DISQUISITIONES ARITHMETICAE

DISQVISITIONES

ARITHMETICAE

AVCTORE

D. CAROLO FRIDERICO GAVSS

LIPSIAE

IN COMMISSIS APVD GERH. FLEISCHER, Jun.

1 8 0 1.

Facsimile of the title page of the first edition of *Disquisitiones Arithmeticae*

DISQUISITIONES
ARITHMETICAE

by Carl Friedrich Gauss

Translated by Arthur A. Clarke, S.J.

NEW HAVEN AND LONDON, YALE UNIVERSITY PRESS, 1966

TRANSLATOR'S PREFACE

IT IS EXTRAORDINARY that one hundred and sixty-four years should have passed between the publication of Gauss' *Disquisitiones Arithmeticae* and its translation into English. No other reason need be offered to justify this enterprise now save the Olympian stature of the author himself, although it is no great presumption to believe that in such a work by such a man there are still hidden profound insights which may yet, after so many years, inspire new discoveries in the field of mathematics.

An apology may be needed for carrying over into English much of the flavor of Gauss' Ciceronian style. I made this decision because I felt that any scholar interested in this work is more concerned with Gauss' thought than in a pithy paraphrase of it. Those fortunate enough to be able to read the original have been satisfied with this for a century and a half. It would be somewhat arbitrary to separate the modern reader too much from the style of the master.

This translation was made from the second edition, edited by Schering for the Königlichen Gesellschaft der Wissenschaften at Göttingen, and printed in 1870 by Dietrich. The reader will find that some footnotes are identified by numerals, others by letters of the alphabet. The former refer to notes that I have inserted, the latter to Gauss' own footnotes. I have also added for the reader's convenience a list of abbreviations of the bibliographical works cited in the text, a list of special symbols used by Gauss with the page numbers where they are defined, and a directory of important terms.

Schering's notes for the second edition read, in part: "In the year 1801 seven sections of the *Disquisitiones Arithmeticae* were published in octavo. The first reprint was published under my direction in 1863 as the first volume of Gauss' *Works*. That edition has been completely sold out, and a new edition is presented here. The eighth section, to which Gauss makes frequent reference and which he had intended to publish with the others, was found among his manuscripts. Since he did not develop it in the same way as the first seven sections, it has been included with his other unpublished arithmetic essays in the second volume of this edition of his *Works....* The form of this edition has been changed to allow for ease of order and summary. I believed that this was justified because Gauss had made such a point of economizing on space. Many formulae which were included in the running text have been displayed to better advantage."

Dr. Herman H. Goldstine first suggested that I undertake this translation, and I am grateful to him for his suggestion and for his continued interest.

ARTHUR A. CLARKE, S.J.

Fordham University
New York, New York
June 1965

TRANSLATOR'S BIBLIOGRAPHICAL ABBREVIATIONS

In the text the dates in brackets represent the years during which the papers were delivered; the unbracketed dates are the years of publication.

Acta acad. Petrop.	*Acta academiae scientiarum imperialis Petropolitanae*, St. Petersburg
Algebra	Leonhard Euler, *Vollständige Anleitung zur Algebra*, St. Petersburg, 1770
Appel au public	Samuel König, *Appel au public du jugement de l'Académie de Berlin sur un fragment de lettre de Mr. de Leibniz, cité par Mr. König*, Leiden, 1752.
Comm. acad. Petrop.	*Commentarii academiae scientiarum imperialis Petropolitanae*, St. Petersburg
Hist. de l'Ac. de Prusse	See below
Hist. Acad. Berlin	*Histoire de l'Académie royale des sciences et belles-lettres avec les mémoires*, Berlin [popularly called *Histoire de l'Académie de Prusse*]
Hist. Acad. Paris	*Histoire de l'Académie royale des sciences avec les mémoires de mathématique et physique*, Paris
Nouv. mém. Acad. Berlin	*Nouveaux mémoires de l'Académie de Berlin*, Berlin
Nova acta erudit.	*Nova acta eruditorum*, Leipzig
Nova acta acad. Petrop.	*Nova acta academiae scientiarum imperialis Petropolitanae*, St. Petersburg

vii

Novi comm. acad. Petrop.	*Novi commentarii academiae scientiarum imperialis Petropolitanae*, St. Petersburg
Opera Mathem.	Pierre de Fermat, *Varia opera Mathematica D. Petri de Fermat, Senatoris Tolosani*, Toulouse, 1679
Opera Mathem. Wall.	*Opera Mathematica*, ed. Johannes Wallis, Oxford, 1693
Opuscula Analytica	Leonhard Euler, St. Petersburg, 1783

CONTENTS

	page
Translator's Preface	v
Bibliographical Abbreviations	vii
Dedication	xv
Author's Preface	xvii
Section I. Congruent Numbers in General	1

Congruent numbers, moduli, residues, and nonresidues, art. 1 ff.
Least residues, art. 4
Elementary propositions regarding congruences, art. 5
Certain applications, art. 12

| Section II. Congruences of the First Degree | 5 |

Preliminary theorems regarding prime numbers, factors, etc., art. 13
Solution of congruences of the first degree, art. 26
The method of finding a number congruent to given residues relative to given moduli, art. 32
Linear congruences with several unknowns, art. 37
Various theorems, art. 38

| Section III. Residues of Powers | 29 |

The residues of the terms of a geometric progression which begins with unity constitute a periodic series, art. 45
If the modulus $= p$ (a prime number), the number of terms in its period is a divisor of the number $p - 1$, art. 49
Fermat's theorem, art. 50
How many numbers correspond to a period in which the number of terms is a given divisor of $p - 1$, art. 52
Primitive roots, bases, indices, art. 57
An algorithm for indices, art. 58
Roots of the congruence $x^n \equiv A$, art. 60
Connection between indices in different systems, art. 69

ix

Bases adapted to special purposes, art. 72

Method of assigning primitive roots, art. 73

Various theorems concerning periods and primitive roots, art. 75

A theorem of Wilson, art. 76

Moduli which are powers of prime numbers, art. 82

Moduli which are powers of the number 2, art. 90

Moduli composed of more than one prime number, art. 92

Section IV. Congruences of the Second Degree   63

Quadratic residues and nonresidues, art. 94

Whenever the modulus is a prime number, the number of residues less than the modulus is equal to the number of nonresidues, art. 96

The question whether a composite number is a residue or nonresidue of a given prime number depends on the nature of the factors, art. 98

Moduli which are composite numbers, art. 100

A general criterion whether a given number is a residue or a nonresidue of a given prime number, art. 106

The investigation of prime numbers whose residues or nonresidues are given numbers, art. 107

The residue -1, art. 108

The residues $+2$ and -2, art. 112

The residues $+3$ and -3, art. 117

The residues $+5$ and -5, art. 121

The residues $+7$ and -7, art. 124

Preparation for a general discussion, art. 125

By induction we establish a general (fundamental) theorem and draw conclusions from it, art. 130

A rigorous demonstration of the fundamental theorem, art. 135

An analogous method of demonstrating the theorem of art. 114, art. 145

A general solution of the problem, art. 146

Linear forms containing all prime numbers for which a given number is a residue or nonresidue, art. 147

The work of other mathematicians concerning these investigations, art. 151

Nonpure congruences of the second degree, art. 152

Section V. Forms and Indeterminate Equations of the Second Degree   108

Plan of our investigation; definition of a form and its symbol, art. 153

Representation of a number; the determinant, art. 154

Values of the expression $\sqrt{(b^2 - ac)}$ (mod. M) to which belongs the representation of the number M by the form (a, b, c), art. 155

One form implying another or contained in it; proper and improper transformation, art. 157

Proper and improper equivalence, art. 158

Opposite forms, art. 159

Neighboring forms, art. 160

Common divisors of the coefficients of forms, art. 161

The connection between all similar transformations of a given form and another given form, art. 162

Ambiguous forms, art. 163

Theorem concerning the case where one form is contained in another both properly and improperly, art. 164

General considerations concerning representations of numbers by forms and their connection with transformations, art. 166

Forms with a negative determinant, art. 171

Special applications for decomposing a number into two squares, into a single and double square, into a single and triple square, art. 182

Forms with positive nonquadratic determinant, art. 183

Forms with quadratic determinant, art. 206

Forms contained in other forms to which, however, they are not equivalent, art. 213

Forms with 0 determinant, art. 215

The general solution by integers of indeterminate equations of the second degree with two unknowns, art. 216

Historical notes, art. 222

Distribution of forms with a given determinant into classes, art. 223

Distribution of classes into orders, art. 226

The partition of orders into genera, art. 228

The composition of forms, art. 234

The composition of orders, art. 245

The composition of genera, art. 246

The composition of classes, art. 249

For a given determinant there are the same number of classes in every genus of the same order, art. 252

Comparison of the number of classes contained in individual genera of different orders, art. 253

The number of ambiguous classes, art. 257

Half of all the characters assignable for a given determinant cannot belong to any properly primitive genus, art. 261

A second demonstration of the fundamental theorem and of other theorems pertaining to the residues -1, $+2$, -2, art. 262

A further investigation of that half of the characters which cannot correspond to any genus, art. 263

A special method of decomposing prime numbers into two squares, art. 265

A digression concerning the treatment of ternary forms, art. 266 ff.

Some applications to the theory of binary forms, art. 286 ff.

How to find the form from whose duplication we get a given binary form of a principal genus, art. 286

Except for those characters for which art. 263, 264 showed it was impossible, all others will belong to some genus, art. 287

The theory of the decomposition of numbers and binary forms into three squares, art. 288

Demonstration of the theorems of Fermat which state that any integer can be decomposed into three trigonal numbers or four squares, art. 293

Solution of the equation $ax^2 + by^2 + cz^2 = 0$, art. 294

The method by which the illustrious Legendre treated the fundamental theorem, art. 296

The representation of zero by ternary forms, art. 299

General solution by rational quantities of indeterminate equations of the second degree in two unknowns, art. 300

The average number of genera, art. 301

The average number of classes, art. 302

A special algorithm for properly primitive classes; regular and irregular determinants etc., art. 305

Section VI. Various Applications of the Preceding Discussions   375

The resolution of fractions into simpler ones, art. 309

The conversion of common fractions into decimals, art. 312

Solution of the congruence $x^2 \equiv A$ by the method of exclusion, art. 319

Solution of the indeterminate equation $mx^2 + ny^2 = A$ by exclusions, art. 323

Another method of solving the congruence $x^2 \equiv A$ for the case where A is negative, art. 327

Two methods for distinguishing composite numbers from primes and for determining their factors, art. 329

Section VII. Equations Defining Sections of a Circle   407

The discussion is reduced to the simplest case in which the number of parts into which the circle is cut is a prime number, art. 336

Equations for trigonometric functions of arcs which are a part or parts of the whole circumference; reduction of trigonometric functions to the roots of the equation $x^n - 1 = 0$, art. 337

Theory concerning the roots of the equation $x^n - 1 = 0$, art. 341 ff.

Except for the root 1, the remaining roots contained in (Ω) are included in the equation $X = x^{n-1} + x^{n-2} +$ etc. $+ x + 1 = 0$; the function X cannot be decomposed into factors in which all the coefficients are rational, art. 341

Declaration of the purpose of the following discussions, art. 342

All the roots in (Ω) are distributed into certain classes (periods), art. 343

Various theorems concerning these periods, art. 344

The solution of the equation $X = 0$ as evolved from the preceding discussions, art. 352

Examples for $n = 19$ where the operation is reduced to the solution of two cubic and one quadratic equation, and for $n = 17$ where the operation is reduced to the solution of four quadratic equations, art. 353, 354

Further discussions concerning periods of roots, art. 355 ff.

Sums in which an even number of terms are real quantities, art. 355

The equation defining the distribution of the roots (Ω) into two periods, art. 356

Demonstration of the theorem mentioned in Section 4, art. 357

The equation for distributing the roots (Ω) into three periods, art. 358

Reduction to pure equations of the equations by which the roots (Ω) are found, art. 359

Application of the preceding to trigonometric functions, art. 361 ff.

Method of finding the angles corresponding to the individual roots of (Ω), art. 361

Derivation of tangents, cotangents, secants, and cosecants from sines and cosines without division, art. 362

Method of successively contracting equations for trigonometric functions, art. 363

Sections of the circle which can be effected by means of quadratic equations or by geometric constructions, art. 365

Additional Notes   461

Tables   463

Gauss' Handwritten Notes   467

List of Special Symbols   470

Directory of Terms   471

TO THE MOST SERENE
PRINCE AND LORD
CHARLES WILLIAM FERDINAND
DUKE OF BRUNSWICK AND LUNEBURG

MOST SERENE PRINCE,

I consider it my greatest good fortune that YOU allow me to adorn this work of mine with YOUR most honorable name. I offer it to YOU as a sacred token of my filial devotion. Were it not for YOUR favor, Most Serene Prince, I would not have had my first introduction to the sciences. Were it not for YOUR unceasing benefits in support of my studies, I would not have been able to devote myself totally to my passionate love, the study of mathematics. It has been YOUR generosity alone which freed me from other cares, allowed me to give myself to so many years of fruitful contemplation and study, and finally provided me the opportunity to set down in this volume some partial results of my investigations. And when at length I was ready to present my work to the world, it was YOUR munificence alone which removed all the obstacles that threatened to delay its publication. Now that I am constrained to acknowledge YOUR remarkable bounty toward me and my work I find myself unable to pay a just and worthy tribute. I am capable only of a secret and ineffable admiration. Well do I recognize that I am not worthy of YOUR gift, and yet everyone knows YOUR extraordinary liberality to all who devote themselves to the higher disciplines. And everyone knows that YOU have never excluded from YOUR patronage those sciences which are commonly regarded as being too recondite and too removed from ordinary life. YOU YOURSELF in YOUR supreme wisdom are well aware of the intimate and necessary bond that unites all sciences among themselves and with whatever pertains to the prosperity of the human society. Therefore I present this book as a witness to my profound regard for YOU and to my dedication to the noblest of sciences. Most Serene Prince, if YOU judge it worthy of that extraordinary favor which YOU have always lavished on me, I will congratulate myself that my work was not in vain and that I have been graced with that honor which I prize above all others.

MOST SERENE PRINCE

Your Highness' most dedicated servant
Brunswick, July 1801   C. F. GAUSS

AUTHOR'S PREFACE

THE INQUIRIES which this volume will investigate pertain to that part of Mathematics which concerns itself with integers. I will rarely refer to fractions and never to surds. The Analysis which is called indeterminate or Diophantine and which discusses the manner of selecting from an infinite set of solutions for an indeterminate problem those that are integral or at least rational (and especially with the added condition that they be positive) is not the discipline to which I refer but rather a special part of it, just as the art of reducing and solving equations (Algebra) is a special part of universal Analysis. And as we include under the heading ANALYSIS all discussion that involves quantity, so integers (and fractions in so far as they are determined by integers) constitute the proper object of ARITHMETIC. However what is commonly called Arithmetic hardly extends beyond the art of enumerating and calculating (i.e. expressing numbers by suitable symbols, for example by a decimal representation, and carrying out arithmetic operations). It often includes some subjects which certainly do not pertain to Arithmetic (like the theory of logarithms) and others which are common to all quantities. As a result it seems proper to call this subject Elementary Arithmetic and to distinguish from it Higher Arithmetic which properly includes more general inquiries concerning integers. We consider only Higher Arithmetic in the present volume.

Included under the heading "Higher Arithmetic" are those topics which Euclid treated with elegance and rigor in Book VII ff., and they can be considered an introduction to this science. The celebrated work of Diophantus, dedicated to the problem of indeterminateness, contains many results which excite a more than ordinary regard for the ingenuity and proficiency of the author because of their difficulty and the subtle devices he uses, especially if we consider the few tools that he had at hand for his work. However, these problems demand a certain dexterity and skillful handling rather than profound principles and, because the questions are too specialized and rarely lead to more general conclusions, Diophantus' book seems to fit into that epoch in the history of Mathematics when scientists were more concerned with creating a characteristic art and a formal Algebraic structure than with attempts to enrich Higher Arithmetic with new discoveries. The really profound discoveries are due to more recent authors like those men of immortal glory P. de Fermat, L. Euler, L. Lagrange, A. M. Legendre (and a few others). They opened the door to what is penetrable in this divine science and enriched it with enormous wealth. I will not recount here the individual discoveries of these geometers since they can be found in the Preface to the appendix which Lagrange added to Euler's *Algebra* and in the recent volume of Legendre (which I shall soon cite). I shall give them their due praise in the proper places in these pages.

The purpose of this volume whose publication I promised five years ago is to present my investigations into the field of Higher Arithmetic. Lest anyone be surprised that the contents here go back over many first principles and that many results had been given energetic attention by other authors, I must explain to the reader that when I first turned to this type of inquiry in the beginning of 1795 I was unaware of the more recent discoveries in the field and was without the means of discovering them. What happened was this. Engaged in other work I chanced on an extraordinary arithmetic truth (if I am not mistaken, it was the theorem of art. 108). Since I considered it so beautiful in itself and since I suspected its connection with even more profound results, I concentrated on it all my efforts in order to understand the principles on which it depended and to obtain a rigorous proof. When I succeeded in this I was so attracted by these questions that I could not let them be. Thus as one result led to another I had completed most of what is presented in the first four sections of this work before I came into contact with similar works of other geometers. Only while studying the writings of these men of genius did I recognize that the greater part of my meditations had been spent on subjects already well developed. But this only increased my interest, and walking in their footsteps I attempted to extend Arithmetic further. Some of these results are embodies in Sections V, VI, and VII. After a while I began to consider publishing the fruits of my new awareness. And I allowed myself to be persuaded not to omit any of the early results, because at that time there was no book that brought together the works of other geometers, scattered as they were among Commentaries of learned Academies. Besides, many of these results are so bound up with one another and with subsequent investigations that new results could not be explained without repeating from the beginning.

Meanwhile there appeared the outstanding work of that man who was already an expert in Higher Arithmetic, Legendre's "Essai d'une théorie des nombres." Here he collected together and systematized not only all that had been discovered up to that time but added many new results of his own. Since this book came to my attention after the greater part of my work was already in the hands of the publishers, I was unable to refer to it in analogous sections of my book. I felt obliged, however, to add some observations in an Appendix and I trust that this understanding and illustrious man will not be offended.

The publication of my work was hindered by many obstacles over a period of four years. During this time I continued investigations which I had already undertaken and deferred to a later date so that the book would not be too large, but I also undertook new investigations. Similarly, many questions which I touched on only lightly because a more detailed treatment seemed less necessary (e.g. the contents of art. 37, 82 ff., and others) have been further developed and have been replaced by more general considerations (cf. what is said in the Appendix about art. 306). Finally, since the book came out much larger than I expected, owing to the size of Section V, I shortened much of what I first intended to do and, especially, I omitted the whole of Section *Eight* (even though I refer to it at times in the present volume; it was to contain a general treatment of algebraic congruences of indeterminate rank). All the treatises which complement the present volume will be published at the first opportunity.

In several difficult discussions I have used synthetic proofs and have suppressed the analysis which led to the results. This was necessitated by brevity, a consideration that must be consulted as much as possible.

The theory of the division of a circle or of a regular polygon treated in Section VII *of itself* does not pertain to Arithmetic but the *principles* involved depend uniquely on Higher Arithmetic. This will perhaps prove unexpected to geometers, but I hope they will be equally pleased with the new results that derive from this treatment.

These are the things I wanted to warn the reader about. It is not my place to judge the work itself. My greatest hope is that it pleases those who have at heart the development of science and that it proposes solutions that they have been looking for or at least opens the way for new investigations.

SECTION I

CONGRUENT NUMBERS IN GENERAL

▶ 1. If a number a divides the difference of the numbers b and c, b and c are said to be *congruent relative to* a; if not, b and c are *noncongruent*. The number a is called the *modulus*. If the numbers b and c are congruent, each of them is called a *residue* of the other. If they are noncongruent they are called *nonresidues*.

The numbers involved must be positive or negative integers,[a] not fractions. For example, -9 and $+16$ are congruent modulo 5; -7 is a residue of $+15$ modulo 11, but a nonresidue modulo 3.

Since every number divides zero, it follows that we can regard any number as congruent to itself relative to any modulus.

▶ 2. Given a, all the residues modulo m are contained in the formula $a + km$ where k is any integer. The easier of the propositions that we will demonstrate follow directly from this, as will be clear to the reader.

Henceforth we shall designate congruence by the symbol $\equiv$, joining to it in parentheses the modulus when it is necessary to do so; e.g. $-7 \equiv 15 \pmod{11}$, $-16 \equiv 9 \pmod{5}$.[b]

▶ 3. THEOREM. *Let* m *successive integers* $a, a + 1, a + 2, \ldots$ $a + m - 1$ *and another integer* A *be given; then one, and only one, of these integers will be congruent to* A *modulo* m.

If $(a - A)/m$ is an integer then $a \equiv A$; if it is a fraction, let k be the next larger integer (or if it is negative, the next *smaller* integer not regarding sign). $A + km$ will fall between a and $a + m$ and will be the desired number. Evidently all the quotients

---

[a] The modulus must obviously be taken *absolutely*, i.e. without sign.

[b] We have adopted this symbol because of the analogy between equality and congruence. For the same reason Legendre, in the treatise which we shall often have occasion to cite, used the same sign for equality and congruence. To avoid ambiguity we have made a distinction.

1

$(a - A)/m$, $(a + 1 - A)/m$, $(a + 2 - A)/m$, etc. lie between $k - 1$ and $k + 1$, so only one of them can be an integer.

▶ 4. Each number therefore will have a residue in the series $0, 1, 2, \ldots m - 1$ and in the series $0, -1, -2, \ldots -(m - 1)$. We will call these the *least residues*, and it is obvious that unless 0 is a residue, they always occur in pairs, one *positive* and one *negative*. If they are unequal in magnitude one will be $< m/2$; otherwise each will $= m/2$ disregarding sign. Thus each number has a residue which is not larger than half the modulus. It will be called the *absolutely least* residue.

For example, relative to the modulus 5, -13 has 2 as least positive residue. It is also the absolutely least residue, whereas -3 is the least negative residue. Relative to the modulus 7, $+5$ is its own least positive residue; -2 is the least negative residue and the absolutely least.

▶ 5. Having established these concepts, let us establish the properties that follow from them.

Numbers that are congruent relative to a composite modulus are also congruent relative to any divisor of the modulus.

If many numbers are congruent to the same number relative to the same modulus, they are congruent to one another (relative to the same modulus).

This identity of moduli is to be understood also in what follows.

Congruent numbers have the same least residues; noncongruent numbers have different least residues.

▶ 6. *Given the numbers A, B, C, etc. and other numbers a, b, c, etc. congruent to each other relative to any modulus whatsoever, i.e. $A \equiv a$, $B \equiv b$, etc., then $A + B + C + $ etc. $\equiv a + b + c + $ etc.*

If $A \equiv a$, $B \equiv b$, then $A - B \equiv a - b$.

▶ 7. *If $A \equiv a$ then also $kA \equiv ka$.*

If k is a positive number, then this is only a particular case of the preceding article (art. 6) letting $A = B = C$ etc., $a = b = c$ etc. If k is negative, $-k$ will be positive. Thus $-kA \equiv -ka$ and so $kA \equiv ka$.

If $A \equiv a$, $B \equiv b$ then $AB \equiv ab$ because $AB \equiv Ab \equiv ba$.

▶ 8. *Given any numbers whatsoever A, B, C, etc. and other numbers a, b, c, etc. which are congruent to them, i.e. $A \equiv a$, $B \equiv b$, etc., the products of each will be congruent; i.e. ABC etc. $\equiv abc$ etc.*

From the preceding article $AB \equiv ab$ and for the same reason $ABC \equiv abc$ and any number of factors may be adjoined.

If all the numbers A, B, C, etc. and the corresponding a, b, c, etc. are assumed equal, then the following theorem holds: *If $A \equiv a$ and k is a positive integer, $A^k \equiv a^k$.*

▶ 9. *Let X be an algebraic function with undetermined x of the form*

$$Ax^a + Bx^b + Cx^c + \text{etc.}$$

where A, B, C, etc. are any integers; a, b, c, etc. are nonnegative integers. Then if x is given values which are congruent relative to some modulus, the resulting values of the function X will also be congruent.

Let f, g be congruent values of x. Then from the preceding article $f^a \equiv g^a$ and $Af^a \equiv Ag^a$, and in the same way $Bf^b \equiv Bg^b$ etc. Thus

$$Af^a + Bf^b + Cf^c + \text{etc.} \equiv Ag^a + Bg^b + Cg^c + \text{etc.} \quad \text{Q.E.D.}$$

It is easy to understand how this theorem can be extended to functions of many undetermined variables.

▶ 10. Thus, if all integers are substituted consecutively for x, and the corresponding values of the function X are reduced to least residues, they will form a sequence in which after an interval of m terms (m being the modulus) the same terms will recur; that is, the sequence will be formed by a *period* of m terms repeated infinitely often. For example, let $X = x^3 - 8x + 6$ and $m = 5$; then for $x = 0, 1, 2, 3$, etc. the values of X produce these least positive residues: 1, 4, 3, 4, 3, 1, 4, etc. where the first five numbers 1, 4, 3, 4, 3 are repeated infinitely often; and if the sequence is continued in a contrary sense, that is, if one gives negative values to x, the same period appears with the order of the terms inverted. From this it follows that no terms other than those that make up this period can occur in the whole sequence.

▶ 11. In this example X cannot become $\equiv 0$, nor $\equiv 2$ (mod. 5) and still less can it $= 0$ or $= 2$. Thus the equations $x^3 - 8x + 6 = 0$ and $x^3 - 8x + 4 = 0$ cannot be solved in integers and consequently, as we know, not by rational numbers. Suppose X is a function in unknown x of the form

$$x^n + Ax^{n-1} + Bx^{n-2} + \text{etc.} + N$$

where A, B, C, etc. are integers, n a positive integer (it is clear that all algebraic equations can be reduced to this form). In general it is clear that in the equation $X = 0$ there exists no rational root unless the congruence $X \equiv 0$ can be satisfied for some modulus. But this omission will be discussed more fully in Section VIII.[1] From this example some small idea of the usefulness of these investigations can be gained.

▶ 12. Many things that are customarily taught in treatises on arithmetic depend on theorems expounded in this section, e.g. rules for deciding whether given numbers are divisible by 9, 11, or any other number. *Relative to the modulus 9* all powers of the number 10 are congruent to unity; and if the number is of the form $a + 10b + 100c +$ etc. it will have, relative to the modulus 9, the same least residue as $a + b + c +$ etc. Thus it is clear that if the single digits of a number expressed in decimal notation are added without regard to position in the number, this sum and the given number will have the same least residue; and thus the latter can be divided by 9 if the former can and vice versa. The same is true of the divisor 3. And since *relative to the modulus 11*, $100 \equiv 1$, in general $10^{2k} \equiv 1$, $10^{2k+1} \equiv 10 \equiv -1$, and a number of the form $a + 10b + 100c +$ etc. will have the same least residue relative to the modulus 11 as $a - b + c$ etc. From this the well-known rule is derived immediately. And from the same principle we can easily deduce all similar rules.

From the preceding argument we can also discover the underlying principles governing the rules that are ordinarily used to verify arithmetic operations. Specifically, if from given numbers others are to be derived by addition, subtraction, multiplication, or by raising to powers, we substitute least residues in place of the given numbers relative to an arbitrary modulus (usually we use 9 or 11 because in our decimal system the residues are easily found). The resulting numbers, as we shall soon see, must be congruent to those deduced from the given numbers, otherwise there is a defect in the calculation.

But since these and similar results are well known, it would be superfluous to dwell on them.

---

[1] Gauss planned eight sections for the *Disquisitiones* and had essentially worked out the eighth section treating of congruences of higher degrees. However, he decided to publish only seven sections in order not to increase the cost of printing the book. See the Author's Preface.

SECTION II

CONGRUENCES OF THE FIRST DEGREE

▶ 13. THEOREM. *The product of two positive numbers each of which is smaller than a given prime number cannot be divided by this prime number.*

Let p be prime, and a positive $< p$: then no positive number b can be found less than p such that $ab \equiv 0$ (mod. p).

Demonstration. If the theorem is false, then we have numbers b, c, d, etc. all $< p$, such that $ab \equiv 0$, $ac \equiv 0$, $ad \equiv 0$, etc. (mod. p). Let b be the smallest of all of these so that no number less than b has the property. Obviously $b > 1$, for if $b = 1$, then $ab = a < p$ (by hypothesis) and so not divisible by p. Now p being prime cannot be divided by b, but lies between two successive multiples of b, mb and $(m + 1)b$. Let $p - mb = b'$; b' will be a positive number and $< b$. Now, since we suppose that $ab \equiv 0$ (mod. p), we also have $mab \equiv 0$ (by art. 7) and subtracting this from $ap \equiv 0$ we get $a(p - mb) = ab' \equiv 0$; i.e. b' must be one of the numbers b, c, d, etc., and it will be smaller than the smallest of them.   Q.E.A.

▶ 14. *If neither a nor b can be divided by a prime number p, the product ab cannot be divided by p.*

Let α, β be the least positive residues of the numbers a, b relative to the modulus p. Neither of them will be 0 (by hypothesis). Now if $ab \equiv 0$ (mod. p) then $\alpha\beta \equiv 0$ because $ab \equiv \alpha\beta$. But this contradicts the previous theorem.

Euclid had already proved this theorem in his *Elements* (Book VII, No. 32). However we did not wish to omit it because many modern authors have employed vague computations in place of proof or have neglected the theorem completely, and because by this very simple case we can more easily understand the nature of the method which will be used later for solving much more difficult problems.

5

▶ 15. *If none of the numbers a, b, c, d, etc. can be divided by a prime p neither can their product abcd etc.*

According to the previous article ab cannot be divided by p; therefore neither can abc; similarly for $abcd$ etc.

▶ 16. THEOREM. *A composite number can be resolved into prime factors in only one way.*

Demonstration. It is clear from elementary considerations that any composite number can be resolved into prime factors, but it is tacitly supposed and generally without proof that this cannot be done in many various ways. Let us suppose that a composite number $A = a^\alpha b^\beta c^\gamma$, etc., with a, b, c, etc. unequal prime numbers, can be resolved in still another way into prime factors. First it is clear that in this second system of factors there cannot appear any other primes except a, b, c, etc., since no other prime can divide A which is composed of these primes. Similarly in this second system of factors none of the prime numbers a, b, c, etc. can be missing, otherwise it would not divide A (preceding article). And so these two resolutions into factors can differ only in that in one of them some prime number appears more often than in the other. Let such a prime by p, which appears in one resolution m times and in the other n times, and let $m > n$. Now remove from each system the factor p, n times. As a result p will remain in one system $m - n$ times and will be missing entirely from the other. That is, we have two resolutions into factors of the number A/p^n. One of them does not contain the factor p, the other contains it $m - n$ times, contradicting what we have just shown.

▶ 17. Thus if a composite number A is the product of B, C, D, etc., it is clear that among the prime factors of B, C, D, etc. none may appear that is not a factor of A. And each of these factors must appear in the resolution of A as many times as it appears altogether in B, C, D, etc. Thus we get a criterion for determining whether a number B divides a number A or not. B will divide A provided it contains no other factor than A and provided it contains none of them more often than A. If either of these conditions is absent, B will not divide A.

It is easy to see by means of the calculus of combinations that if as above a, b, c, etc. are different prime numbers and $A = a^\alpha b^\beta c^\gamma$

etc., then A has $(\alpha + 1)(\beta + 1)(\gamma + 1)$ etc. different divisors including 1 and A itself.

▶ 18. If therefore $A = a^\alpha b^\beta c^\gamma$ etc., $K = k^\kappa l^\lambda m^\mu$ etc. and the primes a, b, c, etc., k, l, m, etc. are all different, then clearly A and K have no common divisor except 1, or in other words that they are prime relative to each other.

Given many numbers A, B, C, etc. the *greatest common divisor* is found as follows. Let all the numbers be resolved into their prime factors, and from these extract the ones which are common to A, B, C, etc. (if there is none, there will be no divisor common to all of them). Then note the number of times each of these prime factors appears in A, in B, in C, etc. or, in other words, note what *dimension* each has in A, in B, in C, etc. Finally, assign to each factor the smallest of all the dimensions that it has in A, in B, in C, etc. Form the product of these, and the result will be the common divisor that we seek.

When we want the *least common multiple*, we proceed as follows. Collect all the prime numbers that divide any of the numbers A, B, C, etc. and assign to each of them the dimension that is the highest in A, B, C, etc. Form the product of these and the result will be the multiple that we seek.

Example. Let $A = 504 = 2^3 3^2 7$, $B = 2880 = 2^6 3^2 5$, $C = 864 = 2^5 3^3$. For the greatest common divisor we have the factors 2, 3 with dimensions 3, 2 respectively; and this becomes $2^3 3^2 = 72$; the least common multiple will be $2^6 3^3 5 \cdot 7 = 60,480$.

We omit the demonstration because of its simplicity. Moreover, we know from elementary considerations how to solve these problems when the resolution of the numbers A, B, C, etc. into factors is not given.

▶ 19. *If the numbers a, b, c, etc. are prime relative to some other number k, then their product abc is also prime relative to k.*

Since none of the numbers a, b, c, etc. has a prime factor in common with k, and since the product abc etc. has no other prime factors but those that belong to one of the numbers a, b, c, etc. the product abc etc. will have no prime factor in common with k. Therefore from the preceding article, k and abc etc. are prime relative to each other.

If the numbers a, b, c, etc. are prime relative to each other, and each of them divides some other number k, then their product divides k.

This follows easily from articles 17, 18. For let p be a prime divisor of the product abc etc. which contains it π times. It is clear that some one of the numbers a, b, c, etc. must contain this same divisor π times. Thus k also, which this number divides, contains p, π times. In like manner for the remaining divisors of the product abc etc.

And so *if two numbers m, n are congruent relative to several moduli a, b, c, etc. which are prime relative to each other, they will also be congruent relative to their product.* For since $m - n$ is divisible by each of the a, b, c, etc., it will be divisible by their product also.

Finally, *if a is prime relative to b and ak is divisible by b, then k is also divisible by b.* For since ak is divisible by both a and b, it is also divisible by ab; i.e. $ak/ab = k/b$ is an integer.

▶ 20. *Suppose a, b, c, etc. are unequal prime numbers and $A = a^{\alpha}b^{\beta}c^{\gamma}$ etc. Then if A is some power, for example $= k^{n}$, all the exponents α, β, γ, etc. will be divisible by n.*

For the number k involves no other prime factors except a, b, c, etc. Let k contain the factor a, α' times; k^{n} or A will contain this factor $n\alpha'$ times; therefore $n\alpha' = \alpha$ and α/n is an integer. In like manner β/n etc. can be shown to be integers.

▶ 21. *When a, b, c, etc. are prime relative to each other and the product abc etc. is some power, for example k^{n}, then each of the numbers a, b, c will be like powers.*

Let $a = l^{\lambda}m^{\mu}p^{\pi}$ etc. with l, m, p, etc. different prime numbers. By hypothesis none of them is a factor of the numbers b, c, etc. So the product abc etc. will contain the factor l, λ times, the factor m, μ times, etc. Thus (preceding article) λ, μ, π, etc. are divisible by n and so

$$\sqrt[n]{a} = l^{\lambda/n}m^{\mu/n}p^{\pi/n} \text{ etc.}$$

is an integer. Likewise for b, c, etc.

Having thus concluded our remarks about prime numbers, we turn now to subjects closer to our purpose.

▶ 22. *Suppose the numbers a, b are divisible by another number k. If they are congruent relative to a modulus m, which is relatively prime to k, then a/k and b/k will be congruent relative to the same modulus.*

It is clear that $a - b$ is divisible by k as well as by m (hypothesis); so (art. 19) $(a - b)/k$ is divisible by m; i.e. $a/k \equiv b/k$ (mod. m).

If, however, other things being equal, m and k have a greatest common divisor e, then $a/k \equiv b/k$ (mod. m/e), for k/e and m/e are relatively prime. But $a - b$ is divisible by k and by m, so $(a - b)/e$ is divisible by k/e and by m/e, hence by km/ee; i.e. $(a - b)/k$ is divisible by m/e, which implies that $a/k \equiv b/k$ (mod. m/e).

▶ 23. *If a is prime relative to m, and e, f are noncongruent relative to the modulus m, then ae, af will be noncongruent relative to m.*

This is simply the inverse of the theorem in the preceding article.

It is clear that if a is multiplied by all integers from 0 to $m - 1$ and the products reduced to their least residues relative to the modulus m, all will be unequal. And since there are m of these residues, none of them $> m$, all the numbers from 0 to $m - 1$ will be included among them.

▶ 24. *Let a, b be given numbers and x an indeterminate or variable number. The expression ax + b can be made congruent to any number relative to a modulus m, provided m is prime relative to a.*

Let the number which is to be made congruent be called c, and let the least positive residue of $c - b$ relative to the modulus m be called e. From the preceding article there is necessarily a value of $x < m$ such that the least residue of the product ax relative to the modulus m will be e; let this value be v, and we have $av \equiv e \equiv c - b$; therefore $av + b \equiv c$ (mod. m). Q.E.F.

▶ 25. Any expression containing two congruent quantities in the manner of an equation will be called a *congruence*. If it involves an unknown, it is said to be *solved* when a value (*root*) is found satisfying the congruence. Hence it is clear what is meant by *solvable and unsolvable congruences*. Obviously distinctions similar to those used when speaking of equations can be used here. Examples of *transcendental* congruences will occur below; with regard to *algebraic* congruences they will be divided according to the highest power of the unknown into congruences of the first, second, and higher *degrees*. Similarly, many congruences involving many unknowns can be proposed, and we can treat of their *elimination*.

▶ 26. According to article 24, a congruence of the first degree $ax + b \equiv c$ always has a solution when the modulus is prime relative to a. Now if v is a suitable value of x, that is, a root of the congruence, it is clear that all numbers congruent to v relative to the modulus involved are also roots (art. 9). It is just as clear that all roots must be congruent to v. For if t is some other root, $av + b \equiv at + b$; thus $av \equiv at$ and $v \equiv t$ (art. 22). We conclude that the congruence $x \equiv v$ (mod. m) gives the complete solution of the congruence $ax + b \equiv c$.

Since all the values of x which are solutions of the congruence are obviously congruent to one another, and since in this respect congruent numbers can be considered as equivalent, we will speak of such solutions as one and the same solution of the congruence. Wherefore, since our congruence $ax + b \equiv c$ admits of no other solutions, we will say that it has one, and only one, solution or that it has one, and only one, root. Thus, e.g., the congruence $6x + 5 \equiv 13$ (mod. 11) admits of no roots other than those that are $\equiv 5$ (mod. 11). This is not at all true in congruences of other degrees or in congruences of the first degree where the unknown is multiplied by a number which is not prime relative to the modulus.

▶ 27. It remains now to add details concerning the manner of solving congruences. We first observe that a congruence of the form $ax + t \equiv u$, where we suppose that the modulus is prime relative to a, depends on $ax \equiv \pm 1$; for if $x \equiv r$ satisfies the latter, $x \equiv \pm(u - t)r$ will satisfy the former. But the undetermined equation $ax = by \pm 1$ is equivalent to the congruence $ax \equiv +1$ with b as its modulus. And we know how to solve this, so it will be sufficient to give the algorithm for calculating it.

If the quantities A, B, C, D, E, etc. depend on $\alpha, \beta, \gamma, \delta$, etc. in the following way

$$A = \alpha, \ B = \beta A + 1, \ C = \gamma B + A, \ D = \delta C + B, \ E = \varepsilon D + C, \text{ etc.}$$

for brevity we will write as follows

$$A = [\alpha], \ B = [\alpha, \beta], \ C = [\alpha, \beta, \gamma], \ D = [\alpha, \beta, \gamma, \delta], \text{ etc.}^{a}$$

Now let us consider the indeterminate equation $ax = by \pm 1$,

---

[a] This relation can be considered more generally, as we do on another occasion. We add more propositions useful for our investigations:

1) $[\alpha, \beta, \gamma, \ldots, \lambda, \mu] \cdot [\beta, \gamma, \ldots \lambda] - [\alpha, \beta, \gamma, \ldots \lambda] \cdot [\beta, \gamma, \ldots \lambda, \mu] = \pm 1$ where the upper sign is taken if the number of terms $\alpha, \beta, \gamma, \ldots \lambda, \mu$ is even and the lower if it is odd.

2) The order of the numbers can be inverted: $[\alpha, \beta, \gamma, \ldots \lambda, \mu] = [\mu, \lambda, \ldots \gamma, \beta, \alpha]$. We omit the simple demonstrations.

where a, b are positive and we assume, as we may, a is not $<b$. Now by the known algorithm for finding the greatest common divisor of two numbers, we form by ordinary division the equations

$$a = \alpha b + c, \qquad b = \beta c + d, \qquad c = \gamma d + e, \qquad \text{etc.}$$

so that α, β, γ, etc. c, d, e, etc. are positive integers and b, c, d, e constantly decreasing until we come to $m = \mu n + 1$. This can always be done, and the result will be

$$a = [n, \mu, \ldots \gamma, \beta, \alpha], \qquad b = [n, \mu, \ldots \gamma, \beta]$$

If we take $x = [\mu, \ldots \gamma, \beta]$, $y = [\mu, \ldots \gamma, \beta, \alpha]$ we will have $ax = by + 1$ when the number of terms $\alpha, \beta, \gamma, \ldots \mu, n$ is even and $ax = by - 1$ when it is odd. Q.E.F.

▶ 28. Euler was the first to give the general solution for indeterminate equations of this type (*Comm. acad. Petrop,˙1* [1734–35], 1740, 46).[1] The method he used consisted in substituting other unknowns for x, y, and it is a method that is well known today. Lagrange treated the problem a little differently. As he noted, it is clear from the theory of continued fractions that if the fraction b/a is converted into the continued fraction

$$\cfrac{1}{\alpha + \cfrac{1}{\beta + \cfrac{1}{\gamma + \text{etc.} + \cfrac{1}{\mu + \cfrac{1}{n}}}}}$$

and if the last part, $1/n$, is deleted and the result reconverted into a common fraction, x/y then $ax = by \pm 1$, provided a is prime relative to b. For the rest, the same algorithm is derived from the two methods. The investigations of Lagrange appear in *Hist. Acad. Berlin*, 1767, p. 173,[2] and with others in an appendix to the French translation of Euler's treatise on algebra.[3]

---

[1] "Solutio problematis arithmetici de inveniendo numero qui per datos numeros divisus, relinquat data residua."

[2] "Sur la solution des problemes indéterminées du second degré."

[3] *Élèmens d'Algèbre par Léonard Euler traduits de l'Allemand avec des Notes et des Additions*, Lyon, 1795. Bernoulli translated the first volume (*De l'Analyse déterminée*), Lagrange the second (*Analyse indéterminée*).

▶ 29. The congruence $ax + t \equiv u$ with its modulus not prime relative to a, reduces easily to the preceding case. Let the modulus be m and δ of the greatest common divisor of a, m. It is clear that any value of x that satisfies the congruence relative to the modulus m also satisfies it relative to the modulus δ (art. 5). But $ax \equiv 0$ (mod. δ) since δ divides a. Therefore, unless $t \equiv u$ (mod. δ), i.e. $t - u$ is divisible by δ, the congruence has no solution.

Now, let $a = \delta e$, $m = \delta f$, $t - u = \delta k$; e will be prime relative to f. Then $ex + k \equiv 0$ (mod. δf) will be equivalent to the proposed congruence $\delta ex + \delta k \equiv 0$ (mod. δf); i.e., whatever value of x satisfies one will satisfy the other and vice versa. For clearly $ex + k$ can be divided by f when $\delta ex + \delta k$ can be divided by δf and vice versa. But we saw above how to solve the congruence $ex + k \equiv 0$ (mod. f); so it is clear that if v is one of the values of x, $x \equiv v$ (mod. f) gives the complete solution of the proposed congruence.

▶ 30. When the modulus is composite, it is sometimes advantageous to use the following method.

Let the modulus $= mn$, and the proposed congruence $ax \equiv b$. First, solve the congruence relative to the modulus m, and suppose that it is satisfied if $x \equiv v$ (mod. m/δ) where δ is the greatest common divisor of the numbers m, a. It is clear that any value of x that satisfies the congruence $ax \equiv b$ relative to the modulus mn also satisfies it relative to the modulus m, and that it will be expressible in the form $v + (m/\delta)x'$ where x' is some undetermined number; the opposite however is not true, since not all numbers of the form $v + (m/\delta)x'$ satisfy the congruence relative to the modulus mn. The manner of determining x' so that $v + (m/\delta)x'$ is a root of the congruence $ax \equiv b$ (mod. mn) can be deduced from the solution of the congruence $(am/\delta)x' + av \equiv b$ (mod. mn) or of the equivalent congruence $(a/\delta)x' \equiv (b - av)/m$ (mod. n). It follows that the solution of any congruence of the first degree relative to a modulus mn can be reduced to the solution of two congruences relative to the moduli m and n. And it is obvious that if n is again the product of two factors, the solution of the congruence relative to the modulus n depends on the solution of two congruences whose moduli are those factors. In general the solution of a congruence relative to a composite modulus depends on the solution of other congruences whose moduli are factors of the composite.

These factors can be taken to be prime numbers if it is convenient to do so.

Example. Suppose we are to solve $19x \equiv 1$ (mod. 140). First solve it relative to the modulus 2 and we get $x \equiv 1$ (mod. 2). Let $x = 1 + 2x'$ and it will become $38x' \equiv -18$ (mod. 140) or equivalently $19x' \equiv -9$ (mod. 70). If this is again solved relative to the modulus 2, it becomes $x' \equiv 1$ (mod. 2), and letting $x' \equiv 1 + 2x''$, this becomes $38x'' \equiv -28$ (mod. 70) or $19x'' \equiv -4$ (mod. 35). Relative to 5, this gives the solution $x'' \equiv 4$ (mod. 5), and substituting $x'' = 4 + 5x'''$, it becomes $95x''' \equiv -90$ (mod. 35) or $19x''' \equiv -18$ (mod. 7). From this it follows that $x''' \equiv 2$ (mod. 7), and letting $x''' = 2 + 7x''''$ we find that $x = 59 + 140x''''$; therefore, $x \equiv 59$ (mod. 140) is the complete solution of the congruence.

▶ 31. In the same way that the root of the equation $ax = b$ can be expressed as b/a, we will designate by b/a a root of the congruence $ax \equiv b$ and join to it the modulus of the congruence to distinguish it. Thus, e.g., $19/17$ (mod. 12) denotes any number that is $\equiv 11$ (mod. 12).[b] From this it is clear that b/a (mod. c) does not signify anything real (or if you prefer, it is imaginary) when a, c have a common divisor which does not divide b. With this exception, the expression b/a (mod. c) will always have real values and indeed an infinite number of them. All of them will be congruent relative to c when a is relatively prime to c, or relative to c/δ when δ is the greatest common divisor of c, a.

These expressions have an algorithm very like that for common fractions. We point out some properties which can be easily deduced from the preceding discussion.

1. If relative to the modulus $c, a \equiv \alpha, b \equiv \beta$ then the expressions a/b (mod. c) and α/β (mod. c) are equivalent.

2. $a\delta/b\delta$ (mod. $c\delta$) and a/b (mod. c) are equivalent.

3. ak/bk (mod. c) and a/b (mod. c) are equivalent when k and c are prime relative to each other.

We could cite many other like propositions but, since they present no difficulty and are not particularly useful for what follows, we will proceed to other considerations.

▶ 32. The problem of *finding all numbers that have given residues relative to any number of given moduli* can easily be solved from

---

[b] By analogy it can be expressed as $11/1$ (mod. 12).

what we have seen and will prove very useful in what follows. Let
two moduli A, B be given. Relative to these we seek the number z
which should be congruent to the numbers a, b respectively. All
such values of z are of the form $Ax + a$ with x indeterminate but
such that $Ax + a \equiv b$ (mod. B). Now if the greatest common
divisor of the numbers A, B is δ, the complete solution of the
congruence will have the form $x \equiv v$ (mod. B/δ) or what comes
to the same thing, $x = v + (kB/\delta)$ with k an arbitrary integer.
Thus the formula $Av + a + (kAB/\delta)$ will include all values of z;
i.e. $z \equiv Av + a$ (mod. AB/δ) will be the complete solution of the
problem. If to the moduli A, B we add a third C, according to
which $z \equiv c$, obviously we proceed in the same way, since the
two previous conditions combine into one. So if the greatest
common divisor of $AB/\delta, C$ is e, and if the solution of the congruence $(AB/\delta)x + Av + a \equiv c$ (mod. C) is $x \equiv w$ (mod. C/e), then
the problem will be completely solved by the congruence
$z \equiv (ABw/\delta) + Av + a$ (mod. $ABC/\delta e$). We observe that AB/δ,
$ABC/\delta e$ are the least common multiples of the numbers A, B and
A, B, C respectively, and it is easily established that no matter
how many moduli we have A, B, C, etc., if their least common
multiple is M, the complete solution will have the form $z \equiv r$
(mod. M). But when none of the auxiliary congruences is solvable,
we conclude that the problem involves an impossibility. But
obviously this cannot happen when all the numbers A, B, C, etc.
are prime relative to each other.

   Example. Let the numbers A, B, C, a, b, c be 504, 35, 16, 17, -4,
33. Here the two conditions $z \equiv 17$ (mod. 504) and $z \equiv -4$
(mod. 35) are equivalent to the one condition $z \equiv 521$ (mod. 2520);
add to this the condition $z \equiv 33$ (mod. 16) and we get finally
$z \equiv 3041$ (mod. 5040).

▶ 33. When all the numbers A, B, C, etc. are prime relative to each
other it is clear that their product is their least common multiple.
In this case, all the congruences $z \equiv a$ (mod. A), $z \equiv b$ (mod. B),
etc. are equivalent to the one congruence $z \equiv r$ (mod. R) where R
is the product of the numbers A, B, C, etc. It follows, in turn, that
the single condition $z \equiv r$ (mod. R) can be resolved into many
conditions, namely that if R is resolved in any manner whatsoever
into factors A, B, C, etc. which are prime relative to each other,
then the conditions $z \equiv r$ (mod. A), $z \equiv r$ (mod. B), $z \equiv r$ (mod. C),

etc. exhaust the proposition. This observation opens to us not only a method of discovering the impossibility when it exists, but also a more satisfactory and elegant way of calculating.

▶ 34. As above let $z \equiv a$ (mod. A), $z \equiv b$ (mod. B), $z \equiv c$ (mod. C). Resolve all moduli into factors which are prime relative to each other: A into $A'A''A'''$ etc.; B into $B'B''B'''$ etc., etc.; and in such a manner that the numbers, A', A'', etc., B', B'', etc., etc. are either primes or powers of primes. If any of the numbers A, B, C, etc. is already a prime or a power of a prime, there is no need for resolving it into factors. It is clear that in place of the conditions proposed we can substitute the following: $z \equiv a$ (mod. A'), $z \equiv a$ (mod. A''), $z \equiv a$ (mod. A'''), etc., $z \equiv b$ (mod. B'), $z \equiv b$ (mod. B''), etc., etc. Now if not all the numbers A, B, C, etc. are prime relative to each other (for example if A is not prime relative to B), it is obvious that not all the prime divisors of A, B can be different. There must be one or another among the factors A', A'', A''', etc. that has an equal or multiple or submultiple among the factors B', B'', B''', etc. Suppose as a *first* possibility that $A' = B'$. Then the conditions $z \equiv a$ (mod. A'), $z \equiv b$ (mod. B') must be identical; that is $a \equiv b$ (mod. A' or B'), and therefore one of them can be rejected. If however $a \equiv b$ (mod. A') is not true, the problem is impossible of solution. Suppose *secondly* that B' is a multiple of A'. The condition of $z \equiv a$ (mod. A') must be contained in the condition $z \equiv b$ (mod. B'); that is, the congruence $z \equiv b$ (mod. A') which is deduced from the latter must be identical with the former. From this it follows that the condition $z \equiv a$ (mod. A') can be rejected unless it is inconsistent with some other condition (in which case the problem is impossible). When all the superfluous conditions have been rejected, all the moduli remaining from the factors A', A'', A''', etc., B', B'', B''', etc., etc. will be prime relative to each other. Then we can be sure of the possibility of the problem and can proceed as described above.

▶ 35. *Example.* If, as above (art. 32), $z \equiv 17$ (mod. 504), $z \equiv -4$ (mod. 35), and $z \equiv 33$ (mod. 16), then these conditions can be reduced to the following: $z \equiv 17$ (mod. 8), $z \equiv 17$ (mod. 9), $z \equiv 17$ (mod. 7), $z \equiv -4$ (mod. 5), $z \equiv -4$ (mod. 7), $z \equiv 33$ (mod. 16). Of these conditions $z \equiv 17$ (mod. 8) and $z \equiv 17$ (mod. 7) can be rejected because the first is contained in the condition $z \equiv 33$ (mod. 16), and the second is identical with $z \equiv -4$ (mod. 7).

There remain:

$$z \equiv \begin{vmatrix} 17 \ (\text{mod. } 9) \\ -4 \ (\text{mod. } 5) \\ -4 \ (\text{mod. } 7) \\ 33 \ (\text{mod. } 16) \end{vmatrix} \quad \text{and from these we get } z \equiv 3041 \ (\text{mod. } 5040)$$

It is clear that it will often be more convenient to collect separately from among the remaining conditions those congruences which derive from one and the same condition, since this can easily be done; e.g. when some of the conditions $z \equiv a$ (mod. A'), $z \equiv a$ (mod. A''), etc. are eliminated, the rest are replaced by $z \equiv a$ relative to the modulus which is the product of all the moduli remaining from the set A', A'', A''', etc. Thus in our example the conditions $z \equiv -4$ (mod. 5), $z \equiv -4$ (mod. 7) are replaced by $z \equiv -4$ (mod. 35). It follows further that it is not a matter of indifference, so far as brevity of calculation is concerned, which superfluous conditions are rejected. But is is not our intention to treat of these details or of other practical artifices that can be learned more easily by usage than by precept.

▶ 36. When all the moduli A, B, C, D, etc. are prime relative to each other it is more often preferable to use the following method. Determine a number α which is congruent to unity relative to the modulus A, and congruent to zero relative to the product of the remaining moduli; that is to say α will be a value (preferably the *least*) of the expression $1/BCD$ etc. (mod. A) multiplied by BCD etc. (see art. 32). Similarly let $\beta \equiv 1$ (mod. B) and $\equiv 0$ (mod. ACD etc.), $\gamma \equiv 1$ (mod. C) and $\equiv 0$ (mod. ABD etc.), etc. Then if we are looking for z which is congruent to a, b, c, d, etc. relative to the moduli A, B, C, D, etc., respectively, we can write

$$z \equiv \alpha a + \beta b + \gamma c + \delta d \text{ etc. (mod. } ABCD \text{ etc.)}$$

Obviously $\alpha a \equiv a$ (mod. A) and the remaining numbers $\beta b, \gamma c$, etc. are all $\equiv 0$ (mod. A), and so $z \equiv a$ (mod. A). A similar demonstration holds for the other moduli. This solution is to be preferred to the former when we are solving more problems of the same type for which the moduli A, B, C, etc. retain their values, for then α, β, γ, etc. have constant values. This usage arises in the problem of chronology when we seek to determine what Julian year it is whose indiction, golden number, and solar cycle are given. Here $A = 15$, $B = 19$, $C = 28$; so, since the value of the expression $1/(19 \cdot 28)$ (mod. 15) or $1/532$ (mod. 15) is 13, α will be 6916. In the same way we find that β is 4200 and γ is 4845, and the number we seek will be the least residue of the number $6916a + 4200b + 4845c$ where a is the indiction, b the golden number, c the solar cycle.

▶ 37. We have said enough about first-degree congruences with a single unknown and will proceed to congruences containing several unknowns. If we were to treat each item with complete rigor, this section would continue interminably. We therefore propose to treat only of those matters that seem worthy of attention, restricting our investigation to a few observations and reserving a full exposition to another occasion.

1) As with equations, it is clear that we must have as many congruences as there are unknowns to be determined.

2) We will propose, therefore, the congruences

$$ax + by + cz \ldots \equiv f \text{ (mod. } m) \qquad (A)$$

$$a'x + b'y + c'z \ldots \equiv f' \qquad (A')$$

$$a''x + b''y + c''z \ldots \equiv f'' \qquad (A'')$$

which are to be as many in number as there are unknowns x, y, z, etc.

Now we determine the numbers ξ, ξ', ξ'', etc. such that

$$b\xi + b'\xi' + b''\xi'' + \text{etc.} = 0$$

$$c\xi + c'\xi' + c''\xi'' + \text{etc.} = 0$$

$$\text{etc.}$$

and such that all the numbers are integers and there is no common factor. Clearly this is possible from the theory of linear equations. In a similar way we determine v, v', v'', etc, ζ, ζ', ζ'', etc., etc. so that

$$av + a'v' + a''v'' + \text{etc.} = 0$$

$$cv + c'v' + c''v'' + \text{etc.} = 0$$

$$\text{etc.}$$

$$a\zeta + a'\zeta' + a''\zeta'' = 0$$
$$b\zeta + b'\zeta' + b''\zeta'' = 0$$

etc.

etc.

3) Clearly if the congruences A, A', A'', etc. are multiplied by ξ, ξ', ξ'', etc. then by v, v', v'', etc., etc. and added, we will get the following congruences:

$$(a\xi + a'\xi' + a''\xi'' + \text{etc.})x \equiv f\xi + f'\xi' + f''\xi'' + \text{etc.}$$
$$(bv + b'v' + b''v'' + \text{etc.})y \equiv fv + f'v' + f''v'' + \text{etc.}$$
$$(c\zeta + c'\zeta' + c''\zeta'' + \text{etc.})z \equiv f\zeta + f'\zeta' + f''\zeta'' + \text{etc.}$$

etc.

which for brevity we shall write as follows:

$$\sum(a\xi)x \equiv \sum(f\xi), \qquad \sum(bv)y \equiv \sum(fv), \qquad \sum(c\zeta)z \equiv \sum(f\zeta), \text{ etc.}$$

4) Various cases must be distinguished.

First, when all the coefficients $\Sigma(a\xi)$, $\Sigma(bv)$, etc. are prime relative to m, the modulus of the congruences, we can solve them according to rules we have already seen, and the complete solution will be found by congruences of the form $x \equiv p$ (mod. m), $y \equiv q$ (mod. m), etc.[c]

For example, given the congruences

$$x + 3y + z \equiv 1, \quad 4x + y + 5z \equiv 7, \quad 2x + 2y + z \equiv 3 \text{ (mod. 8)}$$

we find $\xi = 9$, $\xi' = 1$, $\xi'' = -14$, so $-15x \equiv -26$ and $x \equiv 6$ (mod. 8). In the same way we find $15y \equiv -4$, $15z \equiv 1$ and so $y \equiv 4$, $z \equiv 7$ (mod. 8).

5) *Second*, when not all the coefficients $\Sigma(a\xi)$, $\Sigma(bv)$, etc. are prime relative to the modulus, let α, β, γ, etc. be the greatest common divisors of the modulus m and $\Sigma(a\xi)$, $\Sigma(bv)$, $\Sigma(c\zeta)$, etc.

---

[c] This conclusion needs demonstration, but we have suppressed it here. Nothing more follows from our analysis than that the proposed congruences cannot be solved by other values of the unknowns x, y, etc. We have not shown that these values do satisfy. It is even possible that there is no solution at all. A similar paralogism occurs in treating linear equations.

respectively. It is clear that the problem is impossible unless these divide the numbers $\Sigma(f\xi)$, $\Sigma(fv)$, $\Sigma(f\zeta)$, etc. respectively. If, however, these conditions are fulfilled, the congruences in 3) will be completely solved by congruences of the following type: $x \equiv p$ (mod. m/α), $y \equiv q$ (mod. m/β), $z \equiv r$ (mod. m/γ), etc.; or, if you prefer, there are α different values of x (i.e. not congruent relative to m, say $p, p + m/\alpha, \ldots p + (\alpha - 1)m/\alpha$), β different values of y, etc. satisfying the congruences. Manifestly all solutions of the proposed congruences (if there is any at all) will be found among these. But we cannot invert this conclusion, for in general not all the combinations of all α values of x combined with all those of y and z etc. satisfy the problem, but only those whose connection can be shown by one or more of the conditional congruences. Since, however, the complete solution of this problem is not necessary for what follows, we will not now pursue the argument further but will give some idea of it by an example.

Let these congruences be given:

$$3x + 5y + z \equiv 4, \qquad 2x + 3y + 2z \equiv 7,$$

$$5x + y + 3z \equiv 6 \text{ (mod. 12)}$$

ξ, ξ', ξ''; v, v', v''; ζ, ζ', ζ'' will equal respectively $1, -2, 1$; $1, -1$; $-13, 22, -1$. From this $4x \equiv -4$, $7y \equiv 5$, $28z \equiv 96$. And from this we produce four values of x, say $\equiv 2, 5, 8, 11$; one value of y, say $\equiv 11$; four values of z, say $\equiv 0, 3, 6, 9$ (mod. 12). Now to know which combinations of the values of x can be used with the values of z, we substitute in the proposed congruences $2 + 3t$, $11, 3u$ for x, y, z respectively. This changes the congruences to

$$57 + 9t + 3u \equiv 0, \qquad 30 + 6t + 6u \equiv 0,$$

$$15 + 15t + 9u \equiv 0 \text{ (mod. 12)}$$

and these become

$$19 + 3t + u \equiv 0, \qquad 10 + 2t + 2u \equiv 0,$$

$$5 + 5t + 3u \equiv 0 \text{ (mod. 4)}$$

The first clearly requires that $u \equiv t + 1$ (mod. 4), and when this value is substituted in the remaining congruences it satisfies them also. We conclude that the values of x (2, 5, 8, 11 which are produced by letting $t \equiv 0, 1, 2, 3$), are necessarily combined with the values of $z \equiv 3, 6, 9, 0$ respectively. Altogether we have four solutions:

$$x \equiv 2, \quad 5, \quad 8, 11 \text{ (mod. 12)}$$

$$y \equiv 11, 11, 11, 11$$

$$z \equiv 3, \quad 6, \quad 9, \quad 0$$

Thus we have finished the task set forth in this section. However we will add certain propositions that depend on similar principles and which we will frequently need for what follows.

▶ 38. PROBLEM. *To find how many positive numbers are smaller than a given positive number A and relatively prime to it.*

For brevity we will designate the number of positive numbers which are relatively prime to the given number and smaller than it by the prefix ϕ. We seek therefore ϕA.

I. When A is prime, obviously all numbers from 1 to $A - 1$ are relatively prime to A, so in this case

$$\phi A = A - 1$$

II. When A is a power of a prime number, say $A = p^m$, all numbers divisible by p will not be relatively prime to A, but the rest will be. So of the $p^m - 1$ numbers, these must be rejected: $p, 2p, 3p \ldots (p^m - 1)p$. There remain therefore $p^m - 1 - (p^m - 1)$ or $p^{m-1}(p - 1)$ of them. So

$$\phi p^m = p^{m-1}(p - 1)$$

III. The remaining cases easily reduce to these by means of the following proposition: *If A is resolved into factors M, N, P, etc. which are prime relative to each other, then*

$$\phi A = \phi M \cdot \phi N \cdot \phi P \text{ etc.}$$

To show this, let the numbers that are prime relative to M and less than M be m, m', m'' etc. and let their number be $= \phi M$. Similarly let the numbers that are prime relative to N, P, etc. respectively and which are less than N, P, etc. be n, n', n'', etc., p, p', p'', etc., etc., and let the number of each be ϕN, ϕP, etc. It is clear that all numbers that are prime relative to the product A will also be prime relative to the individual factors M, N, P, etc. and vice versa (art. 19); and further that all numbers congruent to any one of the m, m', m'', etc. relative to the modulus M will be prime relative to M and vice versa. Similarly for N, P, etc. The question then reduces to this: to determine how many numbers there are less than A which are congruent relative to the modulus M among the numbers m, m', m'', etc. and which are congruent relative to the modulus N among the numbers n, n', n'', etc., etc. But from article 32 it follows that all numbers which have given residues relative to each of the moduli M, N, P, etc. will be congruent relative to their product A. Thus there will be only one which is less than A and congruent to the given residues relative to M, N, P, etc. Therefore the number we seek will be equal to the number of combinations of each of the numbers m, m', m'' with each of the n, n', n'' and p, p', p'', etc., etc. It is clear from the theory of combinations that this will be $= \phi M \cdot \phi N \cdot \phi P$ etc.   Q.E.D.

IV. It is easy to see how to apply this to the case we are treating. Let A be resolved into its prime factors; that is, let it be reduced to the form $a^{\alpha} b^{\beta} c^{\gamma}$ etc. where a, b, c, etc. are separate prime numbers. Then we will have

$$\phi A = \phi a^{\alpha} \cdot \phi b^{\beta} \cdot \phi c^{\gamma} \text{ etc.} = a^{\alpha-1}(a-1)b^{\beta-1}(b-1)c^{\gamma-1}(c-1) \text{ etc.}$$

or, more elegantly,

$$\phi A = A \cdot \frac{a-1}{a} \cdot \frac{b-1}{b} \cdot \frac{c-1}{c} \text{ etc.}$$

Example. Let $A = 60 = 2^2 \cdot 3 \cdot 5$; then $\phi A = (1/2) \cdot (2/3) \cdot (4/5) \cdot 60 = 16$. The numbers which are prime relative to 60 are 1, 7, 11, 13, 17, 19, 23, 29, 31, 37, 41, 43, 47, 49, 53, 59.

The first solution of this problem appears in the work of Euler entitled "Theorema arithmetica nova methodo demonstrata" (*Novi comm. acad. Petrop.*, 8 [1760–61], 1763, 74). It was repeated afterward in another dissertation entitled, "Speculationes circa quasdam insignes proprietates numerorum" (*Acta acad. Petrop.* 4, Part 2 [1780], 1784, 18).

▶ 39. If we determine the significance of the character ϕ in such a way that ϕA expresses the number of numbers which are relatively prime to A and *not greater* than A, it is clear that $\phi 1$ will no longer $= 0$, but $= 1$. In no other case will anything be changed. Adopting this definition we have the following theorem:

If a, a', a'', etc. are all the divisors of A (including unity and A itself), we will have

$$\phi a + \phi a' + \phi a'' + \text{etc.} = A$$

Example. Let $A = 30$; then $\phi 1 + \phi 2 + \phi 3 + \phi 5 + \phi 6 + \phi 10 + \phi 15 + \phi 30 = 1 + 1 + 2 + 4 + 2 + 4 + 8 + 8 = 30$.

Demonstration. Multiply by A/a all numbers which are prime relative to a and not greater than a; do the same for a', multiplying by A/a', and we will have $\phi a + \phi a' + \phi a''$ etc. of them, none greater than A. But

1) All these numbers will be unequal. For it is clear that all those generated by the *same* divisor of A will be unequal. Now if two equal numbers were produced from two different divisors M, N and from two numbers μ, v which were respectively prime relative to M, N, i.e. if $(A/M)\mu = (A/N)v$, it would follow that $\mu N = vM$. Suppose $M > N$. Since M is prime relative to μ and since it divides the number μN, it must divide N. A greater number divides a smaller. Q.E.A.

2) All the numbers $1, 2, 3, \ldots A$ are included among these numbers. Let t be any number whatsoever that is not larger than A, and let δ be the greatest common divisor of A, t. A/δ will be the divisor of A which is prime relative to t/δ. Clearly this number t will be found among those produced by the divisor A/δ.

3) It follows from this that the number of these numbers will be A, and therefore

$$\phi a + \phi a' + \phi a'' + \text{etc.} = A \qquad \text{Q.E.D.}$$

▶ 40. *Let the greatest common divisor of the numbers A, B, C, D, etc. $= \mu$. We can always determine the numbers a, b, c, d, etc. such that*

$$aA + bB + cC + \text{etc.} = \mu$$

Demonstration. Consider first only two numbers A, B and let their greatest common divisor $= \lambda$. Then the congruence $Ax \equiv \lambda \pmod{B}$ will be solvable (art. 30). Let the root $\equiv \alpha$ and form $(\lambda - A\alpha)/B = \beta$. Then we will have $\alpha A + \beta B = \lambda$ as we desire.

If there is a third number C, let λ' be the greatest common divisor of λ, C and determine the numbers k, γ such that $k\lambda + \gamma C = \lambda'$, then

$$k\alpha A + k\beta B + \gamma C = \lambda'$$

Clearly λ' is a common divisor of the numbers A, B, C and indeed the greatest of them, for if there were a greater $= \theta$, we would have

$$k\alpha \frac{A}{\theta} + k\beta \frac{B}{\theta} + \gamma \frac{C}{\theta} = \frac{\lambda'}{\theta} \quad \text{an integer} \qquad \text{Q.E.A.}$$

So we have what we set out to show by letting $k\alpha = a$, $k\beta = b$, $\gamma = c$, $\lambda' = \mu$.

If there are more numbers we proceed in the same manner.

And if the numbers A, B, C, D, etc. have no common divisor, clearly

$$aA + bB + cC + \text{etc.} = 1$$

▶ 41. *If p is a prime number and we have p elements, among which as many as we please can be equal so long as not all are equal, then the number of permutations of these elements will be divisible by p.*

Example. Five elements A, A, A, B, B can be arranged in ten different ways.

The demonstration of this theorem can be easily deduced from the well-known theory of permutations. For suppose among these elements there are a elements equal to A, b equal to B, c equal to C, etc. (any of the numbers a, b, c, etc. can be unity), then

$$a + b + c + \text{etc.} = p$$

and the number of permutations will be

$$\frac{1 \cdot 2 \cdot 3 \dots p}{1 \cdot 2 \cdot 3 \dots a \cdot 1 \cdot 2 \dots b \cdot 1 \cdot 2 \dots c \text{ etc.}}$$

Now it is clear that the numerator of this fraction is divisible by the denominator, since the number of permutations must be an integer. But the numerator is divisible by p, whereas the denominator which is composed of factors less than p is not divisible by p (art. 15). Therefore the number of permutations will be divisible by p (art. 19).

We hope now that the alternative demonstration which follows will be acceptable to the reader.

Consider two permutations of the same elements. Suppose that the ordering of elements differs only in that the element which is first in one occupies a different position in the other, but the remaining elements follow the same succession in both; and suppose further that if we consider the first and last elements in one ordering and look at these two in the other ordering, the first follows immediately after the last. Two such permutations we will call *similar permutations*.[d] Thus in our example the permutations *ABAAB* and *ABABA* will be similar because the elements that occupy the first, second, etc. places in the former occupy the third, fourth, etc. places in the latter, and they are arranged in the same succession. Now since any permutation is composed of p elements, it is obvious that we can find $p - 1$ permutations which are similar to it by taking the element in the first place and moving it to the second, third, etc. place. It is clear that the number of all nonidentical permutations is divisible by p, because this number is p times as great as the number of all dissimilar permutations. Suppose therefore we have two permutations

$$PQ \ldots TV \ldots YZ; \qquad V \ldots YZPQ \ldots T$$

one of which is derived from the other by moving terms forward. And suppose further that the two permutations are identical, i.e. $P = V$ etc. Let the term P which is first in the former be the $n + 1$st in the latter. In this latter permutation then the $n + 1$st term will equal the first in the former, the $n + 2$d will equal the second etc., and the $2n + 1$st will again equal the first as will the $3n + 1$st etc.; and in general the $kn + m$th term in the latter will equal the mth term of the former (provided that when $kn + m$ is greater than p, we either consider the series $V \ldots YZPQ \ldots T$ as being continually repeated from the beginning or we subtract from $kn + m$ that multiple of p which is less than $kn + m$ and the nearest to it in magnitude). And so if k is so determined that $kn \equiv 1$ (mod. p), which can be done, since p is prime, it follows

_____

[d] If similar permutations are conceived to be written in a circle so that the last element touches the first, there will be no difference at all since no place can be called first or last.

in general that the mth term is equal to the $m + 1$st or that every term is equal to its successor; i.e. all the terms are equal, contrary to the hypothesis.

▶ 42. *If the coefficients* $A, B, C, \ldots N$; $a, b, c, \ldots n$ *of two functions of the form*

$$x^m + Ax^{m-1} + Bx^{m-2} + Cx^{m-3} \ldots + N \qquad (P)$$

$$x^\mu + ax^{\mu-1} + bx^{\mu-2} + cx^{\mu-3} \ldots + n \qquad (Q)$$

are all rational and not all integers, and if the product of (P) *and* (Q)

$$= x^{m+\mu} + \mathfrak{A}x^{m+\mu-1} + \mathfrak{B}x^{m+\mu-2} + \text{etc.} + \mathfrak{Z}$$

then not all the coefficients $\mathfrak{A}, \mathfrak{B}, \ldots \mathfrak{Z}$ *can be integers.*

Demonstration. Express all fractions among the coefficients A, B, etc. a, b, etc. in their lowest form and select arbitrarily a prime number p which divides one or many of the denominators of these fractions. Suppose p divides the denominator of one of the fractional coefficients in (P). If we divide (Q) by p, it is clear that at least one fractional coefficient in (Q)/p will have p as a factor of its denominator (the coefficient of the first term, $1/p$, for example). It is easy to see that in (P) there will always be one term, a fraction, whose denominator involves *higher* powers of p than the denominators of all fractional coefficients that precede it, and *no lower* powers than the denominators of all succeeding fractional coefficients. Let this term $= Gx^g$ and let the power of p in the denominator of $G = t$. A similar term can be found in (Q)/p. Let it be $= \Gamma x^\gamma$ and let the power of p in the denominator of $\Gamma = \tau$. Obviously the value of $t + \tau$ will at least $= 2$. Now we show that the term $x^{g+\gamma}$ in the product of (P) and (Q) will have a fractional coefficient whose denominator will involve $t + \tau - 1$ powers of p.

Let the terms in (P) which precede Gx^g be $'Gx^{g+1}$, $''Gx^{g+2}$, etc. and those which follow be $G'x^{g-1}$, $G''x^{g-2}$, etc.; in like manner the terms which precede Γx^γ will be $'\Gamma x^{\gamma+1}$, $''\Gamma x^{\gamma+2}$, etc. and the terms which follow will be $\Gamma'x^{\gamma-1}$, $\Gamma''x^{\gamma-2}$, etc. It is clear that in the product of (P) and (Q)/p the coefficient of the term $x^{g+\gamma}$ will

$$= G\Gamma + 'G\Gamma' + ''G\Gamma'' + \text{etc.}$$

$$+ '\Gamma G' + ''\Gamma G'' + \text{etc.}$$

The term $G\Gamma$ will be a fraction, and if it is expressed in lowest terms, it will involve $t + \tau$ powers of p in the denominator. If any of the other terms is a fraction, lower powers of p will appear in the denominators because each of them will be the product of two factors, one of them involving no more than t powers of p, the other involving fewer than τ such powers; or one of them involving no more than τ powers of p, the other involving fewer than t such powers. Thus $G\Gamma$ will be of the form $e/(fp^{t+\tau})$, the others of the form $e'/(f'p^{t+\tau-\delta})$ where δ is positive and e, f, f' are free of the factor p, and the sum will

$$= \frac{ef' + e'fp^{\delta}}{ff'p^{t+\tau}}$$

The numerator is not divisible by p and so there is no reduction that can produce powers of p lower than $t + \tau$. Thus the coefficient of the term $x^{\theta+\gamma}$ in the product of (P) and (Q) will

$$= \frac{ef' + e'fp^{\delta}}{ff'p^{t+\tau-1}}$$

i.e. a *fraction* whose denominator involves $t + \tau - 1$ powers of p.   Q.E.D.

▶ 43. *A congruence of the mth degree*

$$Ax^m + Bx^{m-1} + Cx^{m-2} + \text{etc.} + Mx + N \equiv 0$$

whose modulus is a prime number p which does not divide A, cannot be solved in more than m different ways, that is, it cannot have more than m noncongruent roots relative to p (see art. 25, 26).

Suppose the theorem were false; we would then have congruences of different degrees m, n, etc. which have more than m, n, etc. roots. Let the lowest degree be m. All similar congruences of lower degree will then conform to the statement of our theorem. Since we have already considered the first-degree congruence (art. 26), m will here $= 2$ or greater. Let the congruence

$$Ax^m + Bx^{m-1} + \text{etc.} + Mx + N \equiv 0$$

have at least $m + 1$ roots $x \equiv \alpha$, $x \equiv \beta$, $x \equiv \gamma$, etc.; and let us suppose (this is legitimate) that all the numbers α, β, γ, etc. are positive and less than p, and that the least of them is α. Now in the proposed congruence let $y + \alpha$ be substituted for x. The result will be

$$A'y^m + B'y^{m-1} + C'y^{m-2} + \ldots + M'y + N' \equiv 0$$

Manifestly the congruence is satisfied if $y \equiv 0$ or $y \equiv \beta - \alpha$ or $y \equiv \gamma - \alpha$ etc. All of these are different roots and their number $= m + 1$. But since $y \equiv 0$ is a root, N' is divisible by p. Therefore the expression

$$y(A'y^{m-1} + B'y^{m-2} + \text{etc.} + M') \qquad \text{will be} \qquad \equiv 0 \quad (\text{mod. } p)$$

if y is replaced by one of the m values $\beta - \alpha, \gamma - \alpha$, etc., all of which are >0 and $<p$. So in all these cases we will also have

$$A'y^{m-1} + B'y^{m-2} + \text{etc.} + M' \equiv 0 \qquad (\text{art. 22})$$

That is, this congruence which is of degree $m - 1$ will have m roots, contrary to our theorem (for it is clear that A' will be $= A$ and so not divisible by p as required), but we have supposed that all congruences of degree less than m satisfy the theorem. Q.E.A.

▶ 44. We have supposed here that the modulus p does not divide the coefficient of the highest term, but the theorem is not restricted to this case. For if the first coefficient or any of the others is divisible by p, this term can be safely rejected and the congruence reduced to a lower degree. Now the first coefficient will no longer be divisible by p unless all the original coefficients were. In this case the congruence would have been identity and the unknown completely undetermined.

This theorem was first proposed and demonstrated by Lagrange (*Hist. Acad. Berlin*, 1768, p. 192).[4] It appears also in the dissertation of Legendre, "Recherches d'Analyse indéterminée" (*Hist. Acad. Paris*, 1785, p. 466). In *Novi comm. Acad. Petrop.*, *18* [1773], 1774, 93,[5] Euler showed that the congruence $x^n - 1 \equiv 0$ can have no more than n different roots, and although the result is particular, the method which this illustrious mathematician used is easily adaptable to all congruences. He had previously solved a more limited case in *Novi comm. acad. Petrop.* (5 [1754–55],

---

[4] "Nouvelle Méthode pour résoudre les problèmes indeterminés en nombres entiers."

[5] "Demonstrationes circa residua ex divisione potestatum per numeros primos resultantia."

1760, 6^6), but this method cannot be used in the general case. In Section VIII[7] we will show yet another way of proving the theorem. Although at first glance these methods seem to be different, the expert who wishes to compare them will discover that they all rest on the same principle. However, since this theorem is considered here as no more than a lemma and since the complete exposition does not pertain to this section, we will not stop to treat of composite moduli separately.

[6] "Demonstratio theorematis Fermatiani omnem numerum primum formae $4n + 1$ esse summam duorum quadratorum."

[7] Section VIII was not published.

SECTION III

RESIDUES OF POWERS

▶ 45. THEOREM. *In any geometric progression* $1, a, aa, a^3$, *etc.,
outside of the first term* 1, *there is still another term* a^t *which is
congruent to unity relative to the modulus p when p is prime relative
to a; and the exponent t is* $< p$.

Demonstration. Since the modulus p is prime relative to a and
hence to any power of a, no term of the progression will be
$\equiv 0$ (mod. p) but each of them will be congruent to one of the
numbers $1, 2, 3, \ldots p - 1$. Since the number of these is $p - 1$ it is
clear that if we consider more than $p - 1$ terms of the progression,
not all can have different least residues. So among the terms
$1, a, aa, a^3, \ldots a^{p-1}$ there must be at least one congruent pair. Let
therefore $a^m \equiv a^n$ with $m > n$. Dividing by a^n we get $a^{m-n} \equiv 1$
(art. 22) with $m - n < p$ and > 0. Q.E.D.

Example. In the progression 2, 4, 8, etc. the first term which is
congruent to unity relative to the modulus 13 is $2^{12} = 4096$. But
relative to the modulus 23 in the same progression we have
$2^{11} = 2048 \equiv 1$. Similarly 15625, the sixth power of the number
5, is congruent to unity relative to the modulus 7 but relative to
11 it is 3125, the fifth power. In some cases therefore the power is
less than $p - 1$, but in others it is necessary to go to the $p - 1$st
power itself.

▶ 46. When the progression is continued beyond the term which
is congruent to unity, the same residues that we had from the
beginning will appear again. Thus if $a^t \equiv 1$, we will have $a^{t+1} \equiv a$,
$a^{t+2} \equiv aa$, etc. until we come to the term a^{2t}. Its least residue will
again $\equiv 1$, and the *period* of the residues will begin again. Thus
we have a period of t residues which as long as it is finite will
always be repeated from the beginning; and no residues other
than the ones appearing in this period can occur in the whole

29

progression. In general we will have $a^{mt} \equiv 1$ and $a^{mt+n} \equiv a^n$. Following our notation this can be represented thus:

$$\text{if} \quad r \equiv \rho \quad (\text{mod. } t) \qquad \text{then} \qquad a^r \equiv a^\rho \quad (\text{mod. } p)$$

▶ 47. This theorem facilitates our finding the residues of powers, regardless of the size of the exponents, at the same time that we find the power that is congruent to unity. If, e.g., we want to find the residue resulting from division of 3^{1000} by 13 we have $t \equiv 3$ because $3^3 \equiv 1$ (mod. 13); so since $1000 \equiv 1$ (mod. 3), $3^{1000} \equiv 3$ (mod. 13).

▶ 48. When a^t is the *lowest* power congruent to unity (except $a^0 = 1$, which case we are not considering here) all the t terms that comprise the period of the residues will be different from one another, as is clear from the demonstration in article 45. From article 46 we see that the proposition can be converted; that is, if $a^m \equiv a^n$ (mod. p), we have $m \equiv n$ (mod. t). For if m, n are noncongruent relative to the modulus t, their least residues μ, ν will be different. But $a^\mu \equiv a^m$, $a^\nu \equiv a^n$; so $a^\mu \equiv a^\nu$, i.e. not all the powers less than a^t will be noncongruent, contrary to our hypothesis.

If therefore $a^k \equiv 1$ (mod. p) then $k \equiv 0$ (mod. t); i.e. k is divisible by t.

Thus far we have spoken about moduli which are prime relative to some a. Now we will consider separately moduli which are absolutely prime numbers; we will then develop a more general investigation based on these.

▶ 49. THEOREM. *If p is a prime number that does not divide a, and if a^t is the lowest power of a that is congruent to unity relative to the modulus p, the exponent t will either $= p - 1$ or be a factor of this number.*

See article 45 for examples.

Demonstration. We have already seen that t either $= p - 1$ or $< p - 1$. It remains to show that in the latter case t will always be a factor of $p - 1$.

I. Collect the least positive residues of all the terms $1, a, aa, \ldots$ a^{t-1} and call them $\alpha, \alpha', \alpha''$, etc. Thus $\alpha = 1$, $\alpha' \equiv a$, $\alpha'' \equiv aa$, etc. Obviously all these are different for if two terms a^m, a^n had the same residue, we would have (supposing $m > n$) $a^{m-n} \equiv 1$ with

$m - n < t$, Q.E.A., since by hypothesis no power lower than a^t can be congruent to unity. Further, all the $\alpha, \alpha', \alpha''$, etc. are contained in the series of numbers $1, 2, 3, \ldots p - 1$ without, however, exhausting the series, since $t < p - 1$. Let (A) designate the complex of all the $\alpha, \alpha', \alpha''$, etc. (A) therefore will have t terms.

II. Pick any number β from $1, 2, 3, \ldots p - 1$ which is not contained in (A). Multiply B by all the $\alpha, \alpha', \alpha''$, etc. and produce the least residues of these β, β', β'', etc. There will also be t of them. But all of these will be different from one another as well as from all the $\alpha, \alpha', \alpha''$, etc. If the *former* assertion were false, we would have $\beta a^m \equiv \beta a^n$, and dividing by β, $a^m \equiv a^n$, contrary to what we have just shown. If the *latter* were false, we would have $\beta a^m \equiv a^n$; therefore when $m < n$, $\beta \equiv a^{n-m}$ (i.e. β is congruent to one of the $\alpha, \alpha', \alpha''$, etc., contrary to the hypothesis). If finally $m > n$, multiplying by a^{t-m} we have $\beta a^t \equiv a^{t+n-m}$ or, since $a^t \equiv 1$, $\beta \equiv a^{t+n-m}$, which is the same absurdity. Let (B) designate the complex of numbers β, β', β'', etc. The number of these is t and so we now have $2t$ numbers from $1, 2, 3, \ldots p - 1$. And if (A) and (B) include all these numbers, $(p - 1)/2 = t$, and the theorem is proved.

III. But if there are some numbers left, let one of them be γ. Multiply all the $\alpha, \alpha', \alpha''$, etc. by γ and let the least residues of the products be $\gamma, \gamma', \gamma''$, etc. Designate by (C) the complex of all of these. (C) therefore will have t of the numbers $1, 2, 3, \ldots p - 1$, all of them different from each other and from the numbers contained in (A) and (B). The first two assertions can be demonstrated in the same way as in II. For the third, if we had $\gamma a^m \equiv \beta a^n$, then $\gamma \equiv \beta a^{n-m}$ or $\gamma \equiv \beta a^{t+n-m}$, according as m is $<n$ or $>n$. In either case γ would be congruent to one of the numbers in (B), contrary to the hypothesis. So we have $3t$ numbers of the $1, 2, 3, \ldots p - 1$ and if there is none remaining, $t = (p - 1)/3$ and the theorem is proved.

IV. If, however, there are some remaining, we proceed to a fourth complex of numbers (D) etc. It is clear that since the number of the $1, 2, 3, \ldots p - 1$ is finite, they will be finally exhausted. So $p - 1$ is a multiple of t and t is a factor of the number $p - 1$.   Q.E.D.

▶ 50. Since $(p - 1)/t$ is therefore an integer, it follows that by raising each side of the congruence $a^t \equiv 1$ to the power $(p - 1)/t$ we have $a^{p-1} \equiv 1$, or $a^{p-1} - 1$ *will always be divisible by p when p is a prime and does not divide a.*

This theorem is worthy of attention both because of its elegance and its great usefulness. It is usually called Fermat's theorem after its discoverer. (See Fermat, *Opera Mathem.*, p. 163.[1]) Euler first published a proof in his dissertation entitled "Theorematum quorundam ad numeros primos spectantium demonstratio," (*Comm. acad. Petrop.*, *8* [1736], 1741, 141).[a] The proof is based on the expansion of $(a + 1)^p$. From the form of the coefficients it is easily seen that $(a + 1)^p - a^p - 1$ will always be divisible by p and consequently so will $(a + 1)^p - (a + 1)$ whenever $a^p - a$ is divisible by p. Now since $1^p - 1$ is always divisible by p so also will $2^p - 2$, $3^p - 3$, and in general $a^p - a$. And since p does not divide a, $a^{p-1} - 1$ will be divisible by p. The illustrious Lambert gave a similar demonstration in *Nova acta erudit.*, 1769, 109.[2] But since the expansion of a binomial power seemed quite alien to the theory of numbers, Euler gave another demonstration that appears in *Novi comm. acad. Petrop.*[3] (*8* [1760–61], 1763, 70), which is more in harmony with what we have done in the preceding article. We will offer still others later. Here we will add another deduction based on principles similar to those of Euler. The following proposition of which our theorem is only a particular case will also prove useful for other investigations.

▶ 51. *If* p *is a prime number the pth power of the polynomial* $a + b + c + $ *etc. relative to the modulus* p *is*

$$\equiv a^p + b^p + c^p + \text{etc.}$$

Demonstration. It is clear that the *pth* power of the polynomial $a + b + c + $ etc. is composed of terms of the form $x a^\alpha b^\beta c^\gamma$ etc.

[1] The reference is to a letter of Fermat to Frénicle of 18 October 1640 (Bernard Frénicle de Bessy [1605–75]).

[a] In a previous commentary (*Comm. acad. Petrop.*, 6 [1723–33], 1738, 106 ["Observationes de theoremate quodam Fermatiano aliisque ad numeros primos spectantibus"]), this great man had not yet reached this result. In the famous controversy between Maupertuis and König on the principle of the least action, a controversy that led to strange digressions, König claimed to have at hand a manuscript of Leibniz containing a demonstration of this theorem very like Euler's (König, *Appel au public*, p. 106). We do not wish to deny this testimony, but certainly Leibniz never published his discovery. Cf. *Hist. de l'Ac. de Prusse*, 6 [1750], 1752, 530. [The reference is "Lettre de M. Euler à M. Merian (traduit du Latin)" which was sent from Berlin Sept. 3, 1752.]

[2] "Adnotata quaedam de numeris eorumque anatomia."

[3] "Solutio problematis de investigatione trium numerorum, quorum tan summa quam productum necnon summa productorum ex binis sint numeri quadrati."

where $\alpha + \beta + \gamma +$ etc. $= p$ and x is the number of permutations of p things among which $a, b, c,$ etc. appear respectively $\alpha, \beta, \gamma,$ etc. times. But in article 41 above we showed that this number is always divisible by p unless all the objects are the same; i.e. unless some one of the numbers $\alpha, \beta, \gamma,$ etc. $= p$ and all the rest $= 0$. It follows that all the terms of $(a + b + c +$ etc.$)^p$ except $a^p, b^p, c^p,$ etc. are divisible by p; and when we treat a congruence relative to the modulus p we can safely omit them and thereby get

$$(a + b + c + \text{etc.})^p \equiv a^p + b^p + c^p + \text{etc.} \qquad \text{Q.E.D.}$$

If all the quantities $a, b, c,$ etc. $= 1$, their sum will $= k$ and we will have $k^p \equiv k$, as in the preceding article.

▶ 52. Suppose we are given numbers which are to be made congruent to unity by raising to a power. We know that for the exponent involved to be of lowest degree it must be a divisor of $p - 1$. The question arises whether all the divisors of $p - 1$ enjoy this property. And if we take all numbers not divisible by p and classify them according to the exponent (in the lowest degree) which makes them congruent to unity, how many are there for each exponent? We observe first that it is sufficient to consider all positive numbers from 1 to $p - 1$; for, manifestly, numbers congruent to each other have to be raised to the same power to become congruent to unity, and therefore each number should be related to the same exponent as its least positive residue. Thus we must find out how in this respect the numbers $1, 2, 3, \ldots p - 1$ should be distributed among the individual factors of the number $p - 1$. In brief, if d is one of the divisors of the number $p - 1$ (among these 1 and $p - 1$ itself must be included), we will designate by ψd the number of positive numbers less than p whose dth power is the lowest one congruent to unity.

▶ 53. To make this easier to understand, we give an example. For $p = 19$ the numbers $1, 2, 3, \ldots 18$ will be distributed among the divisors of 18 in the following way:

1	1
2	18
3	7, 11
6	8, 12
9	4, 5, 6, 9, 16, 17
18	2, 3, 10, 13, 14, 15

Thus in this case $\psi 1 = 1$, $\psi 2 = 1$, $\psi 3 = 2$, $\psi 6 = 2$, $\psi 9 = 6$, $\psi 18 = 6$. A little attention shows that with each exponent there are associated as many numbers as there are numbers relatively prime to the exponent not greater than it. In other words in this case (if we keep the symbol of article 39) $\psi d = \phi d$. Now we will show that this observation is true in general.

I. Suppose we have a number a *belonging* to the exponent d (i.e. its dth power is congruent to unity and all lower powers are noncongruent). All its powers $a^2, a^3, a^4, \ldots a^d$ or their least residues will have the same property (their dth power will be congruent to unity). Since this can be expressed by saying that the least residues of the numbers $a, a^2, a^3, \ldots a^d$ (which are all different) are roots of the congruence $x^d \equiv 1$, and since this can have no more than d roots, it is clear that except for the least residues of the numbers $a, a^2, a^3, \ldots a^d$ there can be no other numbers between 1 and $p-1$ inclusively whose power to the exponent d is congruent to unity. So all numbers belonging to the exponent d are found among the least residues of the numbers $a, a^2, a^3, \ldots a^d$. What they are and how many there are, we find as follows. If k is a number relatively prime to d, all the powers of a^k whose exponents are $<d$ will not be congruent to unity; for let $1/k$ (mod. d) $\equiv m$ (see art. 31) and we will have $a^{km} \equiv a$. And if the eth power of a^k were congruent to unity and $e < d$, we would also have $a^{kme} \equiv 1$ and hence $a^e \equiv 1$, contrary to the hypothesis. Manifestly the least residue of a^k belongs to the exponent d. If, however, k has a divisor δ in common with d, the least residue of a^k will not belong to the exponent d, because in that case the (d/δ)th power is already congruent to unity (for kd/δ is divisible by d; that is $\equiv 0$ (mod. d) and $a^{kd/\delta} \equiv 1$). We conclude that there are as many numbers belonging to the exponent d as there are numbers relatively prime to d among $1, 2, 3, \ldots d$. But it must be remembered that this conclusion depends on the supposition that we already have one number a belonging to the exponent d. So the doubt remains as to whether it could happen that no number at all pertains to some exponents. The conclusion is therefore limited to the statement that either $\psi d = 0$ or $\psi d = \phi d$.

▶ 54. II. Let d, d', d'', etc. be all the divisors of the number $p - 1$. Since all the numbers $1, 2, 3, \ldots p - 1$ are distributed among these,

$$\psi d + \psi d' + \psi d'' + \text{etc.} = p - 1$$

But in article 40 we showed that

$$\phi d + \phi d' + \phi d'' + \text{etc.} = p - 1$$

and from the preceding article it follows that ψd is equal to or less than ϕd but not greater. Similarly for $\psi d'$ and $\phi d'$ etc. So if one or more from among the terms $\psi d, \psi d', \psi d''$, etc. were each smaller than the corresponding term from among the $\phi d, \phi d', \phi d''$, etc., the first sum could not be equal to the second. Thus we conclude finally that ψd *is always equal to* ϕd and so does not depend on the size of $p - 1$.

▶ 55. There is a particular case of the preceding proposition which merits special attention. *There always exist numbers with the property that no power less than the $p - 1$st is congruent to unity*, and there are as many of them between 1 and $p - 1$ as there are numbers less than $p - 1$ and relatively prime to $p - 1$. Since the demonstration of this theorem is less obvious than it first seems and because of the importance of the theorem itself we will use a method somewhat different from the preceding. Such diversity of methods is helpful in shedding light on more obscure points. Let $p - 1$ be resolved into its prime factors so that $p - 1 = a^\alpha b^\beta c^\gamma$ etc.; a, b, c, etc. are unequal prime numbers. We will complete the demonstration by means of the following.

I. We can always find a number A (or several of them) which belongs to the exponent a^α, and likewise the numbers B, C, etc. belonging respectively to the exponents b^β, c^γ, etc.

II. The product of all the numbers A, B, C, etc. (or the least residue of this product) belongs to the exponent $p - 1$. We show this as follows.

1) Let g be some number from $1, 2, 3, \ldots p - 1$ which does *not* satisfy the congruence $x^{(p-1)/a} \equiv 1$ (mod. p). Since the degree of the congruence is $< p - 1$ not all these numbers can satisfy it. Now if the $(p - 1)/a^\alpha$ power of $g \equiv h$, then h or its least residue belongs to the exponent a^α.

For it is clear that the a^α power of h is congruent to the $p - 1$ power of g, i.e. to unity. But the $a^{\alpha - 1}$ power of h is congruent to the $(p - 1)/a$ power of g; i.e. it is not congruent to unity. Much less can the $a^{\alpha - 2}, a^{\alpha - 3}$, etc. powers of h be congruent to unity. But the exponent of the smallest power of h which is congruent to unity (that is, the exponent to which h belongs) must divide a^α (art. 48).

And since a^α is divisible only by itself and lower powers of a, a^α will necessarily be the exponent to which h belongs. Q.E.D. By a similar method we show that there exist numbers belonging to the exponents b^β, c^γ, etc.

2) If we suppose that the product of all the A, B, C, etc. does not belong to the exponent $p - 1$ but to a smaller number t, then t divides $p - 1$ (art. 48); that is $(p - 1)/t$ will be an integer greater than unity. It is easily seen that this quotient is one of the prime numbers a, b, c, etc. or at least is divisible by one of them (art. 17), e.g. by a. From this point on the reasoning is the same as above. Thus t divides $(p - 1)/a$ and the product ABC etc. raised to the $(p - 1)/a$ power will also be congruent to unity (art. 46). But it is evident that all the numbers B, C, etc. (A excepted) will become congruent to unity if raised to the $(p - 1)/a$ power because the exponents b^β, c^γ, etc. to which they each belong divide $(p - 1)/a$. Thus we have

$$A^{(p-1)/a}B^{(p-1)/a}C^{(p-1)/a} \text{ etc.} \equiv A^{(p-1)/a} \equiv 1$$

It follows that the exponent to which A belongs ought to divide $(p - 1)/a$ (art. 48); i.e. $(p - 1)/a^{\alpha+1}$ is an integer. But $(p - 1)/a^{\alpha+1}$ $= (b^\beta c^\gamma$ etc.$)/a$ cannot be an integer (art. 15). We conclude therefore that our supposition is inconsistent, i.e. that the product ABC etc. really belongs to the exponent $p - 1$.   Q.E.D.

The second demonstration seems a little longer than the first, but the first is less direct than the second.

▶ 56. This theorem furnishes an outstanding example of the need for circumspection in number theory so that we do not accept fallacies as certainties. Lambert in the dissertation that we praised above, *Nova acta erudit.* 1769, p. 127,[4] makes mention of this proposition but does not mention the need for demonstrating it. No one has attempted the demonstration except Euler in *Novi comm. Acad. Petrop.*, (*18* [1773], 1774, 85), "*Demonstrationes circa residua ex divisione potestatum per numeros primos resultantia.*" See especially his article 37 where he speaks at great length of the need for demonstration. But the demonstration which this shrewdest of men presents has two defects. One is that his article 31 ff. tacitly supposes that the congruence $x^n \equiv 1$ (translating his

---

[4] Cf. p. 32.

argument into our notation) really has n different roots, although nothing more had been shown previously than that it could not have *more* than n roots; the other is that the formula of article 34 was derived only by induction.

▶ 57. Along with Euler we will call numbers belonging to the exponent $p - 1$ *primitive roots*. Therefore if a is a primitive root the least residues of the powers $a, a^2, a^3, \ldots a^{p-1}$ will all be different. It is then easy to deduce that among them we will find all the numbers $1, 2, 3, \ldots p - 1$, since each has the same number of elements. This means that any number not divisible by p is congruent to some power of a. This remarkable property is of great usefulness, and it can considerably reduce the arithmetic operations relative to congruences in much the same way that the introduction of logarithms reduces the operations in ordinary arithmetic. We will arbitrarily choose some primitive root a as a *base* to which we will refer all numbers not divisible by p. And if $a^e \equiv b$ (mod. p), we will call e the *index* of b. For example, if relative to the modulus 19 we take the primitive root 2 as base, we have

numbers 1. 2. 3. 4. 5. 6. 7. 8. 9. 10. 11. 12. 13. 14. 15. 16. 17. 18

indices 0. 1. 13. 2. 16. 14. 6. 3. 8. 17. 12. 15. 5. 7. 11. 4. 10. 9

For the rest, it is clear that with the base unchanged each number has many indices but that they are all congruent relative to the modulus $p - 1$; so when there is question of indices, those that are congruent relative to the modulus $p - 1$ will be regarded as equivalent in the same way that numbers are regarded as equivalent when they are congruent relative to the modulus p.

▶ 58. Theorems pertaining to indices are completely analogous to those that refer to logarithms.

The index of the product of any number of factors is congruent to the sum of the indices of the individual factors relative to the modulus $p - 1$.

The index of the power of a number is congruent relative to the modulus $p - 1$ to the product of the index of the number by the exponent of the power.

Because of their simplicity we omit demonstrations of these theorems.

It is clear from the above that if we wish to construct a table that gives the indices for all numbers relative to different moduli, we can omit from it all numbers larger than the modulus and also all composite numbers. An example of this kind of table appears at the end of this work (Table 1). In the first vertical column are arranged prime numbers and the powers of prime numbers from 3 to 97. These are to be regarded as moduli. Next to each of these in the following column are the numbers chosen as base; then follow the indices of successive prime numbers arranged in blocks of five. At the top of the columns the prime numbers are again arranged in the same order so that it is easy to find the index corresponding to a given prime number relative to a given modulus.

For example, if $p = 67$, the index for the number 60 with 12 as base $\equiv$ 2 Ind. 2 + Ind. 3 + Ind. 5 (mod. 66) $\equiv$ 58 + 9 + 39 $\equiv$ 40

▶ 59. *The index of the value of an expression like a/b (mod. p)* (see art. 31) *is congruent relative to the modulus $p - 1$ to the difference between the index of the numerator a and the index of the denominator b, provided a, b are not divisible by p.*

Let c be such a value. We have $bc \equiv a$ (mod. p) and so

$$\text{Ind. } b + \text{Ind. } c \equiv \text{Ind. } a \text{ (mod. } p - 1)$$

and

$$\text{Ind. } c \equiv \text{Ind. } a - \text{Ind. } b$$

Then if we have two tables, one of which gives the index for any number relative to any prime modulus and the other gives the number belonging to a given index, all congruences of the first degree can be easily solved because they can be reduced to congruences whose modulus is a prime number (art. 30). For example, a given congruence

$$29x + 7 \equiv 0 \text{ (mod. 47)} \qquad \text{becomes} \qquad x \equiv \frac{-7}{29} \text{ (mod. 47)}$$

and so

$$\text{Ind. } x \equiv \text{Ind. } -7 - \text{Ind. } 29 \equiv \text{Ind. } 40 - \text{Ind. } 29 \equiv 15 - 43$$

$$\equiv 18 \text{ (mod. 46)}$$

Now 3 is the number whose index is 18. So $x \equiv 3$ (mod. 47). We have not added the second table, but in Section VI we shall see how to replace it with another.

▶ 60. In article 31 we designated by a special sign roots of congruences of the first degree, so in what follows we will represent simple congruences of higher degrees by a special sign. Just as $\sqrt[n]{A}$ indicates the root of the equation $x^n = A$, so with the modulus added $\sqrt[n]{A}$ (mod. p) will denote any root of the congruence $x^n \equiv A$ (mod. p). We will say that the expression $\sqrt[n]{A}$ (mod. p) has as many values as it has values that are noncongruent relative to p, since all those that are congruent relative to p are considered equivalent (art. 26). It is clear that if A, B are congruent relative to p, the expressions $\sqrt[n]{A}$, $\sqrt[n]{B}$ (mod. p) will be equivalent.

Now if we are given $\sqrt[n]{A} \equiv x$ (mod. p), we will have n Ind. $x \equiv$ Ind. A (mod. $p - 1$). According to the rules of the preceding section, from this congruence we can deduce the values of Ind. x and from these the corresponding values of x. It is easily seen that x will have as many values as there are roots of the congruence n Ind. $x \equiv$ Ind. A (mod. $p - 1$). And manifestly $\sqrt[n]{A}$ will have only one value when n is prime relative to $p - 1$. But when $n, p - 1$ have a greatest common divisor δ, Ind. x will have δ noncongruent values relative to $p - 1$, and $\sqrt[n]{A}$ the same number of noncongruent values relative to p, provided Ind. A is divisible by δ. If this condition is lacking, $\sqrt[n]{A}$ will have no real value.

Example. We are looking for the values of the expression $\sqrt[15]{11}$ (mod. 19). We must therefore solve the congruence 15 Ind. $x \equiv$ Ind. $11 = 6$ (mod. 18), and we find three values of Ind. $x \equiv 4, 10, 16$ (mod. 18). The corresponding values of x are 6, 9, 4.

▶ 61. No matter how expeditious this method is when we have the necessary tables, we should not forget that it is indirect. It will be useful therefore to determine how powerful the direct methods are. We will consider here what we can deduce from previous sections; other considerations that demand more profound investigation will be reserved for Section VIII.[5] We begin with the simplest case where $A = 1$. That is, we will look for the roots of the congruence $x^n \equiv 1$ (mod. p). Here after taking some primitive root as base we must have n Ind. $x \equiv 0$ (mod. $p - 1$). Now if n is prime relative to $p - 1$ this congruence will have only one root; that is, Ind. $x \equiv 0$ (mod. $p - 1$). In this case then $\sqrt[n]{1}$ (mod. p) has a unique value, namely $\equiv 1$. But when the numbers

[5] Section VIII was not published. See Author's Preface.

$n, p - 1$ have a (greatest) common divisor δ, the complete solution of the congruence n Ind. $x \equiv 0$ (mod. $p - 1$) will be Ind. $x \equiv 0$ (mod. $(p - 1)/\delta$) [see art. 29]; i.e. Ind. x relative to the modulus $p - 1$ should be congruent to one of the numbers

$$0, \quad \frac{p - 1}{\delta}, \quad \frac{2(p - 1)}{\delta}, \quad \frac{3(p - 1)}{\delta}, \quad \ldots, \quad \frac{(\delta - 1)(p - 1)}{\delta}$$

that is, it will have δ values which are noncongruent relative to the modulus $p - 1$; and so in this case x also will have δ different values (noncongruent relative to the modulus p). Thus we see that the expression $\sqrt[\delta]{1}$ also has δ different values whose indices are absolutely the same as the previous ones. For this reason the expression $\sqrt[\delta]{1}$ (mod. p) is completely equivalent to $\sqrt[n]{1}$ (mod. p); i.e. the congruence $x^\delta \equiv 1$ (mod. p) has the same roots as $x^n \equiv 1$ (mod. p). The former, however, will be of lower degree if δ and n are not equal.

Example. $\sqrt[15]{1}$ (mod. 19) has three values because 3 is the greatest common divisor of the numbers 15, 18 and they will also be the values of the expression $\sqrt[3]{1}$ (mod. 19). They are 1, 7, 11.

▶ 62. By this reduction we know that other congruences of the form $x^n \equiv 1$ need not be solved unless n is a divisor of the number $p - 1$. We will see later that congruences of this form can always be reduced; what we showed so far is not sufficient to indicate this. There is one case, however, that we can dispose of here, i.e. when $n = 2$. Clearly the values of the expression $\sqrt[2]{1}$ will be $+1$ and -1, since there cannot be more than two, and $+1$ and -1 are always noncongruent unless the modulus $= 2$, in which case $\sqrt[2]{1}$ can have only one value. It follows that $+1$ and -1 will also be values of the expression $\sqrt[2m]{1}$ when m is prime relative to $(p - 1)/2$. This always happens when the modulus is such that $(p - 1)/2$ is an absolutely prime number (except when $p - 1 = 2m$ in which case all the numbers $1, 2, 3, \ldots p - 1$ are roots)—e.g. when $p = 3, 5, 7, 11, 23, 47, 59, 83, 107$, etc. As a corollary we note that the index of -1 is always $\equiv (p - 1)/2$ (mod. $p - 1$), no matter what primitive root is taken as base. This is so because 2 Ind. $(-1) \equiv 0$ (mod. $p - 1$), and so Ind. (-1) will be either $\equiv 0$ or $\equiv (p - 1)/2$ (mod. $p - 1$). However 0 is always an index of $+1$, and $+1$ and -1 must always have different indices (except for the case $p = 2$ which we need not consider here).

▶ 63. We showed in article 60 that the expression $\sqrt[n]{A}$ (mod. p) has δ different values, or none at all if δ is the greatest common divisor of the numbers $n, p - 1$. Now in the same way that $\sqrt[n]{A}$ and $\sqrt[\delta]{A}$ are equivalent when $A \equiv 1$, we will give a more general proof showing that $\sqrt[n]{A}$ can always be reduced to another expression, $\sqrt[\delta]{B}$, to which it will be equivalent. If we denote some value of this by x we will have $x^n \equiv A$; further let t be a value of the expression δ/n (mod. $p - 1$). These values are real as is clear from article 31. Now $x^{tn} \equiv A^t$, but $x^{tn} \equiv x^\delta$ because $tn \equiv \delta$ (mod. $p - 1$). Therefore $x^\delta \equiv A^t$, and so any value of $\sqrt[n]{A}$ will also be a value of $\sqrt[\delta]{A^t}$. As often therefore as $\sqrt[n]{A}$ has real values, it will be completely equivalent to the expression $\sqrt[\delta]{A^t}$. This is true because the former cannot have different values than the latter nor can there be fewer of them except when $\sqrt[n]{A}$ has no real values. In this case, however, it can happen that $\sqrt[\delta]{A^t}$ has real values.

Example. If we are looking for the values of the expression $\sqrt[21]{2}$ (mod. 31), the greatest common divisor of the numbers 21 and 30 is 3, and 3 is a value of the expression 3/21 (mod. 30); so if $\sqrt[21]{2}$ has real values it will be equivalent to the expression $\sqrt[3]{2^3}$ or $\sqrt[3]{8}$; and we find in fact that the values of the latter which are 2, 10, 19 satisfy the former also.

▶ 64. In order to avoid trying this operation in vain, we should investigate the rule for determining whether $\sqrt[n]{A}$ has real values. If we have a table of indices, it is easy, for from article 60 it is clear that we have real values if the index of A with any primitive root taken as base is divisible by δ. In the contrary case there will be no real values. But we can still determine this without such a table. Let the index of A be $= k$. If this is divisible by δ, then $k(p - 1)/\delta$ will be divisible by $p - 1$ and vice versa. But the index of the number $A^{(p - 1)/\delta}$ will be $k(p - 1)/\delta$. So if $\sqrt[n]{A}$ (mod. p) has real values, $A^{(p - 1)/\delta}$ will be congruent to unity; if not, it will be noncongruent. So in the example of the preceding article we have $2^{10} = 1024 \equiv 1$ (mod. 31) and we conclude that $\sqrt[21]{2}$ (mod. 31) has real values. In the same way we see that $\sqrt[2]{-1}$ (mod. p) will always have a pair of real values when p is of the form $4m + 1$ but none when p is of the form $4m + 3$, because $(-1)^{2m} = 1$ and $(-1)^{2m+1} = -1$. This elegant theorem which is commonly enunciated thus: *If p is a prime number of the form $4m + 1$, a square a^2 can be found such that $a^2 + 1$ is divisible by p; but if p is of the form $4m - 1$ such* *a square cannot be found,...* was demonstrated in this manner by Euler in *Novi comm. acad. Petrop.* (*18* [1773], 1774, 112).[6] He had given another demonstration much before that in 1760[7] (*Novi comm. acad. Petrop.*, 5 [1754–55], 5). In a previous dissertation[8] (*4* [1752–53], 1758, 25) he had not yet arrived at the result. Later also Lagrange gave a demonstration of the theorem in *Nouv. mém. Acad. Berlin*, 1775, p. 342.[9] We will give another demonstration in the following section which is properly devoted to this subject.

▶ 65. After discussing how to reduce expressions $\sqrt[n]{A}$ (mod. p) to others in which n is a divisor of $p - 1$, and finding a criterion by which we can determine whether or not they admit of real values, we will consider expressions $\sqrt[n]{A}$ (mod. p) in which n is already a divisor of $p - 1$. First we will show what relation the different values of the expression have among themselves; we will then treat of certain devices by means of which one value can very often be found.

First. When $A \equiv 1$ and r is one of the values for the expressions $\sqrt[n]{1}$ (mod. p) or $r^n \equiv 1$ (mod. p), all the powers of this r will also be values of the expression; there will be as many different ones as there are unities in the exponent to which r belongs (art. 48). Therefore if r is a value belonging to the exponent n, all the powers $r, r^2, r^3, \ldots r^n$ (where *unity* can replace the last) involve values of the expression $\sqrt[n]{1}$ (mod. p). We will explain in Section VIII what aids there are for finding such values that belong to the exponent n.

Second. When A is not congruent to unity and one value z of the expression $\sqrt[n]{A}$ (mod. p) is known, the others can be found as follows. Let the values of the expression $\sqrt[n]{1}$ be

$$1, r, r^2, \ldots, r^{n-1}$$

(as we have just shown). All the values of the expression $\sqrt[n]{A}$ will be

$$z, zr, zr^2, \ldots, zr^{n-1}$$

It is evident that all these satisfy the congruence $x^n \equiv A$ because if

[6] Cf. pp. 27, 36.

[7] Cf. pp. 27, 28.

[8] "De numeris qui sunt aggregata duorum quadratorum."

[9] "Suite des Recherches d'arithmétique imprimées dans le volume de l'année 1773."

one of them is $\equiv zr^k$, its nth power $z^n r^{nk}$ will be congruent to A. This is obvious because $r^n \equiv 1$ and $z^n \equiv A$. And from article 23 it is easy to see that all these values are different. The expression $\sqrt[n]{A}$ can have none except these n values. So, e.g., if one value of the expression $\sqrt[2]{A}$ is z, another will be $-z$. From the preceding we must conclude that it is not possible to find all values of the expression $\sqrt[n]{A}$ without at the same time establishing all the values of the expression $\sqrt[n]{1}$.

▶ 66. The second thing that we proposed to do was to find out when one value of the expression $\sqrt[n]{A}$ (mod. p) can be determined directly (presupposing, of course, that n is a divisor of $p - 1$). This happens when some value is congruent to a power of A. This is a rather frequent occurrence and warrants a moment's consideration. *If this value exists* let it be z, that is $z \equiv A^k$ and $z \equiv z^n$ (mod. p.) As a result $A \equiv A^{kn}$; and so if we can find a number k such that $A \equiv A^{kn}$, A^k will be the value we seek. But this condition is the same as saying that $1 \equiv kn$ (mod. t) where t is the exponent to which A belongs (art. 46, 48). But for such a congruence to be possible it is necessary that n and t be relatively prime. In which case we will have $k \equiv 1/n$ (mod. t); if however t and n have a common divisor, there can be no value of z that is congruent to a power of A.

▶ 67. Since to arrive at this solution it is necessary to know t, let us see how we proceed if it is not known. First, it is obvious that t must divide $(p - 1)/n$ if $\sqrt[n]{A}$ (mod. p) is to have real values as we always suppose here. Let y be some one of these values; we will then have $y^{p-1} \equiv 1$ and $y^n \equiv A$ (mod. p); and by raising both sides of the latter congruence to the $(p - 1)/n$ power we get $A^{(p-1)/n} \equiv 1$; and so $(p - 1)/n$ is divisible by t (art. 48). Now if $(p - 1)/n$ is prime relative to n, the congruence of the preceding article $kn \equiv 1$ can also be solved relative to the modulus $(p - 1)/n$, and obviously any value of k satisfying the congruence relative to this modulus will satisfy it also relative to the modulus t which divides $(p - 1)/n$ (art. 5). Thus we have found what we wanted. If $(p - 1)/n$ is not prime relative to n, eliminate from $(p - 1)/n$ all the prime factors which at the same time divide n. Thus we get the number $(p - 1)/nq$ which is prime relative to n. Here we use q to indicate the product of all the prime factors that have been eliminated. Now if the condition of the previous article holds, that is if t is prime relative to n, then t will also be prime relative to q and so will divide $(p - 1)/nq$. So if we solve the congruence $kn \equiv 1$ (mod. $(p - 1)/nq$), (which can be done because n is prime relative to $(p - 1)/nq$), the value of k will also satisfy the congruence relative to the modulus t, which is what we were looking for. This whole device consists in discovering a number to take the place of the t which we do not know. But we must remember that when $(p - 1)/n$ is not prime relative to n we have supposed the condition of the previous article. If this is not true, all the conclusions will be erroneous; and if by following the given rules without care we find a value for z whose nth power is not congruent to A, we know that the condition is lacking and the method cannot be used.

▶ 68. But even in this case it is often advantageous to have done the work involved, and it is worth investigating the relationship between this false value and the true ones. Suppose therefore that k, z have been duly determined but that z^n is not $\equiv A$ (mod. p). Then if we can only determine the values of the expression $\sqrt[n]{(A/z^n)}$ (mod. p), we will obtain the values of $\sqrt[n]{A}$ by multiplying each of them by z. For if v is a value of $\sqrt[n]{(A/z^n)}$, we will have $(vz)^n \equiv A$. But the expression $\sqrt[n]{(A/z^n)}$ is simpler than $\sqrt[n]{A}$ because very often A/z^n (mod. p) belongs to a lower exponent than A. More precisely, if the greatest common divisor of the numbers t, q is d, A/z^n (mod. p) will belong to the exponent d as we have seen. Substituting for z, we get $A/z^n \equiv 1/A^{kn - 1}$ (mod. p). But $kn - 1$ is divisible by $(p - 1)/nq$, and $(p - 1)/n$ by t (preceding article); that is $(p - 1)/nd$ is divisible by t/d. But t/d and q/d are relatively prime (hypothesis), so also $(p - 1)/nd$ is divisible by tq/dd or $(p - 1)/nq$ by t/d. And thus $kn - 1$ will be divisible by t/d and $(kn - 1)d$ by t. The final result gives us $A^{(kn - 1)d} \equiv 1$ (mod. p), and we deduce that A/z^n raised to the dth power is congruent to unity. It is easy to show that A/z^n cannot belong to any exponent less than d, but since this is not required for our purpose we will not dwell on it. We can be certain therefore that A/z^n (mod. p) always belongs to a smaller exponent than A except when t divides q and so $d = t$.

But what is the advantage in having A/z^n belong to a smaller exponent than A? There are more numbers that can be A than can be A/z^n and when we want to solve many expressions of the form $\sqrt[n]{A}$ relative to the same modulus, we have the advantage of being able to derive several results from the one computation.

Thus, e.g., we will always be able to determine at least one value of the expression $\sqrt[2]{A}$ (mod. 29) if we know the values of the expression $\sqrt[2]{-1}$ (mod. 29) [these are ± 12]. From the preceding article it is easy to see how we can determine one value of this expression directly when t is odd, and that d will be $=2$ when t is even; except for -1 no number belongs to the exponent 2.

Examples. We want to solve $\sqrt[3]{31}$ (mod. 37). Here $p - 1 = 36$, $n = 3$, $(p - 1)/3 = 12$, and so $q = 3$. Now we want $3k \equiv 1$ (mod. 4), and we get this by letting $k = 3$. Thus $z \equiv 31^3$ (mod. 37) $\equiv 6$, and it is true that $6^3 \equiv 31$ (mod. 37). If the values of the expression $\sqrt[3]{1}$ (mod. 37) are known, the remaining values of the expression $\sqrt[3]{6}$ can be determined. The values of $\sqrt[3]{1}$ (mod. 37) are $1, 10, 26$. Multiplying these by 6 we get for the other two $\sqrt[3]{31} \equiv 23, 8$.

If however we are looking for the value of the expression $\sqrt[2]{3}$ (mod. 37), $n = 2$, $(p - 1)/n = 18$, and so $q = 2$. Since we want $2k \equiv 1$ (mod. 9), $k \equiv 5$ (mod. 9). And $z \equiv 3^5 \equiv 21$ (mod. 37); but 21^2 is not $\equiv 3$ whereas it is $\equiv 34$; on the other hand $3/34$ (mod. 37) $\equiv -1$ and $\sqrt[2]{-1}$ (mod. 37) $\equiv \pm 6$; thus we find the true values $\pm 6 \cdot 21 \equiv \pm 15$.

This is practically all we can say about the solution of such expressions. It is clear that direct methods are often rather long, but this is true of almost all direct methods in the theory of numbers; nevertheless their usefulness should be demonstrated. It is beyond the purpose of our investigation, however, to explain one by one the particular artifices that become familiar to anyone working in this field.

▶ 69. We return now to consideration of the roots we call primitive. We have shown that when we take any primitive root as a base all numbers whose indices are prime relative to $p - 1$ will also be primitive roots and that these are the only ones; we therefore know at the same time the number of the primitive roots (see art. 53). In general it is left to our own discretion to decide which root to use as a base. Here, as in logarithmic calculus, we can have many different systems.[b] Let us see how they are connected. Let a, b be two primitive roots and m another number. When we take a as our base, the index of the number $b \equiv \beta$ and the index of the

[b] But it differs in that for logarithms the number of systems is infinite; here there are only as many as there are primitive roots. For it is manifest that congruent bases generate the same system.

number $m \equiv \mu$ (mod. $p - 1$). But when we take b as our base, the index of the number a is $\equiv \alpha$ and of the number m is $\equiv v$ (mod. $p - 1$). Now $\alpha^\beta \equiv b$ and $a^{\alpha\beta} \equiv b^\alpha \equiv a$ (mod. p) [by hypothesis] so $\alpha\beta \equiv 1$ (mod. $p - 1$). By a similar process we find that $v \equiv \alpha\mu$, and $\mu \equiv \beta v$ (mod. $p - 1$). Therefore if we have a table of indices constructed for the base a it is easy to convert it into another in which the base is b. For if for the base a the index of b is $\equiv \beta$, for the base b the index of a will be $\equiv 1/\beta$ (mod. $p - 1$) and by multiplying all the indices of the table by this number, we will find all the indices for the base b.

▶ 70. Although a given number can have varying indices according as different primitive roots are taken as bases, they all agree in this—that the greatest common divisor for each of them and $p - 1$ will be the same. For if for the base a, the index of a given number is m, and for the base b, the index is n and if the greatest common divisors of these numbers and $p - 1$ are μ, v and unequal, then one will be greater; e.g. $\mu > v$ and so μ will not divide n. Now assuming b for the base, let α be the index of a and we will have (preceding article) $n \equiv \alpha m$ (mod. $p - 1$) and μ will divide n. Q.E.A.

We can also see that this greatest divisor common to the indices of a given number and to $p - 1$ does not depend on the base from the fact that it is equal to $(p - 1)/t$. Here t is the exponent to which the number belongs whose indices we are considering. For if the index for any base is k, t will be the smallest number (except zero) which multiplied by k gives a multiple of $p - 1$ (see art. 48, 58), that is, the least value of the expression $0/k$ (mod. $p - 1$) except zero. That this is equal to the greatest common divisor of the numbers k and $p - 1$ is derived with no difficulty from article 29.

▶ 71. It is always permissible to choose a base so that the number pertaining to the exponent t matches some predetermined index whose greatest divisor in common with $p - 1$ is $=(p - 1)/t$. Let us designate by d this divisor, let the index proposed be $\equiv dm$, and let the index of the proposed number be $\equiv dn$. Here a primitive root a is selected for the base; m, n will be prime relative to $(p - 1)/d$ or to t. Then if ε is a value of the expression dn/dm (mod. $p - 1$) and at the same time is prime relative to $p - 1$, a^ε will be a primitive root. With this base the proposed number will produce the index dm just as we wanted (for $a^{\varepsilon dm} \equiv a^{dn} \equiv$ the proposed number).

To prove that the expression dn/dm (mod. $p - 1$) has values relatively prime to $p - 1$ we proceed as follows. This expression is equivalent to n/m (mod. $(p - 1)/d$), or to n/m (mod. t) [see art. 31, 2] and all its values will be prime relative to t; for if any value e should have a common divisor with t, this divisor will also divide me and so also n, since me is congruent to n relative to t. But this is contrary to the hypothesis which demands that n be prime relative to t. When therefore all the prime divisors of $p - 1$ divide t, *all* the values of the expression n/m (mod. t) will be prime relative to $p - 1$ and their number will $= d$. When however $p - 1$ has other divisors f, g, h, etc., which are prime and do not divide t, one value of the expression n/m (mod. t) $\equiv e$. But then since all the numbers t, f, g, h, etc. are prime relative to each other, a number ε can be found which is congruent to e relative to t, and relative to f, g, h, etc. is congruent to other numbers which are prime relative to these (art. 32). Such a number will be divisible by no prime factor of $p - 1$ and so will be prime relative to $p - 1$ as was desired. From the theory of combinations we easily deduce that the number of such values will be equal to

$$\frac{(p - 1) \cdot (f - 1) \cdot (g - 1) \cdot (h - 1) \cdot \text{etc.}}{t \quad \cdot \quad f \quad \cdot \quad g \quad \cdot \quad h \quad \cdot \text{etc.}}$$

but in order not to extend this digression too far, we omit the demonstration. It is not necessary to our purpose in any case.

▶ 72. Although in general the choice of a primitive root for the base is entirely arbitrary, at times some bases will prove to have special advantages over others. In Table 1 we have always taken 10 as the base when it was a primitive root; in other cases we have always chosen the base so that the index of the number 10 would be the smallest possible; i.e. we let it $= p - 1/t$ with t the exponent to which 10 belongs. We will indicate the advantage of this in Section VI where the same table is used for other purposes. But here too there still remains some freedom of choice, as we have seen in the preceding article. And so we have always chosen as base the *smallest* primitive root that satisfies the conditions. Thus for $p = 73$ where $t = 8$ and $d = 9$, a^ε has $72/8 \cdot 2/3$; i.e. 6 values, which are 5, 14, 20, 28, 39, 40. We chose the smallest, 5, as base.

▶ 73. Finding primitive roots is reduced for the most part to trial and error. If we add what we have said in article 55 to what we have indicated above about the solution of the congruence $x^n \equiv 1$, we will have all that can be done by direct methods. Euler (*Opuscula Analytica, 1*, 152[10]) admits that it is extremely difficult to assign these numbers and that their nature is one of the deepest mysteries of numbers. But they can be determined easily enough by the following method. Skillful mathematicians know how to reduce tedious calculations by a variety of devices, and here experience is a better teacher than precept.

1) Arbitrarily select a number a which is prime relative to p (we always use this letter to designate the modulus; usually the calculation is simpler if we choose the smallest possible—e.g. the number 2). Next determine its period (art. 46), i.e. the least residues of its powers, until we come to the power a^t whose least residue is 1.[c] If $t = p - 1$, a is a primitive root.

2) If however $t < p - 1$, choose another number b that is not contained in the period of a and in the same way investigate its period. If we designate by u the exponent to which b belongs, we see that u cannot be equal to t or to a factor of t; for in either case we would have $b^t \equiv 1$ which cannot be, since the period of a includes all numbers whose t power is congruent to unity (art. 53). Now if $u = p - 1$, b would be a primitive root; however if u is not $= p - 1$ but is a multiple of t, we have gained this much— we know we can find a number belonging to a higher exponent and so we will be closer to our goal, which is to find a number belonging to the *maximum* exponent. Now if u does not $= p - 1$ and is not a multiple of t, we can nevertheless find a number belonging to an exponent greater than t, u, namely to the exponent equal to the least common multiple of t, u. Let this number $= y$ and resolve y into two factors m, n which are relatively prime and such that one of them divides t and the other divides u.[d] As a

[10] "Disquisitio accuratior circa residua ex divisione quadratorum altiorumque potestatum per numeros primos relicta."

[c] It is not necessary to know these powers, since the least residue can be easily obtained from the least residue of the preceding power.

[d] From article 18 we see how this can be done without difficulty. Resolve y into factors which are either different prime numbers or powers of different prime numbers. Each of these will divide t or u (or both of them). Assign each of them either to t or to u according to which one it divides. If one of them divides both, it can be arbitrarily assigned. Let the product of those assigned to $t = m$, the product of the others $= n$. Obviously m divides t, n divides u, and $mn = y$.

result, the t/m power of a will $\equiv A$, the u/n power of b will $\equiv B$ (mod. p), and the product AB will be the number belonging to the exponent y. This is clear, since A belongs to the exponent m and B belongs to the exponent n and, since m, n are relatively prime, the product AB will belong to mn. We can prove this in practically the same way as in article 55, Section II.

3) Now if $y = p - 1$, AB will be a primitive root; if not, we proceed as before, using another number not occurring in the period of AB. This will either be a primitive root, or it will belong to an exponent greater than y, or with its help (as before) we can find a number belonging to an exponent greater than y. Since the numbers we get by repeating this operation belong to constantly increasing exponents, we must finally discover a number that belongs to the *maximum* exponent. This will be a primitive root. Q.E.F.

▶ 74. This becomes clearer by an example. Let $p = 73$ and let us look for a primitive root. We will try first the number 2 whose period is

$$1.\ 2.\ 4.\ 8.\ 16.\ 32.\ 64.\ 55.\ 37.\ 1 \text{ etc.}$$

$$0.\ 1.\ 2.\ 3.\ \ 4.\ \ 5.\ \ 6.\ \ 7.\ \ 8.\ 9 \text{ etc.}$$

Since the power of the exponent 9 is therefore congruent to unity, 2 is not a primitive root. We will try another number that does not occur in the period—e.g. 3. Its period will be

$$1.\ 3.\ 9.\ 27.\ 8.\ 24.\ 72.\ 70.\ 64.\ 46.\ 65.\ 49.\ \ 1 \text{ etc.}$$

$$0.\ 1.\ 2.\ \ 3.\ 4.\ \ 5.\ \ 6.\ \ 7.\ \ 8.\ \ 9.\ 10.\ 11.\ 12 \text{ etc.}$$

So 3 is not a primitive root. However the least common multiple of the exponents to which 2, 3 belong (i.e. the numbers 9, 12) is 36 which is resolved into the factors 9 and 4 as we saw in the previous article. Raising 2 to the power 9/9, 3 to the power 3, the product of these is 54 which belongs to the exponent 36. If finally we compute the period of 54 and try a number that is not contained in it (e.g. 5), we find that this is a primitive root.

▶ 75. Before abandoning this line of argument we will present certain propositions which because of their simplicity are worth attention.

The product of all the terms of the period of any number is $\equiv 1$

when their number or the exponent to which the number belongs is odd and $\equiv -1$ when the exponent is even.

Example. For the modulus 13, the period of the number 5 consists of the terms 1, 5, 12, 8, and its product $480 \equiv -1$ (mod. 13).

Relative to the same modulus the period of the number 3 consists of the terms 1, 3, 9, and the product $27 \equiv 1$ (mod. 13).

Demonstration. Let the exponent to which a number belongs be t and the index of the number $(p - 1)/t$. This can always be done if we choose the right base (art. 71). Then the index of the product of all the terms of the period will be

$$\equiv (1 + 2 + 3 + \text{etc.} + t - 1)\frac{p - 1}{t} = \frac{(t - 1)(p - 1)}{2}$$

i.e. $\equiv 0$ (mod. $p - 1$) when t is odd, and $\equiv (p - 1)/2$ when t is even; in the former case the product will be $\equiv 1$ (mod. p) and in the latter $\equiv -1$ (mod. p) (art. 62). Q.E.D.

▶ 76. If the number in the preceding theorem is a primitive root, its period will include all the numbers $1, 2, 3, \ldots, p - 1$. Its product will always $\equiv -1$ (for $p - 1$ will always be even except when $p = 2$; in this case -1 and $+1$ are equivalent). This elegant theorem is usually worded thus: *The product of all numbers less than a given prime number is divisible by this prime number if unity is added to the product.* It was first published by Waring and attributed to Wilson: Waring, *Meditationes Algebraicae* (3d ed., Cambridge, 1782, p. 380).[11] But neither of them was able to prove the theorem, and Waring confessed that the demonstration was made more difficult because no *notation* can be devised to express a prime number. But in our opinion truths of this kind should be drawn from the ideas involved rather than from notations. Afterward Lagrange gave a proof (*Nouv. mém. Acad. Berlin*, 1771[12]). He does this by considering the coefficients that arise in expanding the product

$$(x + 1)(x + 2)(x + 3) \ldots (x + p - 1)$$

By letting this product

$$\equiv x^{p-1} + Ax^{p-2} + Bx^{p-3} + \text{etc.} + Mx + N$$

[11] Page 218 in the first edition of 1770, to which Lagrange refers.

[12] "Démonstration d'un théoreme noveau concernant les nombres premiers." p. 125.

the coefficients A, B, etc., M will be divisible by p, and N will be $= 1 \cdot 2 \cdot 3 \cdot \ldots \cdot p - 1$. If $x = 1$ the product will be divisible by p; but then it will be $\equiv 1 + N$ (mod. p) and so $1 + N$ will necessarily be divisible by p.

Finally, Euler, in *Opuscula Analytica, 1*, 329[13], gave a proof that agrees with what we have shown. Since such distinguished mathematicians did not consider this theorem unworthy of their attention, we are emboldened to add still another proof.

▶ 77. When relative to the modulus p the product of two numbers a, b is congruent to unity we will call the numbers *associates* as Euler did. Then according to the preceding section any positive number less than p will have a positive associate less than p and it will be unique. It is easy to prove that of the numbers $1, 2, 3, \ldots, p - 1$, 1 and $p - 1$ are the unique numbers which are associates of one another, for associate numbers will be roots of the congruence $x^2 \equiv 1$. And since this congruence is of the second degree it cannot have more than two roots; i.e. only 1 and $p - 1$. Suppressing these, the remaining numbers $2, 3, \ldots, p - 2$ will be associated in pairs. Their product then will $\equiv 1$ and the product of all of them $1, 2, 3, \ldots, p - 1$ will $\equiv p - 1$ or $\equiv -1$. Q.E.D.

For example, for $p = 13$ the numbers $2, 3, 4, \ldots, 11$ will be associated as follows: 2 with 7; 3 with 9; 4 with 10; 5 with 8; 6 with 11. That is, $2 \cdot 7 \equiv 1$; $3 \cdot 9 \equiv 1$, etc. Thus $2 \cdot 3 \cdot 4 \ldots 11 \equiv 1$; and so $1 \cdot 2 \cdot 3 \ldots 12 \equiv -1$.

▶ 78. Wilson's theorem can be expressed more generally thus: *The product of all numbers less than a given number A and at the same time relatively prime to it, is congruent relative to A to plus or minus unity.* Minus unity is to be taken when A is of the form p^m or $2p^m$ where p is a prime number different from 2 and also when $A = 4$. Plus unity is taken in all other cases. The theorem as Wilson proposed it is contained in the former case. For example, for $A = 15$ the product of the numbers 1, 2, 4, 7, 8, 11, 13, 14 is $\equiv 1$ (mod. 15). For the sake of brevity we omit the proof and only observe that it can be done as in the preceding article except that the congruence $x^2 \equiv 1$ can have more than two roots because of certain peculiar considerations. We can also look for a proof from a consideration of indices as in article 75 if we include

[13] "Miscellanea Analytica. Theorema a Cl. Waring sine demonstratione propositum."

what we shall soon say about moduli that are not prime.

▶ 79. We return now to the enumeration of other propositions (art. 75).

The sum of all the terms of the period of any number is $\equiv 0$, just as in the example of article 75, $1 + 5 + 12 + 8 = 26 \equiv 0$ (mod. 13).

Demonstration. Let the number whose period we are considering $= a$, and the exponent to which it belongs $= t$, then the sum of all the terms of the period will be

$$\equiv 1 + a + a^2 + a^3 + \text{etc.} + a^{t-1} \equiv \frac{a^t - 1}{a - 1} \text{ (mod. } p\text{)}$$

But $a^t - 1 \equiv 0$; so this sum will always be $\equiv 0$ (art. 22), unless perhaps $a - 1$ is divisible by p, or $a \equiv 1$; and we can except this case if we are willing to consider only one term as a *period*.

▶ 80. *The product of all primitive roots will be* $\equiv 1$, except for the case when $p = 3$; for then there is only one primitive root, 2.

Demonstration. If any primitive root is taken as a base, the indices of all the primitive roots will be prime relative to $p - 1$ and at the same time less than $p - 1$. But the sum of these numbers, i.e. the index of the product of all primitive roots, is $\equiv 0$ (mod. $p - 1$) and so the product $\equiv 1$ (mod. p); for it is easy to see that if k is a number relatively prime to $p - 1$, $p - 1 - k$ will be prime relative to $p - 1$, and so the sum of numbers which are prime relative to $p - 1$ is composed of couples whose sum is divisible by $p - 1$ (k can never be equal to $p - 1 - k$ except for the case when $p - 1 = 2$ or $p = 3$; for clearly in all other cases $(p - 1)/2$ is not prime relative to $p - 1$).

▶ 81. *The sum of all primitive roots is either* $\equiv 0$ (when $p - 1$ is divisible by a *square*), *or* $\equiv \pm 1$ (mod. p) (when $p - 1$ is the product of unequal prime numbers; if the number of these is even the sign is positive but if the number is odd, the sign is negative).

Example. 1. For $p = 13$ the primitive roots are $2, 6, 7, 11$, and the sum is $26 \equiv 0$ (mod. 13).

2. For $p = 11$ the primitive roots are $2, 6, 7, 8$, and the sum is $23 \equiv +1$ (mod. 11).

3. For $p = 31$ the primitive roots are $3, 11, 12, 13, 17, 21, 22, 24$, and the sum is $123 \equiv -1$ (mod. 31).

Demonstration. We showed above (art. 55.II) that if $p - 1 =$

$a^\alpha b^\beta c^\gamma$ etc. (where a, b, c, etc. are unequal prime numbers) and A, B, C, etc. are any numbers belonging to the exponents $a^\alpha, b^\beta, c^\gamma$, etc., respectively, all the products ABC etc. will be primitive roots. It is easy to show that any primitive root can be expressed as a product of this sort and indeed in a unique manner.[ε]

It follows that these products can be taken in place of the primitive roots. But since in these products it is necessary to combine all values of A with all those of B etc., the sum of all these products will be equal to the product of the sum of all the values of A, multiplied by the sum of all the values of B, multiplied by all the values of C etc. as we know from the theory of combinations. Let all the values of $A; B;$ etc. be designated by A, A', A'', etc.; B, B', B'', etc.; etc., and the sum of all the primitive roots will be

$$\equiv (A + A' + \text{etc.})(B + B' + \text{etc.}) \text{ etc.}$$

I now claim that if the exponent $\alpha = 1$, the sum $A + A' + A'' +$ etc. will be $\equiv -1$ (mod. p), but if $\alpha > 1$ the sum will be $\equiv 0$, and in like manner for the remaining β, γ, etc. If we prove these assertions the truth of our theorem will be manifest. For when $p - 1$ is divisible by a square, one of the exponents α, β, γ, etc. will be greater than unity, therefore one of the factors whose sum is congruent to the product of all the primitive roots will $\equiv 0$, and so will the product itself. But when $p - 1$ can be divided by no square, all the exponents α, β, γ, etc. will $= 1$ and so the sum of all the primitive roots will be congruent to the product of factors each of which is $\equiv -1$ and there will be as many of them as there are numbers a, b, c, etc. As a result the sum will be $\equiv \pm 1$ according as there is an even or an odd number of them. We prove this as follows.

1) When $\alpha = 1$ and A is a number belonging to the exponent a, the remaining numbers belonging to this exponent will be

[ε] Let the numbers a, b, c, etc. be so determined that $a \equiv 1$ (mod. a^α) and $\equiv 0$ (mod. $b^\beta c^\gamma$, etc.); $b \equiv 1$ (mod. b^β) and $\equiv 0$ (mod. $a^\alpha c^\gamma$ etc.); etc. (see art. 32); and so $a + b + c$ etc. $\equiv 1$ (mod. $p - 1$) (art. 19). Now if any primitive root r is expressed by a product ABC etc. we will get $A \equiv r^a$, $B \equiv r^b$, $C \equiv r^c$, etc., and A will belong to the exponent a^α, B to the exponent b^β etc.; the product of all the A, B, C, etc. will be $\equiv r$ (mod. p); and it is easily seen that A, B, C, etc. cannot be determined in any other way.

$A^2, A^3, \ldots A^{a-1}$. But

$$1 + A + A^2 + A^3 \ldots + A^{a-1}$$

is the sum of the complete period and thus $\equiv 0$ (art. 79) and

$$A + A^2 + A^3 \ldots + A^{a-1} \equiv -1$$

2) When however $\alpha > 1$ and A is a number belonging to the exponent a^α the remaining numbers belonging to this exponent will be found if from $A^2, A^3, A^4, \ldots A^{a^\alpha - 1}$ we delete A^a, A^{2a}, A^{3a}, etc. (see art. 53); and their sum will be

$$\equiv 1 + A + A^2 \ldots + A^{a^\alpha - 1} - (1 + A^a + A^{2a} \ldots + A^{a^\alpha - a})$$

i.e. congruent to the difference of two periods, and thus $\equiv 0$. Q.E.D.

▶ 82. All that we have said so far presupposes that the modulus is a prime number. It remains to consider the case when a composite number is used as modulus. But since here there are no such elegant properties as in the previous case and there is no need for subtle artifices in order to discover them (because almost everything can be deduced from an application of the preceding principles), it would be superfluous and tedious to treat all the details exhaustively. So we will explain only what this case has in common with the previous one and what is proper to itself.

▶ 83. The propositions in articles 45–48 have already been proven for the general case. But the proposition of article 49 must be changed as follows:

If f designates how many numbers there are relatively prime to m and less than m, i.e. if $f = \phi m$ (art. 38); and if a is a given number relatively prime to m, then the exponent t of the lowest power of a that is congruent to unity relative to the modulus m will be = to f or a factor of f.

The demonstration of the proposition in article 49 has value in this case if we substitute m for p, f for $p - 1$ and if in place of the numbers $1, 2, 3, \ldots p - 1$ we substitute numbers which are relatively prime to m and less than m. Let the reader turn back and do this. But the other demonstrations we considered there (art. 50, 51) cannot be applied to this case without ambiguity. And with respect to the propositions of article 52 et seq. there is a great difference between moduli that are powers of prime numbers and those that are divisible by more than one prime number. We therefore consider moduli of the former type.

▶ 84. If p is a prime number and the modulus $m = p^n$, then we have $f = p^{n-1}(p - 1)$ [art. 38]. Now if we apply what we said in articles 53, 54 to this case, with the necessary changes as in the preceding article, we will discover that everything we said there will have place here provided we first prove that a congruence of the form $x^t - 1 \equiv 0$ (mod. p^n) cannot have more than t different roots. We showed that this was true relative to a prime modulus for a more general proposition in article 43; but this proposition is valid for prime moduli only and cannot be applied to this case. Nevertheless we will show by a special method that the proposition is true for this particular case. In Section VIII we will prove it still more easily.

▶ 85. We propose now to prove this theorem:

If the greatest common divisor of the numbers t and $p^{n-1}(p - 1)$ is e, the congruence $x^t \equiv 1$ (mod. p^n) will have e different roots.

Let $e = kp^v$ so that k does not involve the factor p. As a result it will divide the number $p - 1$. And the congruence $x^t \equiv 1$ relative to the modulus p will have k different roots. If we designate them A, B, C, etc. each root of the same congruence relative to the modulus p^n will be congruent relative to the modulus p to one of the numbers A, B, C, etc. Now we will show that the congruence $x^t \equiv 1$ (mod. p^n) has p^v roots congruent to A and as many congruent to B etc., all relative to the modulus p. From this we find that the number of all the roots will be kp^v or e as we said. We will do this by showing *first* that if α is a root congruent to A relative to the modulus p, then

$$\alpha + p^{n-v}, \qquad \alpha + 2p^{n-v}, \qquad \alpha + 3p^{n-v}, \ldots, \qquad \alpha + (p^v - 1)p^{n-v}$$

will also be roots. *Second*, by showing that numbers congruent to A relative to modulus p cannot be roots unless they are in the form $\alpha + hp^{n-v}$ (h being an integer). As a result there will manifestly be p^v different roots and no more. The same will be true of the roots that are congruent to B, C, etc. *Third*, we will show how we can always find the root that is congruent to A relative to p.

▶ 86. THEOREM. *If as in the preceding article t is a number divisible by p^v, but not by p^{v+1}, we have*

$$(\alpha + hp^\mu)^t - \alpha^t \equiv 0 \;(\text{mod. } p^{\mu+v}) \quad \text{and} \quad \equiv \alpha^{t-1} hp^\mu t \;(\text{mod. } p^{\mu+v+1})$$

The second part of the theorem is not true when $p = 2$ and $\mu = 1$.

The demonstration of this theorem can be shown by expanding a binomial provided that we can show that all the terms after the second are divisible by $p^{\mu+v+1}$. But since a consideration of the denominators of the coefficients leads to many ambiguities, we prefer the following method.

Let us presume *first* that $\mu > 1$ and $v = 1$ and we will then have

$$x^t - y^t = (x - y)(x^{t-1} + x^{t-2}y + x^{t-3}y^2 + \text{etc.} + y^{t-1})$$

$$(\alpha + hp^\mu)^t - \alpha^t = hp^\mu[(\alpha + hp^\mu)^{t-1} + (\alpha + hp^\mu)^{t-2}\alpha + \text{etc.} + \alpha^{t-1}]$$

But

$$\alpha + hp^\mu \equiv \alpha \;(\text{mod. } p^2)$$

and so each term $(\alpha + hp^\mu)^{t-1}, (\alpha + hp^\mu)^{t-2}\alpha$, etc. will be $\equiv \alpha^{t-1}$ (mod. p^2), and the sum of all of them will be $\equiv t\alpha^{t-1}$ (mod. p^2); that is they will be of the form $t\alpha^{t-1} + Vp^2$ where V is any number whatsoever. Thus $(\alpha + hp^\mu)^t - \alpha^t$ will be of the form

$$\alpha^{t-1}hp^\mu t + Vhp^{\mu+2}, \quad \text{i.e.} \quad \equiv \alpha^{t-1}hp^\mu t \;(\text{mod. } p^{\mu+2})$$
$$\text{and} \quad \equiv 0 \;(\text{mod. } p^{\mu+1})$$

And thus the theorem is proven for this case.

Now keeping $\mu > 1$, if the theorem were not true for other values of v, there would necessarily be a limit up to which the theorem would always be true and beyond which it would be false. Let the smallest value of v for which it is false $= \phi$. It is easily seen that if t were divisible by $p^{\phi-1}$ but not by p^ϕ, the theorem would still be true, whereas if we substituted tp for t it would be false. So we have

$$(\alpha + hp^\mu)^t \equiv \alpha^t + \alpha^{t-1}hp^\mu t \;(\text{mod. } p^{\mu+\phi}),$$
$$\text{or} \quad \alpha^t + \alpha^{t-1}hp^\mu t + up^{\mu+\phi}$$

with u an integer. But since the theorem has already been proved for $v = 1$, we get

$$(\alpha^t + \alpha^{t-1}hp^\mu t + up^{\mu+\phi})^p \equiv \alpha^{tp} + \alpha^{tp-1}hp^{\mu+1}$$
$$+ \alpha^{tp-t}up^{\mu+\phi+1} \pmod{p^{\mu+\phi+1}}$$

and so also

$$(\alpha + hp^\mu)^{tp} \equiv \alpha^{tp} + \alpha^{tp-1}hp^\mu tp \pmod{p^{\mu+\phi+1}}$$

i.e. the theorem is true if tp is substituted for t, i.e. for $v = \phi$. But this is contrary to the hypothesis and so the theorem will be true for all values of v.

▶ 87. There remains the case when $\mu = 1$. By a method very similar to the one we used in the preceding article we can show without the help of the binomial theorem that

$$(\alpha + hp)^{t-1} \equiv \alpha^{t-1} + \alpha^{t-2}(t-1)hp \pmod{p^2}$$
$$\alpha(\alpha + hp)^{t-2} \equiv \alpha^{t-1} + \alpha^{t-2}(t-2)hp$$
$$\alpha^2(\alpha + hp)^{t-3} \equiv \alpha^{t-1} + \alpha^{t-2}(t-3)hp$$

etc.

and the sum (since the number of terms $= t$) will be

$$\equiv t\alpha^{t-1} + \frac{(t-1)t}{2}\alpha^{t-2}hp \pmod{p^2}$$

But since t is divisible by p, $(t-1)t/2$ will also be divisible by p in all cases except when $p = 2$, as we pointed out in the preceding article. In the remaining cases however we have $(t-1)t\alpha^{t-2}hp/2 \equiv 0 \pmod{p^2}$ and the sum will be $\equiv t\alpha^{t-1} \pmod{p^2}$ as in the preceding article. The rest of the demonstration proceeds in the same way.

The general result except when $p = 2$ is

$$(\alpha + hp^\mu)^t \equiv \alpha^t \pmod{p^{\mu+v}}$$

and $(\alpha + hp^\mu)^t$ is not $\equiv \alpha^t$ for any modulus which is a higher power of p than $p^{\mu+v}$, provided always that h is not divisible by p and that p^v is the highest power of p that divides the number t.

From this we can immediately derive propositions 1 and 2 which we proposed to show in article 85:

First, if $\alpha^t \equiv 1$ we will also have $(\alpha + hp^{n-v})^t \equiv 1$ (mod. p^n).

Second, if some number α' is congruent relative to the modulus p to A and thus also to α but noncongruent to α relative to the modulus p^{n-v}, and if it satisfies the congruence $x^t \equiv 1$ (mod. p^n), we will let $\alpha' = \alpha + lp^\lambda$ in such a way that l is not divisible by p. It follows that $\lambda < n - v$ and so $(\alpha + lp^\lambda)^t$ will be congruent to α^t relative to the modulus $p^{\lambda+v}$ but not relative to the modulus p^n which is a higher power. As a result α' cannot be a root of the congruence $x^t \equiv 1$.

▶ 88. *Third*, we proposed to find a root of the congruence $x^t \equiv 1$ (mod. p^n) which will be congruent to A. We can show here how this is done only if we already know a root of the same congruence relative to the modulus p^{n-1}. Manifestly this is sufficient, since we can go from the modulus p for which A is a root to the modulus p^2 and from there to all consecutive powers.

Therefore let α be a root of the congruence $x^t \equiv 1$ (mod. p^{n-1}). We are looking for a root of the same congruence relative to the modulus p^n. Let us suppose it $= \alpha + hp^{n-v-1}$. From the preceding article it must have this form (we will consider separately $v = n - 1$, but note that v can never be greater than $n - 1$). We will therefore have

$$(\alpha + hp^{n-v-1})^t \equiv 1 \text{ (mod. } p^{n-1})$$

But

$$(\alpha + hp^{n-v-1})^t \equiv \alpha^t + \alpha^{t-1}htp^{n-v-1} \text{ (mod. } p^n)$$

If, therefore, h is so chosen that $1 \equiv \alpha^t + \alpha^{t-1}htp^{n-v-1}$ (mod. p^n) or [since by hypothesis $1 \equiv \alpha^t$ (mod. p^{n-1}) and t is divisible by p] so that $[(\alpha^t - 1)/p^{n-1}] + \alpha^{t-1}h(t/p^v)$ is divisible by p, we will have the root we want. That this can be done is clear from the preceding section because we presuppose that t cannot be divided by a higher power of p than p^v, and so $\alpha^{t-1}(t/p^v)$ will be prime relative to p.

But if $v = n - 1$, i.e. if t is divisible by p^{n-1} or by a higher power of p, any value A satisfying the congruence $x^t \equiv 1$ relative to the modulus p will satisfy it also relative to the modulus p^n. For if we let $t = p^{n-1}\tau$, we will have $t \equiv \tau$ (mod. $p - 1$); and so, since $A^t \equiv 1$ (mod. p), we will also have $A^\tau \equiv 1$ (mod. p). Let therefore $A^\tau = 1 + hp$ and we will have $A^t = (1 + hp)^{p^{n-1}} \equiv 1$ (mod. p^n) (art. 87).

▶ 89. Everything we have proven in articles 57 ff., with the aid of the theorem stating that the congruence $x^t \equiv 1$ can have no more than t different roots, is also true for a modulus that is the power of a prime number; and if we call *primitive roots* those numbers that belong to the exponent $p^{n-1}(p-1)$, that is to say all numbers containing the numbers not divisible by p in their period, then we have primitive roots here to. Everything we said about indices and their use and about the solution of the congruence $x^t \equiv 1$ can be applied to this case also and, since the demonstrations offer no difficulty, it would be superfluous to repeat them. We showed earlier how to derive the roots of the congruence $x^t \equiv 1$ relative to the modulus p^n from the roots of the same congruence relative to the modulus p. Now we must add some observations for the case when the modulus is some power of the number 2, which we excepted above.

▶ 90. *If some power of the number 2 higher than the second, e.g. 2^n, is taken as modulus, the 2^{n-2}th power of any odd number is congruent to unity.*

For example, $3^8 = 6561 \equiv 1$ (mod. 32).

For any odd number is of the form $1 + 4h$ or of the form $-1 + 4h$ and the proposition follows immediately (theorem in art. 86).

Since, therefore, the exponent belonging to any odd number relative to the modulus 2^n must be a divisor of 2^{n-2}, it is easy to judge to which of the numbers $1, 2, 4, 8, \ldots 2^{n-2}$ it belongs. Suppose the number proposed $= 4h \pm 1$ and the exponent of the highest power of 2 that divides h is $= m$ (m can $= 0$ when h is odd); then the exponent to which the proposed number belongs will $= 2^{n-m-2}$ if $n > m + 2$; if however $n =$ or is $< m + 2$ the proposed number will $\equiv \pm 1$ and so will belong either to the exponent 1 or to the exponent 2. For it is easy to deduce from article 86 that a number of the form $\pm 1 + 2^{m+2}k$ (which is equivalent to $4h \pm 1$) is congruent to unity relative to the modulus 2^n if it is raised to the power 2^{n-m-2} and is noncongruent to unity if it is raised to a lower power of 2. Thus any number of the form $8k + 3$ or $8k + 5$ will belong to the exponent 2^{n-2}.

▶ 91. It follows that here we do not have *primitive roots* in the sense accepted above; that is, there are no numbers which include in their periods all numbers less than the modulus and relatively prime to it. But obviously we have an analogy. For we found that for a number of the form $8k + 3$ the power of an odd exponent is also of the form $8k + 3$ and the power of an even exponent is of the form $8k + 1$; no power then can be of the form $8k + 5$ or $8k + 7$. Since for a number of the form $8k + 3$ the period therefore consists of 2^{n-2} terms of the form $8k + 3$ or $8k + 1$, and since there are no more than 2^{n-2} of these less than the modulus, manifestly any number of the form $8k + 1$ or $8k + 3$ is congruent relative to the modulus 2^n to a power of some number of the form $8k + 3$. In a similar way we can show that the period of a number of the form $8k + 5$ includes all numbers of the form $8k + 1$ and $8k + 5$. Therefore, if we take a number of the form $8k + 5$ as base, we will find real indices for all numbers of the form $8k + 1$ and $8k + 5$ taken positively and for all numbers of the form $8k + 3$ and $8k + 7$ taken negatively. And we must still regard as equivalent indices which are congruent relative to 2^{n-2}. Table 1, in which 5 was always chosen as base for the moduli 16, 32, and 64, should be interpreted in this light (for modulus 8 no table is necessary). For example, the number 19 which is of the form $8n + 3$ and so must be taken *negatively* has the index 7 relative to the modulus 64. This means $5^7 \equiv -19$ (mod. 64). If we took numbers of the form $8n + 1$ and $8n + 5$ negatively and numbers of the form $8n + 3$ and $8n + 7$ positively, it would be necessary to give them imaginary indices, so to speak. If we did that, the calculation of indices could be reduced to a very simple algorithm. But since we would be led too far afield if we wished to treat this point with great rigor, we will save it for another occasion when we may be able to consider in more detail the theory of imaginary quantities. In our opinion no one has yet produced a clear treatment of this subject. Experienced mathematicians will find it easy to develop the algorithm. Those with less training can use our table provided only that they have a good grasp of the principles established above. They can do this in much the same way that people use logarithms even when they know nothing about modern studies of imaginary logarithms.

▶92. Almost all that pertains to residues of powers relative to a modulus composed of many prime numbers can be deduced from the general theory of congruences. We will show later at great length how to reduce congruences relating to moduli composed of several primes to congruences relating to moduli which are primes or powers of primes. For this reason we will not now dwell further on this subject. We only observe that the most elegant property that holds for other moduli is lacking here, namely the property which guarantees that we can always find numbers whose period includes all numbers relatively prime to the modulus. In one case, however, even here we can find such a number. It occurs when the modulus is twice a prime number or twice the power of a prime. For if the modulus m is reduced to the form $A^a B^b C^c$ etc. where A, B, C, etc. are different prime numbers and if we designate $A^{a-1}(A-1)$ by the letter α, $B^{b-1}(B-1)$ by the letter β, etc. and then choose a number z which is relatively prime to m, we get $z^\alpha \equiv 1$ (mod. A^a), $z^\beta \equiv 1$ (mod. B^b), etc. And if μ is the least common multiple of the numbers α, β, γ, etc. we have $z^\mu \equiv 1$ relative to all the moduli A^a, B^b, etc. and so also relative to m which is equal to their product. But except for the case where m is twice a prime or twice the power of a prime, the least common multiple of the numbers α, β, γ, etc. is less than their product (the numbers α, β, γ, etc. cannot be prime relative to each other since they have the common divisor 2). Thus no period can have as many terms as there are numbers which are relatively prime to the modulus and less than it because the number of these is equal to the product of α, β, γ, etc. For example, for $m = 1001$ the 60th power of any number relatively prime to m is congruent to unity because 60 is the least common multiple of the numbers 6, 10, 12. The case where the modulus is twice a prime number or twice the power of a prime number is completely similar to the one where it is a prime or the power of a prime.

▶ 93. We have already mentioned the writings of other geometers concerning the same subjects treated in this section. For those who want a more detailed discussion than brevity has permitted us, we recommend the following treatises of Euler because of the clarity and insight that has placed this great man far ahead of all other commentators: "Theoremata circa residua ex divisione potestatum relicta," *Novi comm. acad. Petrop.*, 7 [1758–59], 1761, 49–82. "Demonstrationes circa residua ex divisione potestatum per numeros primos resultantia," *ibid.*, 18 [1773], 1774, 85–135.

To these we might add *Opuscula Analytica, 1*, Diss. 5, p. 152;
Diss. 8, p. 242.[14]

[14] For title of Dissertation 5 see p. 48; Dissertation 8: "De quisbusdam eximiis pro-
prietatibus circa divisores potestatum occurrentibus."

SECTION IV

CONGRUENCES OF THE SECOND DEGREE

▶ 94. THEOREM. *If we take some number m as modulus, then of the numbers* $0, 1, 2, 3, \ldots m - 1$ *no more than* $(m/2) + 1$ *can be congruent to a square when m is even and not more than* $(m/2) + (1/2)$ *of them when m is odd.*

Demonstration. Since the squares of congruent numbers are congruent to each other, any number that can be congruent to a square will also be congruent to some square whose root is $< m$. It is sufficient therefore to consider the least residues of the squares $0, 1, 4, 9, \ldots (m - 1)^2$. But it is easy to see that $(m - 1)^2 \equiv \cdot 1$, $(m - 2)^2 \equiv 2^2$, $(m - 3)^2 \equiv 3^2$, etc. Thus also when m is even, the squares $[(m/2) - 1]^2$ and $[(m/2) + 1]^2$, $[(m/2) - 2]^2$ and $[(m/2) + 2]^2$, etc. will have the same least residues; and when m is odd, the squares $[(m/2) - (1/2)]^2$ and $[(m/2) + (1/2)]^2$, $[(m/2) - (3/2)]^2$ and $[(m/2) + (3/2)]^2$, etc. will be congruent. It follows that there are no other numbers congruent to a square than those which are congruent to one of the squares $0, 1, 4, 9, \ldots (m/2)^2$ when m is even; when it is odd, any number that is congruent to a square is necessarily congruent to one of the $0, 1, 4, 9, \ldots [(m/2) - (1/2)]^2$. In the former case therefore there will be at most $(m/2) + 1$ different least residues, in the latter case at most $(m/2) + (1/2)$. Q.E.D.

Example. Relative to the modulus 13 the least residues of the squares $0, 1, 2, 3, \ldots 6$ are found to be $0, 1, 4, 9, 3, 12, 10$, and after this they occur in the reverse order $10, 12, 3$, etc. Thus if a number is not congruent to one of these residues, that is, if it is congruent to $2, 5, 6, 7, 8, 11$, it cannot be congruent to a square.

Relative to the modulus 15 we find the residues $0, 1, 4, 9, 1, 10, 6, 4$, after which the same numbers occur in reverse order. Here therefore the number of residues which can be congruent to a square is less than $(m/2) + (1/2)$, since $0, 1, 4, 6, 9, 10$ are the only

63

ones that occur in the list. The numbers 2, 3, 5, 7, 8, 11, 12, 13, 14 and any number congruent to them cannot be congruent to a square relative to the modulus 15.

▶ 95. As a result, for any modulus all numbers can be divided into two classes, one which contains those numbers that can be congruent to a square, the other those that cannot. We will call the former *quadratic residues of the number which we take as modulus*[a] and the *latter quadratic nonresidues of this number.* When no ambiguity can arise we will call them simply *residues* and *nonresidues.* For the rest it is sufficient to classify all the numbers $0, 1, 2, \ldots m - 1$, since numbers that are congruent to these will be put in the same class.

We will begin in this investigation with prime moduli, and this is to be understood even if it is not said in so many words. But we must exclude the prime number 2 and so we will be considering *odd* primes only.

▶ 96. *If we take the prime number p as modulus, half the numbers* $1, 2, 3, \ldots p - 1$ *will be quadratic residues, the rest nonresidues; i.e. there will be* $(p - 1)/2$ *residues and the same number of nonresidues.*

It is easy to prove that all the squares $1, 4, 9, \ldots (p - 1)^2/4$ are not congruent. For if $r^2 \equiv (r')^2$ (mod. p) with r, r' unequal and not greater than $(p - 1)/2$ and $r > r'$, $(r - r')(r + r')$ will be positive and divisible by p. But each factor $(r - r')$ and $(r + r')$ is less than p, so the assumption is untenable (art. 13). We have therefore $(p - 1)/2$ quadratic residues among the numbers $1, 2, 3, \ldots p - 1$. There cannot be more because if we add the residue 0 we get $(p + 1)/2$ of them, and this is larger than the number of all the residues together. Consequently the remaining numbers will be nonresidues and their number will $= (p - 1)/2$.

Since zero is always a residue we will exclude this and all numbers divisible by the modulus from our investigations. This case is clear and would contribute nothing to the elegance of our

---

[a] Actually in this case we give to these expressions a meaning different from that which we have used up to the present. When $r \equiv a^2$ (mod. m) we should say that r is a residue of the square a^2 relative to the modulus m; but for the sake of brevity in this section we will always call r the quadratic residue *of m itself*, and there is no danger of ambiguity. From now on we will not use the expression *residue* to signify the same thing as a congruent number unless perhaps we are treating of *least* residues, where there can be no doubt about what we mean.

theorems. For the same reason we will also exclude 2 as modulus.
▶ 97. Since many of the things we will prove in this section can
be derived from the principles of the preceding section, and since
it is not out of place to discover the same truths by different
methods, we will continue to point out the connection. It can be
easily verified that all numbers congruent to a square have *even*
indices and that numbers not congruent to a square have *odd*
indices. And since $p - 1$ is an even number, there will be as
many even indices as odd ones, namely $(p - 1)/2$ of each, and
thus also there will be the same number of residues and nonresidues.

Examples.

Moduli	Residues
3	1
5	1, 4
7	1, 2, 4
11	1, 3, 4, 5, 9
13	1, 3, 4, 9, 10, 12
17	1, 2, 4, 8, 9, 13, 15, 16

etc.

and all other numbers less than the moduli are nonresidues.

▶ 98. THEOREM. *The product of two quadratic residues of a prime
number p is a residue; the product of a residue and a nonresidue
is a nonresidue; and finally the product of two nonresidues is a
residue.*

Demonstration. I. Let A, B be residues of the squares a^2, b^2;
that is $A \equiv a^2$, $B \equiv b^2$. The product AB will then be congruent
to the square of the number ab; i.e. it will be a residue.

II. When A is a residue, for example $\equiv a^2$, and B a nonresidue,
AB will be a nonresidue. For let us suppose that $AB \equiv k^2$, and
let the value of the expression k/a (mod. p) $\equiv b$. We will have
therefore $a^2 B \equiv a^2 b^2$ and $B \equiv b^2$; i.e. B is a residue contrary to
the hypothesis.

Another way. Take all the numbers among $1, 2, 3, \ldots p - 1$
which are residues [there are $(p - 1)/2$ of them]. Multiply each
by A and all the products will be quadratic residues and all
noncongruent to each other. Now if the nonresidue B is multiplied by A, the product will not be congruent to any of the products

we already have. And so if it were a residue, we would have
$(p + 1)/2$ noncongruent residues, and we have not yet included
the residue 0 among them. This contradicts article 96.

III. Let A, B be nonresidues. Multiply by A all the numbers
among $1, 2, 3, \ldots p - 1$ which are residues. We still have $(p - 1)/2$
nonresidues noncongruent to one another (II); the product AB
cannot be congruent to any of them; so if it were a nonresidue,
we would have $(p + 1)/2$ nonresidues noncongruent to one
another contrary to article 96. So the product etc.   Q.E.D.

These theorems could be derived more easily from the principles
of the preceding section. For since the indices of residues are
always even and the indices of nonresidues odd, the index of the
product of two residues or nonresidues will be even, and the
product itself will be a residue. On the other hand the index of
the product of a residue and a nonresidue will be odd, and so the
product itself will be a nonresidue.

Either method of proof can be used for the following theorems:
*The value of the expression a/b (mod. p) will be a residue when the
numbers a, b are both residues or both nonresidues; it will be a
nonresidue when one of the numbers a, b is a residue and the other
a nonresidue.* They can also be proved by converting the preceding
theorems.

▶ 99. In general the product of any number of factors is a residue
if all the factors are residues or if the number of nonresidues among
them is even; if the number of nonresidues among the factors is
odd, the product will be a nonresidue. It is easy therefore to judge
whether or not a composite number is a residue as long as it is clear
what the individual factors are. That is why in Table 2 we include
only prime numbers. This is the arrangement of the table. Moduli[b]
are listed in the margin, successive prime numbers along the top;
when one of the latter is a residue of some modulus a dash has been
placed in the space corresponding to each of them; when a prime
number is a nonresidue of the modulus, the corresponding space
has been left vacant.

▶ 100. Before we proceed to more difficult subjects, let us add a
few remarks about moduli which are not prime.

If the modulus is p^n, a power of a prime number p (we assume

---

[b] We will soon see how to dispense with composite moduli.

p is not 2), half of all the numbers not divisible by p and less than p^n will be residues, half nonresidues; i.e. the number of each will $= [(p - 1)p^{n-1}]/2$.

For if r is a residue, it will be congruent to a square whose root is not greater than half the modulus (see art. 94). It is easy to see that there are $[(p - 1)p^{n-1}]/2$ numbers that are not divisible by p and less than half the modulus; it remains to show that the squares of all these numbers are noncongruent or that they produce different quadratic residues. Now if the squares of two numbers a, b not divisible by p and less than half the moduli were congruent, we would have $a^2 - b^2$ or $(a - b)(a + b)$ divisible by p^n (we suppose $a > b$). This cannot be unless *either* one of the numbers $(a - b)$, $(a + b)$ is divisible by p^n, but this cannot be because each is $< p^n$, *or* one of them is divisible by p^m, the other by p^{n-m}, i.e. both by p. But this cannot be either. For this means that the sum and the difference $2a$ and $2b$ would be divisible by p and consequently a and b, contrary to the hypothesis. So finally among the numbers not divisible by p and less than the modulus, there are $[(p - 1)p^n]/2$ residues and the others, whose number is the same, will be nonresidues. Q.E.D. This theorem can be proven by considering indices just as in article 97.

▶ 101. *Any number not divisible by p, which is a residue of p, will also be a residue of p^n; if it is a nonresidue of p it will also be a nonresidue of p^n.*

The second part of this theorem is obvious. If the first part were false, among the numbers less than p^n and not divisible by p more would be residues of p than of p^n, i.e. more than $[p^{n-1}(p - 1)]/2$. But clearly the number of residues of p among these numbers is precisely $= [p^{n-1}(p - 1)]/2$.

It is just as easy to find a square congruent to a given residue relative to the modulus p^n if we have a square congruent to this residue relative to the modulus p.

If we have a square a^2 which is congruent to the given residue A relative to the modulus p^μ, we can find the square congruent to A relative to p^ν in the following way (we suppose here that $\nu > \mu$ and $=$ or $< 2\mu$). Suppose that the root of the square we seek $= \pm a + xp^\mu$. It is easy to see that it should have this form. And we should also have $a^2 \pm 2axp^\mu + x^2p^{2\mu} \equiv A$ (mod. p^ν) or if $2\mu > \nu$, $A - a^2 \equiv \pm 2axp^\mu$ (mod. p^ν). Let $A - a^2 = p^\mu d$ and x will be the value of the expression $\pm d/2a$ (mod. $p^{\nu\mu}$) which is equivalent to $\pm$ $(A - a^2)/2ap^\mu$ (mod. p^ν).

Given therefore a square congruent to A relative to p, from it we derive a square congruent to A relative to the modulus p^2; from there we can go to the modulus p^4, then to p^8, etc.

Example. If we are given the residue 6, which is congruent to the square 1 relative to the modulus 5, we find the square 9^2 which is congruent to it relative to 25, 16^2 which is congruent to it relative to 125, etc.

▶ 102. Regarding numbers divisible by p, it is clear that their squares will be divisible by p^2, so that all numbers divisible by p but not by p^2 will be nonresidues of p^n. In general if we have a number $p^k A$ with A not divisible by p we can distinguish the following cases:

1) When $k =$ or is $> n$, we have $p^k A \equiv 0$ (mod. p^n), i.e. a residue.

2) When $k < n$ and odd, $p^k A$ will be a nonresidue.

For if we had $p^k A = p^{2x+1} A \equiv s^2$ (mod. p^n), s^2 would be divisible by p^{2x+1}. This is impossible unless s is divisible by p^{x+1}. But then p^{2x+2} would divide s^2 and also (since $2x + 2$ is certainly not greater than n) $p^k A$, i.e. $p^{2x+1} A$. And this means that p divides A, contrary to the hypothesis.

3) When $k < n$ and even. Then $p^k A$ will be a residue or nonresidue of p^n according as A is a residue or nonresidue of p. For when A is a residue of p, it is also a residue of p^{n-k} But if we suppose that $A \equiv a^2$ (mod. p^{n-k}), we get $Ap^k \equiv a^2 p^k$ (mod. p^n) and $a^2 p^k$ is a square. When on the other hand A is a nonresidue of p, $p^k A$ cannot be a residue of p^n. For if $p^k A \equiv a^2$ (mod. p^n), a^2 would necessarily be divisible by p^k. The quotient will be a square to which A is congruent relative to the modulus p^{n-k} and hence also relative to the modulus p; i.e. A will be a residue of p contrary to the hypothesis.

▶ 103. Since we have excluded the case where $p = 2$ above, we should say a few words about it. When the number 2 is the modulus, every number is a residue and there are no nonresidues. When 4 is the modulus, all odd numbers of the form $4k + 1$ will be residues, and all of the form $4k + 3$ nonresidues. When 8 or a higher power of the number 2 is the modulus, all odd numbers of the form $8k + 1$ are residues, all others of the forms $8k + 3$, $8k + 5$, $8k + 7$ are nonresidues. The last part of this proposition is clear from the fact that the square of any odd number, whether it be of the form $4k + 1$ or $4k - 1$, will be of the form $8k + 1$. We prove the first part as follows.

1) If either the sum or difference of two numbers is divisible by 2^{n-1} the squares of these numbers will be congruent relative to the modulus 2^n. For if one of the numbers $= a$, the other will be of the form $2^{n-1}h \pm a$, and the square of this is $\equiv a^2 \pmod{2^n}$.

2) Any odd number which is a residue of 2^n will be congruent to a square whose root is odd and $< 2^{n-2}$. For let a^2 be a square to which the number is congruent and let $a \equiv \pm \alpha \pmod{2^{n-1}}$ so that α does not exceed half of the modulus (art. 4); $a^2 \equiv \alpha^2$ and so the given number will be $\equiv \alpha^2$. Manifestly both a and α will be odd and $\alpha < 2^{n-2}$.

3) The squares of all odd numbers less than 2^{n-2} will be noncongruent relative to 2^n. For let two such numbers be r and s. If their squares were congruent relative to 2^n, $(r - s)(r + s)$ would be divisible by 2^n (we assume $r > s$). It is easy to see that the numbers $r - s, r + s$ cannot both be divisible by 4. So if only one of them is divisible by 2, the other would be divisible by 2^{n-1} in order to have the product divisible by 2^n; Q.E.A., since each of them is $< 2^{n-2}$.

4) If these squares are reduced to their *least positive residues*, we will have 2^{n-3} different quadratic residues smaller than the modulus,[c] and each of them will be of the form $8k + 1$. But since there are precisely 2^{n-3} numbers of the form $8k + 1$ less than the modulus, all these numbers must be residues.

To find a square which is congruent to a given number of the form $8k + 1$ relative to the modulus 2^n, a method can be used similar to the one in article 101; see also article 88. And finally with regard to even numbers, everything is valid that we said in general in article 102.

▶ 104. If A is a residue of p^n and if we are concerned with the number of different values (i.e. noncongruent relative to the modulus) that the expression $V = \sqrt{A} \pmod{p^n}$ admits, we can easily draw the following conclusions from what has preceded. (We assume that the number p is prime as before, and for brevity's sake we include the case where $n = 1$.)

---

[c] Because the number of odd numbers less than 2^{n-2} is 2^{n-3}.

I. If A is not divisible by p, V will have *one* value for $p = 2$, $n = 1$, that is $V = 1$; *two* values when p is odd or when $p = 2$, $n = 2$, that is if one of them $\equiv v$, the other will $\equiv -v$; *four* values for $p = 2$, $n > 2$, that is if one of them $\equiv v$, the others will be $\equiv -v$, $2^{n-1} + v$, $2^{n-1} - v$.

II. If A is divisible by p but not by p^n, let the highest power of p that divides A be $p^{2\mu}$ (for manifestly this exponent has to be even) and we will have $A = ap^{2\mu}$. It is clear that all the values of V will be divisible by p^μ, and that all the quotients arising from this division will be values of the expression $V' = \sqrt{a}$ (mod. $p^{n-2\mu}$); we will then get all the values of V by multiplying all the values of V' that lie between 0 and $p^{n-\mu}$ by p^μ. Doing this we get

$$vp^\mu, \qquad vp^\mu + p^{n-\mu}, \qquad vp^\mu + 2p^{n-\mu}, \ldots vp^\mu + (p^\mu - 1)p^{n-\mu}$$

where v stands for all the *different* values of V' and so the number of them will be p^μ, $2p^\mu$, or $4p^\mu$ according as the number of values for V (by case I) is 1, 2, or 4.

III. If A is divisible by p^n it is easy to see that by letting $n = 2m$ or $n = 2m - 1$ according as it is even or odd, all numbers divisible by p^m will be values of V; but these will be the numbers $0, p^m, 2p^m$, $\ldots (p^{n-m} - 1)p^m$ and their number will be p^{n-m}.

▶ 105. It remains to consider the case where the modulus m is composed of several prime numbers. Let $m = abc\ldots$ where a, b, c, etc. are different prime numbers or powers of different primes. It is immediately clear that if n is a residue of m, it will also be a residue of each of the a, b, c, etc., and that if it is a nonresidue of any of the numbers a, b, c, etc. it will be a nonresidue of m. And, vice versa, if n is a residue of each of the a, b, c, etc. it will also be a residue of the product m. For suppose $n \equiv A^2, B^2, C^2$, etc. relative to the moduli a, b, c, etc. respectively. If we find the number N which is congruent to A, B, C, etc. relative to the moduli a, b, c, etc., respectively (art. 32), we will have $n \equiv N^2$ relative to all these moduli and so also relative to the product m. From the combination of *any* value of A or of the expression $\sqrt{n}$ (mod. a), together with *any* value of B and *any* value of C etc., we get a value of N or of the expression $\sqrt{n}$ (mod. m), and from different combinations we get different values of N and from all combinations we get all values of N. As a result, the number of all the different values of N will be equal to the product of the various values of A, B, C, etc. which we showed how to determine in the preceding article. And clearly if one value of the expression $\sqrt{n}$ (mod. m) or of N were known, this would also be a value of all the A, B, C, etc.; and since we know from the preceding article how to deduce all the remaining values of these quantities from this one, it follows that from one value of N all the rest can be obtained.

Example. Let the modulus be 315. We want to know whether 46 is a residue or nonresidue. The prime divisors of 315 are 3, 5, 7, and the number 46 is a residue of each of them; therefore it is also a residue of 315. Further, since $46 \equiv 1$ and $\equiv 64$ (mod. 9); $\equiv 1$ and $\equiv 16$ (mod. 5); $\equiv 4$ and $\equiv 25$ (mod. 7), the roots of the squares to which 46 is congruent relative to the modulus 315 will be 19, 26, 44, 89, 226, 271, 289, 296.

▶ 106. From what we have just shown we see that if only we can determine whether a given *prime number* is a residue or a nonresidue of another *given prime*, all other cases can be reduced to this. We will therefore attempt to investigate certain criteria for this case. But before we do this we will show a criterion derived from the preceding section. Although it is of almost no practical use, it is worthy of mention because of its simplicity and generality.

Any number A not divisible by a prime number $2m + 1$ is a residue or nonresidue of this prime according as $A^m \equiv +1$ or $\equiv -1$ (mod. $2m + 1$).

For let a be the index of the number A for the modulus $2m + 1$ in any system whatsoever; a will be even when A is a residue of $2m + 1$, and odd when A is a nonresidue. But the index of the number A^m will be ma; i.e. it will $\equiv 0$ or $\equiv m$ (mod. $2m$), according as a is even or odd. In the former case A^m will be $\equiv +1$, and in the latter case it will $\equiv -1$ (mod. $2m + 1$) (see art. 57, 62).

Example. 3 is a residue of 13 because $3^6 \equiv 1$ (mod. 13), but 2 is a nonresidue of 13 because $2^6 \equiv -1$ (mod. 13).

But as soon as the numbers we are examining are even moderately large this criterion is practically useless because of the amount of calculation involved.

▶ 107. It is very easy, given a modulus, to characterize all the numbers that are residues or nonresidues. If the number $= m$, we determine the squares whose roots do not exceed half of m and also the numbers congruent to these squares relative to m (in practice there are still more expedient methods). All numbers

congruent to any of these relative to m will be residues of m, all
numbers congruent to none of them will be nonresidues. But the
inverse question, *given a number, to assign all numbers of which it
is a residue or a nonresidue*, is much more difficult. The statements
in the preceding article depend on the solution to this problem.
We will investigate it now, beginning with the simplest cases.

▶ 108. THEOREM. -1 *is a quadratic residue of all numbers of
the form* $4n + 1$ *and a nonresidue of all numbers of the form*
$4n + 3$.

Example. -1 is the residue of the numbers 5, 13, 17, 29, 37, 41,
53, 61, 73, 89, 97, etc. arising from the squares of the numbers
2, 5, 4, 12, 6, 9, 23, 11, 27, 34, 22, etc. respectively; on the other hand
it is a nonresidue of the numbers 3, 7, 11, 19, 23, 31, 43, 47, 59, 67,
71, 79, 83, etc.

We mentioned this theorem in article 64, but it is best demonstrated as in article 106. For a number of the form $4n + 1$ we have
$(-1)^{2n} \equiv 1$, for a number of the form $4n + 3$ we get $(-1)^{2n+1} \equiv$
-1. This demonstration agrees with that of article 64, but because
of the elegance and usefulness of the theorem we will show still
another solution.

▶ 109. We will designate by the letter C the complex of all residues
of a prime number p which are less than p and exclusive of the
residue 0. Since the number of these residues is always $=(p - 1)/2$,
C will obviously be even when p is of the form $4n + 1$, and odd
when p is of the form $4n + 3$. By analogy with the language of
article 77 where we spoke of numbers in general, we will call
associated residues those whose product $\equiv 1$ (mod. p); for if r is a
residue then $1/r$ (mod. p) is also a residue. And since the same
residue cannot have more associates among the residues C, it is
clear that all the residues C can be divided into classes each containing a pair of associated residues. Now if there is no residue that
is its own associate, i.e. if each class contains a pair of *unequal*
residues, the number of all residues will be double the number of
classes; but if there are some residues which are their own associates, i.e. there are classes containing only one residue or, if you
prefer, containing the same residue twice, the number of all the
residues C will $=a + 2b$, where a is the number of classes of the
second type and b the number of the first type. So when p is of the

form $4n + 1$, a will be even, and when p is of the form $4n + 3$, a will be odd. But there are no numbers other than 1 and $p - 1$ which can be less than p and associates of themselves (see art. 77); but the first class certainly has 1 among its residues. So in the former case $p - 1$ (or -1, which comes to the same thing) must be a residue, in the latter case, a nonresidue. Otherwise in the first case we would have $a = 1$ and in the second $a = 2$, which is impossible.

▶ 110. This demonstration is due to Euler. He was also the first to discover the previous method (see *Opuscula Analytica, 1*, 135.[1]). It is easy to see that it relies on principles very like those of our second demonstration of Wilson's theorem (art. 77). And if we suppose the truth of this theorem, the previous demonstration becomes much simplified. For among the numbers $1, 2, 3, \ldots p - 1$ there will be $(p - 1)/2$ quadratic residues of p and the same number of nonresidues; so the number of nonresidues will be even when p is of the form $4n + 1$, odd when p is of the form $4n + 3$. Thus the product of all the numbers $1, 2, 3, \ldots p - 1$ will be a residue in the former case, a nonresidue in the latter case (art. 99). But this product will always $\equiv -1$ (mod. p); and so in the former case -1 will also be a residue, in the latter case a nonresidue.

▶ 111. If therefore r is a residue of any prime number of the form $4n + 1$, $-r$ will also be a residue of this prime; and all nonresidues of such a number will remain nonresidues even if the sign is changed.[d] The contrary is true for prime numbers of the form $4n + 3$. Residues become nonresidues when the sign is changed and vice versa (see art. 98).

From what precedes we can derive the general rule: -1 *is a residue of all numbers not divisible by* 4 *or by any prime number of the form* $4n + 3$; *it is a nonresidue of all others* (see art. 103, 105).

▶ 112. Let us consider the residues $+2$ and -2.

If from Table 2 we collect all prime numbers whose residue is $+2$, we have 7, 17, 23, 31, 41, 47, 71, 73, 79, 89, 97. We notice that among these numbers none is of the form $8n + 3$ or $8n + 5$.

Let us see therefore whether this induction can be made a certitude.

[1] Cf. p. 48.

[d] Therefore when we speak of a number in so far as it is a residue or nonresidue of a number of the form $4n + 1$, we can ignore the sign completely or we can use the double sign $\pm$.

First we observe that any composite number of the form $8n + 3$ or $8n + 5$ necessarily involves a prime factor of one or the other of the forms $8n + 3$ or $8n + 5$; for if only prime numbers of the form $8n + 1$, $8n + 7$ were involved, we would get numbers of the form $8n + 1$ or $8n + 7$. So if our induction is true in general, no number of the form $8n + 3$, $8n + 5$ can have $+2$ as a residue. Now certainly there is no number of this form less than 100 whose residue is $+2$. If however there are some such numbers greater than 100, let the least of all of them $= t$. This t will be either of the form $8n + 3$ or $8n + 5$; $+2$ will be a residue of t but a nonresidue of all similar numbers less than t. Let $2 \equiv a^2$ (mod. t). We can always find a such that it is odd and at the same time $< t$ (for a will have at least two positive values less than t whose sum $= t$, one of them being even, the other odd; see art. 104, 105). Let $a^2 = 2 + tu$ (that is $tu = a^2 - 2$); a^2 will be of the form $8n + 1$, tu of the form $8n - 1$; therefore u will be of the form $8n + 3$ or $8n + 5$ according as t is of the form $8n + 5$ or $8n + 3$. But from the equation $a^2 = 2 + tu$ we also have $2 \equiv a^2$ (mod. u); i.e. 2 is a residue of u. It is easy to see that $u < t$ and so t is not the smallest number in our induction, contrary to the hypothesis. And so what we have discovered by induction is proven true for the general case.

Combining this proposition with the propositions of article 111 we deduce the following theorems.

I. *$+2$ will be a nonresidue, -2 a residue of all prime numbers of the form $8n + 3$.*

II. *$+2$ and -2 will both be nonresidues of all prime numbers of the form $8n + 5$.*

▶ 113. By a similar induction from Table 2 we find the prime numbers whose residue is -2: 3, 11, 17, 19, 41, 43, 59, 67, 73, 83, 89, 97.[e] Since none of these is of the forms $8n + 5$, $8n + 7$ we will see whether this induction can lead us to a general theorem. We show as in the preceding article that no composite number of the form $8n + 5$ or $8n + 7$ can involve a prime factor of the form $8n + 5$ or $8n + 7$, so that if our induction is true in general, -2 can be the residue of no number of the form $8n + 5$ or $8n + 7$. If however such numbers did exist, let the least of them be $= t$, and we get $-2 = a^2 - tu$. If, as above, a is taken to be odd and less than t,

---

[e] That is, by taking -2 as the product of $+2$ and -1 (see art. 111).

u will be of the form $8n + 5$ or $8n + 7$, according as t is of the form $8n + 7$ or $8n + 5$. But from the fact that $a^2 + 2 = tu$ and $a < t$, it is easily proven that u also is less than t. Finally -2 will also be a residue of u; i.e. t will not be the smallest number of which -2 is a residue, contrary to the hypothesis. Thus -2 is necessarily a nonresidue of all numbers of the form $8n + 5$, $8n + 7$.

Combining this with the propositions of article 111, we get the following theorems.

I. *Both* -2 *and* $+2$ *are nonresidues of all prime numbers of the form* $8n + 5$, as we have already seen in the preceding article.

II. -2 *is a nonresidue,* $+2$ *a residue of all prime numbers of the form* $8n + 7$.

However in each demonstration we would have been able to take also an even value for a; but then we would have to distinguish the case where a was of the form $4n + 2$ from the one where a was of the form $4n$. The development would then proceed as above without any difficulty.

▶ 114. One case still remains; i.e. when the prime number is of the form $8n + 1$. The preceding method cannot be used here and a special device is needed.

Let $8n + 1$ be a prime modulus and a a primitive root. We will have (art. 62) $a^{4n} \equiv -1 \pmod{8n + 1}$. This congruence can be expressed in the form $(a^{2n} + 1)^2 \equiv 2a^{2n} \pmod{8n + 1}$ or in the form $(a^{2n} - 1)^2 \equiv 2a^{2n}$. It follows that both $2a^{2n}$ and $-2a^{2n}$ are residues of $8n + 1$; but since a^{2n} is a square not divisible by the modulus, both $+2$ and -2 will also be residues (art. 98).

▶ 115. It will be of some use to add here another demonstration of this theorem. This is related to the preceding as the second demonstration (art. 109) of the theorem of article 108 is related to the first (art. 108). Skilled mathematicians can see that the two pairs of proofs are not so different as they seem at first.

I. For any prime modulus of the form $4m + 1$ among the numbers $1, 2, 3, \ldots 4m$ there are m numbers that can be congruent to a biquadratic whereas the other $3m$ cannot be.

This is easily derived from the principles of the preceding section but even without them it is not difficult to see. For we have shown that -1 is always a quadratic residue for such a modulus. Let therefore $f^2 \equiv -1$. Clearly if z is any number not divisible by the modulus, the biquadratics of the four numbers $+z$, $-z$, $+fz$, $-fz$ (two are obviously noncongruent) will be congruent among themselves. And the biquadratic of any number that is congruent to none of these four cannot be congruent to their biquadratics (otherwise the congruence $x^4 \equiv z^4$ which is of the fourth degree would have more than four roots, contrary to art. 43). Thus we deduce that all the numbers $1, 2, 3, \ldots 4m$ furnish only m non-congruent biquadratics and that among the same numbers there are m numbers congruent to these. The others can be congruent to no biquadratic.

II. Relative to a prime modulus of the form $8n + 1$, the number -1 can be made congruent to a biquadratic (-1 will be called the *biquadratic residue* of this prime number).

The number of all biquadratic residues less than $8n + 1$ (zero excluded) will be $= 2n$, i.e. even. It is easy to prove that if r is a biquadratic residue of $8n + 1$, the value of the expression $1/r$ (mod. $8n + 1$) will also be such a residue. So all biquadratic residues can be distributed into classes just as we distributed quadratic residues in article 109, and the rest of the demonstration proceeds in almost the same way as it did there.

III. Let $g^4 \equiv -1$, and h the value of the expression $1/g$ (mod. $8n + 1$). Then we will have

$$(g \pm h)^2 = g^2 + h^2 \pm 2gh \equiv g^2 + h^2 \pm 2$$

(because $gh \equiv 1$). But $g^4 \equiv -1$ and so $-h^2 \equiv g^4h^2 \equiv g^2$. Thus $g^2 + h^2 \equiv 0$ and $(g \pm h)^2 \equiv \pm 2$; i.e. both $+2$ and -2 are quadratic residues of $8n + 1$. Q.E.D.

▶ 116. From what precedes we can easily deduce the following general rule: *$+2$ is a residue of any number that cannot be divided by 4 or by any prime of the form $8n + 3$ or $8n + 5$, and a nonresidue of all others* (e.g. of all numbers of the forms $8n + 3$, $8n + 5$ whether they are prime or composite);

-2 is a residue of any number that cannot be divided by 4 or by any prime of the form $8n + 5$ or $8n + 7$, and a nonresidue of all others.

These elegant theorems were already known to the sagacious Fermat (*Opera Mathem.*, p. 168[2]), but he never divulged the proof

---

[2] See p. 32. The article in question here is a letter of Frénicle to Fermat, 2 August 1641.

which he claimed to have. Euler later searched in vain for it but it was Lagrange who published the first rigorous proof (*Nouv. mém. Acad. Berlin*, 1775, pp. 349–51[3]). Euler seems still to have been unaware of it when he wrote his treatise in *Opuscula Analytica* (*1*, 259).[4]

▶ 117. We pass on to a consideration of the residues $+3$ and -3, and we begin with the second of them.

From Table 2 we find the residues of -3 to be 3, 7, 13, 19, 31, 37, 43, 61, 67, 73, 79, 97. Among them none is of the form $6n + 5$. We show as follows that outside of this table there are no primes of this form with -3 as a residue. First, any composite number of the form $6n + 5$ necessarily involves a prime factor of the same form. Thus when we show that no prime number of the form $6n + 5$ can have -3 as a residue, we also show that no such composite number can have this property. Suppose that there are such numbers outside our table. Let the smallest of them $=t$ and let $-3 = a^2 - tu$. Now if we take a as even and less than t, we will have $u < t$ and -3 a residue of u. But when a is of the form $6n + 2$, tu will be of the form $6n + 1$ and u of the form $6n + 5$. Q.E.A. because we supposed that t was the smallest number that contradicted our induction. Now if a is of the form $6n$, tu will be of the form $36n + 3$ and $tu/3$ of the form $12n + 1$. Thus $u/3$ will be of the form $6n + 5$. But clearly -3 is also a residue of $u/3$ and $u/3 < t$. Q.E.A. Manifestly therefore -3 can be the residue of no number of the form $6n + 5$.

Since every number of the form $6n + 5$ is necessarily contained among those of the form $12n + 5$ or $12n + 11$ and since the first of these is contained among numbers of the form $4n + 1$, the second among numbers of the form $4n + 3$, we have these theorems:

I. *Both -3 and $+3$ are nonresidues of all prime numbers of the form $12n + 5$.*

II. *-3 is a nonresidue, $+3$ a residue, of any prime number of the form $12n + 11$.*

▶ 118. From Table 2 we find that the numbers whose residue is $+3$ are the following: 3, 11, 13, 23, 37, 47, 59, 61, 71, 73, 83, 97. None of these is of the form $12n + 5$ or $12n + 7$. We can use the

[3] See p. 42.
[4] Diss. 8. See p. 62.

same method used in article 112, 113, 117 to show that there are
no numbers at all of the forms $12n + 5$, $12n + 7$ which have $+3$ as
a residue. We omit the details. Combining these results with those
of article 111 we have the following theorems:

I. *Both $+3$ and -3 are nonresidues of all prime numbers of the
form $12n + 5$* (just as we found in the previous article).

II. *$+3$ is a nonresidue, -3 a residue of all prime numbers of the
form $12n + 7$.*

▶ 119. We can learn nothing from this method about numbers of
the form $12n + 1$, and so we must resort to special devices. By
induction it is easy to see that $+3$ and -3 are residues of all prime
numbers of this form. But we need only demonstrate this by showing that -3 is a residue because it follows necessarily that $+3$ will
also be a residue (article 111). However we will show a more general result, i.e. that -3 is a residue of any prime number of the
form $3n + 1$.

Let p be such a prime and a a number belonging to the exponent
3 for the modulus p (that there are such is clear from article 54,
because 3 divides $p - 1$). We have therefore $a^3 \equiv 1$ (mod. p); i.e.
$a^3 - 1$ or $(a^2 + a + 1)(a - 1)$ is divisible by p. But clearly a
cannot be $\equiv 1$ (mod. p) because 1 belongs to the exponent 1;
therefore $a - 1$ will not be divisible by p, but $a^2 + a + 1$ will be.
Therefore $4a^2 + 4a + 4$ also will be; i.e. $(2a + 1)^2 \equiv -3$ (mod. p)
and -3 is a residue of p. Q.E.D.

It is clear that this demonstration (which is independent of the
preceding ones) also includes prime numbers of the form $12n + 7$
which we treated in the preceding article.

We observe further that we could use the method of articles
109 and 115, but for brevity's sake we will not pursue it.

▶ 120. From these results we can easily deduce the following
theorems (see art. 102, 103, 105).

I. *-3 is a residue of any number that cannot be divided by 8 or
by 9 or by any prime number of the form $6n + 5$. It is a nonresidue
of all others.*

II. *$+3$ is a residue of all numbers that cannot be divided by 4 or
by 9 or by any prime of the form $12n + 5$ or $12n + 7$. It is a nonresidue of all others.*

Note this particular case:

-3 is a *residue* of all prime numbers of the form $3n + 1$, or what amounts to the same thing, *of all primes that are residues of* 3. It is a *nonresidue* of all prime numbers of the form $6n + 5$ or, with the exception of the number 2, of all primes of the form $3n + 2$; i.e. *of all primes that are nonresidues of* 3. All other cases follow naturally from this.

The propositions pertaining to the residues $+3$ and -3 were known to Fermat (*Opera Mathem. Wall.*, 2, 857[5]), but Euler was the first to give proofs[6] (*Novi comm. acad. Petrop.*, 8 [1760–61], 1763, 105–28). This is why it is still more astonishing that proof of the propositions relative to the residues $+2$ and -2 kept eluding him, since they depend on similar devices. See also the commentary of Lagrange in *Nouv. mém. Acad. Berlin*[7] (1775, p. 352).

▶ 121. Induction shows that $+5$ is a residue of no odd number of the form $5n + 2$ or $5n + 3$, i.e. of no odd number that is a nonresidue of 5. We show as follows that this rule has no exception. Let the smallest number that might be an exception to this rule $= t$. It is a nonresidue of the number 5, but 5 is a residue of t. Let $a^2 = 5 + tu$ so that a is even and less than t. Then u will be odd and less than t and $+5$ will be a residue of u. Now if a is not divisible by 5, neither will u be. But evidently tu is a residue of 5; but since t is a nonresidue of 5, u will also be a nonresidue. That is, there is a number less than t which is odd, a nonresidue of the number 5, and which has $+5$ as a residue, contrary to the hypothesis. If a is divisible by 5, let $a = 5b$ and $u = 5v$, then $tv \equiv -1 \equiv 4$ (mod. 5); i.e. tv will be a residue of the number 5. From this point the demonstration proceeds as in the previous case.

▶ 122. Both $+5$ and -5 will be nonresidues therefore of all prime numbers which are at the same time nonresidues of 5 and of the form $4n + 1$, i.e. of all prime numbers of the form $20n + 13$ or $20n + 17$; and $+5$ will be a nonresidue, -5 a residue, of all prime numbers of the form $20n + 3$ or $20n + 7$.

In exactly the same way it can be shown that -5 is a nonresidue of all prime numbers of the forms $20n + 11$, $20n + 13$, $20n + 17$, $20n + 19$; and from this it easily follows that $+5$ is a

---

[5] The reference is to a letter: "Epistola XLVI, D. Fermatii ad D. Kenelmum Digby."

[6] "Supplementum quorundam theorematum arithmeticorum, quae in nonnullis demonstrationibus supponuntur."

[7] Cf. p. 42.

residue of all prime numbers of the form $20n + 11$ or $20n + 19$, and a nonresidue of all of the form $20n + 13$ or $20n + 17$. And since any prime number, except 2 and 5 (which have ± 5 as residues), is contained in one of the forms $20n + 1, 3, 7, 9, 11, 13, 17, 19$, it is clear that we are able to judge concerning all primes except those of the form $20n + 1$ or of the form $20n + 9$.

▶ 123. By induction it is easy to establish that $+5$ and -5 are residues of all prime numbers of the form $20n + 1$ or $20n + 9$. And if this is true in general, we have the elegant law that $+5$ *is a residue of all prime numbers that are residues of* 5 (for these numbers are contained in the forms $5n + 1$ or $5n + 4$ or in one of $20n + 1, 9, 11, 19$; we have already considered the third and fourth of these) *and a nonresidue of all odd numbers that are nonresidues of* 5, as we have already demonstrated above. This theorem clearly suffices to judge whether $+5$ (or -5 by considering the product of $+5$ and -1) is a residue or nonresidue of any given number. And finally, observe the analogy between this theorem and the one of article 120 in which we treated the residue -3.

However the verification of this induction is not so easy. When a prime number of the form $20n + 1$, or more generally of the form $5n + 1$, is considered, the problem can be solved in a way similar to that of articles 114 and 119. Let a be some number belonging to the exponent 5 relative to the modulus $5n + 1$. That there are such is clear from the preceding section. Thus we will have $a^5 \equiv 1$ or $(a - 1)(a^4 + a^3 + a^2 + a + 1) \equiv 0$ (mod. $5n + 1$). But since we cannot have $a \equiv 1$ and so not $a - 1 \equiv 0$, we must have $(a^4 + a^3 + a^2 + a + 1) \equiv 0$. Thus also $4(a^4 + a^3 + a^2 + a + 1) = (2a^2 + a + 2)^2 - 5a^2$ will $\equiv 0$; i.e. $5a^2$ will be a residue of $5n + 1$ and so also 5, because a^2 is a residue not divisible by $5n + 1$ (a is not divisible by $5n + 1$ because $a^5 \equiv 1$).   Q.E.D.

The case of a prime number of the form $5n + 4$ demands a more subtle device. But since the propositions we need here will be treated more generally later, we will touch on them only briefly.

I. If p is a prime number and b a given quadratic nonresidue of p, the value of the expression

$$(A) \ldots \frac{(x + \sqrt{b})^{p+1} - (x - \sqrt{b})^{p+1}}{\sqrt{b}}$$

(it is clear that the expansion contains no irrationals) will always be divisible by p, no matter what value is assumed for x. For it is clear from an inspection of the coefficients obtained in the expansion of A that all terms from the second to the penultimate (inclusive) will be divisible by p, and thus $A \equiv 2(p + 1)(x^p + xb^{(p-1)/2})$ (mod. p). But since b is a nonresidue of p we have $b^{(p-1)/2} \equiv -1$ (mod. p) (art. 106); but x^p is always $\equiv x$ (preceding section) and thus $A \equiv 0$.   Q.E.D.

II. In the congruence $A \equiv 0$ (mod. p), the indeterminate x has p dimensions and all the numbers $0, 1, 2, \ldots p - 1$ will be roots. Now let e be a divisor of $p + 1$. The expression

$$\frac{(x + \sqrt{b})^e - (x - \sqrt{b})^e}{\sqrt{b}}$$

(which we will designate by B) will be rational, the indeterminate x will have $e - 1$ dimensions, and it is clear from the fundamentals of analysis that A is divisible by B (indefinitely). Now I say that there are $e - 1$ values of x which, if substituted in B, make B divisible by p. For let $A \equiv BC$ and we find that x will have $p - e - 1$ dimensions in C and thus that the congruence $C \equiv 0$ (mod. p) will have no more than $p - e - 1$ roots. And it follows that the $e - 1$ numbers remaining among $0, 1, 2, 3, \ldots p - 1$ will be roots of the congruence $B \equiv 0$.

III. Now suppose p is of the form $5n + 4$, $e = 5$, b is a nonresidue of p, and the number a is so chosen that

$$\frac{(a + \sqrt{b})^5 - (a - \sqrt{b})^5}{\sqrt{b}}$$

is divisible by p. But this expression becomes

$$= 10a^4 + 20a^2b + 2b^2 = 2[(b + 5a^2)^2 - 20a^4]$$

Therefore we will also have $(b + 5a^2)^2 - 20a^4$ divisible by p; i.e. $20a^4$ is a residue of p, but since $4a^4$ is a residue not divisible by p (for it is easy to see that a cannot be divided by p), 5 will also be a residue of p.   Q.E.D.

Thus it is clear that the theorem enunciated in the beginning of this article is true generally.

We note that the demonstrations for both cases are due to Lagrange[8] (*Nouv. mém. Acad. Berlin*, 1775, p. 352 ff.).

▶ 124. By a similar method we can show:

−7 is a nonresidue of any number that is a nonresidue of 7. By induction we can conclude,

−7 is a residue of any prime number that is a residue of 7.

But no one thus far has proved this rigorously. The demonstration is easy for those residues of 7 that are of the form $4n − 1$; for by the method of the preceding articles it can be shown that $+7$ is always a nonresidue of such prime numbers, and so $−7$ is a residue. But we gain little by this, since the remaining cases cannot be handled in the same way. One case can be solved in the manner of articles 119 and 123. If p is a prime number of the form $7n + 1$, and a belongs to the exponent 7 relative to the modulus p, it is easy to see that

$$\frac{4(a^7 − 1)}{a − 1} = (2a^3 + a^2 − a − 2)^2 + 7(a^2 + a)^2$$

is divisible by p, and thus that $−7(a^2 + a)^2$ is a residue of p. But $(a^2 + a)^2$, as a square, is a residue of p and not divisible by p; for since we supposed that a belongs to the exponent 7, it can neither $\equiv 0$, nor $\equiv −1$ (mod. p); i.e. neither a nor $a + 1$ [nor therefore the square $(a + 1)^2 a^2$] will be divisible by p. Therefore 7 will also be a residue of p.   Q.E.D.   But prime numbers of the form $7n + 2$ or $7n + 4$ cannot be handled by any of the methods thus far considered. A proof however was discovered first by Lagrange in the same work. Below in Section VII we will show that the expression $4(x^p − 1)/(x − 1)$ can always be reduced to the form $X^2 \mp pY^2$ (where the upper sign should be taken when p is a prime number of the form $4n + 1$, the lower when it is of the form $4n + 3$). In this expression X, Y are rational expressions in x, free of fractions. Lagrange did not carry this analysis beyond the case $p = 7$ (see Lagrange, loc. cit. p. 352).

▶ 125. Since the preceding methods are not sufficient to establish general demonstrations, we will now exhibit one that is. We begin with a theorem whose demonstration has long eluded us, although at first glance it seems so obvious that many authors believed there

---

[8] Cf. p. 42.

was no need for demonstration. It reads as follows: *Every number with the exception of squares taken positively is a nonresidue of some prime numbers*. But since we intend to use this theorem only as an aid in proving other considerations, we will explain here only those cases needed for this purpose. Other cases will be established later. We will show therefore that *any prime number of the form $4n + 1$ whether it be taken positively or negatively, is a nonresidue of some prime numbers*[f] and indeed (if it is > 5) of some primes that are less than itself.

First when p a prime number of the form $4n + 1$ (> 17 but $-13 N 3$, $-17 N 5$) is to be taken *negatively*,[9] let $2a$ be the even number immediately greater than $\sqrt{p}$; then $4a^2$ will always be $< 2p$ or $4a^2 - p < p$. But $4a^2 - p$ is of the form $4n + 3$, and p a quadratic residue of $4a^2 - p$ (since $p \equiv 4a^2$ (mod. $4a^2 - p$)); and if $4a^2 - p$ is a prime number, $-p$ will be a nonresidue; if not, some factor of $4a^2 - p$ will necessarily be of the form $4n + 3$; and since $+p$ has to be a residue, $-p$ will be a nonresidue. Q.E.D.

For *positive* prime numbers we distinguish two cases. *First* let p be a prime number of the form $8n + 5$. Let a be any positive number $< \sqrt{(p/2)}$. Then $8n + 5 - 2a^2$ will be a positive number of the form $8n + 5$ or $8n + 3$ (according as a is even or odd) and so necessarily divisible by some prime number of the form $8n + 3$ or $8n + 5$, for a product of numbers of the form $8n + 1$ and $8n + 7$ cannot have the form $8n + 3$ or $8n + 5$. Let it here be denoted by q and we will have $8n + 5 \equiv 2a^2$ (mod. q). But 2 will be a nonresidue of q (art. 112) and so also $2a^2$ and $8n + 5$. Q.E.D.[8]

▶ 126. We have no such obvious devices to demonstrate that any prime number of the form $8n + 1$ taken positively is always a nonresidue of some prime number less than itself. But since this truth is of such great importance, we cannot omit a rigorous demonstration even though it is somewhat long. We begin as follows:

---

[f] Obviously we must except $+1$.

[9] $-13 N 3$ means that -13 is a nonresidue of 3; $-17 N 5$ means that -17 is a nonresidue of 5. See p. 88.

[8] Article 98. It is clear that a^2 is a residue of q not divisible by q, for otherwise the prime number p would be divisible by q. Q.E.A.

LEMMA. *Suppose we have two series of numbers*

A, B, C, etc. ...   (I) A', B', C', etc. ...   (II)

(whether the number of terms in each is the same or not, is a matter of indifference) *so arranged that if p is a prime number or the power of a prime number that divides one or more terms of the second series, there are at least as many terms of the first series divisible by p. I claim that the product of all numbers* (I) *is divisible by the product of all numbers* (II).

Example. Let (I) consist of the numbers $12, 18, 45$; (II) of the numbers $3, 4, 5, 6, 9$. Then if we take successively the numbers $2, 4, 3, 9, 5$ we find that there are $2, 1, 3, 2, 1$ terms in (I) and $2, 1, 3, 1, 1$ terms in (II) which are, respectively, divisible by these numbers; and the product of all the terms (I) $= 9720$ which is divisible by 3240, the product of all the terms (II).

Demonstration. Let the product of all the terms (I) $= Q$, the product of all the terms of the series (II) $= Q'$. It is clear that any prime number which is a divisor of Q' is also a divisor of Q. Now we will show that any prime factor of Q' has at least as many dimensions in Q as it has in Q'. Let such a divisor be p and let us suppose that in series (I) there are a terms divisible by p, b terms divisible by p^2, c terms divisible by p^3, etc. In a similar way determine the letters a', b', c', etc. for series (II). In Q, p will have $a + b + c +$ etc. dimensions and $a' + b' + c' +$ etc. dimensions in Q'. But a' is certainly not greater than a, b' not greater than b, etc. (by hypothesis); and therefore $a' + b' + c' +$ etc. will certainly not be $> a + b + c +$ etc. Since therefore no prime number can have higher dimensions in Q' than in Q, Q will be divisible by Q' (art. 17). Q.E.D.

▶ 127. LEMMA. *In the progression* $1, 2, 3, 4, \ldots n$ *there can not be more terms divisible by a number h than in the progression* $a,\ a + 1,\ a + 2, \ldots a + n - 1$, *which has the same number of terms.*

We see without any trouble that if n is a multiple of h, there are n/h terms in each progression divisible by h; if n is not a multiple of h, let $n = eh + f$ with $f < h$. In the first series there will be e terms divisible by h, in the second either e or $e + 1$.

As a corollary of this we have that well-known proposition

from the theory of figurative numbers, namely that

$$\frac{a(a + 1)(a + 2)\ldots(a + n - 1)}{1 \cdot 2 \cdot 3 \ldots n}$$

is always an integer. But until now, as far as we know, no one has demonstrated it directly.

Finally, the lemma could have been expressed more generally as follows:

In the progression $a, a + 1, a + 2, \ldots, a + n - 1$ there are at least as many terms congruent relative to the modulus h to any given number r as there are terms in the progression $1, 2, 3, \ldots n$ divisible by h.

▶ 128. THEOREM. *Let a be any number of the form $8n + 1$; p some number which is prime relative to a having $+a$ as a residue; and let m be an arbitrary number. I claim that in the progression*

$a, \frac{1}{2}(a - 1), 2(a - 4), \frac{1}{2}(a - 9), 2(a - 16), \ldots 2(a - m^2)$ or $\frac{1}{2}(a - m^2)$

according as m is even or odd, there are at least as many terms divisible by p as there are in the progression

$$1, 2, 3, \ldots 2m + 1$$

We will designate the first of these by (I), the second by (II).

Demonstration. I. When $p = 2$, all the terms except the first, i.e. m terms, will be divisible in (I); the same number will be divisible in (II).

II. Let p be an odd number or double or four times an odd number and let $a \equiv r^2$ (mod. p). Then in the progression $-m, -(m - 1), -(m - 2), \ldots + m$ [it has as many terms as (II) and we will call it (III)] relative to the modulus p there will be as many terms congruent to r as there are in series (II) divisible by p (preceding article). But among them there can be no pairs which differ only in sign.[h] And each of them will have a corresponding value in series (I) which is divisible by p. This means that if $\pm b$ were a term of series (III) congruent to r relative to p, $a - b^2$

---

[h] For if $r \equiv -f \equiv +f$ (mod. p), $2f$ would be divisible by p and so also $2a$ [since $f^2 \equiv a$ (mod. p)]. This cannot be unless $p = 2$, since by hypothesis a and p are relatively prime. But we have already treated this case by itself.

would be divisible by p. And so if b is even, the term $2(a - b^2)$ of series (I) will be divisible by p. But if b is odd, the term $(a - b^2)/2$ will be divisible by p; for manifestly $(a - b^2)/p$ is an *even* integer because $a - b^2$ is divisible by 8, whereas p is divisible at most by 4 (by hypothesis, a is of the form $8n + 1$, and b^2 which is the square of an odd number will be of the same form, and the difference will be of the form $8n$). Thus we conclude that in series (I) there are as many terms divisible by p as there are terms in (III) congruent to r relative to p, i.e. as many or more than there are in (II) divisible by p. Q.E.D.

III. Let p be of the form $8n$ and $a \equiv r^2$ (mod. $2p$). It is easy to see that a, which is by hypothesis a residue of p, is also a residue of $2p$. Then there will be at least as many terms in series (III) congruent to r relative to p as there are in (II) divisible by p, and they will all be unequal in value. But for each of them there will be a corresponding term in (I) which is divisible by p. For if $+b$ or $-b \equiv r$ (mod. p), we will have $b^2 \equiv r^2$ (mod. $2p$)[i] and the term $(a - b^2)/2$ will be divisible by p. Wherefore there will be at least as many terms in (I) divisible by p as there are in (II). Q.E.D.

▶ 129. THEOREM. *If a is a prime number of the form $8n + 1$ there is necessarily some prime number below $2\sqrt{a} + 1$ of which a is a nonresidue.*

Demonstration. If it is possible, let a be a residue of all primes less than $2\sqrt{a} + 1$. Then a would be a residue of all composite numbers less than $2\sqrt{a} + 1$ (cf. the rules for judging whether a given number is a residue of a composite number or not: art. 105). Let the number next smaller than $\sqrt{a}$ be $= m$. Then in the series

(I) $\ldots a, \frac{1}{2}(a - 1), 2(a - 4), \frac{1}{2}(a - 9), \ldots 2(a - m^2)$ or $\frac{1}{2}(a - m^2)$

there will be as many or more terms divisible by some number smaller than $2\sqrt{a} + 1$ as in

(II) $\ldots 1, 2, 3, 4, \ldots 2m + 1$ (preceding article)

And it follows that the product of all terms (I) will be divisible by the product of all terms (II) (art. 126). But the former product

---

[i] That is, we will have $b^2 - r^2 = (b - r)(b + r)$ composed of two factors, one of which is divisible by p (hypothesis), the other divisible by 2 (since both b and r are odd); and so $b^2 - r^2$ is divisible by $2p$.

is either equal to $a(a - 1)(a - 4)\ldots(a - m^2)$ or to half of this product (according as m is even or odd). And so the product $a(a - 1)(a - 4)\ldots(a - m^2)$ will certainly be divisible by the product of all the terms (II) and since all these terms are prime relative to a, so will the product if we delete the factor a. But the product of all the terms (II) can be expressed as follows:

$$[m + 1][(m + 1)^2 - 1][(m + 1)^2 - 4]\ldots[(m + 1)^2 - m^2]$$

And therefore

$$\frac{1}{m + 1}\cdot\frac{a - 1}{(m + 1)^2 - 1}\cdot\frac{a - 4}{(m + 1)^2 - 4}\cdots\frac{a - m^2}{(m + 1)^2 - m^2}$$

will be an integer although it is the product of fractions which are less than unity; for since $\sqrt{a}$ is necessarily irrational, $m + 1 > \sqrt{a}$ and $(m + 1)^2 > a$. We conclude that our supposition cannot be true. Q.E.D.

Now because a is certainly > 9, we will have $2\sqrt{a} + 1 < a$, and there exists a prime number $< a$, of which a is a nonresidue.

▶ 130. Now that we have rigorously established that any prime number of the form $4n + 1$, taken both positively and negatively, is a nonresidue of some prime less than itself, we proceed to a more exact and more general comparison of prime numbers, one of which is a residue or nonresidue of the other.

We have shown above that -3 and $+5$ are residues or nonresidues of all prime numbers that are residues or nonresidues of 3, 5 respectively.

By induction it is found that the numbers $-7, -11, +13, +17, -19, -23, +29, -31, +37, +41, -43, -47, +53, -59$, etc. are residues or nonresidues of all prime numbers which, taken positively, are residues or nonresidues of these primes respectively. This induction can be accomplished easily with the help of Table 2.

It can be seen that among these primes those of the form $4n + 1$ are positive, those of the form $4n + 3$ are negative.

▶ 131. We will soon show that what we have discovered by induction is true for the general case. But before we do this it will be necessary to discover all the consequences of this theorem, supposing it to be true:

If p is a prime number of the form $4n + 1$, $+p$ will be a residue or nonresidue of any prime number which taken positively is a *residue or nonresidue of p. If p is of the form* $4n + 3$, $-p$ *will have the same property.*

Since almost everything that can be said about quadratic residues depends on this theorem, the term *fundamental theorem* which we will use from now on should be acceptable.

To indicate our reasoning as briefly as possible, we will denote prime numbers of the form $4n + 1$ by the letters a, a', a'', etc. and prime numbers of the form $4n + 3$ by the letters b, b', b'', etc.; any numbers of the form $4n + 1$ will be denoted by A, A', A'', etc., any numbers of the form $4n + 3$ by B, B', B'', etc.; finally the letter R placed between two quantities will indicate that the former is a residue of the latter, and the letter N will indicate the contrary. For example, $+5\,R\,11$, $\pm 2\,N\,5$ will indicate that $+5$ is a residue of 11, and $+2$ or -2 is a nonresidue of 5. Now with the help of the theorems in article 111 we can easily deduce the following propositions from the fundamental theorem.

	If	we have
1.	$\pm a\,R\,a'$	$\pm a'\,R\,a$
2.	$\pm a\,N\,a'$	$\pm a'\,N\,a$
3.	$+a\,R\,b\ \}$ $-a\,N\,b\ \}$	$\pm b\,R\,a$
4.	$+a\,N\,b\ \}$ $-a\,R\,b\ \}$	$\pm b\,N\,a$
5.	$\pm b\,R\,a$	$\{\,+a\,R\,b$ $\{\,-a\,N\,b$
6.	$\pm b\,N\,a$	$\{\,+a\,N\,b$ $\{\,-a\,R\,b$
7.	$+b\,R\,b'\ \}$ $-b\,N\,b'\ \}$	$\{\,+b'\,N\,b$ $\{\,-b'\,R\,b$
8.	$+b\,N\,b'\ \}$ $-b\,R\,b'\ \}$	$\{\,+b'\,R\,b$ $\{\,-b'\,N\,b$

▶ 132. This table includes all the cases when two prime numbers are compared; what follows pertains to any numbers, but the demonstrations of these is less obvious.

	If	we have
9.	$\pm a\,R\,A$	$\pm A\,R\,a$
10.	$\pm b\,R\,A$	$\{\,+A\,R\,b$ $\{\,-A\,N\,b$

11.	$+ a \, R \, B$	$\pm B \, R \, a$
12.	$- a \, R \, B$	$\pm B \, N \, a$
13.	$+ b \, R \, B$	$\begin{cases} - B \, R \, b \\ + B \, N \, b \end{cases}$
14.	$- b \, R \, B$	$\begin{cases} + B \, R \, b \\ - B \, N \, b \end{cases}$

Since the same principles lead to the demonstration of all these propositions, it will not be necessary to develop them all. A demonstration of proposition 9 which we will do as an example should be sufficient. First we observe that any number of the form $4n + 1$ has either no factor of the form $4n + 3$ or else two or four etc. of them; i.e. the number of such factors (including any that may be equal) will always be even; and that any number of the form $4n + 3$ involves an odd number of factors of the form $4n + 3$ (i.e. one or three or five etc.). The number of factors of the form $4n + 1$ remains undetermined.

Proposition 9 can be demonstrated as follows. Let A be the product of prime factors a', a'', a''', etc. b, b', b'', etc.; the number of factors b, b', b'', etc. will be even (or there may be none of them, which reduces to the same thing). Now if a is a residue of A, it will also be a residue of all the factors a', a'', a''', etc. b, b', b'', etc. By propositions 1, 3 of the preceding article, each of these factors will be residues of a and so will the product A as well as $- A$. And if $- a$ is a residue of A, by that very fact it is a residue of all the factors a', a'', etc. b, b', etc.; each of the a', a'', etc. will be residues of a, each of the b, b', etc. nonresidues. But since the latter are even in number, the product of all of them, i.e. A, will be a residue of a, and so also will $- A$.

▶ 133. We begin now a more general investigation. Consider any two odd, signed numbers P and Q which are relatively prime. Let us conceive of P without respect to its sign as resolved into its prime factors, and designate by p the number of these factors for which Q is a nonresidue. If any prime number of which Q is a nonresidue occurs many times among the factors of P, it is to be counted many times. Similarly let q be the number of prime factors of Q for which P is a nonresidue. The numbers p, q will have a certain mutual relation depending on the nature of the numbers P, Q. That is, if one of the numbers p, q is even or odd, the form of the numbers P, Q will indicate whether the other is even or odd. We show this relation in the following table.

The numbers p, q will both be even or both odd when the numbers P, Q have the forms:

1.	$+A$,	$+A'$
2.	$+A$,	$-A'$
3.	$+A$,	$+B$
4.	$+A$,	$-B$
5.	$-A$,	$-A'$
6.	$+B$,	$-B'$

On the other hand one of the numbers p, q will be even, the other odd, when the numbers P, Q have the forms:

7.	$-A$,	$+B$
8.	$-A$,	$-B$
9.	$+B$,	$+B'$
10.	$-B$,	$-B'$ [j]

Example. Let the given numbers be -55 and $+1197$ which is the fourth case; 1197 is a nonresidue of one prime factor of 55, namely the number 5. But -55 is a nonresidue of three of the prime factors of 1197, namely 3, 3, 19.

If P and Q are prime numbers, these propositions reduce to those which we considered in article 131. Here p and q cannot be greater than 1, and so when p is even, it necessarily $=0$; i.e. Q will be a residue of P. But when p is odd, Q will be a nonresidue of P, and vice versa. And so writing a, b in place of A, B it follows from 8 that if $-a$ is a residue or nonresidue of b, $-b$ will be a nonresidue or residue of a, which agrees with 3 and 4 of article 131.

In general Q cannot be a residue of P unless $p = 0$; if therefore p is odd, Q will certainly be a nonresidue of P.

The propositions of the preceding article can be derived from this fact without any difficulty.

It is apparent that this general representation is more than idle speculation because the demonstration of the fundamental theorem would be incomplete without it.

▶ 134. We attempt now to deduce these propositions.

I. Let us as before conceive of P as resolved into its prime factors neglecting all signs. Let Q be resolved into factors in any

[j] Let $l = 1$ if both $P, Q \equiv 3$ (mod. 4), otherwise $l = 0$. Let $m = 1$ if both P, Q are negative, otherwise $m = 0$. The relation then depends on $l + m$.

way whatsoever, but here we take into account the sign of Q. Now combine each of the former with each of the latter. Then if s designates the number of all combinations in which a factor of Q is a nonresidue of the factor of P, p and s will either both be even or both odd. Let the prime factors of P be f, f', f'', etc. And among the factors into which Q is resolved, let m of them be nonresidues of f, m' of them nonresidues of f', m'' of them nonresidues of f'', etc. Then obviously,

$$s = m + m' + m'' + \text{etc.}$$

and p will indicate how many numbers among m, m', m'', etc. are odd. Thus s will be even when p is even, odd when p is odd.

II. This is true in general no matter how Q is resolved into factors. Now to particular cases. For the first case, let one of the numbers, P, be positive and the other, Q, either of the form $+A$ or $-B$. Resolve P, Q into their prime factors, assigning to each of the factors of P a positive sign and to each of the factors of Q a positive or negative sign according as they are of the form a or b; manifestly Q will be either of the form $+A$ or $-B$, as required. Combine each of the factors of P with each of the factors of Q and designate as before s, the number of the combinations in which the factor of Q is a nonresidue of the factor of P. Similarly let t be the number of combinations in which the factor of P is a nonresidue of the factor of Q. But from the fundamental theorem it follows that these combinations must be identical and therefore $s = t$. Finally, from what we have just shown, $p \equiv s$ (mod. 2), $q \equiv t$ (mod. 2) and so $p \equiv q$ (mod. 2).

Thus we have propositions 1, 3, 4, and 6 of article 133.

The other propositions can be demonstrated directly in a similar manner but they demand one new consideration; it is easier to derive them from the preceding in the following way.

III. Let us denote again P, Q any odd numbers which are relatively prime, p, q the number of prime factors of P, Q for which Q, P are respectively nonresidues. And let p' be the number of prime factors of P for which $-Q$ is a nonresidue (when Q is negative, manifestly $-Q$ will be positive). Now let all the prime factors of P be distributed into four classes:

1) factors of the form a of which Q is a residue

2) factors of the form b for which Q is a residue; let the number of these be χ

3) factors of the form a for which Q is a nonresidue; let the number of these be ψ

4) factors of the form b for which Q is a nonresidue; let the number of these be ω

It is easy to see that $p = \psi + \omega$, $p' = \chi + \psi$.

Now when P is of the form $+A$, $\chi + \omega$ and $\chi - \omega$ will be even; and thus $p' = p + \chi - \omega \equiv p$ (mod. 2). When P is of the form $\pm B$, we find by a similar computation that the numbers p, p' will be noncongruent relative to the modulus 2.

IV. Let us apply this to individual cases. Let both P and Q be of the form $+A$. From proposition 1 we have $p \equiv q$ (mod. 2); but $p' \equiv p$ (mod. 2) so also $p' \equiv q$ (mod. 2). This agrees with proposition 2. Similarly if P is of the form $-A$, Q of the form $+A$, we have $p \equiv q$ (mod. 2) from proposition 2 which we have just seen; then since $p' \equiv p$, we get $p' \equiv q$. Proposition 5 is thus demonstrated.

In the same way proposition 7 can be deduced from 3; proposition 8 from either 4 or 7; proposition 9 from 6; proposition 10 from 6.

▶ 135. We have not demonstrated the propositions of article 133 in the preceding article but we nevertheless showed that their truth depends on the truth of the fundamental theorem. By the method we followed it is clear that these propositions are true for the numbers P, Q if only the fundamental theorem is true for all the prime factors of these numbers compared among themselves, even if it were not generally true. Now we come to a demonstration of the fundamental theorem. We preface it with the following explanation:

We will say that the fundamental theorem is true up to some number M, if it is true for any two prime numbers neither of which is larger than M.

It should be understood in the same way if we say that the theorems of articles 131, 132, and 133 are true up to some term. If the fundamental theorem is true up to some term, it is clear that these propositions will be true up to the same term.

▶ 136. It is easy to confirm that the fundamental theorem is true for small numbers by induction and so the limit can be determined up to which it certainly applies. Let us suppose this induction established; how far we have carried it is a matter of indifference. Thus it would be sufficient to confirm it up to the number 5. This can be done by a single observation since $+5\,N\,3$, $\pm3\,N\,5$.

Now if the fundamental theorem were not generally true, there would be some limit T up to which it would be valid, but it would not be valid for the next greater number $T + 1$. This is the same as saying that there are two prime numbers, the larger of which is $T + 1$, such that they contradict the fundamental theorem when compared together. It would imply however that any other pair of prime numbers would satisfy the theorem if only both were less than $T + 1$. From this it would follow that the propositions of articles 131, 132, and 133 would also be valid up to T. We will show now that this supposition is inconsistent. There are various cases we must distinguish according to the different forms that $T + 1$ and the prime number less than $T + 1$ can have. We will call that prime number p.

When both $T + 1$ and p are of the form $4n + 1$, the fundamental theorem can be false in two ways, i.e. if we had at the same time

either	$\pm p\,R(T + 1)$	and	$\pm(T + 1)N\,p$
or	$\pm p\,N(T + 1)$	and	$\pm(T + 1)R\,p$

When both $T + 1$ and p are of the form $4n + 3$, the fundamental theorem will be false if we had at the same time

	either	$+p\,R(T + 1)$	and	$-(T + 1)N\,p$
[or what is the same thing	$-p\,N(T + 1)$	and	$+(T + 1)R\,p$]
	or	$+p\,N(T + 1)$	and	$-(Y + 1)R\,p$
[or what is the same thing	$-p\,R(T + 1)$	and	$+(T + 1)N\,p$]

When $T + 1$ is of the form $4n + 1$, and p of the form $4n + 3$, the fundamental theorem will be false if we had

either	$\pm p\,R(T + 1)$	and	$+(T + 1)N\,p$	
				[or $-(T + 1)R\,p$]
or	$\pm p\,N(T + 1)$	and	$-(T + 1)N\,p$	
				[or $+(T + 1)R\,p$]

When $T + 1$ is of the form $4n + 3$, and p of the form $4n + 1$, the fundamental theorem will be false if we had

either	$+p\,R(T + 1)$[or $-p\,N(T + 1)$]	and	$\pm(T + 1)N\,p$
or	$+p\,N(T + 1)$[or $-p\,R(T + 1)$]	and	$\pm(T + 1)R\,p$

If it can be shown that none of these eight cases is valid, it will be certain likewise that the truth of the fundamental theorem is circumscribed by no limits. We proceed to this exposition now but, since some of the cases depend on others, we cannot preserve the same order we used to enumerate them.

▶ 137. *First case. When $T + 1$ is of the form $4n + 1$ ($=a$) and p is of the same form, we cannot have $\pm p \, R \, a$ and $\pm a \, N \, p$ or we cannot have $\pm a \, N \, b$ if $\pm p \, R \, a$.*

Let $+p \equiv e^2$ (mod. a) where e is even and $< a$ (this is always possible). We can distinguish two cases.

I. When e is not divisible by p. Let $e^2 = p + af$. Here f will be positive, of the form $4n + 3$ (form B), $< a$ and not divisible by p. Further $e^2 \equiv p$ (mod. f); i.e. $p \, R \, f$ and so from proposition 11 of article 132 $\pm f \, R \, p$ (since $p, f < a$). But we have also $af \, R \, p$ and therefore $\pm a \, R \, p$.

II. When e is divisible by p, let $e = gp$ and $e^2 = p + aph$ or $pg^2 = 1 + ah$. Then h will be of the form $4n + 3$ (B) and relatively prime to p and g^2. Further we have $pg^2 R \, h$ and so also $p \, R \, h$ and from this (prop. 11, art. 132) $\pm h \, R \, p$. But we also have $-ah \, R \, p$ because $-ah \equiv 1$ (mod. p); therefore $\mp a \, R \, p$.

▶ 138. *Second case. When $T + 1$ is of the form $4n + 1$ ($=a$), p of the form $4n + 3$, and $\pm p \, R(T + 1)$ we cannot have $+(T + 1)N \, p$ or $-(T + 1)R \, p$.* This was the fifth case above.

Let as above $e^2 = p + fa$ and e even and $< a$.

I. When e is not divisible by p, f will also not be divisible by p. Further, f will be positive, of the form $4n + 1$ (A), and $< a$; but $+p \, R \, f$ and therefore (prop. 10, art. 132) $+f \, R \, p$. But also $+fa \, R \, p$; therefore $+a \, R \, p$ or $-a \, N \, p$.

II. When e is divisible by p, let $e = pg$ and $f = ph$. Therefore $g^2 p = 1 + ha$. Then h will be positive, of the form $4n + 3$ (B), and relatively prime to p and g^2. Further, $+g^2 p \, R \, h$ and so $+p \, R \, h$; as a result (prop. 13, art. 132) $-h \, R \, p$. But we have $-ha \, R \, p$ and so $+a \, R \, p$ and $-a \, N \, p$.

▶ 139. *Third case. When $T + 1$ is of the form $4n + 1$ ($=a$), p of the same form and $\pm p \, N \, a$, we cannot have $\pm a \, R \, p$* (second case above).

Take any prime number less than a of which $+a$ is a nonresidue. We have shown that there are such (art. 125, 129). But here we must consider two cases separately according as the prime number is of the form $4n + 1$ or $4n + 3$, for we did not show that there are such prime numbers of *each* form.

I. Let the prime number be of the form $4n + 1$ and $= a'$. Then we will have $\pm a' N a$ (art. 131) and therefore $\pm a'p R a$. Let now $e^2 \equiv a'p$ (mod. a) and e even and $< a$. Then again we have to distinguish four cases.

1) When e is not divisible by p or by a'. Let $e^2 = a'p \pm af$, taking whichever sign makes f positive. Then we have $f < a$, relatively prime to a' and p and, for the upper sign, of the form $4n + 3$, for the lower of the form $4n + 1$. For brevity's sake we will designate by $[x, y]$ the number of prime factors of the number y for which x is a nonresidue. Then we have $a'p R f$ and so $[a'p, f] = 0$. Thus $[f, a'p]$ will be an even number (prop. 1, 3, art. 133), i.e. either $= 0$ or $= 2$. So f will be a residue of each of the numbers a', p or of neither. The former case is impossible, since $\pm af$ is a residue of a' and $\pm a N a'$ (by hypothesis); therefore $\pm f N a'$. So f must be a nonresidue of each of the numbers a', p. But since $\pm af R p$, we will have $\pm a N p$. Q.E.D.

2) When e is divisible by p but not by a', let $e = gp$ and $g^2 = a' \pm ah$ with the sign so determined that h is positive. Then we will have $h < a$, relatively prime to a', g and p, and for the upper sign of the form $4n + 3$, for the lower of the form $4n + 1$. If we multiply the equation $g^2p = a' \pm ah$ by p and a', we can deduce easily that $pa' R h \ldots (\alpha)$; $\pm ahp R a' \ldots (\beta)$; $aa'h R p \ldots (\gamma)$. It follows from (α) that $[pa', h] = 0$ and so (prop. 1, 3, art. 133) $[h, pa']$ is even; i.e. h is a nonresidue of both p, a' or of neither. *In the former case* it follows from (β) that $\pm ap N a'$ and, since by hypothesis $\pm a N a'$, we have $\pm p R a'$. Thus by the fundamental theorem which is valid for the numbers p, a', since they are less than $T + 1$, we get $\pm a' R p$. Now $h N p$, therefore by (γ) $\pm a N p$. Q.E.D. *In the latter case* it follows from (β) that $\pm ap R a'$, so $\pm p N a'$, $\pm a' N p$. But because $h R p$ we get from (γ) that $\pm a N p$. Q.E.D.

3) When e is divisible by a' but not by p. For this case the demonstration is almost the same as in the preceding and it will cause no difficulty to anyone who has understood it.

4) When e is divisible both by a' and by p and thus also by the product $a'p$ (we have supposed that a', p are *unequal* because otherwise the hypothesis $a N a'$ would contain $a N p$). Let $e = ga'p$ and $g^2a'p = 1 \pm ah$. Then we will have $h < a$, relatively prime to a'

and p, and of the form $4n + 3$ when the upper sign applies, of the form $4n + 1$ when the lower sign applies. From this equation we can easily deduce the following, $a'p \, R \, h \ldots (\alpha)$; $\pm ah \, R \, a' \ldots (\beta)$; $\pm ah \, R \, p \ldots (\gamma)$. From (α) which agrees with (α) in 2) the same result follows. That is we have either $h \, R \, p$, $h \, R \, a'$ or $h \, N \, p$, $h \, N \, a'$. But in the former case (β) would give us $a \, R \, a'$ contrary to the hypothesis. So therefore $h \, N \, p$ and by (γ) $a \, N \, p$ also.

II. When the prime number is of the form $4n + 3$, the demonstration is so like the preceding that it seems superfluous to include it. We only observe for those who wish to do it themselves (which we highly recommend), that it would be advantageous to consider each sign separately after developing the equation $e^2 = bp \pm af$ (b designates the prime number).

▶ 140. *Fourth case. When $T + 1$ is of the form $4n + 1$ ($= a$), p of the form $4n + 3$ and $\pm p \, N \, a$, we cannot have $+a \, R \, p$ or $-a \, N \, p$* (sixth case above).

Since the demonstration of this is just like that of the third case, we omit it for the sake of brevity.

▶ 141. *Fifth case. When $T + 1$ is of the form $4n + 3$ ($= b$), p of the same form and $+p \, R \, b$ or $-p \, N \, b$, we cannot have $+b \, R \, p$ or $-b \, N \, p$* (third case above).

Let $p = e^2$ (mod. b) with e even and $< b$.

I. When e is not divisible by p. Let $e^2 = p + bf$ where f is positive, of the form $4n + 3$, $< b$, and relatively prime to p. Further $p \, R \, f$ and so (prop. 13, art. 132) $-f \, R \, p$. Thus also since $+bf R \, p$ we get $-b \, R \, p$ and $+b \, N \, p$. Q.E.D.

II. When e is divisible by p, let $e = pg$ and $g^2 p = 1 + bh$. Then h will be of the form $4n + 1$ and relatively prime to p; also $p \equiv g^2 p^2$ (mod. h), therefore $p \, R \, h$; from this we have $+h \, R \, p$ (prop. 10, art. 132) and then because $-bh \, R \, p$ it follows that $-b \, R \, p$ or $+b \, N \, p$. Q.E.D.

▶ 142. *Sixth case. When $T + 1$ is of the form $4n + 3$ ($= b$), p of the form $4n + 1$, and $p \, R \, b$, we cannot have $\pm b \, N \, p$* (seventh case above).

We omit the demonstration which is completely like the preceding.

▶ 143. *Seventh case. When $T + 1$ is of the form $4n + 3$ ($= b$), p of the same form, and $+p \, N \, b$ or $-p \, R \, b$, we cannot have $+b \, N \, p$ or $-b \, R \, p$* (fourth case above).

Let $-p \equiv e^2$ (mod. b) with e even and $< b$.

I. When e is not divisible by p. Let $-p = e^2 - bf$ where f is positive, of the form $4n + 1$, relatively prime to p, and less than b (for certainly e is not greater than $b - 1$, $p < b - 1$, and so $bf = e^2 + p < b^2 - b$; i.e. $f < b - 1$). Further we have $-p \, R \, f$ and from this (prop. 10, art. 132) $+f \, R \, p$. And since $+bf \, R \, p$ we get $+b \, R \, p$ or $-b \, N \, p$.

II. When e is divisible by p, let $e = pg$ and $g^2 p = -1 + bh$. Then h will be positive, of the form $4n + 3$, relatively prime to p, and $< b$. Further we get $-p \, R \, h$ and so (prop. 14, art. 132) $+h \, R \, p$. And since $bh \, R \, p$ it follows that $+b \, R \, p$ or $-b \, N \, p$. Q.E.D.

▶ 144. *Eighth case. When $T + 1$ is of the form $4n + 3$ $(=b)$, p of the form $4n + 1$, and $+p \, N \, b$ or $-p \, R \, b$, we cannot have $\pm b \, R \, p$* (last case above).

The demonstration is the same as in the preceding case.

▶ 145. In the preceding demonstration we always took an even value for e (art. 137, 144); an odd value could have been used just as well but then many more distinctions would have to be introduced. (Those who enjoy these investigations will find it very useful to apply themselves to this task.) Furthermore the theorems pertaining to the residues $+2$ and -2 would have to be presupposed. But since our demonstration was accomplished without these theorems we will present them by a new method which should not be disdained, since it is more direct than the methods used above to show that ± 2 is a residue of any prime number of the form $8n + 1$. We will suppose the other cases (regarding prime numbers of the form $8n + 3$, $8n + 5$, $8n + 7$) have been established by the above methods and that the theorem has been established only by induction; in the following reflections this induction will be raised to a certainty.

If ± 2 is not a residue of all prime numbers of the form $8n + 1$, let the smallest prime for which ± 2 is a nonresidue $= a$, so that the theorem is valid for all primes less than a. Now take some prime number $< a/2$ for which a is a nonresidue (it is clear from article 129 that this can be done). Let this number be $= p$ and by the fundamental theorem $p \, N \, a$. And so $\pm 2p \, R \, a$. Let therefore $e^2 \equiv 2p$ (mod. a) so that e is odd and $< a$. Then two cases must be distinguished.

I. When e is not divisible by p, let $e^2 = 2p + aq$; q will be positive, of the form $8n + 7$ or $8n + 3$ (according as p is of the form $4n + 1$ or $4n + 3$), $< a$, and not divisible by p. Now all prime factors of q are divided into four classes: e of the form $8n + 1$, f of the form $8n + 3$, g of the form $8n + 5$, h of the form $8n + 7$; let the product of the factors of the first class be E, of the second, third, and fourth classes F, G, H respectively.[k] Let us consider *first* the case where p is of the form $4n + 1$ and q of the form $8n + 7$. Clearly we will have $2\,R\,E$, $2\,R\,H$ and therefore $p\,R\,E$, $p\,R\,H$, and finally $E\,R\,p$, $H\,R\,p$. Further, 2 will be a nonresidue of any factor of the form $8n + 3$ or $8n + 5$ and so also p; thus such a factor will be a nonresidue of p; and in conclusion FG will be a residue of p if $f + g$ is even, a nonresidue if $f + g$ is odd. But $f + g$ cannot be odd; for regardless of what e, f, g, h are individually, if $f + g$ is odd $EFGH$ or q will be of the form $8n + 3$ or $8n + 5$ in every case. And this is contrary to the hypothesis. We get therefore $FG\,R\,p$, $EFGH\,R\,p$, or $q\,R\,p$, but since $aq\,R\,p$ this implies that $a\,R\,p$, contrary to the hypothesis. *Second*, when p is of the form $4n + 3$ we can show in a similar way that $p\,R\,E$ and so $E\,R\,p$, $-p\,R\,F$ and consequently $F\,R\,p$; since $g + h$ is even we get $GH\,R\,p$ and it follows finally that $q\,R\,p$, $a\,R\,p$, contrary to the hypothesis.

II. When e is divisible by p, a demonstration can be developed in a similar way. The skillful mathematician (for whom this article is written) will be able to accomplish this with no difficulty. For brevity's sake we shall omit it.

▶ 146. By the fundamental theorem and the propositions pertaining to the residues -1 and ± 2 it can always be determined whether a given number is a residue or a nonresidue of any other given number. But it will be very useful to restate the conclusions arrived at above in order to bring together all that is necessary for the solution.

PROBLEM. *Given any two numbers P, Q, to discover whether Q is a residue or a nonresidue of P.*

Solution. I. Let $P = a^\alpha b^\beta c^\gamma$ etc. where a, b, c, etc. are unequal prime numbers taken positively (for obviously we must consider P absolutely). For brevity in this article we shall speak simply of a *relation* of two numbers x, y meaning that the former x is a residue

---

[k] If there are no factors from any one of these classes, the number 1 should be written in place of the product.

or nonresidue of the latter y. Thus the relation of Q, P depends on the relations of Q, a^α; Q, b^β; etc. (art. 105).

II. In order to determine the relation of Q, a^α (and of Q, b^β, etc. as well) it is necessary to distinguish two cases.

1. When Q is divisible by a. Set $Q = Q'a^e$ where Q' is not divisible by a. Then if $e = \alpha$ or $e > \alpha$ we have $Q\,R\,a^\alpha$; but if $e < \alpha$ and odd, we get $Q\,N\,a^\alpha$; and if $e < \alpha$ and even, Q will have the same relation to a^α as Q' has to $a^{\alpha-e}$. This reduces to case.

2. When Q is not divisible by a. Two cases must be again distinguished.

(A) When $a = 2$. Then always when $\alpha = 1$ we have $Q\,R\,a^\alpha$; when $\alpha = 2$ it is required that Q be of the form $4n + 1$; and when $\alpha = 3$ or > 3, Q will have to be of the form $8n + 1$. If this condition holds $Q\,R\,a^\alpha$.

(B) When a is any other prime number. Then Q will have the same relation to a^α as it has to a (see art. 101).

III. We will investigate the relation of any number Q to a prime number a (odd) as follows. When $Q > a$ substitute in place of Q its least positive residue relative to the modulus a.[1] This will have the same relation to a that Q has.

Resolve Q, or the number taken in its place, into its prime factors p, p', p'', etc. adjoining the factor -1 when Q is negative. Then clearly the relation of Q to a depends on the relations of the single numbers p, p', p'', etc. to a. That is, if among those factors there are $2m$ nonresidues of a we will have $Q\,R\,a$; if there are $2m + 1$ of them, we will have $Q\,N\,a$. For it is easy to see that if among the factors p, p', p'', etc. a pair or four or six or in general $2k$ are equal, these can be safely disregarded.

IV. If -1 and 2 appear among the factors p, p', p'', etc. their relation to a can be determined from articles 108, 112, 113, 114. The relation of the remaining factors to a depends on that of a to them (fundamental theorem and the propositions of article 131). Let p be one of them and we will find (by treating a, p as we treated Q and a which are respectively greater) that the relation between a and p can be determined by articles 108–114 [provided the least residue of a (mod. p) contains no odd prime factors] or that the

___
[1] *Residue* in accordance with the meaning in article 4. It will be especially useful to take the *absolutely* least residue.

relation depends on that of p to prime numbers that are less than p. The same holds for the other factors p', p'', etc. By continuing this operation we will come finally to numbers whose relations can be determined by the propositions of articles 108–114. This can be seen more clearly by example.

Example. We want to find the relation of the number $+453$ to 1236: $1236 = 4 \cdot 3 \cdot 103$; $+453 \, R \, 4$ by II.2(A); $+453 \, R \, 3$ by II.1. Now to explore the relation of $+453$ to 103. It will be the same as that of $+41$ ($\equiv 453$, mod. 103) to 103; or the same as that of $+103$ to 41 (fundamental theorem) or of -20 to 41. But $-20 \, R \, 41$; for $-20 = -1 \cdot 2 \cdot 2 \cdot 5$; $-1 \, R \, 41$ (art. 108); and $+5 \, R \, 41$ and so $41 \equiv 1$ and is thus a residue of 5 (fundamental theorem). From this it follows that $+453 \, R \, 103$, and then $+453 \, R \, 1236$. It is true that $453 \equiv 297^2$ (mod. 1236).

▶ 147. If we are given a number A, certain *formulae* can be shown that contain all numbers relatively prime to A of which A is a residue, or all numbers that can be *divisors* of numbers of the form $x^2 - A$ (where x^2 is an undetermined square).[m] For brevity we will consider only these divisors which are odd and relatively prime to A, since all others can be easily reduced to these cases.

First let A be either a positive prime number of the form $4n + 1$ or negative of the form $4n - 1$. Then according to the fundamental theorem all prime numbers which taken positively are residues of A, will be divisors of $x^2 - A$; and all prime numbers (except 2 which is always a divisor) which are nonresidues of A, will be nondivisors of $x^2 - A$. Let all residues of A (excluding zero) which are less than A be denoted by r, r', r'', etc., all nonresidues by n, n', n'', etc. Then any prime number contained in one of the forms $Ak + r$, $Ak + r'$, $Ak + r''$, etc. will be a divisor of $x^2 - A$; but any prime contained in one of the forms $Ak + n$, $Ak + n'$, etc. will be a nondivisor. In these formulae k is an indeterminate integer. We will call the first set *forms of the divisors of $x^2 - A$*, the second set *forms of the nondivisors*. The number of members in each set will be $(A - 1)/2$. Further if C is an odd composite number and $A \, R \, B$, all the prime factors of B will be contained in one of the above forms and so also B. Therefore *any* odd number contained in a form of the nondivisors will be a nondivisor of the form $x^2 - A$.

---

[m] We will call these numbers simply *divisors of $x^2 - A$*. It is obvious what we mean by *nondivisors*.

But this theorem cannot be inverted; for if B is an odd composite nondivisor of the form $x^2 - A$, some of the prime factors of B will be nondivisors and there will be an *even* number of them. But B itself will be found in a form of a divisor (see art. 99).

Example. For $A = -11$ the forms of the divisors of $x^2 + 11$ will be: $11k + 1, 3, 4, 5, 9$; the forms of the nondivisors will be $11k + 2, 6, 7, 8, 10$. Thus -11 will be a nonresidue of all odd numbers which are contained in the latter forms, a residue of all primes belonging to the former forms.

We will have similar forms for divisors and nondivisors of $x^2 - A$ no matter what number A is. Obviously we should consider only those values of A that are not divisible by some square; for if $A = a^2 A'$, all divisors[n] of $x^2 - A$ will also be divisors of $x^2 - A'$, and so for the nondivisors. But we must distinguish three cases: (1) when A is of the form $+(4n + 1)$ or $-(4n - 1)$; (2) when A is of the form $-(4n + 1)$ or $+(4n - 1)$; (3) when A is of the form $\pm(4n + 2)$, i.e. even.

▶ 148. The *first case*, when A is of the form $+(4n + 1)$ or $-(4n - 1)$. Resolve A into its prime factors, and to those of the form $4n + 1$ ascribe a positive sign, and to those of the form $4n - 1$ a negative sign (the product of all these will $= A$). Let these factors be a, b, c, d, etc. Now distribute all numbers which are less than A and relatively prime to A into two classes. In the first class will be all numbers that are nonresidues of none of the numbers a, b, c, d, etc., or of two of them, or of four of them, or in general of an even number of them; in the second class will be those numbers that are nonresidues of one of the numbers a, b, c, etc., or of three of them, or in general of an odd number of them. Designate the former by r, r', r'', etc., the latter by n, n', n'', etc. Then the forms $Ak + r$, $Ak + r'$, $Ak + r''$, etc. will be forms of the divisors of $x^2 - A$, the forms $Ak + n$, $Ak + n'$, etc. will be forms of the non-divisors of $x^2 - A$ (i.e. *every prime number except 2 will be a divisor or a nondivisor of $x^2 - A$ according as it is contained in one of the former or latter forms*). For if p is a positive prime number and a residue or nonresidue of one of the numbers a, b, c, etc., this number will be a residue or a nonresidue of p (fundamental theorem). So if among the numbers a, b, c, etc. there are m numbers of which

[n] That is, those relatively prime to A.

p is a nonresidue, the same number will be nonresidues of p. Therefore if p is contained in one of the former forms, m will be even and $A R p$. But if it is contained in one of the latter forms, m will be odd and $A N p$.

Example. Let $A = +105 = (-3)(+5)(-7)$. Then the numbers r, r', r'', etc. will be these: 1, 4, 16, 46, 64, 79 (nonresidues of none of the numbers 3, 5, 7); 2, 8, 23, 32, 53, 92 (nonresidues of the numbers 3, 5); 26, 41, 59, 89, 101, 104 (nonresidues of the numbers 3, 7); 13, 52, 73, 82, 97, 103 (nonresidues of the numbers 5, 7). The numbers n, n', n'', etc. will be these: 11, 29, 44, 71, 74, 86; 22, 37, 43, 58, 67, 88; 19, 31, 34, 61, 76, 94; 17, 38, 47, 62, 68, 83. The first six are nonresidues of 3, the next six nonresidues of 5, then follow the nonresidues of 7, and finally those that are nonresidues of all three at the same time.

From the theory of combinations and articles 32 and 96 we can easily see that the number of numbers r, r', r'', etc. will

$$= t\left(1 + \frac{l(l-1)}{1 \cdot 2} + \frac{l(l-1)(l-2)(l-3)}{1 \cdot 2 \cdot 3 \cdot 4} + \dots\right)$$

and of the numbers n, n', n'', etc.

$$= t\left(l + \frac{l(l-1)(l-2)}{1 \cdot 2 \cdot 3} + \frac{l(l-1)\dots(l-4)}{1 \cdot 2 \dots 5} + \dots\right)$$

where l determines the number of numbers a, b, c, etc.;

$$t = 2^{-l}(a-1)(b-1)(c-1) \text{ etc.}$$

and each series is to be continued until it breaks off. (There will be t numbers which are residues of a, b, c, etc., $tl(l-1)/(1 \cdot 2)$ which are nonresidues of two of these etc., but brevity does not permit a fuller explanation.) The sum° of each of the series $= 2^{l-1}$. That is, the former is derived from

$$1 + (l-1) + \frac{(l-1)(l-2)}{1 \cdot 2} + \dots$$

by adding the second and third term, the fourth and fifth etc., and the latter is derived from the same equation by adding the first and second, third and fourth, etc. There will therefore be as many

---

° Disregarding the factor t.

forms for the divisors of $x^2 - A$ as there are forms for the non-divisors, namely $(a - 1)(b - 1)(c - 1)$ etc./2.

▶ 149. We can consider the *second* and *third cases* together. We can express A as $= (-1)Q$, or $= (+2)Q$, or $= (-2)Q$ where Q is a number of the form $+(4n + 1)$ or $-(4n - 1)$, such as those we considered in the preceding article. In general let $A = \alpha Q$ where $\alpha = -1$ or ± 2. Then A will be a residue of all numbers for which both or neither α and Q are residues; and a nonresidue of all numbers for which only one of the numbers α, Q is a nonresidue. From this the forms of the divisors and nondivisors of $x^2 - A$ can be easily derived. If $\alpha = -1$ distribute all numbers less than $4A$ and relatively prime to it into two classes, putting into the first those numbers which are at the same time in some form of the divisors of $x^2 - Q$ and in the form $4n + 1$, and also those numbers which are at the same time in some form of the nondivisors of $x^2 - Q$ and in the form $4n + 3$; all other numbers will be put in the second class. Let the members of the former class be r, r', r'', etc. and those of the latter class, n, n', n'', etc. A will be a residue of all prime numbers contained in any of the forms $4Ak + r, 4Ak + r'$, $4Ak + r''$, etc.; a nonresidue of all primes contained in any of the forms $4Ak + n, 4Ak + n'$, etc.

If $\alpha = \pm 2$ distribute all numbers less than $8Q$ and relatively prime to it into two classes, putting into the first class those numbers which are at the same time contained in some form of the divisors of $x^2 - Q$ and in one of the forms $8n + 1, 8n + 7$ when the upper sign holds, or in one of the forms $8n + 1, 8n + 3$ when the lower sign holds; and also those numbers contained in some form of the nondivisors of $x^2 - Q$ and at the same time in one of the forms $8n + 3, 8n + 5$ when the upper sign holds, or in one of the forms $8n + 5, 8n + 7$ when the lower sign holds. All other numbers will be put into the second class. Then if we designate numbers in the former class as r, r', r'', etc. and numbers of the latter class as n, n', n'', etc., $\pm 2Q$ will be a residue of all prime numbers contained in any of the forms $8Qk + r, 8Qk + r'$, etc., and a nonresidue of all primes in any of the forms $8Qk + n, 8Qk + n'$, $8Qk + n''$, etc. And it is easy to show here too that there are as many forms of the divisors of $x^2 - A$ as there are nondivisors.

Example. By this method we find that $+10$ is a residue of all prime numbers in any of the forms $40k + 1, 3, 9, 13, 27, 31, 37, 39$,

a nonresidue of all primes in any of the forms $40k + 7, 11, 17, 19,$
$21, 23, 29, 33$.

▶ 150. These forms have many remarkable properties, but we
indicate only one of them. If B is a composite number relatively
prime to A, and if among its prime factors there are $2m$ of them
contained in some form of the nondivisors of $x^2 - A$, B will be
contained in some form of the divisors of $x^2 - A$; and if the
number of prime factors of B contained in some form of the nondivisors of $x^2 - A$ is odd, B also will be contained in a form of
the nondivisors. We omit the demonstration, which is not difficult.
From all of this it follows that not only any prime number but
also any odd composite number relatively prime to A contained
in some form of the nondivisors will be itself a nondivisor; for
some prime factor of such a number is necessarily a nondivisor.

▶ 151. The fundamental theorem must certainly be regarded as
one of the most elegant of its type. No one has thus far presented
it in as simple a form as we have done above. What is more
surprising is that Euler already knew other propositions which
depend on it and which should have led to its discovery. He was
aware that certain forms existed which contain all the prime
divisors of numbers of the form $x^2 - A$, that there were others
containing all prime nondivisors of numbers of the same form,
and that the two sets were mutually exclusive. And he knew further
the method of finding these forms, but all his attempts at demonstration were in vain, and he succeeded only in giving a greater
degree of verisimilitude to the truth that he had discovered by
induction. In a memoir entitled "*Novae demonstrationes circa
divisores numerorum formae $xx + nyy$*" which he read in the St.
Petersburg Academy (Nov. 20, 1775) and which was published
after his death,[10] he seems to believe that he had fulfilled his
resolve. But an error did creep in, for on p. 65 he *tacitly* presupposed the existence of such forms of the divisors[p] and nondivisors,
and from this it was not difficult to discover *what* their *form* should
be. But the method he used to prove this supposition does not

[p] Namely that there do exist numbers r, r', r'', etc., n, n', n'', etc., all divisors and $< 4A$
such that all prime divisors of $x^2 - A$ are contained in one of the forms $4Ak + r, 4Ak + r'$,
etc., and all prime nondivisors in one of the forms $4Ak + n, 4Ak + n'$, etc. (k being indeterminate).

[10] *Nova acta acad. Petrop.*, 1 [1783], 1787, 47–74.

seem to be suitable. In another paper, "De criteriis aequationis $fxx + gyy = hzz$ utrumque resolutionem admittat necne,"[11] *Opuscula Analytica*, *1*, 211 (f, g, h are given, x, y, z are indeterminate) he finds by induction that if the equation is solvable for one value of $h = s$, it will also be solvable for any other value congruent to s relative to the modulus $4fg$ provided it is a prime number. From this proposition the supposition we spoke of can easily be demonstrated. But the demonstration of this theorem also eluded his efforts.[q] This is not remarkable because in our judgment it is necessary to start from the fundamental theorem. The truth of the proposition will flow automatically from what we will show in the following section.

After Euler, the renowned Legendre worked zealously on the same problem in his excellent tract, "Recherches d'analyse indétérminée," *Hist. Acad. Paris*, 1785, p. 465 ff. He arrived at a theorem basically the same as the fundamental theorem. He showed that if p, q are two positive prime numbers, the absolutely least residues of the powers $p^{(q-1)/2}$, $q^{(p-1)/2}$ relative to the moduli q, p respectively, are either both $+1$ or both -1 when either p or q is of the form $4n + 1$; but when both p and q are of the form $4n + 3$ one least residue will be $+1$, the other -1 (p. 516). From this, according to article 106, we derive the fact that the *relation* (taken according to the meaning in article 146) of p to q and of q to p will be the *same* when either p or q is of the form $4n + 1$, *opposite* when both p and q are of the form $4n + 3$. This proposition is contained among the propositions of article 131 and follows also from 1, 3, 9 of article 133; on the other hand the fundamental theorem can be derived from it. Legendre also attempted a demonstration and, since it is extremely ingenious, we will speak of it at some length in the following section. But since he

---

[11] The original article reads "utrum ea" instead of "utrumque."

[q] As he himself confesses (*Opuscula Analytica*, *1*, 216): "A demonstration of this most elegant theorem is still sought even though it has been investigated in vain for so long and by so many.... And anyone who succeeds in finding such a demonstration must certainly be considered most outstanding." With what ardor this great man searched for a proof of this theorem and of others which are only special cases of the fundamental theorem can be seen in many other places, e.g. *Opuscula Analytica*, *1*, 268 (Additamentum ad Diss. 8) and *2*, 275 (Diss. 13) and in many dissertations in *Comm. acad. Petrop.* which we have praised from time to time. [For Dissertation 8 see p. 62. Dissertation 13 is entitled "De insigni promotione scientiae numerorum" *Ed. Note.*]

presupposed many things without demonstration (as he himself confesses on p. 520: "*Nous avons supposé seulement ...*"), some of which have not been demonstrated by anyone up till now, and some of which cannot be demonstrated in our judgment without the help of the fundamental theorem itself, the road he has entered upon seems to lead to an impasse, and so our demonstration must be regarded as the first. Below we shall give *two other demonstrations* of this most important theorem, which are totally different from the preceding and from each other.

▶ 152. Thus far we have treated the pure congruence $x^2 \equiv A$ (mod. m) and we have learned to recognize whether or not it is solvable. By article 105 the investigation of the *roots themselves* was reduced to the case where m is either a prime or a power of a prime, and afterward by applying article 101 to the case where m is prime. For this case what we said in article 61 along with what we will show in sections V and VIII[12] embraces almost all that can be derived by direct methods. But in the cases where they are applicable they are infinitely more prolix than the indirect methods which we will show in section VI and so they are memorable not for their usefulness in practice but for their beauty. *Congruences of the second degree that are not pure* can be reduced to pure congruences easily. Suppose we are given the congruence

$$ax^2 + bx + c \equiv 0$$

which is to be solved relative to the modulus m. The following congruence is equivalent:

$$4a^2x^2 + 4abx + 4ac \equiv 0 \,(\text{mod. } 4am)$$

i.e. any number that satisfies one will satisfy the other. The second congruence can be put in the form

$$(2ax + b)^2 \equiv b^2 - 4ac \,(\text{mod. } 4am)$$

and from this all the values of $2ax + b$ less than $4m$ can be found, if any exists. If we designate them by r, r', r'', etc. all solutions of the proposed congruence can be deduced from the solutions of the congruences

$$2ax \equiv r - b, \qquad 2ax \equiv r' - b, \text{ etc. (mod. } 4am)$$

---

[12] Section VIII was not published. See Author's Preface.

which we showed how to find in section II. But we observe that the solution can be shortened by various artifices; e.g. in place of the given congruence another can be found

$$a'x^2 + 2b'x + c' \equiv 0$$

which is equivalent to it and in which a' divides m. Brevity will not allow us to take up these considerations here, but see the last section.

SECTION V

FORMS AND INDETERMINATE EQUATIONS OF THE SECOND DEGREE

▶ 153. In this section we shall treat particularly of functions in two unknowns x, y of the form

$$ax^2 + 2bxy + cy^2$$

where a, b, c are given integers. We will call these functions *forms of the second degree* or simply *forms*. On this investigation depends the solution of the famous problem of finding all the solutions of any indeterminate equation of the second degree in two unknowns as long as these unknown values are integers or rational numbers. This problem has already been solved in all generality by Lagrange, and many things pertaining to the nature of the *forms* were proved by this great geometer. Still other properties were discovered in part first by Euler and later by Fermat. However, a careful inquiry into the nature of the forms revealed so many new results that we decided it would be worthwhile to review the whole argument from the beginning—the more so because what these men have discovered is scattered in so many different places that few scholars are aware of them, further because the method we will use is almost entirely our own, and finally because the new things we add could not be understood without a new exposition of their discoveries. We have no doubt that many remarkable results still lie hidden and are a challenge to the talents of others. In the proper place we will report on the history of important truths.

We represent the form $ax^2 + 2bxy + cy^2$ by the symbol (a, b, c) when we are not concerned with the unknowns x, y. Thus this expression will denote in an indefinite manner a sum of three terms: the first a product of a given number a by the square of some indeterminate number; the second a doubled product of the number b by this indeterminate number times another indeterminate number; the third a product of the number c by the square of this second indeterminate number. For example, $(1, 0, 2)$ indicates the sum of a square and double a square. And although (a, b, c) and (c, b, a) designate the same thing if we look at the *terms alone*, nevertheless they are different if we heed the *order* of the terms. And so we will carefully distinguish between them. We shall see clearly from what follows the advantage to be gained from this.

▶ 154. We shall say that a given number is *represented* by a given form if we can find integral values of the indeterminate form so that its value is equal to the given number. Thus we have the following:

THEOREM. *If the number M can so be represented by the form (a, b, c) that the values of the unknowns that accomplish this are prime relative to each other, $b^2 - ac$ will be a quadratic residue of the number M.*

Demonstration. Let the values of the unknowns be m, n; i.e.

$$am^2 + 2bmn + cn^2 = M$$

and choose the numbers μ, ν so that $\mu m + \nu n = 1$ (art. 40). Then

$$(am^2 + 2bmn + cn^2)(a\nu^2 - 2b\mu\nu + c\mu^2)$$

$$= [\mu(mb + nc) - \nu(ma + nb)]^2 - (b^2 - ac)(m\mu + n\nu)^2$$

or

$$M(a\nu^2 - 2b\mu\nu + c\mu^2) = [\mu(mb + nc) - \nu(ma + nb)]^2 - (b^2 - ac)$$

and therefore

$$b^2 - ac \equiv [\mu(mb + nc) - \nu(ma + nb)]^2 \qquad (\text{mod. } M)$$

i.e. $b^2 - ac$ is a quadratic residue of M.

We shall see in what follows that the properties of the form (a, b, c) depend in a special way on the nature of this number $b^2 - ac$. We shall call it the *determinant* of this form.

▶ 155. Now

$$\mu(mb + nc) - \nu(ma + nb)$$

will be the value of the expression

$$\sqrt{(b^2 - ac)} \qquad (\text{mod. } M)$$

But it is clear that the numbers μ, v can be determined in infinitely many ways so that $\mu m + vn = 1$, so various values of this expression will be produced. Let us examine the connection they have with one another. Let not only $\mu m + vn = 1$ but also $\mu'm + v'n = 1$ and derive

$$\mu(mb + nc) - v(ma + nb) = v, \qquad \mu'(mb + nc) - v'(ma + nb) = v'$$

Multiplying the equation $\mu m + vn = 1$ by μ' and $\mu'm + v'n = 1$ by μ and subtracting, we get $\mu' - \mu = n(\mu'v - \mu v')$; and similarly multiplying the first by v', the second by v, and subtracting, we get $v' - v = m(\mu v' - \mu'v)$. From this we find directly

$$v' - v = (\mu'v - \mu v')(am^2 + 2bmn + cn^2) = (\mu'v - \mu v')M$$

or $v' \equiv v$ (mod. M). Therefore, no matter how μ, v are determined, the formula $\mu(mb + nc) - v(ma + nb)$ cannot give *different* (i.e. noncongruent) values of the expression $\sqrt{(b^2 - ac)}$ (mod. M). If therefore v is a value of this formula, we will say that the representation of the number M by the form $ax^2 + 2bxy + cy^2$ in which $x = m$, $y = n$ *belongs to the value v of the expression* $\sqrt{(b^2 - ac)}$ (mod. M). And it is easy to show that if a value of this formula is v and $v' \equiv v$ (mod. M), in place of the numbers μ, v which give v, others can be found μ', v' which give v'. It is sufficient to let

$$\mu' = \mu + \frac{n(v' - v)}{M}, \qquad v' = v - \frac{m(v' - v)}{M}$$

and we get

$$\mu'm + v'n = \mu m + vn = 1$$

Thus the value of the formula found by using μ', v' will exceed that found by using μ, v by the quantity $(\mu'v - \mu v')M$, which is equal to $(\mu m + vn)(v' - v) = v' - v$, so this value will $= v'$.

▶ 156. If we have two representations of the same number M by the same form (a, b, c) in which the values of the unknowns are relatively prime, they can belong to the same value of the expression $\sqrt{(b^2 - ac)}$ (mod. M) or to different ones. Let

$$M = am^2 + 2bmn + cn^2 = am'm' + 2bm'n' + cn'n'$$

and

$$\mu m + vn = 1, \qquad \mu'm' + v'n' = 1$$

It is clear that if

$$\mu(mb + nc) - v(ma + nb) \equiv \mu'(m'b + n'c) - v'(m'a + n'b) \text{ (mod. } M)$$

the congruence will always stay the same no matter what suitable values for $\mu, v; \mu', v'$ are chosen. In this case we will say that both representations belong to the *same* value of the expression $\sqrt{(b^2 - ac)}$ (mod. M); but if the congruence does not hold for some values of $\mu, v; \mu', v'$ it will not hold for any values at all, and the representations will belong to *different* values. Now if

$$\mu(mb + nc) - v(ma + nb) \equiv -[\mu'(m'b + n'c) - v'(m'a + n'b)]$$

the representations are said to belong to *opposite* values of the expression $\sqrt{(b^2 - ac)}$. We will use all these designations when we treat of many representations of the same number by *different* forms which, however, have the same determinant.

Example. Let the proposed form be $(3, 7, -8)$ and the determinant $= 73$. By this form we have the following representations of the number 57:

$$3 \cdot 13^2 + 14 \cdot 13 \cdot 25 - 8 \cdot 25^2; \qquad 3 \cdot 5^2 + 14 \cdot 5 \cdot 9 - 8 \cdot 9^2$$

For the first we can put $\mu = 2, v = -1$, and the value of the expression $\sqrt{73}$ (mod. 57) to which the representation belongs will

$$= 2(13 \cdot 7 - 25 \cdot 8) + (13 \cdot 3 + 25 \cdot 7) = -4$$

In a similar way if we let $\mu = 2$ and $v = -1$, the second representation will be found to belong to the value $+4$. Thus the two representations belong to opposite values.

Before we go any farther we observe that the forms whose determinant $= 0$ will be excluded from the following investigations because they only upset the elegance of the theorems and so require separate treatment.

▶ 157. If the form F with unknown x, y can be transformed into another form F' with unknown x', y' by substitutions like

$$x = \alpha x' + \beta y', \qquad y = \gamma x' + \delta y'$$

where $\alpha, \beta, \gamma, \delta$ are integers, we will say that the former *implies* the latter or that the latter is *contained in the former*. Let the form F be

$$ax^2 + 2bxy + cy^2$$

and the form F'

$$a'x'x' + 2b'x'y' + c'y'y'$$

and we have the three following equations

$$a' = a\alpha^2 + 2b\alpha\gamma + c\gamma^2$$

$$b' = a\alpha\beta + b(\alpha\delta + \beta\gamma) + c\gamma\delta$$

$$c' = a\beta^2 + 2b\beta\delta + c\delta^2$$

Multiplying the second equation by itself, the first by the third, and subtracting we get

$$b'b' - a'c' = (b^2 - ac)(\alpha\delta - \beta\gamma)^2$$

From this it follows that the determinant of the form F' is divisible by the determinant of the form F and that the quotient is a square; manifestly therefore these determinants will have the *same sign*. And if further the form F' could be transformed by a similar substitution into the form F, i.e. if F' is contained in F and F is contained in F', the determinants of the forms will be equal,[a] and $(\alpha\delta - \beta\gamma)^2 = 1$. In this case we call the forms *equivalent*. Thus equality of determinants is a necessary condition for equivalence of forms, although equivalence does not follow from equality of determinants alone. We will call the substitution $x = \alpha x' + \beta y'$, $y = \gamma x' + \delta y'$ a *proper transformation* if $\alpha\delta - \beta\gamma$ is a positive number, *improper* if $\alpha\delta - \beta\gamma$ is negative. And we will say that the form F' is contained *properly* or *improperly* in the form F, if F can be transformed into the form F' by a proper or improper transformation. So if the forms F, F' are equivalent, we have $(\alpha\delta - \beta\gamma)^2 = 1$, and if the transformation is proper $\alpha\delta - \beta\gamma = +1$; if it is improper $\alpha\delta - \beta\gamma = -1$. If many transformations are all proper or all improper, we will say they are *similar*. A proper and an improper form will be called *dissimilar*.

▶ 158. *If the determinants of the forms F, F' are equal and F' is contained in F, then F will also be contained in F' properly or improperly according as F' is contained in F properly or improperly.* Transform F into F' by

$$x = \alpha x' + \beta y', \qquad y = \gamma x' + \delta y'$$

---

[a] It is clear from the preceding analysis that this proposition is applicable to forms whose determinant $= 0$. But the equation $(\alpha\delta - \beta\gamma)^2 = 1$ must not be extended to this case.

and F' will be transformed into F by

$$x' = \delta x - \beta y, \qquad y' = -\gamma x + \alpha y$$

For it is clear that by this substitution the same thing results from F' as from F by putting

$$x = \alpha(\delta x - \beta y) + \beta(-\gamma x + \alpha y),$$
$$y = \gamma(\delta x - \beta y) + \delta(-\gamma x + \alpha y)$$

or

$$x = (\alpha\delta - \beta\gamma)x, \qquad y = (\alpha\delta - \beta\gamma)y$$

Manifestly F becomes $(\alpha\delta - \beta\gamma)^2 F$, i.e. F itself (preceding article). It is obvious that the latter transformation will be proper or improper according as the former is proper or improper.

If F' is *properly* contained in F, and F properly contained in F', we will call the forms *properly equivalent;* if they are *improperly* contained in each other we will call the forms *improperly equivalent.* The usefulness of these distinctions will soon be made clear.

Example. The form $2x^2 - 8xy + 3y^2$, by use of the substitutions $x = 2x' + y'$, $y = 3x' + 2y'$, is transformed into the form $-13x'x' - 12x'y' - 2y'y'$. The latter is transformed into the former by the substitutions $x' = 2x - y$, $y' = -3x + 2y$. Therefore the forms $(2, -4, 3)$, $(-13, -6, -2)$ will be properly equivalent.

We will now turn our attention to the following problems: I. Given any two forms having the same determinant, we want to know whether or not they are equivalent, whether or not they are properly or improperly equivalent, or both (for this can happen too). And when they have unequal determinants we want to find whether only one implies the other or whether the implication is mutual. If there is an implication, is it proper or improper or both? Finally we want to find all the transformations of one form into the other, both proper and improper. II. Given a form we want to find whether a given number can be represented by it and to determine all the representations. But, since forms with a negative determinant require a method different from forms with a positive determinant, we will first consider what is common to both and then consider each type of form separately.

▶ 159. *If the form F implies the form F' and this in turn implies the form F'', the form F will also imply the form F''.*

Let the unknowns of the forms F, F', F'' be respectively x, y; x', y'; x'', y'' and let F be transformed into F' by the equations

$$x = \alpha x' + \beta y', \qquad y = \gamma x' + \delta y'$$

F' into F'' by

$$x' = \alpha' x'' + \beta' y'', \qquad y' = \gamma' x'' + \delta' y''$$

It is clear that F will be transformed into F'' by

$$x = \alpha(\alpha' x'' + \beta' y'') + \beta(\gamma' x'' + \delta' y'),$$
$$y = \gamma(\alpha' x'' + \beta' y'') + \delta(\gamma' x'' + \delta' y'')$$

or

$$x = (\alpha\alpha' + \beta\gamma')x'' + (\alpha\beta' + \beta\delta')y'',$$
$$y = (\gamma\alpha' + \delta\gamma')x'' + (\gamma\beta' + \delta\delta')y''$$

Thus F implies F''.
Since

$$(\alpha\alpha' + \beta\gamma')(\gamma\beta' + \delta\delta') - (\alpha\beta' + \beta\delta')(\gamma\alpha' + \delta\gamma') =$$
$$(\alpha\delta - \beta\gamma)(\alpha'\delta' - \beta'\gamma')$$

will be positive if both $\alpha\delta - \beta\gamma$ and $\alpha'\delta' - \beta'\gamma'$ are positive or both negative, and since it will be negative if one of these numbers is positive the other negative, the form F will *properly* imply the form F'' if F implies F' and F' implies F'' in the same way, and *improperly* if the two implications are different.

From this it follows that if we have any number of forms F, F', F'', F''', etc., each of which implies its successor, the first will imply the last. If the number of forms which improperly imply their successors is even, the implication will be *proper;* if this number is odd, the implication will be *improper.*

If the form F is equivalent to the form F' and F' is equivalent to the form F'', the form F will be equivalent to the form F'', and they will be properly equivalent if the form F is equivalent to the form F' in the same way that the form F' is to the form F'' and improperly equivalent otherwise.

For, since the forms F, F' are respectively equivalent to F', F'', the former imply the latter; thus F implies F'', and also the latter imply the former. So F, F'' will be equivalent. From the preceding it follows that F properly or improperly implies F'' according as F and F', F' and F'' are equivalent in the same or different manner, and the same holds for F'' relative to F. In the former case F, F'' will be properly equivalent, in the latter case improperly equivalent.

The forms $(a, -b, c)$, (c, b, a), $(c, -b, a)$ *are equivalent to the form* (a, b, c), *the first two improperly, the last properly.*

For $ax^2 + 2bxy + cy^2$ is transformed into $ax'x' - 2bx'y' + cy'y'$ by letting $x = x' + 0 \cdot y'$, $y = 0 \cdot x' - y'$. This transformation is improper because $1(-1 - 0 \cdot 0) = -1$. The transformation to the form $cx'x' + 2bx'y' + ay'y'$ is done by the improper transformation $x = 0 \cdot x' + y'$, $y = x' + 0 \cdot y'$ and to the form $cx'x' - 2bx'y' + ay'y'$ by the proper transformation $x = 0 \cdot x' - y'$, $y = x' + 0 \cdot y'$.

Thus it is manifest that any form equivalent to the form (a, b, c) is *properly* equivalent either to (a, b, c) itself or to the form $(a, -b, c)$. Similarly if any form implies the form (a, b, c) or is contained in it, that form will *properly* imply the form (a, b, c) or the form $(a, -b, c)$ or will be *properly* contained in one of them. We will call the forms (a, b, c), $(a, -b, c)$ *opposites*.

▶160. If the forms (a, b, c), (a', b', c') have the same determinant and further $c = a'$ and $b \equiv -b'$ (mod. c) or $b + b' \equiv 0$ (mod. c) we shall call them *neighboring* forms, and when we want a more accurate determination we will say that the former is a neighbor to the latter *by the first part*, the latter a neighbor to the former *by the last part*.

Thus, e.g., the form $(7, 3, 2)$ is a neighbor to the form $(3, 4, 7)$ *by the last part*, the form $(3, 1, 3)$ is a neighbor to its opposite $(3, -1, 3)$ by both parts.

Neighboring forms are always properly equivalent. For the form $ax^2 + 2bxy + cy^2$ is transformed into the neighboring form $cx'x' + 2b'x'y' + c'y'y'$ by the substitution $x = -y'$, $y = x' + [(b + b')y'/c]$ [which is proper because $0[(b + b')/c] - 1(-1) = 1]$. The transformation can be easily proven by expanding and using the equation $b^2 - ac = b'b' - cc'$. By hypothesis $(b + b')/c$ is an integer. But these definitions and conclusions do not hold if

$c = a' = 0$. This case cannot occur except in forms whose determinant is a square.

The forms (a, b, c), (a', b', c') are properly equivalent if $a = a'$, $b \equiv b'$ (mod. a). For the form (a, b, c) is properly equivalent to the form $(c, -b, a)$ (preceding article) but this form is a neighbor to the form (a', b', c') by the first part.

▶ 161. *If the form (a, b, c) implies the form (a', b', c'), any common divisor of the numbers a, b, c will also divide the numbers a', b', c', and any common divisor of the numbers $a, 2b, c$ will divide $a', 2b', c'$.*

For if the form $ax^2 + 2bxy + cy^2$ is transformed into the form $a'x'x' + 2b'x'y' + c'y'y'$, by the substitutions $x = \alpha x' + \beta y'$, $y = \gamma x' + \delta y'$ we get the following equations

$$a\alpha^2 + 2b\alpha\gamma + c\gamma^2 = a'$$
$$a\alpha\beta + b(\alpha\delta + \beta\gamma) + c\gamma\delta = b'$$
$$a\beta^2 + 2b\beta\delta + c\delta^2 = c'$$

From this the proposition follows immediately [for the second part of the proposition in place of the second equation we use $2a\alpha\beta + 2b(\alpha\delta + \beta\gamma) + 2c\gamma\delta = 2b'$].

It follows then that the greatest common divisor of the numbers $a, b(2b), c$ also divides the greatest common divisor of the numbers $a', b'(2b'), c'$. If, further, the form (a', b', c') implies the form (a, b, c), i.e. the forms are equivalent, the greatest common divisor of the numbers $a, b(2b), c$ will be equal to the greatest common divisor of the numbers $a', b'(2b'), c'$ because each must divide the other. If therefore in this case $a, b(2b), c$ do not have a common divisor, i.e. if the greatest divisor $= 1$, $a', b'(2b'), c'$ will not have a common divisor either.

▶ 162. PROBLEM. *If the form $AX^2 + 2BXY + CY^2 \ldots F$ implies the form $ax^2 + 2bxy + cy^2 \ldots f$ and if any transformation of the former to the latter is given, deduce all other similar transformations from this one.*

Solution. Let the given transformation be $X = \alpha x + \beta y$, $Y = \gamma x + \delta y$ and suppose first that we know another similar to it: $X = \alpha' x + \beta' y$, $Y = \gamma' x + \delta' y$. Let us investigate what follows from this. Denote the determinants of the forms F, f by D, d and let $\alpha\delta - \beta\gamma = e$, $\alpha'\delta' - \beta'\gamma' = e'$. We have (art. 157) $d = De^2 = De'e'$

and since by hypothesis e, e' have the same sign, $e = e'$. We get also the following six equations:

$$A\alpha^2 + 2B\alpha\gamma + C\gamma^2 = a \qquad [1]$$

$$A\alpha'\alpha' + 2B\alpha'\gamma' + C\gamma'\gamma' = a \qquad [2]$$

$$A\alpha\beta + B(\alpha\delta + \beta\gamma) + C\gamma\delta = b \qquad [3]$$

$$A\alpha'\beta' + B(\alpha'\delta' + \beta'\gamma') + C\gamma'\delta' = b \qquad [4]$$

$$A\beta^2 + 2B\beta\delta + C\delta^2 = c \qquad [5]$$

$$A\beta'\beta' + 2B\beta'\delta' + C\delta'\delta' = c \qquad [6]$$

If for the sake of brevity we use $a', 2b', c'$ to designate the numbers

$$A\alpha\alpha' + B(\alpha\gamma' + \gamma\alpha') + C\gamma\gamma'$$

$$A(\alpha\beta' + \beta\alpha') + B(\alpha\delta' + \beta\gamma' + \gamma\beta' + \delta\alpha') + C(\gamma\delta' + \delta\gamma')$$

$$A\beta\beta' + B(\beta\delta' + \delta\beta') + C\delta\delta'$$

we will deduce the following new equations[b] from the preceding equations:

$$a'a' - D(\alpha\gamma' - \gamma\alpha')^2 = a^2 \qquad [7]$$

$$2a'b' - D(\alpha\gamma' - \gamma\alpha')(\alpha\delta' + \beta\gamma' - \gamma\beta' - \delta\alpha') = 2ab \qquad [8]$$

$$4b'b' - D[(\alpha\delta' + \beta\gamma' - \gamma\beta' - \delta\alpha')^2 + 2ee'] = 2b^2 + 2ac$$

from this we get by adding $2Dee' = 2d = 2b^2 - 2ac$

$$4b'b' - D(\alpha\delta' + \beta\gamma' - \gamma\beta' - \delta\alpha')^2 = 4b^2 \qquad [9]$$

$$a'c' - D(\alpha\delta' - \gamma\beta')(\beta\gamma' - \delta\alpha') = b^2$$

and subtracting $D(\alpha\delta - \beta\gamma)(\alpha'\delta' - \beta'\gamma') = b^2 - ac$ we get

$$a'c' - D(\alpha\gamma' - \gamma\alpha')(\beta\delta' - \delta\beta') = ac \qquad [10]$$

$$2b'c' - D(\alpha\delta' + \beta\gamma' - \gamma\beta' - \delta\alpha')(\beta\delta' - \delta\beta') = 2bc \qquad [11]$$

$$c'c' - D(\beta\delta' - \delta\beta')^2 = c^2 \qquad [12]$$

[b] This is the origin of these equations: 7 comes from $1 \cdot 2$ (i.e. if equation [1] is multiplied by equation [2] or, rather, the first part of the former is multiplied by the first part of the latter, and the last part of the former by the last part of the latter and the products equated); 8 from $1 \cdot 4 + 2 \cdot 3$; the next (which is not numbered) from $1 \cdot 6 + 2 \cdot 5 + 3 \cdot 4 + 3 \cdot 4$; the next, not numbered, from $3 \cdot 4$; 11 from $3 \cdot 6 + 4 \cdot 5$; 12 from $5 \cdot 6$. We shall always use a similar designation in what follows. We leave the derivation to the reader.

We will suppose that the greatest common divisor of the numbers a, $2b$, c is m and that the numbers $\mathfrak{A}$, $\mathfrak{B}$, $\mathfrak{C}$ are so determined that

$$\mathfrak{A}a + 2\mathfrak{B}b + \mathfrak{C}c = m$$

(art. 40). Multiply the equations [7], [8], [9], [10], [11], [12] respectively by $\mathfrak{A}^2$, $2\mathfrak{A}\mathfrak{B}$, $\mathfrak{B}^2$, $2\mathfrak{A}\mathfrak{C}$, $2\mathfrak{B}\mathfrak{C}$, $\mathfrak{C}^2$ and sum the products. Now if for brevity's sake we put

$$\mathfrak{A}a' + 2\mathfrak{B}b' + \mathfrak{C}c' = T \quad [13]$$

$$\mathfrak{A}(\alpha\gamma' - \gamma\alpha') + \mathfrak{B}(\alpha\delta' + \beta\gamma' - \gamma\beta' - \delta\alpha') + \mathfrak{C}(\beta\delta' - \delta\beta') = U \quad [14]$$

where T, U are manifestly integers, we will get

$$T^2 - DU^2 = m^2$$

We are led therefore to this elegant conclusion, *that the solution of the indeterminate equation $t^2 - Du^2 = m^2$ in integers depends on two similar transformations of the form F into f; that is $t = T$, $u = U$.* But since in our reasoning we have not supposed that the transformations are *different*, *one* transformation taken twice must give the solution. But since $\alpha' = \alpha$, $\beta' = \beta$, etc., $a' = a$, $b' = b$, $c' = c$ in such a case, $T = m$, $U = 0$ and the solution is obvious.

Now let us assume that the first transformation and the solution of the indeterminate equation are known. We will investigate how the other transformation can be deduced from this or how α', β', γ', δ' depend on $\alpha, \beta, \gamma, \delta, T, U$. First we multiply equation [1] by $\delta\alpha' - \beta\gamma'$, [2] by $\alpha\delta' - \gamma\beta'$, [3] by $\alpha\gamma' - \gamma\alpha'$, [4] by $\gamma\alpha' - \alpha\gamma'$ and add the products. As a result we get

$$(e + e')a' = (\alpha\delta' - \beta\gamma' - \gamma\beta' + \delta\alpha')a \quad [15]$$

In a similar way from

$$(\delta\beta' - \beta\delta')([1] - [2]) + (\alpha\delta' - \beta\gamma' - \gamma\beta' + \delta\alpha')([3] + [4]) +$$
$$(\alpha\gamma' - \gamma\alpha')([5] - [6])$$

we get

$$2(e + e')b' = 2(\alpha\delta' - \beta\gamma' - \gamma\beta' + \delta\alpha')b \quad [16]$$

Finally from

$$(\delta\beta' - \beta\delta')([3] - [4]) + (\alpha\delta' - \gamma\beta')[5] + (\delta\alpha' - \beta\gamma')[6]$$

we get

$$(e + e')c' = (\alpha\delta' - \beta\gamma' - \gamma\beta' + \delta\alpha')c \qquad [17]$$

Substituting these values [15], [16], [17] in [13] we get

$$(e + e')T = (\alpha\delta' - \beta\gamma' - \gamma\beta' + \delta\alpha')(\mathfrak{A}a + 2\mathfrak{B}b + \mathfrak{C}c)$$

or

$$2eT = (\alpha\delta' - \beta\gamma' - \gamma\beta' + \delta\alpha')m \qquad [18]$$

From this it is much easier to calculate T than from [13]. By combining this equation with [15], [16], [17] we find $ma' = Ta$, $2mb' = 2Tb$, $mc' = Tc$. Substituting these values of a', $2b'$, c' in equations [7]—[12] and writing $m^2 + DU^2$ in place of T^2, with suitable manipulation we get

$$(\alpha\gamma' - \gamma a')^2 m^2 = a^2 U^2$$

$$(\alpha\gamma' - \gamma\alpha')(\alpha\delta' + \beta\gamma' - \gamma\beta' - \delta\alpha')m^2 = 2abU^2$$

$$(\alpha\delta' + \beta\gamma' - \gamma\beta' - \delta\alpha')^2 m^2 = 4b^2 U^2$$

$$(\alpha\gamma' - \gamma\alpha')(\beta\delta' - \delta\beta')m^2 = acU^2$$

$$(\alpha\delta' + \beta\gamma' - \gamma\beta' - \delta\alpha')(\beta\delta' - \delta\beta')m^2 = 2bcU^2$$

$$(\beta\delta' - \delta\beta')^2 m^2 = c^2 U^2$$

With the help of equation [14] and $\mathfrak{A}a + 2\mathfrak{B}b + \mathfrak{C}c = m$ we easily deduce the following equations (by multiplying the first, second, fourth; second, third, fifth; fourth, fifth, sixth respectively by $\mathfrak{A}, \mathfrak{B}, \mathfrak{C}$ and adding the product):

$$(\alpha\gamma' - \gamma\alpha')Um^2 = maU^2$$

$$(\alpha\delta' + \beta\gamma' - \gamma\beta' - \delta\alpha')Um^2 = 2mbU^2$$

$$(\beta\delta' - \delta\beta')Um^2 = mcU^2$$

Dividing these by mU^c

$$aU = (\alpha\gamma' - \gamma\alpha')m \qquad [19]$$

$$2bU = (\alpha\delta' + \beta\gamma' - \gamma\beta' - \delta\alpha')m \qquad [20]$$

$$cU = (\beta\delta' - \delta\beta')m \qquad [21]$$

^{c} This would not be allowed if $U = 0$; but then the truth of [19], [20], [21] would follow immediately from [1], [3], [6].

and from these equations any U can be decided much more easily than from [14]. It follows also from this that no matter how $\mathfrak{A}, \mathfrak{B}, \mathfrak{C}$ are determined (and there is an infinite number of different ways of doing it) we will always have the same values for T and U.

Now if [18] is multiplied by α, [19] by 2β, [20] by $-\alpha$ we get by addition

$$2\alpha e T + 2(\beta a - \alpha b)U = 2(\alpha\delta - \beta\gamma)\alpha' m = 2e\alpha' m$$

Similarly from $\beta[18] + \beta[20] - 2\alpha[21]$

$$2\beta e T + 2(\beta b - \alpha c)U = 2(\alpha\delta - \beta\gamma)\beta' m = 2e\beta' m$$

From $\gamma[18] + 2\delta[19] - \gamma[20]$

$$2\gamma e T + 2(\delta a - \gamma b)U = 2(\alpha\delta - \beta\gamma)\gamma' m = 2e\gamma' m$$

Finally from $\delta[18] + \delta[20] - 2\gamma[21]$

$$2\delta e T + 2(\delta b - \gamma c)U = 2(\alpha\delta - \beta\gamma)\delta' m = 2e\delta' m$$

If we substitute the values from [1], [3], [5] for a, b, c in these formulas, we get

$$\alpha' m = \alpha T - (\alpha B + \gamma C)U$$
$$\beta' m = \beta T - (\beta B + \delta C)U$$
$$\gamma' m = \gamma T + (\alpha A + \gamma B)U$$
$$\delta' m = \delta T + (\beta A + \delta B)U^{\text{d}}$$

From the preceding analysis it follows that there is no transformation of the form F into f similar to the given one which is not contained in the formula

$$X = \frac{1}{m}[\alpha t - (\alpha B + \gamma C)u]x + \frac{1}{m}[\beta t - (\beta B + \delta C)u]y$$

$$Y = \frac{1}{m}[\gamma t + (\alpha A + \gamma B)u]x + \frac{1}{m}[\delta t + (\beta A + \delta B)u]y \qquad (\text{I})$$

Here the indeterminate numbers t, u are all the integers satisfying the equation $t^2 - Du^2 = m^2$. We cannot yet conclude that all

---

[d] From this we easily deduce: $AeU = (\delta\gamma' - \gamma\delta')m$
$$2BeU = (\alpha\delta' - \delta\alpha' + \gamma\beta' - \beta\gamma')m$$
$$CeU = (\beta\alpha' - \alpha\beta')m$$

values of t, u satisfying this equation will give suitable transformations when substituted in formula (I). But

1. It is easy to show by calculation with the help of equations [1], [3], [5] and $t^2 - Du^2 = m^2$; that the form F will always be transformed into f by substituting any values of t, u. We omit the calculation because it is long rather than difficult.

2. Any transformation deduced from the formula will be similar to the proposed one. For

$$\frac{1}{m}[\alpha t - (\alpha B + \gamma C)u] \cdot \frac{1}{m}[\delta t + (\beta A + \delta B)u]$$

$$-\frac{1}{m}[\beta t - (\beta B + \delta C)u] \cdot \frac{1}{m}[\gamma t + (\alpha A + \gamma B)u]$$

$$= \frac{1}{m^2}(\alpha\delta - \beta\gamma)(t^2 - Du^2) = \alpha\delta - \beta\gamma$$

3. If the forms F, f have unequal determinants, it can happen that formula (I) for certain values of t, u will produce substitutions that imply *fractions*. These must be rejected. But all the others will be suitable transformations and they will be the only ones.

4. Now if the forms F, f have the same determinant and are thus *equivalent*, formula (I) will produce no transformations implying fractions, and in this case it will give the complete solution of the problem. We show this as follows.

From the theorem of the preceding article it follows in this case that m will be a common divisor of the numbers $A, 2B, C$. We know $t^2 - Du^2 = m^2$; $t^2 - B^2u^2 = m^2 - ACu^2$ and so $t^2 - B^2u^2$ is divisible by m^2. Since this is so, $4t^2 - 4B^2u^2$ is also divisible by m^2 and therefore (since $2B$ is divisible by m) $4t^2$ by m^2 and so $2t$ by m. As a result of this $2(t + Bu)/m$, $2(t - Bu)/m$ will be integers and indeed (since their difference $4Bu/m$ is even) they will both be even or both odd. If they are both odd, the product will be odd but since the square of the number $(t^2 - B^2u^2)/m^2$, an integer as we have just shown, is necessarily even, this case is impossible. Thus $2(t + Bu)/m$, $2(t - Bu)/m$ are always even, and $(t + Bu)/m$, $(t - Bu)/m$ are integers. From this we can conclude easily that all four coefficients in (I) are always integers.   Q.E.D.

We conclude from the preceding that if we have all solutions of the equation $t^2 - Du^2 = m^2$, we can derive all transformations of the form (A, B, C) to the form (a, b, c) similar to the given transformation. We will show later how to find these solutions. We only observe here that the number of solutions is always finite when D is negative or a positive square. When D is positive but not a square there is an infinite number of solutions. When this case occurs and we do not have $D = d$ (see 3 above), we must further find out how to distinguish a priori the values of t, u that give integral substitutions from those that do not. For this case we will give in article 214 another method which is free of this difficulty.

Example. The form $x^2 + 2y^2$ is transformed into the form $(6, 24, 99)$ by the proper substitution $x = 2x' + 7y'$, $y = x' + 5y'$. We want *all* proper transformations of the former form into the latter. Here $D = -2$, $m = 3$, and so the equation to be solved is $t^2 + 2u^2 = 9$. This can be satisfied in six ways, namely $t = 3, -3,$ $1, -1, 1, -1$; $u = 0, 0, 2, 2, -2, -2$ respectively. The third and sixth solutions give fractional substitutions and thus must be rejected; the rest give the following four substitutions:

$$
x = \begin{vmatrix} 2x' + 7y' \\ -2x' - 7y' \\ -2x' - 9y' \\ 2x' + 9y' \end{vmatrix}
\qquad
y = \begin{vmatrix} x' + 5y' \\ -x' - 5y' \\ x' + 3y' \\ -x' - 3y' \end{vmatrix}
$$

The first of these is the given one.

▶ 163. We have already remarked above that a form F can imply another form F' both properly and improperly. This can happen if another form G can be interposed between F and F' so that F implies G and G implies F', and G is such that it is improperly equivalent to itself. For if we suppose that F implies G properly or improperly: since G implies G improperly, F will imply G improperly or properly (respectively) and thus in either case both properly and improperly (art. 159). In the same way, no matter how G is presumed to imply F', F will always imply F' properly and improperly. That there are such forms equivalent to themselves improperly we see in the most obvious case when the middle term of the form $= 0$. Such a form will be opposite to itself (art. 159) and so improperly equivalent. More generally any form $(a, b,$

c) in which 2*b* is divisible by *a* will have this property. The form (*c*, *b*, *a*) will be a neighbor to this by the first part (art. 160) and thus properly equivalent to it; but (*c*, *b*, *a*) by article 159 is improperly equivalent to the form (*a*, *b*, *c*); and so (*a*, *b*, *c*) is improperly equivalent to itself. Such forms (*a*, *b*, *c*) in which 2*b* is divisible by *a* we will call *ambiguous* forms. We have therefore the following theorem:

Form F will imply another form F' both properly and improperly if we can find an ambiguous form contained in F which implies F'. And this proposition can also be converted.

▶ 164. THEOREM. *If the form $Ax^2 + 2Bxy + Cy^2 \dots (F)$ implies the form $A'x'x' + 2B'x'y' + C'y'y' \dots (F')$ both properly and improperly, an ambiguous form can be found which is contained in F and which implies the form F'.*

Let us suppose that the form F is transformed into the form F' by the substitution

$$x = \alpha x' + \beta y', \qquad y = \gamma x' + \delta y'$$

and by another dissimilar substitution

$$x = \alpha' x' + \beta' x', \qquad y = \gamma' x' + \delta' y'$$

Then designating the numbers $\alpha\delta - \beta\gamma$, $\alpha'\delta' - \beta'\gamma'$ by e, e' we will have $B'B' - A'C' = e^2(B^2 - AC) = e'e'(B^2 - AC)$; thus $ee = e'e'$ and because by hypothesis e, e' have opposite signs, $e = -e'$ or $e + e' = 0$. Now it is evident that if in F', $\delta'x'' - \beta'y''$ is substituted for x' and $-\gamma'x'' + \alpha'y''$ for y', the same form will be produced. If further we write in F

either   1) for x $\alpha(\delta'x'' - \beta'y'') + \beta(-\gamma'x'' + \alpha'y'')$

i.e. $(\alpha\delta' - \beta\gamma')x'' + (\beta\alpha' - \alpha\beta')y''$

and   for y $\gamma(\delta'x'' - \beta'y'') + \delta(-\gamma'x'' + \alpha'y'')$

i.e. $(\gamma\delta' - \delta\gamma')x'' + (\delta\alpha' - \gamma\beta')y''$

or   2) for x $\alpha'(\delta'x'' - \beta'y'') + \beta'(-\gamma'x'' + \alpha'y'')$ i.e. $e'x''$

and   for y $\gamma'(\delta'x'' - \beta'y'') + \delta'(-\gamma'x'' + \alpha'y'')$ i.e. $e'y''$

and if we designate the numbers $\alpha\delta' - \beta\gamma'$, $\beta\alpha' - \alpha\beta'$, $\gamma\delta' - \delta\gamma'$, $\delta\alpha' - \gamma\beta'$ by a, b, c, d, the form F will be transformed into the same form by two substitutions

$$x = ax'' + by'', \qquad y = cx'' + dy''; \qquad x = e'x'', \qquad y = e'y''$$

and we will obtain the three following equations:

$$Aa^2 + 2Bac + Cc^2 = Ae'e' \qquad [1]$$

$$Aab + B(ad + bc) + Ccd = Be'e' \qquad [2]$$

$$Ab^2 + 2Bbd + Cd^2 = Ce'e' \qquad [3]$$

From the values of a, b, c, d, however, we find

$$ad - bc = ee' = -e^2 = -e'e' \qquad [4]$$

And from $d[1] - c[2]$

$$(Aa + Bc)(ad - bc) = (Ad - Bc)e'e'$$

and thus

$$A(a + d) = 0$$

Further, from $(a + d)[2] - b[1] - c[3]$ we get

$$[Ab + B(a + d) + Cc][ad - bc] = [-Ab + B(a + d) - Cc]e'e'$$

and therefore

$$B(a + d) = 0$$

Finally from $a[3] - b[2]$

$$(Bb + Cd)(ad - bc) = (-Bb + Ca)e'e'$$

and thus

$$C(a + d) = 0$$

Therefore since not all A, B, C can $= 0$, necessarily $a + d = 0$ or $a = -d$.

From $a[2] - b[1]$ we get

$$(Ba + Cc)(ad - bc) = (Ba - Ab)e'e'$$

and from this

$$Ab - 2Ba - Cc = 0 \qquad [5]$$

From the equations $e + e' = 0$, $a + d = 0$ or

$$\alpha\delta - \beta\gamma + \alpha'\delta' - \beta'\gamma' = 0, \qquad \alpha\delta' - \beta\gamma' - \gamma\beta' + \delta\alpha' = 0$$

it follows that $(\alpha + \alpha')(\delta + \delta') = (\beta + \beta')(\gamma + \gamma')$ or

$$(\alpha + \alpha'):(\gamma + \gamma') = (\beta + \beta'):(\delta + \delta')$$

Let this proportion[e] in lowest terms be equal to the proportion $m:n$ where m, n are relatively prime and select μ, ν in such a way that $\mu m + \nu n = 1$. Further let r be the greatest common divisor of the numbers a, b, c. Its square will divide $a^2 + bc$ or $bc - ad$ or e^2, so therefore r will also divide e. Having done this, if we suppose that by the substitution

$$x = mt + \frac{\nu e}{r}u, \qquad y = nt - \frac{\mu e}{r}u$$

the form F is transformed into the form $Mt^2 + 2Ntu + Pu^2$ (G), this will be ambiguous and will imply the form F'.

Demonstration. I. To show that the form G is ambiguous we will show that

$$M(b\mu^2 - 2a\mu\nu - c\nu^2) = 2Nr$$

and since r divides a, b, c, $(b\mu^2 - 2a\mu\nu - c\nu^2)/r$ is an integer and therefore $2N$ is a multiple of M.

$$M = Am^2 + 2Bmn + Cn^2$$

$$Nr = [Am\nu - B(m\mu - n\nu) - Cn\mu]e \qquad\qquad [6]$$

And by calculation it is easy to confirm that

$$2e + 2a = e - e' + a - d = (\alpha - \alpha')(\delta + \delta') - (\beta - \beta')(\gamma + \gamma')$$

$$2b = (\alpha + \alpha')(\beta - \beta') - (\alpha - \alpha')(\beta + \beta')$$

Thus since $m(\gamma + \gamma') = n(\alpha + \alpha')$, $m(\delta + \delta') = n(\beta + \beta')$ we have $m(2e + 2a) = -2nb$ or

$$me + ma + nb = 0 \qquad\qquad [7]$$

---

[e] If all $\alpha + \alpha', \gamma + \gamma', \beta + \beta', \delta + \delta' = 0$, the ratio would be indeterminate and the method not applicable. But a little attention shows that this would not be consistent with our presuppositions. For we would have $\alpha\delta - \beta\gamma = \alpha'\delta' - \beta'\gamma'$, i.e. $e = e'$, and since $e = -e'$, $e = e' = 0$. But then also $B'B' - A'C'$, i.e. the determinant of the form F', would $= 0$. We excluded such forms entirely.

In the same way we find

$$2e - 2a = e - e' - a + d = (\alpha + \alpha')(\delta - \delta') - (\beta + \beta')(\gamma - \gamma')$$

$$2c = (\gamma - \gamma')(\delta + \delta') - (\gamma + \gamma')(\delta - \delta')$$

and from this $n(2e - 2a) = -2mc$ or

$$ne - na + mc = 0 \qquad\qquad [8]$$

Now to $m^2(b\mu^2 - 2a\mu v - cv)$ we add

$$[1 - m\mu - nv][mv(e - a) + (m\mu + 1)b]$$

$$+ (me + ma + nb)(m\mu v + v) + (ne - na + mc)mv^2$$

which manifestly equals 0, since

$$1 - \mu m - vn = 0, \qquad me + ma + nb = 0, \qquad ne - na + mc = 0$$

If we multiply this out and cancel terms we get $2mve + b$. Therefore

$$m^2(b\mu^2 - 2a\mu v - cv^2) = 2mve + b \qquad\qquad [9]$$

In the same manner by adding to $mn(b\mu^2 - 2a\mu v - cv^2)$ the following:

$$[1 - m\mu - nv][(nv - m\mu)e - (1 + m\mu + nv)a]$$

$$-(me + ma + nb)m\mu^2 + (ne - na + mc)nv^2$$

we find

$$mn(b\mu^2 - 2a\mu v - cv^2) = (nv - m\mu)e - a \qquad\qquad [10]$$

Finally by adding to $n^2(b\mu^2 - 2a\mu v - cv^2)$ the following:

$$[m\mu + nv - 1][n\mu(e + a) + (nv + 1)c]$$

$$-(me + ma + nb)n\mu^2 - (ne - na + mc)(n\mu v + \mu)$$

we get

$$n^2(b\mu^2 - 2a\mu v - cv^2) = -2n\mu e - c \qquad\qquad [11]$$

And now from [9], [10], [11] we deduce

$$(Am^2 + 2Bmn + Cn^2)(b\mu^2 - 2a\mu v - cv^2)$$

$$= 2e[Amv + B(nv - m\mu) - Cn\mu] + Ab - 2Ba - Cc$$

or because of [6]

$$M(b\mu^2 - 2a\mu v - cv^2) = 2Nr \quad \text{Q.E.D.}$$

II. In order to prove that the form G implies the form F' we will show *first* that G is transformed into F' by letting

$$t = (\mu\alpha + v\gamma)x' + (\mu\beta + v\delta)y'$$

$$u = \frac{r}{e}(n\alpha - m\gamma)x' + \frac{r}{e}(n\beta - m\delta)y' \tag{S}$$

second that $r(n\alpha - m\gamma)/e$, $r(n\beta - m\delta)/e$ are integers.

1. Since F is transformed into G by letting

$$x = mt + \frac{ve}{r}u, \qquad y = nt - \frac{\mu e}{r}u$$

the form G will be changed by the substitution (S) into the same form as that of F by letting

$$x = m[(\mu\alpha + v\gamma)x' + (\mu\beta + v\delta)y'] + v[(n\alpha - m\gamma)x' + (n\beta - m\delta)y']$$

i.e.

$$= \alpha(m\mu + nv)x' + \beta(m\mu + nv)y' \quad or \quad = \alpha x' + \beta y'$$

and

$$y = n[(\mu\alpha + v\gamma)x' + (\mu\beta + v\delta)y'] - \mu[(n\alpha - m\gamma)x' + (n\beta - m\delta)y']$$

i.e.

$$= \gamma(nv + m\mu)x' + \delta(nv + m\mu)y' \quad or \quad = \gamma x' + \delta y'$$

By this substitution F is transformed into F', and therefore by the substitution (S) G will be transformed into F'.

2. From the values of e, b, d we can find $\alpha'e + \gamma b - \alpha d = 0$ or, since $d = -a$, $n\alpha'e + n\alpha a + n\gamma b = 0$; so using [7], $n\alpha'e + n\alpha a = m\gamma e + m\gamma a$ or

$$(n\alpha - m\gamma)a = (m\gamma - n\alpha')e \tag{12}$$

Further, $\alpha nb = -\alpha m(e + a)$, $\gamma mb = -m(\alpha'e + \alpha a)$ and so

$$(n\alpha - m\gamma)b = (\alpha' - \alpha)me \tag{13}$$

Finally $\gamma'e - \gamma a + ac = 0$; then multiplying by n and substituting for na its value in [8] we get

$$(n\alpha - m\gamma)c = (\gamma - \gamma')ne \tag{14}$$

In a similar manner $\beta'e + \delta b - \beta d = 0$ or $n\beta'e + n\delta b + n\beta a = 0$ and thus by [7], $n\beta'e + n\beta a = m\delta e + m\delta a$ or

$$(n\beta - m\delta)a = (m\delta - n\beta')e \qquad [15]$$

Further, $\beta nb = -\beta m(e + a)$, $\delta mb = -m(\beta'e + \beta a)$ and so

$$(n\beta - m\delta)b = (\beta' - \beta)me \qquad [16]$$

Finally $\delta'e - \delta a + \beta c = 0$; then multiplying by n and substituting for na its value in [8] we get

$$(n\beta - m\delta)c = (\delta - \delta')ne \qquad [17]$$

Now since the greatest common divisor of the numbers a, b, c is r, integers $\mathfrak{A}, \mathfrak{B}, \mathfrak{C}$ can be found such that

$$\mathfrak{A}a + \mathfrak{B}b + \mathfrak{C}c = r \qquad \bullet$$

And from [12], [13], [14]; [15], [16], [17]

$$\mathfrak{A}(m\gamma - n\alpha') + \mathfrak{B}(\alpha' - \alpha)m + \mathfrak{C}(\gamma - \gamma')n = \frac{r}{e}(n\alpha - m\gamma)$$

$$\mathfrak{A}(m\delta - n\beta') + \mathfrak{B}(\beta' - \beta)m + \mathfrak{C}(\delta - \delta')n = \frac{r}{e}(n\beta - m\delta)$$

and so $r(n\alpha - m\gamma)/e$, $r(n\beta - m\delta)/e$ are integers. Q.E.D.

▶ 165. *Example.* The form $3x^2 + 14xy - 4y^2$ can be transformed into $-12x'x' - 18x'y' + 39y'y'$ properly by

$$x = 4x' + 11y', \qquad y = -x' - 2y'$$

and improperly by

$$x = -74x' + 89y', \qquad y = 15x' - 18y'$$

Here therefore $\alpha + \alpha', \beta + \beta', \gamma + \gamma', \delta + \delta'$ are $-70, 100, 14, -20$; and $-70:14 = 100:-20 = 5:-1$. We therefore let $m = 5$, $n = -1, \mu = 0, \nu = -1$. The numbers a, b, c are $-237, -1170, 48$; and the greatest common divisor $= 3 = r$; and finally $e = 3$. Thus the transformation (S) will be $x = 5t - u, y = -t$. By this the form $(3, 7, -4)$ is transformed into the ambiguous form $t^2 - 16tu + 3u^2$.

If the forms F, F' are equivalent, the form G contained in F will also be contained in F'. But since it implies F', it will be equivalent to it and thus also to the form F. In this case we will enunciate the theorem as follows:

If F, F' are properly and improperly equivalent, an ambiguous form can be found which is equivalent to both of them. But in this case $e = \pm 1$, and so r which divides e will $= 1$.

What we have said suffices for transformations of forms in general; now we will go on to consider *representations*.

▶ 166. *If the form F implies the form F', any number that can be represented by F' can also be represented by F.*

Let the unknowns of the forms F, F' be $x, y; x', y'$, respectively, and let us suppose that the number M can be represented by F' by letting $x' = m$, $y' = n$ and, further, that the form F is transformed into F' by the substitution

$$x = \alpha x' + \beta y', \qquad y = \gamma x' + \delta y'$$

It is manifest that if we let

$$x = \alpha m + \beta n, \qquad y = \gamma m + \delta n$$

F will be transformed into M.

If M can be represented by the form F' in various ways, e.g. by also letting $x' = m', y' = n'$, then various representations of M by F will follow. For then we would have

$$\alpha m + \beta n = \alpha m' + \beta n' \quad \text{and} \quad \gamma m + \delta n = \gamma m' + \delta n'$$

But if this were so, we would have $\alpha\delta - \beta\gamma = 0$, and thus also the determinant of the form $F = 0$ contrary to the hypothesis or else $m = m', n = n'$. From this it follows that M can be represented in at least as many ways by F as by F'.

If therefore F implies F' and F' implies F, i.e. if F, F' are equivalent, the number M can be represented by either of them and in as many ways by one as by the other.

Finally we observe that in this case the greatest common divisor of the numbers m, n is equal to the greatest common divisor of the numbers $\alpha m + \beta n, \gamma m + \delta n$. Let it $= \Delta$, and choose the numbers μ, ν such that $\mu m + \nu n = \Delta$. Then we will have

$$(\delta\mu - \gamma\nu)(\alpha m + \beta n) - (\beta\mu - \alpha\nu)(\gamma m + \delta n)$$

$$= (+\alpha\delta - \beta\gamma)(\mu m + \nu n) = \pm\Delta$$

Thus the greatest common divisor of the numbers $\alpha m + \beta n$,

$\gamma m + \delta n$ divides Δ, and Δ divides this divisor because manifestly it divides $\alpha m + \beta n, \gamma m + \delta n$. Wherefore Δ necessarily equals the greatest common divisor. When therefore m, n are relatively prime, $\alpha m + \beta n, \gamma m + \delta n$ will be relatively prime.

▶167. THEOREM. *If the forms*

$$ax^2 + 2bxy + cy^2 \tag{F}$$

$$a'x'x' + 2b'x'y' + c'y'y' \tag{F'}$$

are equivalent, if their determinant $= D$, *and if the latter is transformed into the former by letting*

$$x' = \alpha x + \beta y, \qquad y' = \gamma x + \delta y$$

if further the number M *is represented by* F *by letting* $x = m$, $y = n$ *and by* F' *by letting*

$$x' = \alpha m + \beta n = m', \qquad y' = \gamma m + \delta n = n'$$

in such a manner that m *and* n *and ipso facto* m' *and* n' *are relatively prime, then both representations belong either to the same value of the expression* $\sqrt{D}$ *(mod. M) or to opposite values according as the transformation of the form* F' *into* F *is proper or improper.*

Demonstration. Let the numbers μ, ν be so determined that $\mu m + \nu n = 1$, and let

$$\frac{\delta\mu - \gamma\nu}{\alpha\delta - \beta\gamma} = \mu', \qquad \frac{-\beta\mu + \alpha\nu}{\alpha\delta - \beta\gamma} = \nu'$$

(which are integers since $\alpha\delta - \beta\gamma = \pm 1$). We will then have $\mu'm' + \nu'n' = 1$ (cf. end of preceding article).

Further let

$$\mu(bm + cn) - \nu(am + bn) = V,$$

$$\mu'(b'm' + c'n') - \nu'(a'm' + b'n') = V'$$

and V, V' will be values of the expression $\sqrt{D}$ (mod. M), to which the first and second representation belong. If for μ', ν', m', n' their values are substituted, and in V

for a, $a'\alpha^2 + 2b'\alpha\gamma + c'\gamma^2$

for b, $a'\alpha\beta + b'(\alpha\delta + \beta\gamma) + c'\gamma\delta$

for c, $\quad a'\beta^2 + 2b'\beta\delta + c'\delta^2$

we find by calculation that $V = V'(\alpha\delta - \beta\gamma)$.

Therefore either $V = V'$ or $V = -V'$ according as $\alpha\delta - \beta\gamma$ $= +1$ or $= -1$; i.e. the representations belong to the same value of the expression $\sqrt{D}$ (mod. M) or to opposite values according as the transformation of the form F' into F is proper or improper. Q.E.D.

If therefore we have many representations of the number M by the form (a, b, c) by means of relatively prime values of the unknowns x, y and if they give *different* values for the expression $\sqrt{D}$ (mod. M), representations corresponding to the form (a', b', c') will belong to the same values respectively. And if there is no representation by any form of the number M belonging to a given determined value, there will be none belonging to this value by any equivalent form.

▶ 168. THEOREM. *If the number M is represented by the form $ax^2 + 2bxy + cy^2$, and if by giving to x, y relatively prime values m, n the value of the expression $\sqrt{D}$ (mod. M), to which this expression belongs, is N, then the forms (a, b, c), $[M, N, (N^2 - D)/M]$ will be properly equivalent.*

Demonstration. From article 155 it is clear that we can find integers μ, ν such that

$$m\mu + n\nu = 1, \qquad \mu(bm + cn) - \nu(am + bn) = N$$

Using this, by the substitution $x = mx' - \nu y'$, $y = nx' + \mu y'$ which is manifestly proper, the form (a, b, c) is transformed into a form whose determinant $= D(m\mu + n\nu)^2 = D$, i.e. into an equivalent form. If we presume that this form $= [M', N', (N'N' - D)/M']$ we get

$$M' = am^2 + 2bmn + cn^2 = M$$

$$N' = -m\nu a + (m\mu - n\nu)b + n\mu c = N$$

Thus the form into which (a, b, c) will be changed by the transformation will be $[M, N, (N^2 - D)/M]$. Q.E.D.

But from the equations

$$m\mu + n\nu = 1, \qquad \mu(mb + nc) - \nu(ma + nb) = N$$

we deduce

$$\mu = \frac{nN + ma + nb}{am^2 + 2bmn + cn^2} = \frac{nN + ma + nb}{M}, \qquad \nu = \frac{mb + nc - mN}{M}$$

and these numbers will be integers.

We must point out that this proposition does not hold if $M = 0$, for then the term $(N^2 - D)/M$ will be *indeterminate*.[f]

▶ 169. If we have many representations of the number M by (a, b, c) belonging to the same value N of the expression $\sqrt{D}$ (mod. M) (we always presuppose that the values of x, y are relatively prime), then we can derive from them many proper transformations of the form (a, b, c) into $[M, N, (N^2 - D)/M]\ldots$ (G). For if such a representation results from the values $x = m'$, $y = n'$, (F) will also be transformed into (G) by the substitution

$$x = m'x' + \frac{m'N - m'b - n'c}{M}y', \qquad y = n'x' + \frac{n'N + m'a + n'b}{M}y'$$

Reciprocally, from every proper transformation of the form (F) into (G) will follow a representation of the number M by the form (F) belonging to the value N. That is, if by letting $x = mx' - \nu y'$, $y = nx' + \mu y'$ (F) is transformed into (G), then M will be represented by (F) by letting $x = m$, $y = n$. And since $m\mu + n\nu = 1$, the value of the expression $\sqrt{D}$ (mod. M) to which the representation belongs will be $\mu(bm + cn) - \nu(am + bn)$; i.e. N. From many different proper transformations will follow just as many different representations belonging to N.[g] As a result, if

<hr>

[f] If we wish to extend the terminology to this case, we can say that if N is the value of the expression $\sqrt{D}$ (mod. M) or $N^2 \equiv D$ (mod. M), it will signify that $N^2 - D$ is a multiple of M and therefore $= 0$.

[g] If we suppose that the same representation comes from two different proper transformations, they will have to be:

$$(1) \ x = mx' - \nu y', \ y = nx' + \mu y'; \quad (2) \ x = mx' - \nu'y', \ y = nx' + \mu'y'.$$

But from the two equations

$$m\mu + n\nu = m\mu' + n\nu', \ \mu(mb + nc) - \nu(ma + nb) = \mu'(mb + nc) - \nu'(ma + nb)$$

it is easy to deduce that either $M = 0$ or $\mu = \mu'$, $\nu = \nu'$. But we have already excluded $M = 0$.

we have all the proper transformations of the form (F) into (G), from them will follow all representations of M by (F) belonging to the value N. When we want to investigate the question of the representations of a given number by a given form (in which the indeterminate values are relatively prime), we are reduced to finding all the proper transformations of the form into another given equivalent form.

In applying here what we have seen in article 162, it is easy to conclude as follows: If a representation of the number M by the form (F) belonging to the value N is $x = \alpha$, $y = \gamma$, then the general formula embracing all representations of the same number by the form (F) belonging to the value N will be

$$x = \frac{\alpha t - (\alpha b + \gamma c)u}{m}, \qquad y = \frac{\gamma t + (\alpha^2 + \gamma b)u}{m}$$

where m is the greatest common divisor of the numbers $a, 2b, c$; and t, u all the pairs of numbers satisfying the equation $t^2 - Du^2 = m^2$.

▶170. If the form (a, b, c) is equivalent to an ambiguous form and thus equivalent to the form $[M, N, (N^2 - D)/M]$ properly and improperly, or properly to the forms $[M, N, (N^2 - D)/M]$ and $[M, -N, (N^2 - D)/M]$, we will have representations of the number M by the form (F) belonging both to the value N and to the value $-N$. And vice versa, if we have many representations of the number M by the same form (F) belonging to N, and $-N$ [*opposite* values of the expression $\sqrt{D}$ (mod. M)], the form (F) will be properly and improperly equivalent to the form (G), and we can find an ambiguous form equivalent to (F).

These general considerations concerning representations will suffice for now. We will proceed below to representations in which the indeterminate values are not relatively prime. With respect to other properties, forms whose determinant is negative must be treated quite differently from forms with positive determinants. We will consider each of these separately and will begin with the former type since they are easier.

▶171. PROBLEM. *Given a form* (a, b, a') *whose negative determinant* $= -D$ *with D a positive number, to find a form* (A, B, C)

which is properly equivalent to this in which A is not greater than
$\sqrt{4D/3}$, *C or less than 2B.*

Solution. We will suppose that in the given form not all three conditions hold simultaneously, for then there would be no need to find another form. Let b' be the absolutely least residue of the number $-b$ relative to the modulus a'[h] and $a'' = (b'b' + D)/a'$ which is an integer because $b'b' \equiv b^2$, $b'b' + D \equiv b^2 + D \equiv aa' \equiv 0 \pmod{a'}$. Now if $a'' < a'$, let b'' be the absolutely least residue of $-b'$ relative to the modulus a'', and $a''' = (b''b'' + D)/a''$. If again $a''' < a''$, let again b''' be the absolutely least residue of $-b''$ relative to the modulus a''' and $a'''' = (b'''b''' + D)/a'''$. Continue this operation until in the progression a', a'', a''', a'''', etc. we come to the term a^{m+1} which is not smaller than its predecessor a^m. This will happen eventually because otherwise the progression would have an infinite number of continuously decreasing integers. The form (a^m, b^m, a^{m+1}) will satisfy all the conditions.

Demonstration. I. In the progression of the forms (a, b, a'), (a', b', a''), (a'', b'', a'''), etc. each form is neighbor to its predecessor and so the last will be properly equivalent to the first (art. 159, 160).

II. Since b^m is the absolutely least residue of $-b^{m-1}$ relative to the modulus a^m it will not be greater than $a^m/2$ (art. 4).

III. Since $a^m a^{m+1} = D + b^{2m}$, and a^{m+1} is not $< a^m$, a^{2m} will not be $> D + b^{2m}$, and since b^m is not $> a^m/2$, a^{2m} will not be $> D + 1/4a^{2m}$, $3a^{2m}/4$ will not be $> D$, and finally a^m is not $> \sqrt{4D/3}$.

Example. Given the form $(304, 217, 155)$ with the determinant $= -31$, we find the following progression of forms:

$(304, 217, 155), (155, -62, 25), (25, 12, 7), (7, 2, 5), (5, -2, 7)$

The last is the one we are looking for. In the same way, given the form $(121, 49, 20)$ with the determinant $= -19$ we find the equivalent forms: $(20, -9, 5), (5, -1, 4), (4, 1, 5)$; and $(4, 1, 5)$ is the form we seek.

Forms like (A, B, C) whose determinant is negative and in

---

[h] It is worthwhile to observe that if a or a', the first or last term of a form (a, b, a'), should $= 0$, its determinant will be a positive square, but this cannot happen in the present case. For a similar reason, the outer terms a, a' of a form with a negative determinant cannot have opposite signs.

which A is not greater than $\sqrt{4D/3}$, C or less than $2B$ we shall call *reduced forms*. Thus a reduced form can be found which is properly equivalent to any form that has a negative determinant.

▶ 172. PROBLEM. *To find conditions under which two nonidentical reduced forms (a, b, c), (a', b', c') with the same determinant $-D$ can be properly equivalent.*

Solution. Let us suppose (as is legitimate) that a' is not $> a$ and that the form $ax^2 + 2bxy + cy^2$ is transformed into $a'x'x' + 2b'x'y' + c'y'y'$ by the proper substitution $x = \alpha x' + \beta y'$, $y = \gamma x' + \delta y'$. Then we will have the equations

$$a\alpha^2 + 2b\alpha\gamma + c\gamma^2 = a' \qquad [1]$$

$$a\alpha\beta + b(\alpha\delta + \beta\gamma) + c\gamma\delta = b' \qquad [2]$$

$$\alpha\delta - \beta\gamma = 1 \qquad [3]$$

From [1] it follows that $aa' = (a\alpha + b\gamma)^2 + D\gamma^2$; so aa' will be positive; and since $ac = D + b^2$, $a'c' = D + b'b'$ both $ac, a'c'$ will be positive; therefore a, a', c, c' will all have the same sign. But neither a nor a' is $> \sqrt{4D/3}$, so aa' will not be $> 4D/3$; and much less can $D\gamma^2$ [$= aa' - (a\alpha + b\gamma)^2$] be greater than $4D/3$. Therefore γ will either $= 0$ or $= \pm 1$.

I. If $\gamma = 0$, it follows from [3] that either $\alpha = 1$, $\delta = 1$ or $\alpha = -1$, $\delta = -1$. In either case we get from [1] $a' = a$ and from [2] $b' - b = \pm\beta a$. But b is not $> a/2$, and b' is not $> a'/2$ and so not $> a/2$. Therefore the equation $b' - b = \pm\beta a$ is inconsistent unless:

either $b = b'$ from which would follow $c' = (b'b' + D)/a' = (b^2 + D)/a = c$ and the forms (a, b, c), (a', b', c') would be identical, contrary to the hypothesis,

or $b = -b' = \pm a/2$. In this case also $c' = c$, and the form (a', b', c') will be $(a, -b, c)$, i.e. opposite to the form (a, b, c). It is clear that these forms are ambiguous, since $2b = \pm a$.

II. If $\gamma = \pm 1$, we get from [1] $a\alpha^2 + c - a' = \pm 2b\alpha$. But c is not less than a and therefore not less than a'; so $a\alpha^2 + c - a'$ or $2b\alpha$ will certainly not be less than $a\alpha^2$. So since $2b$ is not greater than a, α will not be less than α^2; therefore necessarily $\alpha = 0$ or $= \pm 1$.

1) If $\alpha = 0$ we get from [1] $a' = c$ and, since a is neither greater than c nor less than a', we have necessarily $a' = a = c$. Further, from [3] we get $\beta\gamma = -1$; therefore from [2] $b + b' = \pm \delta c = \pm \delta a$. And as in I above it follows that:

either $b = b'$ in which case the forms (a, b, c), (a', b', c') would be identical contrary to the hypothesis,

or $b = -b'$ in which case the forms (a, b, c), (a', b', c') will be opposites.

2) If $\alpha = \pm 1$ it follows from [1] that $\pm 2b = a + c - a'$. And since neither a nor c is $< a'$, $2b$ will not be $< a$ nor $< c$. But it is also true that $2b$ is not $> a$ nor $> c$, so necessarily $\pm 2b = a = c$, and thus from the equation $\pm 2b = a + c - a'$ it will also $= a'$. From [2] therefore

$$b' = a(\alpha\beta + \gamma\delta) + b(\alpha\delta + \beta\gamma)$$

or, since $\alpha\delta - \beta\gamma = 1$

$$b' - b = a(\alpha\beta + \gamma\delta) + 2b\beta\gamma = a(\alpha\beta + \gamma\delta + \beta\gamma)$$

wherefore necessarily, as before

either $b = b'$ and the forms (a, b, c), (a', b', c') are identical, contrary to the hypothesis,

or $b = -b'$ and the forms are opposite. Also since $a = \pm 2b$ the forms will be ambiguous in this case.

From all of this it follows that the forms (a, b, c), (a', b', c') cannot be properly equivalent unless they are opposite and at the same time *either* ambiguous *or* $a = c = a' = c'$. In these cases it was evident a priori that the forms (a, b, c), (a', b', c') were properly equivalent. For if the forms are opposite they must be improperly equivalent and further, if they are ambiguous, they must also be properly equivalent. If $a = c$, the form $([D + (a - b)^2]/2, a - b, a)$ will be a neighbor to and therefore equivalent to the form (a, b, c). But since $D + b^2 = ac = a^2$, we have $[D + (a - b)^2]/a = 2a - 2b$, but the form $(2a - 2b, a - b, a)$ is ambiguous and therefore also properly equivalent to its opposite (a, b, c).

It is also easy to judge when two reduced forms that are not opposite (a, b, c), (a', b', c') can be improperly equivalent. They will be improperly equivalent if (a, b, c), $(a', -b', c')$ are not identical and are properly equivalent. And the inverse of this proposition also holds. Thus the condition under which the given forms are improperly equivalent is that they be identical and, further, either ambiguous or $a = c$. Reduced forms which are neither identical nor opposite cannot be properly or improperly equivalent.

▶ 173. PROBLEM. *Given two forms F and F' with the same negative determinant, to discover whether they are equivalent.*

Solution. We will seek two reduced forms f, f' properly equivalent to the forms F, F' respectively. If the forms f, f' are equivalent properly or improperly or both, then F, F' will be also. But if f, f' are in no way equivalent, neither will F, F' be.

From the preceding article we distinguish four cases:

1) If f, f' are neither identical nor opposite, F, F' will not be equivalent in any way.

2) If f, f' are *first* either identical or opposite, and *second* either ambiguous or have their outer terms equal, F, F' will be equivalent properly and improperly.

3) If f, f' are identical but are not ambiguous and do not have outer terms equal, F, F' will be only properly equivalent.

4) If f, f' are opposite but are not ambiguous and do not have outer terms equal, F, F' will be only improperly equivalent.

Example. The forms $(41, 35, 30)$, $(7, 18, 47)$ have a determinant $= -5$; $(1, 0, 5)$, $(2, 1, 3)$ are reduced forms and since these are not equivalent, the former will not be equivalent either. The same reduced form $(2, 1, 3)$ is equivalent to the forms $(23, 38, 63)$, $(15, 20, 27)$. Since the reduced form is ambiguous, the forms $(23, 38, 63)$, $(15, 20, 27)$ will be properly and improperly equivalent. The reduced forms $(9, 2, 9)$, $(9, -2, 9)$ are equivalent to the forms $(37, 53, 78)$, $(53, 73, 102)$. And since the reduced forms are opposite and their outer terms are equal, the given forms are properly and improperly equivalent.

▶ 174. The number of reduced forms having a given determinant $-D$ is always finite and quite small relative to the number D, and there are two methods of finding the forms themselves. We will designate by (a, b, c) the reduced forms with determinant $-D$. We therefore wish to determine all the values of a, b, c.

First method. Take for a all numbers positive and negative which are not greater than $\sqrt{4D/3}$ and whose quadratic residue is $-D$. And for each a let b be successively equal to all values of the expression $\sqrt{-D}$ (mod. a) not greater than $a/2$. Both positive and negative values are to be used. For each pair of values for a, b let $c = (D + b^2)/a$. If any forms gotten this way make $c < a$, they are to be rejected. The rest are manifestly reduced forms.

Second method. Take for b all numbers both positive and negative which are not greater than $(1/2)\sqrt{4D/3}$ or $\sqrt{D/3}$. For each b decompose $b^2 + D$ in every way possible into pairs of factors neither of which is less than $2b$ (account should be taken of the sign involved). Let one of the factors (the smaller when they are unequal) $= a$ and let the other $= c$. Since a will not be $> \sqrt{4D/3}$, all such forms will manifestly be reduced. Finally it is clear that no reduced form can be found that cannot be found by both methods.

Example. Let $D = 85$. The limit of the values of a is $\sqrt{340/3}$, which lies between 10 and 11. The numbers 1 through 10 whose residue is -85 are $1, 2, 5, 10$. Thus we have twelve forms: $(1, 0, 85)$, $(2, 1, 43)$, $(2, -1, 43)$, $(5, 0, 17)$, $(10, 5, 11)$, $(10, -5, 11)$; $(-1, 0, -85)$, $(-2, 1, -43)$, $(-2, -1, -43)$, $(-5, 0, -17)$, $(-10, 5, -11)$, $(-10, -5, -11)$.

By the second method the limit of the values of b is $\sqrt{85/3}$, which lies between 5 and 6. For $b = 0$ the following forms result:

$$(1, 0, 85), (-1, 0, -85), (5, 0, 17), (-5, 0, -17)$$

for $b = \pm 1$ these result: $(2, \pm 1, 43)$, $(-2, \pm 1, -43)$.

For $b = \pm 2$ there is none, since 89 cannot be resolved into two factors which are both not < 4. The same is true for $\pm 3, \pm 4$. For $b = \pm 5$ we get

$$(10, \pm 5, 11), (-10, \pm 5, -11)$$

▶ 175. If among all the reduced forms of a given determinant, we reject one or the other of the pairs of forms which are properly equivalent without being identical, the remaining forms have the following remarkable property: that any form of the determinant will be properly equivalent to one of them and indeed only to one (otherwise some of them would be properly equivalent among themselves). Thus it is clear that *all forms of the same determinant can be distributed into as many classes as there are remaining forms*, that is, by putting forms which are properly equivalent to the same reduced form into the same class. Thus for $D = 85$, the following forms remain:

$$(1, 0, 85), (2, 1, 43), (5, 0, 17), (10, 5, 11)$$

$$(-1, 0, -85), (-2, 1, -43), (-5, 0, -17), (-10, 5, -11)$$

so all the forms of the determinant -85 can be distributed into eight classes according as they are properly equivalent to the first or second etc. of these. Obviously forms of the same class will be properly equivalent, and forms of different classes cannot be properly equivalent. We will treat of this argument concerning the classification of forms more in detail later, adding here only one observation. We showed above that if the determinant of the form (a, b, c) was negative, $= -D$, a and c would have the same sign (because $ac = b^2 + D$ and is therefore positive). By the same reasoning it is clear that if the forms (a, b, c), (a', b', c') are equivalent, all the terms a, c, a', c' will have the same sign. For if the former is transformed into the latter by the substitution $x = \alpha x' + \beta y'$, $y = \gamma x' + \delta y'$, we have $a\alpha^2 + 2b\alpha\gamma + c\gamma^2 = a'$. From this $aa' = (a\alpha + b\beta)^2 + D\gamma^2$ which is certainly not negative. And since neither a nor a' can $= 0$, aa' will be positive and the signs of a, a' must be the same. Thus, forms whose outer terms are positive must be completely separated from those whose outer terms are negative, and it is sufficient to consider only those of the reduced forms that have positive outer terms because the others are of the same number and they derive from these by changing the signs of the outer terms. The same thing holds for the forms that are to be rejected or retained from among the reduced forms.

▶176. The following table shows forms with negative determinants according to which all other forms of the same determinant can be distributed into classes. We remark according to the note

in the preceding article that we have only half of them listed, that
is, those whose outer terms are positive.

D				
1	$(1, 0, 1)$			
2	$(1, 0, 2)$			
3	$(1, 0, 3)$,	$(2, 1, 2)$		
4	$(1, 0, 4)$,	$(2, 0, 2)$		
5	$(1, 0, 5)$,	$(2, 1, 3)$		
6	$(1, 0, 6)$,	$(2, 0, 3)$		
7	$(1, 0, 7)$,	$(2, 1, 4)$		
8	$(1, 0, 8)$,	$(2, 0, 4)$,	$(3, 1, 3)$	
9	$(1, 0, 9)$,	$(2, 1, 5)$,	$(3, 0, 3)$	
10	$(1, 0, 10)$,	$(2, 0, 5)$		
11	$(1, 0, 11)$,	$(2, 1, 6)$,	$(3, 1, 4)$,	$(3, -1, 4)$
12	$(1, 0, 12)$,	$(2, 0, 6)$,	$(3, 0, 4)$,	$(4, 2, 4)$

It would be superfluous to continue this table further here,
since we will show later a more suitable way of arranging it.

It is clear from examining the table that any form with determinant -1 is properly equivalent to the form $x^2 + y^2$ if the outer
terms are positive, and to the form $-x^2 - y^2$ if they are negative;
that any form with determinant -2 whose outer terms are positive
is equivalent to the form $x^2 + 2y^2$ etc.; that any form with determinant -11 whose outer terms are positive is equivalent to one
of the following: $x^2 + 11y^2$, $2x^2 + 2xy + 6y^2$, $3x^2 + 2xy + 4y^2$,
$3x^2 - 2xy + 4y^2$, etc.

▶ 177. PROBLEM. *We have a series of forms in which each form is
a neighbor by the last part to the preceding form: we want a proper
transformation of the first into any form of the series.*

Solution. Let these forms be given: $(a, b, a') = F$; $(a', b', a'') = F'$;
$(a'', b'', a''') = F''$; $(a''', b''', a'''') = F'''$; etc. Designate $(b + b')/a'$,
$(b' + b'')/a''$, $(b'' + b''')/a'''$, etc. by h', h'', h''', etc. respectively. Let the
unknowns of the forms F, F', F'', etc. be x, y; x', y'; x'', y''; etc.
Suppose F is transformed into

F' by letting $\quad x = \alpha'x' + \beta'y', \qquad y = \gamma'x' + \delta'y'$

F'' by letting $\quad x = \alpha''x'' + \beta''y'', \qquad y = \gamma''x'' + \delta''y''$

•

F''' by letting $\quad x = \alpha'''x''' + \beta'''y''', \qquad y = \gamma'''x''' + \delta'''y'''$

etc.

Then since

F becomes F' by letting $x = -y', \qquad y = x' + h'y'$

F' becomes F'' by letting $x' = -y'', \qquad y' = x'' + h''y''$

F'' becomes F''' by letting $x'' = -y''', \qquad y'' = x''' + h''y''',$

etc. (art. 160)

we get the following algorithm easily (art. 159)

$\alpha' = 0 \qquad \beta' = -1 \qquad\qquad \gamma' = 1 \qquad \delta' = h'$

$\alpha'' = \beta' \qquad \beta'' = h''\beta' - \alpha' \qquad \gamma'' = \delta' \qquad \delta'' = h''\delta' - \gamma'$

$\alpha''' = \beta'' \qquad \beta''' = h'''\beta'' - \alpha'' \qquad \gamma''' = \delta'' \qquad \delta''' = h'''\delta'' - \gamma''$

$\alpha'''' = \beta''' \qquad \beta'''' = h''''\beta''' - \alpha''' \qquad \gamma'''' = \delta''' \qquad \delta'''' = h''''\delta''' - \gamma'''$

etc.

or

$\alpha' = 0 \qquad \beta' = -1 \qquad\qquad \gamma' = 1 \qquad \delta' = h'$

$\alpha'' = \beta' \qquad \beta'' = h''\beta' \qquad\qquad \gamma'' = \delta' \qquad \delta'' = h''\delta' - 1$

$\alpha''' = \beta'' \qquad \beta''' = h'''\beta'' - \beta' \qquad \gamma''' = \delta'' \qquad \delta''' = h'''\delta'' - \delta'$

$\alpha'''' = \beta''' \qquad \beta'''' = h''''\beta''' - \beta'' \qquad \gamma'''' = \delta''' \qquad \delta'''' = h''''\delta''' - \delta''$

etc.

It is not hard to see that all these transformations are proper both because of the way in which they were constructed and from article 159.

This very simple algorithm is extremely well suited for calculating, and is analogous to the algorithm in article 27 and can be reduced to it.[i] This solution is not restricted to forms with a negative determinant but is applicable to all cases as long as none of the numbers a', a'', a''', etc. $= 0$.

[i] We have, using the notation of article 27, $\beta^n = \pm[-h'', h''', -h''''', \ldots, \pm h^n]$. The undetermined signs should be $- - ; - + ; + - ; + +$; according as n is of the form $4k + 0$; 1; 2; 3; and $\delta^n = \pm[h', -h'', h''', \ldots, \pm h^n]$ where the undetermined signs should be $+ -$; $+ +$; $- -$; $- +$; according as n is of the form $4k + 0$; 1; 2; 3. But space does not permit us to explain this more fully. Each one can confirm it easily for himself.

▶ 178. PROBLEM. *Given two properly equivalent forms F, f with the same negative determinant: to find a proper transformation of one into the other.*

Solution. Let us suppose that the form F is (A, B, A') and that by the method of article 171 we have found a series of forms (A', B', A''), (A'', B'', A''') etc. up to the reduced form (A^m, B^m, A^{m+1}); and similarly that f is (a, b, a') and that by the same method we have found the series (a', b', a''), (a'', b'', a''') up to the reduced form (a^n, b^n, a^{n+1}). We can identify two cases.

I. If the forms (A^m, B^m, A^{m+1}), (a^n, b^n, a^{n+1}) are either identical or opposite and ambiguous at the same time, then the forms (A^{m-1}, B^{m-1}, A^m), $(a^n, -b^{n-1}, a^{n-1})$ will be neighbors (here A^{m-1} designates the penultimate term of the progression $A, A', A'', \ldots, A^m$. Similarly for $B^{m-1}, a^{n-1}, b^{n-1}$). For $A^m = a^n$, $B^{m-1} \equiv -B^m$ (mod. A^m), $b^{n-1} \equiv -b^n$ (mod. a^n or A^m). As a result $B^{m-1} - b^{n-1} \equiv b^n - B^m$. But if the forms (A^m, B^m, A^{m+1}), (a^n, b^n, a^{n+1}) are identical this will be $\equiv 0$; if they are opposite and ambiguous it will be $\equiv 2b^n$ and therefore $\equiv 0$. So in the series of forms

$$(A, B, A'), (A', B', A''), \ldots, (A^{m-1}, B^{m-1}, A^m)$$
$$(a^n, -b^{n-1}, a^{n-1}), (a^{n-1}, -b^{n-2}, a^{n-2}), \ldots, (a', -b, a), (a, b, a')$$

each form is a neighbor to its predecessor, and by the preceding article a proper transformation of the first F into the last f can be found.

II. If the forms (A^m, B^m, A^{m+1}), (a^n, b^n, a^{n+1}) are not identical but opposite and at the same time $A^m = A^{m+1} = a^n = a^{n+1}$, then the series of forms

$$(A, B, A'), (A', B', A''), \ldots, (A^m, B^m, A^{m+1})$$
$$(a^n, -b^{n-1}, a^{n-1}), (a^{n-1}, -b^{n-2}, a^{n-2}), \ldots, (a', -b, a), (a, b, a')$$

will have the same property. For $A^{m+1} = a^n$, and $B^m - b^{n-1} = -(b^n + b^{n-1})$ is divisible by a^n. And so by the preceding article a proper transformation of the first form F into the last f can be found.

Example. For the forms $(23, 38, 63)$, $(15, 20, 27)$ we have the series $(23, 38, 63)$, $(63, 25, 10)$, $(10, 5, 3)$, $(3, 1, 2)$, $(2, -7, 27)$, $(27, -20,$

15), (15, 20, 27). Therefore

$$h' = 1, h'' = 3, h''' = 2, h'''' = -3, h''''' = -1, h'''''' = 0$$

Thus the transformation of the form $23x^2 + 76xy + 63y^2$ into $15t^2 + 40tu + 27u^2$ is $x = -13t - 18u, y = 8t + 11u$.

From the solution of this problem follows the solution of the next one: *If the forms F, f are improperly equivalent, to find an improper transformation of the form F into f.* For if $f = at^2 + 2btu + A'u^2$, the form $ap^2 - 2bpq + A'q^2$ which is opposite to f will be properly equivalent to F. We have only to find a proper transformation of the form F into this. Let it be $x = \alpha p + \beta q$, $y = \gamma p + \delta q$ and clearly F will be transformed into f by $x = \alpha t - \beta u, y = \gamma t - \delta u$. And this is an improper transformation.

If the forms F, f are properly and improperly equivalent, we can find a proper and an improper transformation.

▶ 179. PROBLEM. *If the forms F, f are equivalent, to find all transformations of the form F into f.*

Solution. If the forms F, f are equivalent in only one way, i.e. only properly or only improperly, according to the preceding article we can find a transformation of the form F into f. It is clear that there can be no others except those that are similar to this one. If the forms F, f are properly and improperly equivalent, we can find two transformations, one proper, the other improper. Now let the form $F = (A, B, C)$, $B^2 - AC = -D$ and let the greatest common divisor of the numbers $A, 2B, C = m$. Then from article 162 it is clear that in the former case all transformations of the form F into f can be derived from one transformation and that in the latter case all proper transformations can be derived from a proper transformation and all improper transformations from an improper one—provided we have all solutions of the equation $t^2 + Du^2 = m^2$. When we find these, the problem will be solved.

We have $D = AC - B^2$, $4D = 4AC - 4B^2$ therefore $4D/m^2 = (4AC/m^2) - (2B/m)^2$ will be an integer. Now

1) If $4D/m^2 > 4$ then $D > m^2$. So in $t^2 + Du^2 = m^2$, u necessarily $= 0$ and t can have no values except $+m$ and $-m$. Thus if F, f are equivalent in only one way and if we have a transformation

$$x = \alpha x' + \beta y', \qquad y = \gamma x' + \delta y'$$

then there can be no other transformations except this one, which results from letting $t = m$ (art. 162) and the following one:

$$x = -\alpha x' - \beta y', \qquad y = -\gamma x' - \delta y'$$

But if F, f are properly and improperly equivalent and we have a proper transformation

$$x = \alpha x' + \beta y', \qquad y = \gamma x' + \delta y'$$

and an improper one

$$x = \alpha' x' + \beta' y', \qquad y = \gamma' x' + \delta' y'$$

then there can be no other transformations except these two (which result from letting $t = m$) and the following two (letting $t = -m$) which are proper and improper respectively:

$$x = -\alpha x' - \beta y', \qquad y = -\gamma x' - \delta y'$$
$$x = -\alpha' x' - \beta' y', \qquad y = -\gamma' x' - \delta' y'$$

2) If $4D/m^2 = 4$ or $D = m^2$, the equation $t^2 + Du^2 = m^2$ admits of four solutions: $t, u = m, 0; = -m, 0; = 0, 1; = 0, -1$. Thus if F, f are equivalent in only one way and we have the transformation

$$x = \alpha x' + \beta y', \qquad y = \gamma x' + \delta y'$$

there will be *four* transformations altogether,

$$x = \pm \alpha x' \pm \beta y', \qquad y = \pm \gamma x' \pm \delta y'$$
$$x = \mp \frac{\alpha B + \gamma C}{m} x' \mp \frac{\beta B + \delta C}{m} y',$$
$$y = \pm \frac{\alpha A + \gamma B}{m} x' \pm \frac{\beta A + \delta B}{m} y'$$

If F, f are equivalent in two ways, that is to say if besides the given transformation there is another one dissimilar to it, this one will produce four more, making eight in all. And if there are eight such transformations it is easy to show that F, f are equivalent in two ways. For since $D = m^2 = AC - B^2$, m will divide B. The determinant of the form $(A/m, B/m, C/m)$ will be $= -1$ and so the form $(1, 0, 1)$ or $(-1, 0, -1)$ will be equivalent to it. It is easy to see further that the same transformation that carries $(A/m, B/m, C/m)$ into $(\pm 1, 0, \pm 1)$ will carry the form (A, B, C) into $(\pm m, 0, \pm m)$, which is ambiguous. And so the form (A, B, C) being equivalent to an ambiguous form will be both properly and improperly equivalent to any form to which it is equivalent at all.

3) If $4D/m^2 = 3$ or $4D = 3m^2$, then m will be even and there will be six solutions of the equation $t^2 + Du^2 = m^2$,

$$t, u = m, 0; = -m, 0; = \tfrac{1}{2}m, 1; = -\tfrac{1}{2}m, -1; = \tfrac{1}{2}m, -1; = -\tfrac{1}{2}m, 1$$

If therefore we have two dissimilar transformations of the form F into f

$$x = \alpha x' + \beta y', \qquad y = \gamma x' + \delta y'$$
$$x = \alpha' x' + \beta' y', \qquad y = \gamma' x' + \delta' y'$$

there will be twelve transformations in all, that is, six similar to the first transformation,

$$x = \pm \alpha x' \pm \beta y', \qquad y = \pm \gamma x' \pm \delta y'$$

$$x = \pm \left(\frac{1}{2}\alpha - \frac{\alpha B + \gamma C}{m} \right) x' \pm \left(\frac{1}{2}\beta - \frac{\beta B + \delta C}{m} \right) y'$$

$$y = \pm \left(\frac{1}{2}\gamma + \frac{\alpha A + \gamma B}{m} \right) x' \pm \left(\frac{1}{2}\delta + \frac{\beta A + \delta B}{m} \right) y'$$

$$x = \pm \left(\frac{1}{2}\alpha + \frac{\alpha B + \gamma C}{m} \right) x' \pm \left(\frac{1}{2}\beta + \frac{\beta B + \delta C}{m} \right) y'$$

$$y = \pm \left(\frac{1}{2}\gamma - \frac{\alpha A + \gamma B}{m} \right) x' \pm \left(\frac{1}{2}\delta - \frac{\beta A + \delta B}{m} \right) y'$$

and six similar to the second transformation, which can be derived from these six by substituting $\alpha', \beta', \gamma', \delta'$ for $\alpha, \beta, \gamma, \delta$.

To show that in this case F, f are always equivalent in both ways, consider the following. The determinant of the form $(2A/m, 2B/m, 2C/m)$ will $= -4D/m^2 = -3$ and therefore this form (art. 176) will be equivalent either to the form $(\pm 1, 0, \pm 3)$ or $(\pm 2, \pm 1, \pm 2)$. And the form (A, B, C) will be equivalent either to the form $(\pm m/2, 0, \pm 3m/2)$ or to $(\pm m, m/2, \pm m)$.[j] Since both these forms are ambiguous, it will be equivalent in both ways to whichever of them it is equivalent at all.

[j] It can be shown that the form (A, B, C) is necessarily equivalent to the second of these, but this is not needed here.

4) If we suppose that $4D/m^2 = 2$, we have $(2B/m)^2 = (4AC/m^2)$ -2 and therefore $(2B/m^2)^2 \equiv 2$ (mod. 4). But since no square can be $\equiv 2$ (mod. 4) this case cannot occur.

5) By supposing that $4D/m^2 = 1$, we get $(2B/m)^2 = (4AC/m^2) - 1$ $\equiv -1$ (mod. 4). But since this is impossible, neither can this case occur.

And since D cannot be ≤ 0, there are no other cases.

▶ 180. PROBLEM. *To find all representations of a given number M by the form $ax^2 + 2bxy + cy^2 \dots F$, in which the determinant $-D$ is negative and the values x, y are relatively prime.*

Solution. From article 154 it is clear that M cannot be represented in this way unless $-D$ is a quadratic residue of M. Let us therefore first investigate all the different (i.e. noncongruent) values of the expression $\sqrt{-D}$ (mod. M). Let them be N, $-N$, N', $-N'$, N'', $-N''$, etc. To make the calculation simpler, all the N, N', etc. can be determined so that they are not $> M/2$. Now since each representation should belong to one of these values, we will consider each of them separately.

If the forms F and $[M, N, (D + N^2)/M]$ are not properly equivalent, there can be no representation of M belonging to the value N (art. 168). If they are properly equivalent, we will want a proper transformation of the form F into

$$Mx'x' + 2Nx'y' + \frac{D + N^2}{M}y'y'$$

Let it be

$$x = \alpha x' + \beta y', \qquad y = \gamma x' + \delta y'$$

and we get $x = \alpha$, $y = \gamma$ as the representation of the number M by the form F which belongs to the value N. Let the greatest common divisor of the numbers $A, 2B, C = m$, and we will distinguish three cases (art. 179):

1) If $4D/m^2 > 4$, there can be no other representation belonging to N except these *two* (art. 169, 179):

$$x = \alpha, y = \gamma; \qquad x = -\alpha, y = -\gamma$$

2) If $4D/m^2 = 4$, we have *four* representations:

$$x = \pm\alpha: y = \pm\gamma; \qquad x = \mp\frac{\alpha B + \gamma C}{m}, y = \pm\frac{\alpha A + \gamma B}{m}$$

3) If $4D/m^2 = 3$, we have *six* representations:

$$x = \pm\alpha \qquad\qquad\qquad y = \pm\gamma$$

$$x = \pm\left(\frac{1}{2}\alpha - \frac{\alpha B + \gamma C}{m}\right), \qquad y = \pm\left(\frac{1}{2}\gamma + \frac{\alpha A + \gamma B}{m}\right)$$

$$x = \pm\left(\frac{1}{2}\alpha + \frac{\alpha B + \gamma C}{m}\right), \qquad y = \pm\left(\frac{1}{2}\gamma - \frac{\alpha A + \gamma B}{m}\right)$$

In the same manner we can find representations belonging to the values $-N$, N', $-N$, etc.

▶ 181. If we are looking for representations of the number M by the form F when the values of x, y are not relatively prime, we can reduce it to the case which we have just considered. Suppose we have this representation by letting $x = \mu e$, $y = \mu f$ in such a way that μ is the greatest common divisor of $\mu e, \mu f$ or, in other words, that e, f are relatively prime. Then we will have $M = \mu^2(Ae^2 + 2Bef + Cf^2)$, and so M is divisible by μ^2. The substitution $x = e$, $y = f$ will be the representation of the number M/μ^2 by the form F in which x, y are relatively prime values. If therefore M is divisible by no square (except 1), e.g. if it is a prime number, there will be no representation of M. But if M has quadratic divisors, let them be μ^2, ν^2, π^2, etc. First we will look for all representations of the number M/μ^2 by the form (A, B, C) in which the values x, y are relatively prime. If these values are multiplied by μ they will give all representations of M in which the greatest common divisor of the numbers x, y is μ. In a similar way all representations of M/ν^2, in which the values of x, y are relatively prime, will give all representations of M in which the greatest common divisor of the values of x, y is ν etc.

And thus by the preceding rules we can find all representations of a given number by a given form that has a negative determinant.

▶ 182. Let us now consider certain particular cases both because of their remarkable elegance and because of the painstaking work done on them by Euler, who endowed them with an almost classical distinction.

I. No number can be so represented by $x^2 + y^2$ that x, y are relatively prime (or decomposed into two squares which are relatively prime) unless its quadratic residue is -1. But all numbers which enjoy this property can be so represented when taken positively. Let M be such a number, and let all the values of the expression $\sqrt{-1}$ (mod. M) be $N, -N, N', -N', N'', -N''$, etc. Then by article 176 the form $[M, N, (N^2 + 1)/M]$ will be properly equivalent to the form $(1, 0, 1)$. Let a proper transformation of the latter into the former be $x = \alpha x' + \beta y'$, $y = \gamma x' + \delta y'$, and the representations of the number M by the form $x^2 + y^2$ belonging to N will be these four:[k] $x = \pm\alpha$, $y = \pm\gamma$; $x = \mp\gamma$, $y = \pm\alpha$.

Since the form $(1, 0, 1)$ is ambiguous, the form $[M, -N, (N^2 + 1)/M]$ will also be properly equivalent to it, and the first can be properly transformed into the second by letting $x = ax' - \beta y'$, $y = -\gamma x' + \delta y'$. From this we derive the four representations of M belonging to $-N : x = \pm\alpha, y = \mp\gamma; x = \pm\gamma, y = \pm\alpha$. Thus there are eight representations of M, half of which belong to N, half to $-N$; but all of these specify *only one* separation of the number M into two squares, $M = \alpha^2 + \gamma^2$ as long as we consider only the squares themselves and not the order or signs of the roots.

If then there are no other values of the expression $\sqrt{-1}$ (mod. M) except N and $-N$, which happens, for example, when M is a prime number, M can be resolved into two relatively prime squares in only one way. Now since -1 is a quadratic residue of any prime number of the form $4n + 1$ (art. 108), manifestly a prime number cannot be separated into two squares that are not relatively prime, and we have the following theorem.

Any prime number of the form $4n + 1$ can be decomposed into two squares and in only one way.

$1 = 0 + 1$, $5 = 1 + 4$, $13 = 4 + 9$, $17 = 1 + 16$, $29 = 4 + 25$, $37 = 1 + 36$, $41 = 16 + 25$, $53 = 4 + 49$, $61 = 25 + 36$, $73 = 9 + 64$, $89 = 25 + 64$, $97 = 16 + 81$, etc.

This extremely elegant theorem was already known by Fermat but was first demonstrated by Euler, *Novi comm. acad. Petrop.*, 5

[k] It is obvious that this case is contained in article 180. 2.

[1754–55], 3 ff.[1] In Volume 4 there exists a dissertation on the same subject (p. 3 ff.),[2] but he had not yet found a solution (see especially art. 27).

If therefore a number of the form $4n + 1$ can be resolved into two squares in many ways or in no way, it will certainly not be prime.

On the other hand, however, if the expression $\sqrt{-1}$ (mod. M) should have other values than N and $-N$, there would be still other representations of M belonging to them. In this case M could be resolved into two quadratics in more than one way; e.g. $65 = 1 + 64 = 16 + 49$, $221 = 25 + 196 = 100 + 121$.

Other representations in which x, y have values which are not relatively prime can easily be found by our general method. We only observe that if a number involves factors of the form $4n + 3$ and if it cannot be freed from these by division by squares (which happens when one or more of these factors has an *odd dimension*), then this number *cannot* be resolved into two squares *by any method*.[1]

II. No number having -2 as a residue can be represented by the form $x^2 + 2y^2$ with x, y relatively prime. All others can. Let -2 be a residue of the number M, and N the value of the expression $\sqrt{-2}$ (mod. M). Then by article 176 the forms $(1, 0, 2)$, $[M, N, (N^2 + 2)/M]$ will be properly equivalent. Transform the former into the latter properly by letting $x = \alpha x' + \beta y'$, $y = \gamma x' + \delta y'$, and we will get $x = \alpha$, $y = \gamma$ as the representation of the number M belonging to N. Besides this one we also have $x = -\alpha$, $y = -\gamma$, and there are no others belonging to N (art. 180).

As before, we see that the representations $x = \pm\alpha$, $y = \pm\gamma$ belong to the value $-N$. All four of these representations produce

[1] Cf. p. 27.

[2] Cf. p. 42.

---

[1] If the number $M = 2^\mu S a^\alpha b^\beta c^\gamma \ldots$, where a, b, c, etc. are unequal prime numbers of the form $4n + 1$, and S is the product of all the prime factors of M of the form $4n + 3$ (any positive number can be reduced to this form, for if M is odd we let $\mu = 0$ and if M has no factors of the form $4n + 3$ we let $S = 1$), M cannot in any way be resolved into two squares if S is not a square. If S is a square we will have $(\alpha + 1)(\beta + 1)(\gamma + 1)$etc./2 decompositions of M when any of the numbers α, β, γ, etc. is odd or $(\alpha + 1)(\beta + 1)(\gamma + 1)$etc./2 $+ 1/2$ when all the numbers α, β, γ, etc. are even (as long as we pay attention only to the squares themselves). Those who are well versed in the calculus of combinations will be able to derive this theorem (we cannot dwell on this or on other particular cases) from our general theory without any difficulty (cf. art. 105).

only one decomposition of M into a square and the double of a square. And if the expression $\sqrt{-2}$ (mod. M) has no values but N and $-N$, there will be no other decompositions. From this fact and with the help of the propositions of article 116 we easily deduce the following theorem.

Any prime number of the form $8n + 1$ or $8n + 3$ can be decomposed into a square and the double of a square, and in only one way.

$1 = 1 + 0, 3 = 1 + 2, 11 = 9 + 2, 17 = 9 + 8, 19 = 1 + 18,$
$\quad 41 = 9 + 32, 43 = 25 + 18, 59 = 9 + 50, 67 = 49 + 18,$
$\quad 73 = 1 + 72, 83 = 81 + 2, 89 = 81 + 8, 97 = 25 + 72,$ etc.

This theorem and many like it were known to Fermat, but Lagrange was the first to demonstrate it: "Suite des recherches d'Arithmétique," *Nouv. mém. Acad. Berlin*, 1775, p. 323 ff.[3] Euler had already discovered much pertaining to the same subject: "Specimen de usu observationum in mathesi pura," *Novi comm. acad. Petrop.*, 6 [1756–57], 1761, 185 ff. But a complete demonstration of the theorem always eluded his efforts (see p. 220). Compare also his dissertation in Volume 8 (for the years 1760, 1761) "Supplementum quorundam theorematum arithmeticorum" toward the end.[4]

III. By a similar method it can be shown that any number whose quadratic residue is -3 can be represented either by the form $x^2 + 3y^2$ or by $2x^2 + 2xy + 2y^2$ with the values x, y relatively prime. Thus since -3 is a residue of all prime numbers of the form $3n + 1$ (art. 119), and since only *even* numbers can be represented by the form $2x^2 + 2xy + 2y^2$ we have, just as above, the following theorem.

Any prime number of the form $3n + 1$ can be decomposed into a square and the triple of a square, and in only one way.

$1 = 1 + 0, \quad 7 = 4 + 3, \quad 13 = 1 + 12, \quad 19 = 16 + 3,$
$31 = 4 + 27, \quad 37 = 25 + 12, \quad 43 = 16 + 27, \quad 61 = 49 + 12,$
$\quad\quad\quad 67 = 64 + 3, \quad 73 = 25 + 48,$ etc.

Euler first gave a demonstration of this theorem in the commentary we have already commended (*Novi comm. acad. Petrop.*, 8, 105 ff.[4]).

We could continue in the same way and show for example that

[3] Cf. p. 42.
[4] Cf. p. 79.

any prime number of the form $20n + 1$, $20n + 3$, $20n + 7$, or $20n + 9$ (numbers which have -5 as a residue) can be represented by one of the forms $x^2 + 5y^2, 2x^2 + 2xy + 3y^2$, and indeed that prime numbers of the form $20n + 1$ and $20n + 9$ can be represented by the first form, and those of the form $20n + 3$, $20n + 7$ by the second form; and it can be further shown that doubles of prime numbers of the form $20n + 1, 20n + 9$ can be represented by the form $2x^2 + 2xy + 3y^2$ and doubles of primes of the form $20n + 3, 20n + 7$ by the form $x^2 + 5y^2$. But the reader can derive this proposition and an infinite number of other particular ones from the preceding and the following discussions. We will pass on now to a consideration of *forms with a positive determinant* and, since the properties are quite different when the determinant is a square and when it is not a square, we will first exclude forms with a square determinant and consider them separately later.

▶ 183. PROBLEM. *Given a form (a, b, a') whose positive nonquadratic determinant $= D$: find a form (A, B, C) properly equivalent to it in which B is positive and $< \sqrt{D}$; A, if it is positive, $-A$, if A is negative, is to lie between $\sqrt{D} + B$ and $\sqrt{D} - B$.*

Solution. We will suppose that neither condition yet holds in the given form; otherwise there would be no need to look for another form. We observe further that in a form with a *non-quadratic* determinant neither the first nor last term can $= 0$ (art. 171). Let $b' \equiv -b$ (mod. a') and let it be between the limits $\sqrt{D}$ and $\sqrt{D} \mp a'$ (the upper sign is to be used when a' is positive, the lower when it is negative). This can be done as in article 3. Let $(b'b' - D)/a' = a''$, which is an integer because $b'b' - D \equiv b^2 - D \equiv aa' \equiv 0$ (mod. a'). Now if $a'' < a'$ let b'' again $\equiv -b'$ (mod. a'') and let it lie between $\sqrt{D}$ and $\sqrt{D} \mp a''$ (according as a'' is positive or negative) and let $(b''b'' - D)/a'' = a'''$. If here again $a''' < a''$ let again $b''' \equiv b''$ (mod. a''') and lying between $\sqrt{D}$ and $\sqrt{D} \mp a'''$ and let $(b'''b''' - D)/a''' = a''''$. Continue this operation until in the series a', a'', a''', a'''', etc. we come to a term a^{m+1} which is not less than the preceding term a^m. This will happen eventually because otherwise we would have a continuously decreasing infinite series of integers. Now let $a^m = A$, $b^m = B$, $a^{m+1} = C$, and the form (A, B, C) will satisfy all conditions.

Demonstration. I. Since in the series of forms (a, b, a'), (a', b', a''), (a'', b'', a'''), etc. each is neighbor to the preceding, the last (A, B, C) will be properly equivalent to the first (a, b, a').

II. Since B lies between $\sqrt{D}$ and $\sqrt{D} \mp A$ (always taking the upper sign when A is positive, the lower when A is negative), it is clear that if we let $\sqrt{D} - B = p$, $B - (\sqrt{D} \mp A) = q$, these numbers p, q will be positive. Now it is easy to confirm that $q^2 + 2pq + 2p\sqrt{D} = D + A^2 - B^2$; thus $D + A^2 - B^2$ is a positive number which we will let $= r$. Now since $D = B^2 - AC$, $r = A^2 - AC$, and $A^2 - AC$ is a positive number. Since by hypothesis A is not greater than C, this cannot happen unless AC is negative, and so the signs of A, C must be opposite. Therefore $B^2 = D + AC < D$ and $B < \sqrt{D}$.

III. Further, since $-AC = D - B^2$, $AC < D$; and so (since A is not $> C$) $A < \sqrt{D}$. Therefore $\sqrt{D} \mp A$ will be positive and so also B which lies between the limits $\sqrt{D}$ and $\sqrt{D} \mp A$.

IV. Because of the above, $\sqrt{D} + B \mp A$ is, a fortiori, positive and since $\sqrt{D} - B \mp A = -q$ is negative, $\pm A$ will lie between $\sqrt{D} + B$ and $\sqrt{D} - B$.   Q.E.D.

Example. Let us be given the form $(67, 97, 140)$ whose determinant $= 29$. We find the series of forms $(67, 97, 140)$, $(140, -97, 67)$, $(67, -37, 20)$, $(20, -3, -1)$, $(-1, 5, 4)$. The last is the one we are looking for.

Such forms (A, B, C) with a nonquadratic positive determinant D in which the positive value of A lies between $\sqrt{D} + B$ and $\sqrt{D} - B$ (B positive and $< \sqrt{D}$), we will call *reduced forms*. Thus reduced forms with nonquadratic positive determinant differ somewhat from reduced forms with negative determinant, but because of the great analogy that exists between the two, we prefer not to introduce different designations for them.

▶ 184. If we could demonstrate the equivalence of two *reduced* forms with a positive determinant as easily as equivalence for forms with a negative determinant (art. 172), we could determine the equivalence of *any* two forms with the same positive determinant without any difficulty. But this is far from the situation, and it can happen that many reduced forms are equivalent among themselves. Before we come to this problem we must investigate more thoroughly the nature of reduced forms (always understood, with nonquadratic positive determinant).

1) If (a, b, c) is a reduced form, a and c will have opposite signs. For letting the determinant $= D$, we will have $ac = b^2 - D$ and thus negative since $b < \sqrt{D}$.

2) The number c, just as a, if taken positively, will lie between $\sqrt{D} + b$ and $\sqrt{D} - b$. For $-c = (D - b^2)/a$; so ignoring sign, c will lie between $(D - b^2)/\sqrt{D} + b$ and $(D - b^2)/\sqrt{D} - b$; i.e. between $\sqrt{D} - b$ and $\sqrt{D} + b$.

3) From this it is clear that (c, b, a) will also be a reduced form.

4) Both a and c will be $< 2\sqrt{D}$. For each is $< \sqrt{D} + b$ and therefore, a fortiori, $< 2\sqrt{D}$.

5) The number b will lie between $\sqrt{D}$ and $\sqrt{D} \mp a$ (taking the upper sign when a is positive, the lower when it is negative). For since $\pm a$ lies between $\sqrt{D} + b$ and $\sqrt{D} - b$, $\pm a - (\sqrt{D} - b)$ or $b - (\sqrt{D} \mp a)$ will be positive; $b - \sqrt{D}$ however is negative; so b will lie between $\sqrt{D}$ and $\sqrt{D} \mp a$. In the same way it can be demonstrated that b lies between $\sqrt{D}$ and $\sqrt{D} \mp c$ (according as c is positive or negative).

6) *For any reduced form (a, b, c) there is one, and only one, reduced form which is neighbor to it by either side.*

Let $a' = c$, $b' \equiv -b$ (mod. a') so that b' lies between $\sqrt{D}$ and $\sqrt{D} \mp a'$,[m] $c' = (b'b' - D)/a'$ and the form (a', b', c') will be neighbor to the form (a, b, c) by the last part. And it is likewise manifest that if we have any other reduced form neighbor by the last part to (a, b, c), it cannot be different from (a', b', c'). We will now show that this is really a reduced form.

A) If we let

$$\sqrt{D} + b \mp a' = p, \quad \pm a' - (\sqrt{D} - b) = q, \quad \sqrt{D} - b = r$$

it follows from 2) above and the definition of a reduced form that p, q, r will be positive. Further let

$$b' - (\sqrt{D} \mp a') = q', \quad \sqrt{D} - b' = r'$$

and q', r' will be positive because b' lies between $\sqrt{D}$ and $\sqrt{D} \mp a'$. Finally let $b + b' = \pm ma'$ and m will be an integer. Now it is clear that $p + q' = b + b'$ and so $b + b'$ or $+ma'$ is positive and

---

[m] Where the signs are ambiguous the upper is to be taken when a' is positive, the lower when a' is negative.

therefore also m; and it follows that $m - 1$ is certainly not negative. Further

$$r + q' \pm ma' = 2b' \pm a' \quad \text{or} \quad 2b' = r + q' \pm (m - 1)a'$$

and $2b'$ and b' will necessarily be positive. And since $b' + r' = \sqrt{D}$ we have $b' < \sqrt{D}$.

B) Further we have

$$r \pm ma' = \sqrt{D} + b' \quad \text{or} \quad r \pm (m - 1)a' = \sqrt{D} + b' \mp a'$$

and so $\sqrt{D} + b' \mp a'$ will be positive. Thus also since $\pm a' - (\sqrt{D} - b') = q'$ and so positive, $\pm a'$ will lie between $\sqrt{D} + b'$ and $\sqrt{D} - b'$. Therefore (a', b', c') will be a reduced form.

In the same way if we have $'c = a$, $'b \equiv -b$ (mod. $'c$) and $'b$ lying between $\sqrt{D}$ and $\sqrt{D} \pm 'c$, then $'a = ('b'b - D)/'c$ and the form $('a, 'b, 'c)$ will be reduced. Manifestly this form is neighbor to the form (a, b, c) by the first part and no other reduced form except $('a, 'b, 'c)$ can have this property.

Example. Given the reduced form $(5, 11, -14)$ with determinant $= 191$, the reduced form $(-14, 3, 13)$ will be neighbor by the last part, $(-22, 9, 5)$ by the first part.

7) If the reduced form (a', b', c') is neighbor by the last part to the reduced form (a, b, c), the form (c', b', a') will be neighbor by the first part to the reduced form (c, b, a); and if the form $('a, 'b, 'c)$ is neighbor to the reduced form (a, b, c) by the first part, the reduced form $('c, 'b, 'a)$ will be neighbor by the last part to the reduced form (c, b, a). Further the forms $(-'a, -'b, -'c)$, $(-a, b, -c)$, $(-a', b', -c')$ will be reduced forms and the second will be neighbor to the first by the last part, and the third to the second; the first will be neighbor by the first part to the second, and the second to the third. In like manner for the three forms $(-c', b', -a')$, $(-c, b, -a)$, $(-'c, 'b, -'a)$. This is obvious and needs no explanation.

▶ 185. The number of all reduced forms of a given determinant D is always finite, and they can be found in two different ways. We will designate all the reduced forms of the determinant D by the indefinite symbol (a, b, c) and seek to determine all values of a, b, c.

First method. Take for a all numbers (both positive and negative) which are less than $2\sqrt{D}$ and which have D as a quadratic residue. For each a let b equal all positive values of the expression

$\sqrt{D}$ (mod. a) lying between $\sqrt{D}$ and $\sqrt{D} \mp a$; and for each of the determined values of a, b let $c = (b^2 - D)/a$. If any form arises by this method in which $\pm a$ lies outside $\sqrt{D} + b$ and $\sqrt{D} - b$, it is to be rejected.

Second method. Take for b all positive numbers less than $\sqrt{D}$. For each b resolve $b^2 - D$ into pairs of factors in every way possible so that each factor, neglecting sign, lies between $\sqrt{D} + b$ and $\sqrt{D} - b$. Set one of these $= a$, the other $= c$. Manifestly each resolution into factors furnishes two forms because each factor can be put $= a$ or $= c$.

Example. Let $D = 79$ and there will be twenty-two values of a: $\mp 1, 2, 3, 5, 6, 7, 9, 10, 13, 14, 15$. From these we find nineteen forms:

$$(1, 8, -15), \quad (2, 7, -15), \quad (3, 8, -5), \quad (3, 7, -10), \quad (5, 8, -3),$$
$$(5, 7, -6), \quad (6, 7, -5), \quad (6, 5, -9), \quad (7, 4, -9), \quad (7, 3, -10),$$
$$(9, 5, -6), \quad (9, 4, -7), \quad (10, 7, -3), \quad (10, 3, -7), \quad (13, 1, -6),$$
$$(14, 3, -5), \quad (15, 8, -1), \quad (15, 7, -2), \quad (15, 2, -5)$$

and just as many others if we change the signs of the outer terms, e.g. $(-1, 8, 15), (-2, 7, 15)$, etc., so that there are thirty-eight in all. But of these the six $(\pm 13, 1, \mp 6), (\pm 14, 3, \mp 5), (\pm 15, 2, \mp 5)$ must be rejected, leaving thirty-two reduced forms. By the second method the same forms will arise in the following order:[n]

$$(\pm 7, 3, \mp 10), \quad (\pm 10, 3, \mp 7), \quad (\pm 7, 4, \mp 9), \quad (\pm 9, 4, \mp 7),$$
$$(\pm 6, 5, \mp 9), \quad (\pm 9, 5, \mp 6), \quad (\pm 2, 7, \mp 15), \quad (\pm 3, 7, \mp 10),$$
$$(\pm 5, 7, \mp 6), \quad (\pm 6, 7, \mp 5), \quad (\pm 10, 7, \mp 3), \quad (\pm 15, 7, \mp 2),$$
$$(\pm 1, 8, \mp 15), \quad (\pm 3, 8, \mp 5), \quad (\pm 5, 8, \mp 3), \quad (\pm 15, 8, \mp 1)$$

▶ 186. Let F be the reduced form of the determinant D, and F' a reduced form neighbor to it by the last part; F'' a reduced form neighbor to F'' by the last part; F''' a reduced form neighbor to F'' by the last part, etc. Then it is clear that all the forms, F', F'', F''', etc. are completely determined and are properly equivalent among themselves and to the form F. Since the number of all reduced forms of a given determinant is finite, manifestly all forms in the infinite progression F, F', F'', etc. cannot be different. Suppose F^m and F^{m+n} are identical and F^{m-1}, F^{m+n-1} reduced forms,

---

[n] For $b = 1$, -78 cannot be resolved into two factors which, disregarding sign, fall between $\sqrt{79} + 1$ and $\sqrt{79} - 1$; so this value must be disregarded and for the same reason the values 2 and 6.

neighbors by the first part to the reduced form and thus identical; then in the same way F^{m-2}, F^{m+n-2}, etc., and finally F and F^n will be identical. So in the series F, F', F'', etc., if only it is continued long enough, the first form F will necessarily recur; and if we suppose that F^n is the first which is identical with F or that all the $F', F'', \ldots, F^{n-1}$ are different from the form F, it is easy to see that all the forms $F, F', F'', \ldots, F^{n-1}$ will be different. We will call the complex of these forms the *period of the form F*. If therefore the series is continued beyond the last form of the period, the same forms F, F', F'', etc. will appear again, and the whole infinite progression F, F', F'', etc. will be made up of this period of the form F repeated infinitely often.

The series F, F', F'', etc. can also be continued backward by putting ahead of the form F the reduced form $'F$ which is neighbor to it by the first part; and by again putting before the form $'F$ the reduced form $''F$ which is neighbor to $'F$ by the first part, etc. In this way we get a series of forms that is infinite *in both directions*.

$$\ldots, {}'''F, {}''F, {}'F, F, F', F'', F''', \ldots$$

and obviously $'F$ is identical with F^{n-1}, $''F$ with F^{n-2}, etc. and so on the left side as well, the series is made up of the period of the form F repeated infinitely often.

If we give indices $0, 1, 2$, etc., $-1, -2$, etc., to the forms F, F', F'', etc., $'F, ''F$, etc. and in general the index m to the form F^m and the index $-m$ to the form $''F$, it is clear that *forms of the series will be identical or different according as their indices are congruent or noncongruent relative to the modulus n*.

Example. The period of the form $(3, 8, -5)$ whose determinant $= 79$ is $(3, 8, -5)$, $(-5, 7, 6)$, $(6, 5, -9)$, $(-9, 4, 7)$, $(7, 3, -10)$, $(-10, 7, 3)$. After the last one we again get $(3, 8, -5)$. Therefore here $n = 6$.

▶ 187. Here are some general observations regarding these periods.

1) If the forms F, F', F'', etc.; $'F, ''F, '''F$, etc. are designated as follows: $(a, b, -a')$, $(-a', b', a'')$, $(a'', b'', -a''')$, etc.; $(-'a, 'b, a)$, $('a, ''b, -'a)$, $(-'''a, '''b, ''a)$, all the a, a', a'', a''', etc., $'a, ''a, '''a$ etc. will have the *same sign* (art. 184.1), and all the b, b', b'', etc., $'b, ''b$, etc. will be positive.

2) It follows from this that the number n (the number of forms which make up the period of the form F) is always *even*. For the first term of the form of any F^m of the period manifestly will have the same sign as the first term a of the form F if m is even, opposite if m is odd. Therefore since F^n and F are identical, n will necessarily be even.

3) Using article 184.6 we have the algorithm for finding the numbers b', b'', b''', etc., a'', a''', etc.:

$$b' \equiv -b \,(\text{mod. } a') \quad \text{between the limits } \sqrt{D} \text{ and } \sqrt{D} \mp a'; \quad a'' = \frac{D - b'b'}{a'}$$

$$b'' \equiv -b' \,(\text{mod. } a'') \qquad \ldots \qquad \sqrt{D} \mp a''; \; a''' = \frac{D - b''b''}{a''}$$

$$b''' \equiv -b'' \,(\text{mod. } a''') \qquad \ldots \qquad \sqrt{D} \mp a'''; \; a'''' = \frac{D - b'''b'''}{a'''}$$

$$\text{etc.}$$

In the second column the upper or lower sign is to be taken according as a, a', a'', etc. are positive or negative. In place of the formulae in the third column the following can also be used and they will prove more convenient when D is a large number:

$$a'' = \frac{b + b'}{a'}(b - b') + a$$

$$a''' = \frac{b' + b''}{a''}(b' - b'') + a'$$

$$a'''' = \frac{b'' + b'''}{a'''}(b'' - b''') + a''$$

$$\text{etc.}$$

4) Any form F^m contained in the period of the form F will have the same period as F. That is, the period will be $F^m, F^{m+1}, \ldots, F^{n-1}$, $F, F', \ldots, F^{m-1}$, and the same forms will occur in the same order as in the period of the form F and differ from it only with respect to the beginning and end.

5) It is clear from the above that all reduced forms of the same determinant D can be *distributed* into periods. Let us take any of these forms F at random and investigate its period $F, F', F'', \ldots,$

F^{n-1}, which we will designate as P. If this does not yet contain all the reduced forms of the determinant D, let G be one not contained in it and let its period be Q. It is clear that P and Q cannot have any form in common; for otherwise G also would be contained in P and the periods would coincide completely. If P and Q have not yet exhausted all the reduced forms, let H be one of the missing ones and we will have a third period R which has no form in common with either P or Q. We can continue in this way until all reduced forms are exhausted. Thus, e.g., all the reduced forms of the determinant 79 are distributed into six periods:

I. $(1, 8, -15), (-15, 7, 2), (2, 7, -15), (-15, 8, 1)$
II. $(-1, 8, 15), (15, 7, -2), (-2, 7, 15), (15, 8, -1)$
III. $(3, 8, -5),\ (-5, 7, 6),\ (6, 5, -9),\ (-9, 4, 7),\ (7, 3, -10), (-10, 7, 3)$
IV. $(-3, 8, 5),\ (5, 7, -6),\ (-6, 5, 9),\ (9, 4, -7),\ (-7, 3, 10), (10, 7, -3)$
V. $(5, 8, -3),\ (-3, 7, 10), (10, 3, -7), (-7, 4, 9),\ (9, 5, -6),\ (-6, 7, 5)$
VI. $(-5, 8, 3),\ (3, 7, -10), (-10, 3, 7), (7, 4, -9),\ (-9, 5, 6),\ (6, 7, -5)$

6) We will call *associated forms* those which are made up of the same terms but in reverse order, as $(a, b, -a')$, $(-a', b, a)$. From article 184.7 it is easy to see that if the period of the reduced form F is $F, F', F'', \ldots, F^{n-1}$ and if f is associated with the form F, and the forms $f', f'', \ldots, f^{n-2}, f^{n-1}$ are associated with the forms $F^{n-1}, F^{n-2}, \ldots, F'', F'$ respectively: then the period of the form f will be $f, f', f'', \ldots, f^{n-2}, f^{n-1}$ and thus it will be made up of the same number of forms as the period of the form F. We will call periods of associated forms *associated periods*. Thus in our example periods III and VI, IV and V are associated.

7) But it can happen that the form f occurs in the period of its associate F, as in period I and II of our example, and thus the period of the form F coincides with the period of the form f or *the period of the form F is an associate of itself.* When this happens this period will have two ambiguous forms. For suppose the period of the form F is made up of $2n$ forms, that is F and F^{2n} are identical; further let $2m + 1$ be the index of the form f in the period of the form F;[o] that is F^{2m+1} and F are associates. Then it is clear that F' and F^{2m} will also be associates; also F'' and F^{2m-1}, etc., and thus also F^m and F^{m+1}. Let $F^m = (a^m, b^m, -a^{m+1})$,

[o] The index here will necessarily be odd because manifestly the first terms of the forms F, f will have opposite signs (see observation 2 above).

$F^{m+1} = (-a^{m+1}, b^{m+1}, a^{m+2})$. Then we will have $b^m + b^{m+1} \equiv 0$ (mod. a^{m+1}); from the definition of associated forms we have $b^m = b^{m+1}$ and thus $2b^{m+1} \equiv 0$ (mod. a^{m+1}); that is, the form F^{m+1} is ambiguous. By the same reasoning F^{2m+1} and F^{2n} will be associates; and so also F^{2m+2} and F^{2n-1}; F^{2m+3} and F^{2n-2}; etc.; and finally F^{m+n} and F^{m+n+1}, the second one of each pair being ambiguous, as can be proven in a similar way. Since $m + 1$ and $m + n + 1$ are noncongruent relative to the modulus $2n$, the forms F^{m+1} and F^{m+n+1} will not be identical (art. 186 where n means what $2n$ does here). Thus in I the ambiguous forms are $(1, 8, -15)$, $(2, 7, -15)$; in II $(-1, 8, 15)$, $(-2, 7, 15)$.

8) Reciprocally, *any period in which an ambiguous form occurs is an associate of itself.* Obviously if F^m is an ambiguous reduced form, a form that is associated with it (which is also a reduced form) will at the same time be neighbor by the first part; i.e. F^{m-1} and F^m will be associates. But then the whole period will be associated with itself. From this it is clear that *it is not possible to have only one ambiguous form in any period.*

9) But it is also true that *there cannot be more than two in the same period.* For suppose that in the period of the form F consisting of $2n$ forms there are three ambiguous forms F^λ, F^μ, F^ν belonging to the indices λ, μ, ν, respectively, where λ, μ, ν are unequal numbers lying between the limits 0 and $2n - 1$ (inclusive). Then the forms $F^{\lambda-1}$ and F^λ will be associates; similarly $F^{\lambda-2}$ and $F^{\lambda+1}$, etc., and finally F and $F^{2\lambda-1}$. By the same reasoning F and $F^{2\mu-1}$ will be associates and also F and $F^{2\nu-1}$; therefore $F^{2\lambda-1}$, $F^{2\mu-1}$, $F^{2\nu-1}$ will be identical and the indices $2\lambda - 1$, $2\mu - 1$, $2\nu - 1$ will be congruent relative to the modulus $2n$, and thus also $\lambda \equiv \mu \equiv \nu$ (mod. n). Q.E.A. because manifestly there cannot be three numbers between the limits 0 and $2n - 1$ that differ from one another and are congruent relative to the modulus n.

▶ 188. Since all forms from the same period are properly equivalent, the question arises whether forms from different periods can also be properly equivalent. But before we show that *this is impossible* we should say something about the transformation of reduced forms.

The transformation of forms will frequently be treated below, and so we will avoid prolixity as much as possible by using the following shorter method of writing. If the form $LX^2 + 2MXY + NY^2$ is transformed into the form $lx^2 + 2mxy + ny^2$ by the substitution $X = \alpha x + \beta y$, $Y = \gamma x + \delta y$, we will say simply that (L, M, N) is transformed into (l, m, n) by the substitution $\alpha, \beta, \gamma, \delta$. In this way it will not be necessary to denote by proper characters the unknowns of each of the forms that is being treated. It is obvious that the *first* unknown must be carefully distinguished from the *second* in any form.

Let $(a, b, -a')\ldots f$ be a given reduced form with determinant D. Just as in article 186 we form a series of reduced forms which is infinite in both directions, $\ldots, ''f, 'f, f, f', f'', \ldots$ and we let

$$f' = (-a', b', a''), \quad f'' = (a'', b'', -a'''), \text{ etc.}$$

$$'f = (-'a, 'b, a), \quad ''f = (''a, ''b, -'a), \text{ etc.}$$

and let

$$\frac{b + b'}{-a'} = h', \quad \frac{b' + b''}{a''} = h'', \quad \frac{b'' + b'''}{-a'''} = h''', \text{ etc.}$$

$$\frac{'b + b}{a} = h, \quad \frac{''b + 'b}{-'a} = 'h, \quad \frac{'''b + ''b}{''a} = ''h, \text{ etc.}$$

Then it is clear that if (as in art. 177) the numbers $\alpha', \alpha'', \alpha'''$, etc. $\beta', \beta'', \beta'''$, etc., etc. are formed according to the following algorithm

$$\alpha' = 0 \quad \beta' = -1 \qquad\qquad \gamma' = 1 \quad \delta' = h'$$

$$\alpha'' = \beta' \quad \beta'' = h''\beta' \qquad\qquad \gamma'' = \delta' \quad \delta'' = h''\delta' - 1$$

$$\alpha''' = \beta'' \quad \beta''' = h'''\beta'' - \beta' \qquad \gamma''' = \delta'' \quad \delta''' = h'''\delta'' - \delta'$$

$$\alpha'''' = \beta''' \quad \beta'''' = h''''\beta''' - \beta'' \quad \gamma'''' = \delta''' \quad \delta'''' = h''''\delta''' - \delta''$$

etc.

f will be transformed

into f' by the substitution $\alpha', \beta', \gamma', \delta'$

into f'' $\qquad \ldots \qquad \alpha'', \beta'', \gamma'', \delta''$

into f''' $\qquad \ldots \qquad \alpha''', \beta''', \gamma''', \delta'''$

etc.

and all these transformations will be proper.

Since $'f$ is transformed into f by the proper substitution $0, -1, 1, h$ (art. 158), f will be transformed into $'f$ by the proper substitution $h, 1, -1, 0$. By similar reasoning $'f$ will be transformed into $''f$ by the proper substitution $'h, 1, -1, 0$; $''f$ into $'''f$ by the proper substitution $''h, 1, -1, 0$ etc. From this by article 159 we conclude in the same way as in article 177 that if the numbers $'\alpha, ''\alpha, '''\alpha$, etc., $'\beta, ''\beta, '''\beta$, etc., etc. are formed according to the following algorithm

$$'\alpha = h \qquad\qquad '\beta = 1 \qquad '\gamma = -1 \qquad\qquad '\delta = 0$$

$$''\alpha = 'h'\alpha - 1 \qquad ''\beta = '\alpha \qquad ''\gamma = 'h'\gamma \qquad\qquad ''\delta = '\gamma$$

$$'''\alpha = ''h''\alpha - '\alpha \qquad '''\beta = ''\alpha \qquad '''\gamma = ''h''\gamma - '\gamma \qquad '''\delta = ''\gamma$$

$$''''\alpha = '''h'''\alpha - ''\alpha \qquad ''''\beta = '''\alpha \qquad ''''\gamma = '''h'''\gamma - ''\gamma \qquad ''''\delta = '''\gamma$$

etc.

then f will be transformed

> into $'f$ by the substitution $'\alpha, '\beta, '\gamma, '\delta$
>
> into $''f$ $\qquad \ldots \qquad$ $''\alpha, ''\beta, ''\gamma, ''\delta$
>
> into $'''f$ $\qquad \ldots \qquad$ $'''\alpha, '''\beta, '''\gamma, '''\delta$

etc.

and all these transformations are proper.

If we let $\alpha = 1, \beta = 0, \gamma = 0, \delta = 1$ these numbers will have the same relation to the form f as $\alpha', \beta', \gamma', \delta'$ have to f'; $\alpha'', \beta'', \gamma'', \delta''$ have to f'' etc.; $'\alpha, '\beta, '\gamma, '\delta$ to $'f$ etc. That is to say, by the substitution $\alpha, \beta, \gamma, \delta$ the form f will be transformed into f. Then the infinite series $\alpha', \alpha'', \alpha'''$, etc., $'\alpha, ''\alpha, '''\alpha$, etc. will be neatly joined together by the insertion of the term α so that they can be conceived of as one continuous series infinite in both directions according to the one law of procession $\ldots, '''\alpha, ''\alpha, '\alpha, \alpha, \alpha', \alpha'', \alpha''', \ldots$ The following is the law of procession:

$$'''\alpha + '\alpha = ''h''\alpha, \quad ''\alpha + \alpha = 'h'\alpha, \quad '\alpha + \alpha' = h\alpha, \quad \alpha + \alpha'' = h'\alpha',$$
$$\alpha' + \alpha''' = h'\alpha'', \text{ etc.}$$

or in general (if we suppose that a negative index indicates the same thing written on the right as a positive index written on the left)

$$\alpha^{m-1} + \alpha^{m+1} = h^m \alpha^m$$

In a similar way the series $\ldots, {''}\beta, {'}\beta, \beta, \beta', \beta'', \ldots$ will be continuous, and its law is

$$\beta^{m-1} + \beta^{m+1} = h^{m+1} \beta^m$$

This series is identical with the preceding if each term is moved up one place $''\beta = '\alpha$, $'\beta = \alpha$, $\beta = \alpha'$, etc. The law for the continuous series $\ldots, {''}\gamma, {'}\gamma, \gamma, \gamma', \gamma'', \ldots$ will be

$$\gamma^{m-1} + \gamma^{m+1} = h^m \gamma^m$$

and the law for $\ldots, {''}\delta, {'}\delta, \delta, \delta', \delta'', \ldots$ will be

$$\delta^{m-1} + \delta^{m+1} = h^{m+1} \delta^m$$

and generally $\delta^m = \gamma^{m+1}$.

Example. Let the given form f be $(3, 8, -5)$. It is transformed into the form

$''''''f$	$(-10, 7, 3)$ by the substitution	$-805,$	$-152,$	$+143,$	$+27$
$'''''f$	$(3, 8, -5)$	$-152,$	$+45,$	$+27,$	-8
$''''f$	$(-5, 7, 6)$	$+45,$	$+17,$	$-8,$	-3
$'''f$	$(6, 5, -9)$	$+17,$	$-11,$	$-3,$	$+2$
$''f$	$(-9, 4, 7)$	$-11,$	$-6,$	$+2,$	$+1$
$'f$	$(7, 3, -10)$	$-6,$	$+5,$	$+1,$	-1
$'f$	$(-10, 7, 3)$	$+5,$	$+1,$	$-1,$	0
f	$(3, 8, -5)$	$+1,$	$0,$	$0,$	$+1$
f'	$(-5, 7, 6)$	$0,$	$-1,$	$+1,$	-3
f''	$(6, 5, -9)$	$-1,$	$-2,$	$-3,$	-7
f'''	$(-9, 4, 7)$	$-2,$	$+3,$	$-7,$	$+10$
f''''	$(7, 3, -10)$	$+3,$	$+5,$	$+10,$	$+17$
f'''''	$(-10, 7, 3)$	$+5,$	$-8,$	$+17,$	-27
f''''''	$(3, 8, -5)$	$-8,$	$-45,$	$-27,$	-152
f'''''''	$(-5, 7, 6)$	$-45,$	$+143,$	$-152,$	$+483$

etc.

▶ 189. The following should be noted with regard to this algorithm.

1) All the numbers a, a', a'', etc., $'a, ''a$, etc. will have the same sign; all the numbers b, b', b'', etc., $'b, ''b$, etc. will be positive; the

series ..., $''h, 'h, h, h', h''$,... will have alternate signs, that is if all the a, a', etc. are positive, h^m or $^m h$ will be positive when m is even, negative when m is odd; but if a, a', etc. are negative h^m or $^m h$ will be negative for m even, positive for m odd.

2) If a is positive and thus h' negative, h'' positive etc., we will have $\alpha'' = -1$ negative, $\alpha''' = h''\alpha''$ negative and $> \alpha''$ (or $= \alpha''$ if $h'' = 1$); $\alpha'''' = h'''\alpha''' - \alpha''$ positive and $> \alpha''$ (because $h'''\alpha'''$ positive, α'' negative); $\alpha''''' = h''''\alpha'''' - \alpha'''$ positive and $> \alpha''''$ (because $h''''\alpha''''$ positive); etc. Thus it is easy to conclude that the series $\alpha', \alpha'', \alpha'''$, etc. will increase continuously and that there will always be two positive and two negative signs, so that α^m has the sign $+, +, -, -$ according as $m \equiv 0, 1, 2, 3$ (mod. 4). If a is negative, by a like reasoning we find α'' negative, α''' positive and either $>$ or $= \alpha''$; α'''' positive $> \alpha'''$; α''''' negative $> \alpha''''$; etc., so that the series $\alpha', \alpha'', \alpha'''$, etc. continually increases, and the sign of the term α^m will be $+, -, -, +$ according as $m \equiv 0, 1, 2, 3$ (mod. 4).

3) In this way we find that all four progressions $\alpha', \alpha'', \alpha'''$, etc.; $\gamma, \gamma', \gamma''$, etc.; $\alpha', \alpha, '\alpha, ''\alpha$, etc.; $\gamma, '\gamma, ''\gamma$, etc. increase continuously and thus all the following, which are identical with them: β, β', β'', etc.; $'\delta, \delta, \delta', \delta''$, etc.; $\beta, '\beta, ''\beta$, etc.; $'\delta, ''\delta$, etc.; and according as $m \equiv 0, 1, 2, 3$ (mod. 4) the sign

$$\text{of } \alpha^m, + \pm - \mp; \quad \text{of } \beta^m, \pm - \mp +$$
$$\text{of } \gamma^m, \pm + \mp -; \quad \text{of } \delta^m, + \mp - \pm$$
$$\text{of } {}^m\alpha, + \pm - \mp; \quad \text{of } {}^m\beta, \mp + \pm -$$
$$\text{of } {}^m\gamma, \mp - \pm +; \quad \text{of } {}^m\delta, + \mp - \pm$$

with the upper sign being used when a is positive, the lower when a is negative. It is especially important to note the following: if we designate any positive index by m, α^m and γ^m will have the same sign when a is positive, opposite signs when a is negative, and similarly for β^m and δ^m; on the other hand $^m\alpha$ and $^m\gamma$ or $^m\beta$ and $^m\delta$ will have the same sign when a is negative, opposite when a is positive.

4) Using the notation of article 27 we can show the size of the numbers α^m etc. by letting

$$\mp h' = k', \quad \pm h'' = k'', \quad \mp h''' = k''', \text{ etc.}$$
$$\pm h = k, \quad \mp 'h = 'k, \quad \pm ''h = ''k, \text{ etc.}$$

so that all the numbers k', k'', etc.; $k, 'k$, etc. will be positive:

$$\alpha^m = \pm[k'', k''', k'''', \ldots, k^{m-1}]; \quad \beta^m = \pm[k'', k''', k'''', \ldots, k^m]$$

$$\gamma^m = \pm[k', k'', k''', \ldots, k^{m-1}]; \quad \delta^m = \pm[k', k'', k''', \ldots, k^m]$$

$$^m\alpha = \pm[k, 'k, ''k, \ldots, {}^{m-1}k]; \quad {}^m\beta = \pm[k, 'k, ''k, \ldots, {}^{m-2}k]$$

$$^m\gamma = \pm['k, ''k, \ldots, {}^{m-1}k]; \quad {}^m\delta = \pm['k, ''k, \ldots, {}^{m-2}k]$$

As for the *signs*, they must be determined by what we have just said above. By means of these formulae, whose proof we will omit because it is very easy, the calculation involved can be done very quickly.

▶ 190. LEMMA. *Let us designate any integers by the letters m, μ, m', n, v, n' in such a way however that no one of the last three $= 0$. I say that if μ/v lies between the limits m/n and m'/n' exclusively and if $mn' - nm' = \mp 1$, then the denominator v will be greater than n and n'.*

Demonstration. Manifestly $\mu nn'$ will lie between vmn' and vnm', and thus the difference between this number and either limit will be less than the difference between one limit and the other; i.e. we have $vmn' - vnm' > \mu nn' - vmn'$ and $> \mu nn' - vnm'$, or $v > n'(\mu n - vm)$ and $> n(\mu n' - vn')$. Thus it follows that since $\mu n - vm$ certainly does not $= 0$ (for otherwise we would have $\mu/v = m/n$ contrary to the hypothesis) nor does $\mu n' - vm' = 0$ (for a similar reason), but each will at least $= 1$, therefore $v > n'$ and $> n$. Q.E.D.

It is therefore clear that v cannot $= 1$; i.e. if $mn' - nm' = \pm 1$ no integer can lie between the fractions m/n, m'/n'. Nor can zero lie between them, i.e. the fractions cannot have opposite signs.

▶ 191. THEOREM. *If the reduced form $(a, b, -a')$ with determinant D is transformed by the substitution $\alpha, \beta, \gamma, \delta$ into the reduced form $(A, B, -A')$ with the same determinant: first, $\pm\sqrt{(D-b)}/a$ will lie between α/γ and β/δ (as long as neither γ nor $\delta = 0$, i.e. if each limit is finite). The upper sign is to be used when neither of these limits has a sign opposite to that of a (or more clearly, when either both have the same sign or one has the same sign, the other $= 0$). The lower sign is to be used when neither has the same sign as a; second, $\pm(\sqrt{D}+b)/a'$ will lie between γ/α and δ/β (as long as* *neither α nor β = 0). The upper sign is to be used when neither limit has a sign opposite to the sign of a' (or a), the lower sign when neither has the same sign as a'.*[P]

Demonstration. We have the equations

$$a\alpha^2 + 2b\alpha\gamma - a'\gamma^2 = A \qquad [1]$$

$$a\beta^2 + 2b\beta\delta - a'\delta^2 = -A' \qquad [2]$$

From this we deduce

$$\frac{\alpha}{\gamma} = \frac{\pm\sqrt{(D + aA/\gamma^2)} - b}{a} \qquad [3]$$

$$\frac{\beta}{\delta} = \frac{\pm\sqrt{(D - aA'/\delta^2)} - b}{a} \qquad [4]$$

$$\frac{\gamma}{\alpha} = \frac{\pm\sqrt{(D - a'A/\alpha^2)} + b}{a'} \qquad [5]$$

$$\frac{\delta}{\beta} = \frac{\pm\sqrt{(D + a'A'/\beta^2)} + b}{a'} \qquad [6]$$

Equations [3], [4], [5], [6] must be discarded if γ, δ, α, β respectively = 0. But there is still a doubt about which *signs* to give to the radical quantities. We will decide this in the following manner.

It is immediately clear that in [3] and [4] the upper sign is to be used when neither α/γ nor β/δ has a sign opposite to that of a; because if the lower sign were used, $a\alpha/\gamma$ and $a\beta/\delta$ would be negative quantities. Now since A and A' have the same sign, $\sqrt{D}$ falls between $\sqrt{(D + aA/\gamma^2)}$ and $\sqrt{(D - aA'/\delta^2)}$ and so in this case $(\sqrt{D} - b)/a$ between α/γ and β/δ. Thus the first part of the theorem must be demonstrated for the former case.

In the same way we see that in [5] and [6] the lower sign must be taken when neither γ/α nor δ/β has the same sign as a' or a, because if we took the upper sign, $a'\gamma/\alpha$, $a'\delta/\beta$ would necessarily be positive quantities. Then it would happen in this case that $-(\sqrt{D} + b)/a'$ would lie between γ/α and δ/β. So the second part

[P] Manifestly no other cases can be involved here since, according to the preceding article, $\alpha\delta - \beta\gamma = \pm 1$ and thus the pair of limits cannot have opposite signs nor can they both = 0.

of the theorem has to be shown for the latter case. Now if it were equally easy to show that in [3] and [4] the lower signs should be taken when neither of the quantities α/γ, β/δ has the same sign as a, and the upper sign in [5] and [6] when neither γ/α nor δ/β has opposite sign, then it would follow for the former case that $-(\sqrt{D} - b)/a$ lies between α/γ and β/δ and for the latter case that $(\sqrt{D} + b)/a'$ lies between γ/α and δ/β; that is, the first part of the theorem has to be demonstrated for the latter case and the second part for the former case. But even though this is not diffi-cult, it cannot be done without certain ambiguities and we prefer the following method.

When none of the numbers $\alpha, \beta, \gamma, \delta = 0$, then α/γ and β/δ will have the same signs as γ/α, δ/β. When therefore neither of these quantities has the same sign as a' or a, and thus $-(\sqrt{D} + b)/a'$ lies between γ/α and δ/β, neither of the quantities $\alpha/\gamma, \beta/\delta$ will have the same sign as a and $a'/-(\sqrt{D} + b) = -(\sqrt{D} - b)/a$ (since $aa' = D - b^2$) will lie between α/γ and β/δ. Therefore for the case when neither α nor $\beta = 0$, the first part of the theorem covers the second case (for the condition that neither γ nor $\delta = 0$ has already been taken care of in the theorem itself). In a similar way when no one of the numbers $\alpha, \beta, \gamma, \delta = 0$ and neither α/γ nor β/δ has a sign opposite that of a or a', and so $(\sqrt{D} - b)/a'$ lies between α/γ and β/δ, neither γ/α nor δ/β will have a sign opposite that of a', and $a/(\sqrt{D} - b) = (\sqrt{D} + b)/a'$ will lie between γ/α and δ/β. Therefore, where neither γ nor $\delta = 0$, the second part of the theorem demonstrates the second case.

It thus remains to show that the first part of the theorem also applies to the second case if either of the numbers $\alpha, \beta = 0$, and that the second part applies to the first case if either γ or $\delta = 0$. But *all these cases are impossible.* For suppose for the *first* part of the theorem that neither γ nor $\delta = 0$; that α/γ, β/δ do not have the same sign as a and that (1) $\alpha = 0$. Then from the equation $\alpha\delta - \beta\gamma = \pm 1$ we have $\beta = \pm 1$, $\gamma = \pm 1$. And from [1] $A = -a'$, so A and a' and therefore also a and A' will have opposite signs and $\sqrt{(D - aA'/\delta^2)} > \sqrt{D} > b$. From this it is clear that in [4] the lower sign is necessarily taken because, if we took the upper sign, β/δ would manifestly have the same sign as a. We have therefore $\beta/\delta > -(\sqrt{D} - b)/a > 1$ (since $a < \sqrt{D} + b$ from the definition of a reduced form). Q.E.A., since $\beta = \pm 1$, and δ does not $= 0$.

(2) Let $\beta = 0$. Then from the equation $\alpha\delta - \beta\gamma = \pm 1$ we have $\alpha = \pm 1$, $\delta = \pm 1$. From [2] $- A' = -a'$, so a' and a and A will have the same sign and $\sqrt{(D + aA/\alpha^2)} > \sqrt{D} > b$. Thus clearly in [3] we have to take the lower sign because if we took the upper sign, α/γ would have the same sign as a. We get therefore $\alpha/\gamma > -(\sqrt{D} - b)/a > 1$. Q.E.A., for the same reason as before.

For the *second* part if we suppose that neither α nor $\beta = 0$; that γ/α, δ/β do not have signs opposite that of a' and that (1) $\gamma = 0$: from the equation $\alpha\delta - \beta\gamma = \pm 1$ we get $\alpha = \pm 1$, $\delta = \pm 1$. Thus by [1] $A = a$ and a' and A' will have the same sign so $\sqrt{(D + a'A'/\beta^2)} > \sqrt{D} > b$. Therefore in [6] the upper sign has to be taken because if we took the lower sign, δ/β would have a sign opposite that of a'. We get therefore $\delta/\beta > (\sqrt{D} + b)/a' > 1$. Q.E.A., because $\delta = \pm 1$ and β does not $= 0$. Finally (2), if we had $\delta = 0$, from $\alpha\delta - \beta\gamma = \pm 1$ we have $\beta = \pm 1$, $\gamma = \pm 1$ and so from [2] $- A' = a$. Therefore $\sqrt{(D - a'A/\alpha^2)} > \sqrt{D} > b$, and the upper sign must be taken in [5]; $\gamma/\alpha > (\sqrt{D} + b)/a' > 1$. Q.E.A. And the theorem is demonstrated in all its generality.

Since the difference between α/γ and β/δ is $= 1/\gamma\delta$, the difference between $\pm(\sqrt{D} - b)/a$ and α/γ or β/δ will be $< 1/\gamma\delta$; however, between $\pm(\sqrt{D} - b)/a$ and α/γ or between that quantity and β/δ there can be no fraction whose denominator is not greater than γ or δ (preceding lemma). In the same way the difference between the quantity $\pm(\sqrt{D} + b)/a$ and the fraction γ/α or δ/β will be less than $1/\alpha\beta$, and no fraction can lie between that quantity and either of these fractions unless the denominator is less than α and β.

▶ 192. By applying the preceding theorem to the algorithm of article 188 it follows that the quantity $(\sqrt{D} - b)/a$, which we shall designate by L, will lie between α'/γ' and β'/δ'; between α''/γ'' and $\beta''\delta''$; between α'''/γ''' and β'''/δ''', etc. (for it is easy to see from article 189.3, toward the end, that none of these limits has a sign opposite to that of a, so a positive sign must be given to the radical quantity $\sqrt{D}$); or between α'/γ' and α''/γ''; between α''/γ'' and α'''/γ''', etc. Therefore all the fractions α'/γ', α'''/γ''', α''''/γ'''', etc. will lie on one side of L, and all the fractions α''/γ'', α''''/γ'''', $\alpha''''''/\gamma''''''$, etc. on the other side. But since $\gamma' < \gamma'''$, α'/γ' will lie *outside* α'''/γ''' and L, and for a similar reason α''/γ'' will lie outside L and

α''''/γ'''' ; α'''/γ''' outside L and α''''/γ''''; etc. Thus these quantities lie in the following order:

$$\frac{\alpha'}{\gamma'}, \frac{\alpha'''}{\gamma'''}, \frac{\alpha''''}{\gamma''''}, \ldots L \ldots, \frac{\alpha'''''}{\gamma'''''}, \frac{\alpha''''}{\gamma''''}, \frac{\alpha''}{\gamma''}$$

The difference between α'/γ' and L will be less than the difference between α'/γ' and α''/γ''; i.e. $< 1/\gamma'\gamma''$ and for a similar reason the difference between α''/γ'' and L will be $< 1/\gamma''\gamma'''$ etc. Therefore the fractions α'/γ', α''/γ'', α'''/γ''', etc. will continuously approach the limit L closer and closer and, since γ', γ'', γ''' keep increasing indefinitely, the difference between the fractions and the limit can be made less than any given quantity.

From article 189 none of the quantities γ/α, $'\gamma/'\alpha$, $''\gamma/''\alpha$, etc. will have the same sign as a; thus, by the reasoning above, these numbers and the quantity $-(\sqrt{D} + b)/a'$, which we will designate as L', will lie in the following order:

$$\frac{\gamma}{\alpha}, \frac{''\gamma}{''\alpha}, \frac{''''\gamma}{''''\alpha}, \ldots L' \ldots, \frac{'''''\gamma}{'''''\alpha}, \frac{'''\gamma}{'''\alpha}, \frac{'\gamma}{'\alpha}$$

The difference between γ/α and L' will be less than $1/'\alpha\alpha$, the difference between $'\gamma/'\alpha$ and L' less than $1/''\alpha'\alpha$, etc. Therefore the fractions γ/α, $'\gamma/'\alpha$, etc. will continuously approach L' closer and closer and the difference can be made less than any given quantity.

In the example of article 188 we have $L = (\sqrt{79} - 8)/3 = 0.2960648$, and the approaching fractions are $0/1, 1/3, 2/7, 3/10, 5/17, 8/27, 45/152, 143/483$, etc. And $143/483 = 0.2960662$. In the same example $L' = -(\sqrt{79} + 8)/5 = -0.1776388$, and the approximating fractions are $0/1, -1/5, -1/6, -2/11, -3/17, -8/45, -27/152, -143/805$, etc. And $143/805 = 0.1776397$.

▶ 193. THEOREM. *If the reduced forms f, F are properly equivalent, each of them will be contained in the period of the other.*

Let $f = (a, b, -a')$, $F = (A, B, -A')$ and the determinant of these forms D, and let the first be transformed into the second by the proper substitution $\mathfrak{A}, \mathfrak{B}, \mathfrak{C}, \mathfrak{D}$. If we look for the period of the form f and calculate the two-way infinite series of the reduced forms and the transformations of the form f into these as we did in article 188, then *either* $+\mathfrak{A}$ will be equal to some term of the series $\ldots, ''\alpha, '\alpha, \alpha, \alpha', \alpha'', \ldots$ (and if we let this $= \alpha^m$, $+\mathfrak{B}$ will

$= \beta^m, +\mathfrak{C} = \gamma^m, +\mathfrak{D} = \delta^m$); *or* $-\mathfrak{A}$ will be equal to some term α^m and $-\mathfrak{B}, -\mathfrak{C}, -\mathfrak{D}$, respectively, will $= \beta^m, \gamma^m, \delta^m$ (where m can also designate a negative index). In either case F will manifestly be identical with f^m.

Demonstration. I. We will have the four equations

$$a\mathfrak{A}^2 + 2b\mathfrak{A}\mathfrak{C} - a'\mathfrak{C}^2 = \quad A \qquad [1]$$

$$a\mathfrak{A}\mathfrak{B} + b(\mathfrak{A}\mathfrak{D} + \mathfrak{B}\mathfrak{C}) - a'\mathfrak{C}\mathfrak{D} = \quad B \qquad [2]$$

$$a\mathfrak{B}^2 + 2b\mathfrak{B}\mathfrak{D} + a'\mathfrak{D}^2 = -A' \qquad [3]$$

$$\mathfrak{A}\mathfrak{D} - \mathfrak{B}\mathfrak{C} = \quad 1 \qquad [4]$$

Let us *first* consider the case where one of the numbers $\mathfrak{A}, \mathfrak{B}, \mathfrak{C}, \mathfrak{D} = 0$.

1) If $\mathfrak{A} = 0$ we get from [4] $\mathfrak{B}\mathfrak{C} = -1$, and so $\mathfrak{B} = \pm 1$, $\mathfrak{C} = \mp 1$. Then from [1], $-a' = A$; from [2], $-b \pm a'\mathfrak{D} = B$ or $B \equiv -b$ (mod. a' or A); therefore it follows that the form $(A, B, -A')$ is neighbor by the last part to the form $(a, b, -a')$. Since the first of these is reduced, it is necessarily identical with f'. Therefore $B = b'$, and thus from [2] $b + b' = -a'\mathfrak{C}\mathfrak{D} = \pm a'\mathfrak{D}$; from this, since $(b + b')/-a' = h'$, we have $\mathfrak{D} = \mp h'$. And finally it follows that $\mp \mathfrak{A}, \mp \mathfrak{B}, \mp \mathfrak{C}, \mp \mathfrak{D}$ respectively $= 0, -1, +1, h'$ or $= \alpha', \beta', \gamma', \delta'$.

2) If $\mathfrak{B} = 0$ we get from [4] $\mathfrak{A} = \pm 1$, $\mathfrak{D} = \pm 1$; from [3] $a' = A'$; from [2] $b \mp a'\mathfrak{C} = B$, or $b \equiv B$ (mod. a'). But since both f and F are reduced forms, both b and B will lie between $\sqrt{D}$ and $\sqrt{D} \mp a'$ (according as a' is positive or negative, by art. 184.5). So necessarily $b = B$ and $\mathfrak{C} = 0$. Thus the forms f, F are identical and $\pm \mathfrak{A}, \pm \mathfrak{B}, \pm \mathfrak{C}, \pm \mathfrak{D} = 1, 0, 0, 1 = \alpha, \beta, \gamma, \delta$ respectively.

3) If $\mathfrak{C} = 0$ we get from [4] $\mathfrak{A} = \pm 1, \mathfrak{D} = \pm 1$; from [1] $a = A$; from [2] $\pm a\mathfrak{B} + b = B$ or $b \equiv B$ (mod. a). Since both b and B lie between $\sqrt{D}$ and $\sqrt{D} \mp a$, we have necessarily that $B = b$ and $\mathfrak{B} = 0$. So this case does not differ from the preceding one.

4) If $\mathfrak{D} = 0$ we get from [4] $\mathfrak{B} = \pm 1$, $\mathfrak{C} = \mp 1$; from [3] $a = -A'$; from [2] $\pm a\mathfrak{A} - b = B$ or $B \equiv -b$ (mod a). Thus the form F will be neighbor by the first part to the form f and so identical with the form $'f$. Therefore since $('b + b)/a = h$, and $B = 'b$ we have $\pm \mathfrak{A} = h$. And finally, $\pm \mathfrak{A}, \pm \mathfrak{B}, \pm \mathfrak{C}, \pm \mathfrak{D}$ respectively $= h, 1, -1, 0 = '\alpha, '\beta, '\gamma, '\delta$.

There remains now the case where none of the numbers $\mathfrak{A}$, $\mathfrak{B}$, $\mathfrak{C}$, $\mathfrak{D} = 0$. By the lemma of article 190 the quantities $\mathfrak{A}/\mathfrak{C}$, $\mathfrak{B}/\mathfrak{D}$, $\mathfrak{C}/\mathfrak{A}$, $\mathfrak{D}/\mathfrak{B}$ will have the same sign and there will result two cases according as this sign is the same or opposite to the sign of a, a'.

II. If $\mathfrak{A}/\mathfrak{C}$, $\mathfrak{B}/\mathfrak{D}$ have the same sign as a the quantity $(\sqrt{D} - b)/a$ (which we will designate by L) will lie between these fractions (art. 191). We will show now that $\mathfrak{A}/\mathfrak{C}$ is equal to one of the fractions α''/γ'', α'''/γ''', α''''/γ'''', etc. and $\mathfrak{B}/\mathfrak{D}$ to the next following; that is if $\mathfrak{A}/\mathfrak{C}$ were $= \alpha^m/\gamma^m$ then $\mathfrak{B}/\mathfrak{D}$ would $= \alpha^{m+1}/\gamma^{m+1}$. In the preceding article we showed that the quantities α'/γ', α''/γ'', α'''/γ''', etc. [for brevity's sake we will denote them by (1), (2), (3), etc.] and L follow this order (I): (1), (3), (5), $\ldots L \ldots$, (6), (4), (2); the first of these quantities $= 0$ (since $\alpha' = 0$); the rest have the same sign as L or a. But since by hypothesis $\mathfrak{A}/\mathfrak{C}$, $\mathfrak{B}/\mathfrak{D}$ (for which we will write $\mathfrak{M}$, $\mathfrak{N}$) have the same sign, it is clear that these quantities lie to the right of (1) (or if you prefer on the same side as L) and, indeed, since L lies between them, one to the right of L, the other to the left. It is easy to see that $\mathfrak{M}$ cannot lie to the right of (2) for otherwise $\mathfrak{N}$ would lie between (1) and L, and it would follow *first* that (2) lies between $\mathfrak{M}$ and $\mathfrak{N}$, and the denominator of the fraction (2) would be greater than the denominator of the fraction $\mathfrak{N}$ (art. 190); *second* that $\mathfrak{N}$ would lie between (1) and (2) and the denominator of the fraction $\mathfrak{N}$ would be greater than the denominator of the fraction (2). Q.E.A.

Suppose $\mathfrak{M}$ is equal to none of the fractions (2), (3), (4), etc. Let us see what results. Manifestly if the fraction $\mathfrak{M}$ lies to the left of L, it will necessarily lie between (1) and (3), or between (3) and (5), or between (5) and (7), etc. (because L is irrational, and thus certainly not equal to $\mathfrak{M}$, the fractions (1), (3), (5), etc. can approach closer and closer to any given quantity unequal to L). Now if $\mathfrak{M}$ lies to the right of L, it will necessarily lie between (2) and (4), or between (4) and (6), or between (6) and (8), etc. Suppose therefore that $\mathfrak{M}$ lies between (m) and $(m + 2)$; it is obvious that the quantities $\mathfrak{M}, (m), (m + 1), (m + 2), L$ lie in the following order (II): (m), $(\mathfrak{M})$, $(m + 2)$, L, $(m + 1)$.[q]

Then necessarily $\mathfrak{N} = (m + 1)$. For $\mathfrak{N}$ will lie to the right of L;

---

[q] It makes no difference whether the order in (II) is the same as in (I) or opposite to it, i.e. whether in (I) (m) lies to the left of L or to the right of it.

but if it also lay to the right of $(m + 1)$, $(m + 1)$ would lie between $\mathfrak{M}$ and $\mathfrak{N}$ making $\gamma^{m+1} > \mathfrak{C}$, and $\mathfrak{M}$ would lie between (m) and $(m + 1)$ making $\mathfrak{C} > \gamma^{m+1}$ (art. 190), Q.E.A. But if $\mathfrak{N}$ should lie to the left of $(m + 1)$, that is between $(m + 2)$ and $(m + 1)$ we would have $\mathfrak{D} > \gamma^{m+2}$, and because $(m + 2)$ is between $\mathfrak{M}$ and $\mathfrak{N}$, we would have $\gamma^{m+2} > \mathfrak{D}$. Q.E.A. We have therefore $\mathfrak{N} = (m + 1)$; that is $\mathfrak{B}/\mathfrak{D} = \alpha^{m+1}/\gamma^{m+1} = \beta^m/\delta^m$.

Since $\mathfrak{A}\mathfrak{D} - \mathfrak{B}\mathfrak{C} = 1$, $\mathfrak{B}$ will be relatively prime to $\mathfrak{D}$, and for a similar reason β^m will be relatively prime to δ^m. Thus the equation $\mathfrak{B}/\mathfrak{D} = \beta^m/\delta^m$ will be inconsistent unless either $\mathfrak{B} = \beta^m$, $\mathfrak{D} = \delta^m$ or $\mathfrak{B} = -\beta^m$, $\mathfrak{D} = -\delta^m$. Now since the form f is transformed by the proper substitution $\alpha^m, \beta^m, \gamma^m, \delta^m$ into the form f^m, which is $(\pm a^m, b^m, \mp a^{m+1})$, we will have the equations

$$a\alpha^{2m} + 2b\alpha^m\gamma^m - a'\gamma^{2m} = \pm a^m \qquad [5]$$

$$a\alpha^m\beta^m + b(\alpha^m\delta^m + \beta^m\gamma^m) - a'\gamma^m\delta^m = b^m \qquad [6]$$

$$a\beta^{2m} + 2b\beta^m\delta^m - a'\delta^{2m} = \mp a^{m+1} \qquad [7]$$

$$\alpha^m\delta^m - \beta^m\gamma^m = 1 \qquad [8]$$

We get (from equations [7] and [3]): $\mp a^{m+1} = -A'$. Further, by multiplying equation [2] by $\alpha^m\delta^m - \beta^m\gamma^m$, equation [6] by $\mathfrak{A}\mathfrak{D} - \mathfrak{B}\mathfrak{C}$ and subtracting we have by easy computation:

$$B - b^m = [\mathfrak{C}\alpha^m - \mathfrak{A}\gamma^m][a\mathfrak{B}\beta^m + b(\mathfrak{D}\beta^m + \mathfrak{B}\delta^m) - a'\mathfrak{D}\delta^m]$$

$$+ [\mathfrak{B}\delta^m - \mathfrak{D}\beta^m][a\mathfrak{A}\alpha^m + b(\mathfrak{C}\alpha^m + \mathfrak{A}\gamma^m) - a'\mathfrak{C}\gamma^m] \qquad [9]$$

or since either $\beta^m = \mathfrak{B}$, $\delta^m = \mathfrak{D}$ or $\beta^m = -\mathfrak{B}$, $\delta^m = -\mathfrak{D}$,

$$B - b^m = \pm(\mathfrak{C}\alpha^m - \mathfrak{A}\gamma^m)(a\mathfrak{B}^2 + 2b\mathfrak{B}\mathfrak{D} - a'\mathfrak{D}^2)$$

$$= \mp(\mathfrak{C}\alpha^m - \mathfrak{A}\gamma^m)A'$$

Thus $B \equiv b^m$ (mod. A'); and because both B and b^m lie between $\sqrt{D}$ and $\sqrt{D \mp A'}$ we must have $B = b^m$, and so $\mathfrak{C}\alpha^m - \mathfrak{A}\gamma^m = 0$ or $\mathfrak{A}/\mathfrak{C} = \alpha^m/\gamma^m$; i.e. $\mathfrak{M} = (m)$.

In this way, therefore, starting with the supposition that $\mathfrak{M}$ is not equal to any of the quantities (2), (3), (4), etc., we deduced that it is actually equal to one of them. But if we suppose from the beginning that $\mathfrak{M} = (m)$ we will clearly have either $\mathfrak{A} = \alpha^m$, $\mathfrak{C} = \gamma^m$ or $-\mathfrak{A} = \alpha^m$, $-\mathfrak{C} = \gamma^m$. In either case we get from [1] and

[5] $A = \pm a^m$ and from [9] $B - b^m = \pm(\mathfrak{B}\delta^m - \mathfrak{D}\beta^m)A$ or $B \equiv b^m$ (mod. A). From this we conclude in the same way as above that $B = b^m$, and thus $\mathfrak{B}\delta^m = \mathfrak{D}\beta^m$; therefore since $\mathfrak{B}$ and $\mathfrak{D}$ are relatively prime and β^m and δ^m are relatively prime, we have either $\mathfrak{B} = \beta^m$, $\mathfrak{D} = \delta^m$ or $-\mathfrak{B} = \beta^m$, $-\mathfrak{D} = \delta^m$, and from [7] $-A' = \mp a^{m+1}$. Thus the forms F, f^m are identical. With the help of the equation $\mathfrak{A}\mathfrak{D} - \mathfrak{B}\mathfrak{C} = \alpha^m\delta^m - \beta^m\gamma^m$ it is no trouble to show that when $+\mathfrak{A} = \alpha^m$, $+\mathfrak{C} = \gamma^m$ we should take $+\mathfrak{B} = \beta^m$, $+\mathfrak{D} = \delta^m$; and on the other hand when $-\mathfrak{A} = \alpha^m$, $-\mathfrak{C} = \gamma^m$ we should take $-\mathfrak{B} = \beta^m$, $-\mathfrak{D} = \delta^m$.   Q.E.D.

III. If the sign of the quantities $\mathfrak{A}/\mathfrak{C}$, $\mathfrak{B}/\mathfrak{D}$ is opposite to that of a, the demonstration is so similar to the preceding one that it is sufficient to add only the principal points. The quantity $-(\sqrt{D} + b)/a'$ will lie between $\mathfrak{C}/\mathfrak{A}$ and $\mathfrak{D}/\mathfrak{B}$. The fraction $\mathfrak{D}/\mathfrak{B}$ will be equal to one of the fractions

$$\frac{'\delta}{'\beta}, \frac{''\delta}{''\beta}, \frac{'''\delta}{'''\beta}, \text{ etc.} \tag{I}$$

and if we let this be $= {}^m\delta/{}^m\beta$

$$\frac{\mathfrak{C}}{\mathfrak{A}} \text{ will be } = \frac{{}^m\gamma}{{}^m\alpha} \tag{II}$$

We demonstrate (I) as follows: If we suppose that $\mathfrak{D}/\mathfrak{B}$ is equal to none of these fractions, it will lie between two of them ${}^m\delta/{}^m\beta$ and ${}^{m+2}\delta/{}^{m+2}\beta$. And in the same way as above we show that

$$\frac{\mathfrak{C}}{\mathfrak{A}} = \frac{{}^{m+1}\delta}{{}^{m+1}\beta} = \frac{{}^m\gamma}{{}^m\alpha}$$

and either $\mathfrak{A} = {}^m\alpha$, $\mathfrak{C} = {}^m\gamma$ or $-\mathfrak{A} = {}^m\alpha$, $-\mathfrak{C} = {}^m\gamma$. But since by the proper substitution ${}^m\alpha, {}^m\beta, {}^m\gamma, {}^m\delta$, f is transformed into the form

$${}^mf = (\pm {}^ma, {}^mb, \mp {}^{m-1}a)$$

we can derive three equations. From these and the equations [1], [2], [3], [4] and the equation ${}^m\alpha {}^m\delta - {}^m\beta {}^m\gamma = 1$ we deduce in the same way as above that the first term A of the form F is equal to the first term of the form ${}^mf$ and that the middle term of the former is congruent relative to the modulus A to the middle term of the latter. It follows from this that because both forms are reduced, thus requiring the middle term of each to lie between $\sqrt{D}$ and $\sqrt{D \mp A}$, these middle terms will be equal. From this we conclude that $^m\delta/^m\beta = \mathfrak{D}/\mathfrak{B}$. The truth of assertion (I) is derived by supposing that this is false.

Suppose that $^m\delta/^m\beta = \mathfrak{D}/\mathfrak{B}$ in exactly the same way. By using the same equations we can also show that $^m\gamma/^m\alpha = \mathfrak{C}/\mathfrak{A}$, which was the second assertion (II). Now with the help of the equations $\mathfrak{A}\mathfrak{D} - \mathfrak{B}\mathfrak{C} = 1$, $^m\alpha^m\delta - ^m\beta^m\gamma = 1$ we deduce that either

$$\mathfrak{A} = {}^m\alpha, \qquad \mathfrak{B} = {}^m\beta, \qquad \mathfrak{C} = {}^m\gamma, \qquad \mathfrak{D} = {}^m\delta$$

or

$$-\mathfrak{A} = {}^m\alpha, \qquad -\mathfrak{B} = {}^m\beta, \qquad -\mathfrak{C} = {}^m\gamma, \qquad -\mathfrak{D} = {}^m\delta$$

and the forms F, ^mf are identical.   Q.E.D.

▶ 194. Since the forms which we have called associates above (art. 187. 6) are always improperly equivalent (art. 159), it is clear that if the reduced forms F, f are improperly equivalent, and if the form G is an associate of the form F, then the forms f, G will be properly equivalent and the form G will be contained in the period of the form f. And if the forms F, f are properly and improperly equivalent, it is clear that both F and G should be found in the period of the form f. Therefore this period will be an associate of itself and will contain two ambiguous forms (art. 187. 7). Thus we admirably confirm the theorem of article 165, which has already guaranteed that we can find an ambiguous form equivalent to the forms F, f.

▶ 195. PROBLEM. *Given any two forms* Φ, ϕ *with the same determinant: to determine whether or not they are equivalent.*

Solution. We will look for two reduced forms F, f properly equivalent respectively to the given forms Φ, ϕ (art. 183). Now according as these will be only properly equivalent or only improperly equivalent or both or neither, then the given forms will be properly equivalent only, or only improperly, or both or neither. Calculate the period of one of the reduced forms, e.g. the period of the form f. If the form F occurs in this period and the form of its associate does not, then manifestly the *first* case holds; on the other hand, if the associate appears here but F does not, the *second* case holds; if both appear, the *third* case holds; if neither, the *fourth* case holds.

Example. Let the given forms be (129, 92, 65), (42, 59, 81) with determinant 79. For these we have the properly equivalent reduced forms (10, 7, −3), (5, 8, −3). The period of the first of these is: (10, 7, −3), (−3, 8, 5), (5, 7, −6), (−6, 5, 9), (9, 4, −7), (−7, 3, 10). Since the form (5, 8, −3) does not appear here but only its associate (−3, 8, 5), we conclude that the given forms are only improperly equivalent.

If all the reduced forms of a given determinant are distributed in the same way as above (art. 187. 5) into periods P, Q, R, etc., and if we select a form from each period at random, F from P; G from Q; H from R, etc., no two of the forms F, G, H, etc. can be properly equivalent. And any other form of the same determinant will be properly equivalent to one and *only one* of these. Thus, *all the forms of this determinant can be distributed into as many classes as it has periods*, that is by putting the forms which are properly equivalent to the form F into the first class, those which are properly equivalent to G in the second class, etc. In this way all the forms contained in the same class will be properly equivalent, and forms contained in different classes cannot be properly equivalent. But we will not dwell here on a subject that will be treated more in detail later.

▶ 196. PROBLEM. *Given two properly equivalent forms* Φ, ϕ: *to find a proper transformation of one into the other.*

Solution. By the method of article 183 we can find two series of forms

$$\Phi, \Phi', \Phi'', \ldots, \Phi^n \quad \text{and} \quad \phi, \phi', \phi'', \ldots, \phi^v$$

such that every following form is properly equivalent to its predecessor and the last Φ^n, ϕ^v are reduced forms. And since we supposed that Φ, ϕ are properly equivalent, Φ^n is necessarily contained in the period of the form ϕ^v. Let $\phi^v = f$ and let its period up to the form Φ^n be

$$f, f', f'', \ldots, f^{m-1}, \Phi^n$$

so that in this period the index of the form Φ^n is m; and let us designate the forms which are opposite to the associates of the forms

$$\Phi, \Phi', \Phi'', \ldots, \Phi^n \quad \text{by} \quad \Psi, \Psi', \Psi'', \ldots, \Psi^n \text{ respectively}^r$$

Then in the series

$$\phi, \phi', \phi'', \ldots, f, f', f'', \ldots, f^{m-1}, \Psi^{n-1}, \Psi^{n-2}, \ldots, \Psi, \Phi$$

each form will be neighbor to the preceding one by the last part, and by article 177 we can find a proper transformation of the first ϕ into the last Φ. This is easily seen for the other terms of the series; for the terms f^{m-1}, Ψ^{n-1} we prove it as follows: let

$$f^{m-1} = (g, h, i); f^m \text{ or } \Phi^m = (g', h', i'); \Phi^{n-1} = (g'', h'', i'')$$

The form (g', h', i') will be neighbor by the last part to each of the forms (g, h, i) and (g'', h'', i''); therefore $i = g' = i''$ and $-h \equiv h' \equiv -h''$ (mod. i or g' or i''). From this it is manifest that the form $(i'', -h'', g'')$, i.e. the form Ψ^{n-1}, is neighbor by the last part to the form (g, h, i), i.e. to the form f^{m-1}.

If the forms Φ, ϕ are improperly equivalent, the form ϕ will be properly equivalent to the form that is opposite to Φ. We can therefore find the proper transformation of the form ϕ into the form opposite to Φ; if we suppose that this can be done by the transformation $\alpha, \beta, \gamma, \delta$, it is easy to see that ϕ will be transformed improperly into Φ by the substitution $\alpha, -\beta, \gamma, -\delta$.

And if the forms Φ, ϕ are properly and improperly equivalent, we can find *two* transformations, one proper and the other improper.

Example. We saw in the preceding article that (129, 92, 65) and (42, 59, 81) are improperly equivalent. Let us try to find the improper transformation of the first into the second. We must first look at the proper transformation of the form (129, 92, 65) into the form (42, −59, 81). For this purpose we calculate the series of forms (129, 92, 65), (65, −27, 10), (10, 7, −3), (−3, 8, 5), (5, 22, 81), (81, 59, 42), (42, −59, 81). From this we deduce the proper transformation −47, 56, 73, −87, by means of which (129, 92, 65) is transformed into (42, −59, 81); therefore it will transform into (42, 59, 81) by the improper transformation −47, −56, 73, 87.

---

^{r} ψ is derived from Φ by interchanging the first and last terms and by assigning the opposite sign to the middle term. The same is true for the other members of the series.

▶ 197. If we have one transformation of a form $(a, b, c) \ldots \phi$ into an equivalent form Φ, from this we can deduce *all* similar transformations of the form ϕ into Φ, if only we can determine all the solutions of the indeterminate equation $t^2 - Du^2 = m^2$. D indicates the determinant of the forms Φ, ϕ; m is the greatest common divisor of the numbers $a, 2b, c$ (art. 162). We solved this problem above for a negative value of D. Now we will consider a positive value. But since obviously any value of t or u that satisfies the equation will also satisfy it if we change the sign, it will be sufficient to assign *positive* values to t, u, and any solution by positive values will furnish four solutions actually. To do this we will first find the *smallest* values of t, u (except the obvious ones where $t = m$, $u = 0$) and then from these we will see how to derive the others.

▶ 198. PROBLEM. *Given a form (M, N, P) whose determinant is D, and m the greatest common divisor of the numbers $M, 2N, P$: to find the smallest numbers t, u satisfying the indeterminate equation $t^2 - Du^2 = m^2$.*

Solution. Let us take at will the reduced form $(a, b, -a') \ldots f$ with determinant D, in which the greatest common divisor of the numbers $a, 2b, a'$ is m. That there is one is clear from the fact that a reduced form equivalent to the form (M, N, P) can be found and that by article 161 it is endowed with this property. But for our present purpose any reduced form that fulfills this condition can be used. Calculate the period of the form f, which we will suppose consists of n forms. Retaining all the signs that we used in article 188, f^n will be $(+a^n, b^n, -a^{n+1})$ because n is even and f will be transformed into this form by the proper substitution $\alpha^n, \beta^n, \gamma^n, \delta^n$. But since f and f^n are identical, f will also be transformed into f^n by the proper substitution $1, 0, 0, 1$. From these two similar transformations of the form f into f^n, by article 162 we can deduce an *integral* solution of the equation $t^2 - Du^2 = m^2$, namely $t = (\alpha^n + \delta^n)m/2$ (equation 18, art. 162), $u = \gamma^n m/a$ (equation 19).[5] Take these values positively if they are not already so and designate them by T, U. These values T, U will be the smallest values of t, u except for $t = m$, $u = 0$ (they must be different from these, since clearly γ^n cannot $= 0$).

---

[5] The quantities which in article 162 were α, β, γ, δ; α', β', γ', δ'; A, B, C; a, b, c; e are here $1, 0, 0, 1$; $\alpha^n, \beta^n, \gamma^n, \delta^n$; $a, b, -a'$; $a, b, -a'$; 1.

Suppose there are still smaller values of t, u, namely t', u', which are both positive and u' does not $= 0$. Then by article 162 the form f will be transformed into a form which is identical with itself by the proper transformation $(t' - bu')/m$, $a'u'/m$, au'/m, $(t' + bu')/m$. Now from article 193. II it follows that *either* $(t' - bu')/m$ or $-(t' - bu')/m$ must be equal to one of the numbers $\alpha'', \alpha''', \alpha''''$, etc., for example to α^μ (for since $t't' = Du'u' + m^2 = b^2u'u' + aa'u'u' + m^2$ we have $t't' > b^2u'u'$ and thus $t' - bu'$ positive; the fraction $(t' - bu')/au'$ which corresponds to the fraction $\mathfrak{A}/\mathfrak{C}$ in article 193 will have the same sign as a or a'); and in the former case $a'u'/m$, au'/m, $(t' + bu')/m$ will $= \beta^\mu, \gamma^\mu, \delta^\mu$ respectively. In the latter case they will equal the same quantities but with changed signs. Since we have $u' < U$, i.e. $u' < \gamma^n m/a$ and > 0, we get $\gamma^\mu < \gamma^n$ and > 0; and since the series $\gamma, \gamma', \gamma''$, etc. is continuously increasing, μ will necessarily lie between 0 and n exclusively. And the corresponding form f^μ will be identical with the form f. Q.E.A. since all the forms f, f', f'', etc. up to f^{n-1} are supposed to be different. From this we conclude that the smallest values of t, u (except the values $m, 0$) will be T, U.

Example. If $D = 79$, $m = 1$ we can use the form $(3, 8, -5)$ for which $n = 6$, and $\alpha^n = -8$, $\gamma^n = -27$, $\delta^n = -152$ (art. 188). Thus $T = 80$, $U = 9$ which are the smallest values of the numbers t, u satisfying the equation $t^2 - 79u^2 = 1$.

▶ 199. In practice, more suitable formulae can be developed. We have $2b\gamma^n = -a(\alpha^n - \delta^n)$, which is easy to deduce from article 162 by multiplying equation [19] by $2b$, [20] by a, and changing the symbols used there into the ones we are using here. From this we get $\alpha^n + \delta^n = 2\delta^n - (2b\gamma^n)/a$ and thus

$$\pm T = m(\delta^n - \frac{b}{a}\gamma^n), \quad \pm U = \frac{\gamma^n m}{a}$$

By a similar method we obtain the following values

$$\pm T = m(\alpha^n + \frac{b}{a}\beta^n), \quad \pm U = \frac{\beta^n m}{a'}$$

These two sets of formulae are very convenient because $\gamma^n = \delta^{n-1}$, $\alpha^n = \beta^{n-1}$, so if we use the second set it is sufficient to compute the series $\beta', \beta'', \beta''', \ldots, \beta^n$; and if we use the first set the series $\delta', \delta'', \delta'''$, etc. will be sufficient. Furthermore from article 189. 3 we can easily deduce that, since n is even, α^n and $b\beta^n/a'$ have the same sign. This is also true of δ^n and $b\gamma^n/a$ so that in the first formula we can take for T the absolute difference, and in the second the absolute sum without having to pay attention to the sign. Using the symbols in article 189. 4 we get the following from the first formula

$$T = m[k', k'', k''', \ldots, k^n] - \frac{mb}{a}[k', k'', k''', \ldots, k^{n-1}],$$

$$U = \frac{m}{a}[k', k'', k''', \ldots, k^{n-1}]$$

and from the second formula

$$T = m[k'', k''', \ldots, k^{n-1}] + \frac{mb}{a'}[k'', k''', \ldots, k^n],$$

$$U = \frac{m}{a'}[k'', k''', \ldots, k^n]$$

where we can also write $m[k'', k''', \ldots, k^n, b/a']$ for the value of T.

Example. For $D = 61$, $m = 2$ we can use the form $(2, 7, -6)$. From this we find $n = 6$; $k', k'', k''', k'''', k''''', k''''''$ respectively $= 2, 2, 7, 2, 2, 7$.
Thus

$$T = 2[2, 2, 7, 2, 2, 7] - 7[2, 2, 7, 2, 2] = 2888 - 1365 = 1523$$

from the first formula; the same thing results from the second formula

$$T = 2[2, 7, 2, 2] + \frac{7}{3}[2, 7, 2, 2, 7]$$

and

$$U = [2, 2, 7, 2, 2] = \frac{1}{3}[2, 7, 2, 2, 7] = 195$$

There are many other devices for simplifying the calculation, but brevity does not permit us to speak of them in detail here.

▶ 200. In order to derive *all* the values of t, u from the smallest values we present the equation $T^2 - DU^2 = m^2$ in the following form

$$\left(\frac{T}{m} + \frac{U}{m}\sqrt{D}\right)\left(\frac{T}{m} - \frac{U}{m}\sqrt{D}\right) = 1$$

From this we have also

$$\left(\frac{T}{m} + \frac{U}{m}\sqrt{D}\right)^e \left(\frac{T}{m} - \frac{U}{m}\sqrt{D}\right)^e = 1 \qquad [1]$$

where e can be any number. Now for the brevity we will designate the values of the quantities

$$\frac{m}{2}\left(\frac{T}{m} + \frac{U}{m}\sqrt{D}\right)^e + \frac{m}{2}\left(\frac{T}{m} - \frac{U}{m}\sqrt{D}\right)^e,$$

$$\frac{m}{2\sqrt{D}}\left(\frac{T}{m} + \frac{U}{m}\sqrt{D}\right)^e - \frac{m}{2\sqrt{D}}\left(\frac{T}{m} - \frac{U}{m}\sqrt{D}\right)^e {}^\text{t}$$

in general by t^e, u^e respectively; i.e. for $e = 0$ they will be t^0, u^0 (these values are $m, 0$); for $e = 1$ they will be t', u' (these values are T, U); for $e = 2$ they will be t'', u''; for $t = 3$ they will be t''', u''', etc. And we will show that if for e we take all non-negative integers, i.e. 0 and all positive integers from 1 to ∞, these expressions will produce all the positive values of t, u: that is (I) all values of these expressions are truly values of t, u; (II) all these values are integers; (III) there are no positive values of t, u which are not contained in these formulae.

I. If we substitute for t^e, u^e their values and use equation [1] it is easy to find that

$$(t^e + u^e\sqrt{D})(t^e - u^e\sqrt{D}) = m^2; \quad \text{i.e.} \quad t^{2e} - Du^{2e} = m^2$$

II. In the same way it is easy to confirm that in general

$$t^{e+1} + t^{e-1} = \frac{2T}{m}t^e, \quad u^{e+1} + u^{e-1} = \frac{2T}{m}u^e$$

Thus it is manifest that the two series t^0, t', t'', t''', etc., u^0, u', u'', u''', etc. are recurrent and that the scale of relationship for each is $2T/m, -1$, that is

$$t'' = \frac{2T}{m}t' - t^0, \quad t''' = \frac{2T}{m}t'' - t', \text{etc.}, \quad u'' = \frac{2T}{m}u', \text{etc.}$$

Now since by hypothesis we are given a form (M, N, P) with

---

¹ Only in these four expressions and in equation [1] does e denote the *exponent* of the power; in all others, letters written above designate the *index*.

determinant D in which $M, 2N, P$ are divisible by m, we will have

$$T^2 = (N^2 - MP)U^2 + m^2$$

and manifestly $4T^2$ will be divisible by m^2. Thus $2T/m$ will be an integer and positive. And since $t^0 = m$, $t' = T$, $u^0 = 0$, $u' = U$, and are thus integers: all the numbers t'', t''', etc. u'', u''', etc. will also be integers. Further since $T^2 > m^2$ all the numbers t^0, t', t'', t''', etc. will be positive and continuously increasing; the same is true of the numbers u^0, u', u'', u''', etc.

III. Suppose there are other positive values of t, u not contained in the series t^0, t', t'', etc. u^0, u', u'', etc. We will call them $\mathfrak{T}, \mathfrak{U}$. Manifestly since the series u^0, u', etc. increases from 0 to infinity, $\mathfrak{U}$ will necessarily lie between two neighboring terms u^n and u^{n+1} so that $\mathfrak{U} > u^n$ and $\mathfrak{U} < u^{n+1}$. In order to show the absurdity of this supposition, we observe

1) The equation $t^2 - Du^2 = m^2$ will also be satisfied if we let

$$t = \frac{1}{m}(\mathfrak{T}t^n - D\mathfrak{U}u^n), \quad u = \frac{1}{m}(\mathfrak{U}t^n - \mathfrak{T}u^n)$$

This can be shown with no difficulty by substitution. We will show that these values, which for brevity we let $= \tau, v$, are always *integers*. If (M, N, P) is a form with determinant D, and m the greatest common divisor of the numbers $M, 2N, P$, both $\mathfrak{T} + N\mathfrak{U}$ and $t^n + Nu^n$ will be divisible by m and so also $\mathfrak{A}(t^n + Nu^n) - u^n(\mathfrak{T} + N\mathfrak{U})$ or $\mathfrak{U}t^n - \mathfrak{T}u^n$. Therefore v will be an integer and so also τ because $\tau^2 = Dv^2 + m^2$.

2) Clearly v cannot $= 0$; for from this would follow

$$\mathfrak{U}^2 t^{2n} = \mathfrak{T}^2 u^{2n}$$

or

$$\mathfrak{U}^2(Du^{2n} + m^2) = u^{2n}(D\mathfrak{U}^2 + m^2)$$

or $\mathfrak{U}^2 = u^{2n}$ contrary to the hypothesis that $\mathfrak{U} > u^n$. Since therefore except for the value 0, the smallest value of u is U, v will certainly not be less than U.

3) From the values of $t^n, t^{n+1}, u^n, u^{n+1}$ it is easy to confirm that

$$mU = u^{n+1}t^n - t^{n+1}u^n$$

And so $\mathfrak{U}t^n - \mathfrak{T}u^n$ will certainly not be less than $u^{n+1}t^n - t^{n+1}u^n$.

4) Now from the equation $\mathfrak{T}^2 - D\mathfrak{U}^2 = m^2$ we have

$$\frac{\mathfrak{T}}{\mathfrak{U}} = \sqrt{\left(D + \frac{m^2}{\mathfrak{U}^2}\right)}$$

and similarly

$$\frac{t^{n+1}}{u^{n+1}} = \sqrt{\left(D + \frac{m^2}{u^{2n+2}}\right)}$$

From this it is easy to see $\mathfrak{T}/\mathfrak{U} > t^{n+1}/u^{n+1}$. This along with the conclusion in 3) gives us

$$(\mathfrak{U}t^n - \mathfrak{T}u^n)\left(t^n + u^n\frac{\mathfrak{T}}{\mathfrak{U}}\right) > (u^{n+1}t^n - t^{n+1}u^n)\left(t^n + u^n\frac{t^{n+1}}{u^{n+1}}\right)$$

or by multiplying out and in place of $\mathfrak{T}^2, t^{2n}, t^{2n+2}$ substituting their values $D\mathfrak{U}^2 + m^2$, $Du^{2n} + m^2$, $Du^{2n+2} + m^2$ we have

$$\frac{1}{\mathfrak{U}}(\mathfrak{U}^2 - u^{2n}) > \frac{1}{u^{n+1}}(u^{2n+2} - u^{2n})$$

Now since each quantity is positive we get by transposing $\mathfrak{U} + (u^{2n}/u^{n+1}) > u^{n+1} + (u^{2n}/\mathfrak{U})$. Q.E.A. because the first part of the former quantity is *less* than the first part of the second quantity and the second part of the former is less than the second part of the latter. Therefore the supposition is inconsistent and the progression t^0, t', t'', etc. u^0, u', u'', etc. will exhibit all possible values of t, u.

Example. For $D = 61$, $m = 2$ we find that the smallest positive values of t, u are $1523, 195$, so *all* positive values will be expressed by the formulae

$$t = \left(\frac{1523}{2} + \frac{195}{2}\sqrt{61}\right)^e + \left(\frac{1523}{2} - \frac{195}{2}\sqrt{61}\right)^e$$

$$u = \frac{1}{\sqrt{61}}\left[\left(\frac{1523}{2} + \frac{195}{2}\sqrt{61}\right)^e - \left(\frac{1523}{2} - \frac{195}{2}\sqrt{61}\right)^e\right]$$

And we find

$$t^0 = 2, \, t' = 1523, \, t'' = 1523t' - t^0 = 2319527,$$
$$t''' = 1523t'' - t' = 3523618098, \text{ etc.}$$
$$u^0 = 0, \, u' = 195, \, u'' = 1523u' - u^0 = 296985,$$
$$u''' = 1523u'' - u' = 452307960, \text{ etc.}$$

▶ 201. We would like to add the following observations relative to the problem treated in the preceding articles.

1) Since we have shown how to solve the equation $t^2 - Du^2 = m^2$ for all cases when m is the greatest common divisor of the three numbers $M, 2N, P$ such that $N^2 - MP = D$, it will be useful to assign all numbers that can be such divisors, that is, all the values of m for a given value of D. Let $D = n^2 D'$ so that D' is entirely free from quadratic factors. This can be done by letting n^2 be the largest square that divides D and, if D has no quadratic factor, by letting $n = 1$. Then,

First, if D' is of the form $4k + 1$, any divisor of $2n$ will be a value of m and vice versa. For if g is a divisor of $2n$ we will have the form $[g, n, n^2(1 - D')/g]$ whose determinant is D and in which the greatest common divisor of the numbers $g, 2n, n^2(D' - 1)/g$ will obviously be g (for it is clear that $n^2(D' - 1)/g^2 = [4n^2/g^2] \cdot [(D' - 1)/4]$ is an integer). If on the other hand we suppose that g is a value of m, that is, the greatest common divisor of the numbers $M, 2N, P$ and that $N^2 - MP = D$, manifestly $4D$ or $4n^2D'$ will be divisible by g^2. It follows that $2n$ is divisible by g. For if g did not divide $2n$, g and $2n$ would have a greatest common divisor less than g. Suppose it were $= \delta$, and $2n = \delta n'$, $g = \delta g'$; $n'n'D$ would be divisible by $g'g'$. So n' and g' as well as $n'n'g'$ and $g'g'$ would be relatively prime and D' would be divisible by $g'g'$, contrary to the hypothesis which says that D' is free from all quadratic factors.

Second, if D' were of the form $4k + 2$ or $4k + 3$, any divisor of n would be a value of m and, inversely, any value of m would divide n. For if g is a divisor of n we will have a form $(g, 0, -n^2D'/g)$ whose determinant $= D$. Clearly the greatest common divisor of the numbers $g, 0, n^2D'/g$ will be g. Now if we suppose g is a value of m, that is, the greatest common divisor of the numbers $M, 2N, P$ and that $N^2 - MP = D$, in the same way as above g will divide $2n$ and $2n/g$ will be an integer. If this quotient is odd: the square $4n^2/g^2$ would be $\equiv 1$ (mod. 4), and thus $4n^2D'/g^2$ would either $\equiv 2$, or $\equiv 3$ (mod. 4). But $4n^2D'/g^2 = 4D/g^2 = 4N^2/g^2 - 4MP/g^2 \equiv 4N^2/g^2$ (mod. 4) and thus $4N^2/g^2$ either $\equiv 2$ or $\equiv 3$ (mod. 4). Q.E.A., because every square must be congruent either to zero or to unity relative to the modulus 4. Therefore $2n/g$ is necessarily even, and so n/g is an integer or g is a divisor of n.

Thus it is clear that 1 is always a value of m, that is to say the equation $t^2 - Du^2 = 1$ is solvable in the preceding manner for any positive nonquadratic value of D; 2 will be a value of m only if D is of the form $4k$ or $4k + 1$.

2) If m is greater than 2 but yet a suitable number, the solution of the equation $t^2 - Du^2 = m^2$ can be reduced to the solution of a similar equation in which m is either 1 or 2. So letting $D = n^2D'$, if m divides n, m^2 will divide D. Then if we suppose that the smallest values of p, q in the equation $p^2 - (Dq^2/m^2) = 1$ are $p = P$, $q = Q$, the smallest values of t, u in the equation $t^2 - Du^2 = m^2$ will be $t = mP$, $u = Q$. But if m does not divide n it will at least divide $2n$ and will certainly be even; and $4D/m^2$ will be an integer. And if then the smallest values of p, q in the equation $p^2 - (4Dq^2/m^2) = 4$ are found to be $p = P$, $q = Q$, the smallest values of t, u in the equation $t^2 - Du^2 = m^2$ will be $t = mP/2$, $u = Q$. In either case, however, we can deduce not only the smallest values of t, u from a knowledge of the smallest values of p, q but by this method we can deduce *all* values of the former from *all* values of the latter.

3) Suppose we designate by t^0, u^0; t', u'; t'', u'', etc. all the positive values of t, u in the equation $t^2 - Du^2 = m^2$ (as in the preceding article). If it happens that any values in the series are congruent to the first values relative to any given modulus r, for example if $t^\rho \equiv t^0$ (or $\equiv m$), $u^\rho \equiv u^0$ or $\equiv 0$ (mod. r), and if at the same time the succeeding values are congruent to the second values, i.e.

$$t^{\rho+1} \equiv t', \quad u^{\rho+1} \equiv u' \text{ (mod. } r\text{)}$$

we will also have

$$t^{\rho+2} \equiv t'', u^{\rho+2} \equiv u''; \quad t^{\rho+3} \equiv t''', u^{\rho+3} \equiv u'''; \text{ etc.}$$

This can easily be seen from the fact that each series t^0, t', t'', etc., u^0, u', u'', etc. is a recurrent series; that is since

$$t'' = \frac{2T}{m}t' - t^0, \quad t^{\rho+2} = \frac{2T}{m}t^{\rho+1} - t^\rho$$

we will have

$$t'' \equiv t^{\rho-2}$$

and similarly for the rest. Thus it follows that in general

$$t^{h+\rho} \equiv t^h, \quad u^{h+\rho} \equiv u^h \text{ (mod. } r\text{)}$$

where h is any number; and even more generally, if

$$\mu \equiv v \;(\text{mod. } \rho) \text{ then } t^\mu \equiv t^v, \quad u^\mu \equiv u^v \;(\text{mod. } r)$$

4) We can always satisfy the conditions required by the preceding observation; that is we can always find an index ρ (for any given modulus r) for which we will have

$$t^\rho \equiv t^0, \quad t^{\rho+1} \equiv t', \quad u^\rho \equiv u^0, \quad u^{\rho+1} \equiv u' \qquad \bullet$$

To show this we observe

First, that the third condition can always be satisfied. For by the criteria given in 1) it is clear that the equation $p^2 - r^2 D q^2 = m^2$ is solvable; and if we suppose that the smallest positive values of p, q (except $m, 0$) are P, Q, manifestly $t = P$, $u = rQ$ will be among the values of t, u. Therefore P, rQ will be contained in the series t^0, t', etc., u^0, u', etc., and if $P = t^\lambda$, $rQ = u^\lambda$ we will have $u^\lambda \equiv 0 = u^0$ (mod. r). Further it is clear that between u^0 and u^λ there will be no term that is congruent to u^0 relative to the modulus r.

Second, if the other three conditions are also fulfilled, for example if $u^{\lambda+1} \equiv u'$, $t^\lambda \equiv t^0$, $t^{\lambda+1} \equiv t'$, then we should let $\rho = \lambda$. But if one or another of these conditions does not hold, we can certainly let $\rho = 2\lambda$. For from equation [1] and the general formulae for t^e, u^e in the preceding article we can deduce

$$t^{2\lambda} = \frac{1}{m}(t^{2\lambda} + D u^{2\lambda}) = \frac{1}{m}(m^2 + 2D u^{2\lambda})$$

and thus

$$\frac{t^{2\lambda} - t^0}{r} = \frac{2D u^{2\lambda}}{mr}$$

This quantity will be an integer because by hypothesis r divides u^λ, and m^2 divides $4D$ and so, a fortiori, m divides $2D$. Further, we have $u^{2\lambda} = 2t^\lambda u^\lambda / m$, and since

$$4t^{2\lambda} = 4D u^{2\lambda} + 4m^2$$

and thus divisible by m^2, $2t^\lambda$ will be divisible by m and then $u^{2\lambda}$ by r or

$$u^{2\lambda} \equiv u^0 \;(\text{mod. } r)$$

In the third place we find

$$t^{2\lambda+1} = t' + \frac{2Du^{2\lambda+1}}{m}$$

and since for a similar reason $2Du^{\lambda}/mr$ is an integer, we have

$$t^{2\lambda+1} \equiv t' \pmod{r}$$

Finally we find

$$u^{2\lambda+1} = u' + \frac{2t^{\lambda+1}u^{\lambda}}{m}$$

and since $2t^{\lambda+1}$ is divisible by m and u^{λ} by r, we have

$$u^{2\lambda+1} \equiv u' \pmod{r} \quad \text{Q.E.D.}$$

The usefulness of the latter two observations will appear in what follows.

▶ 202. A particular case of the problem of solving the equation $t^2 - Du^2 = 1$ had already been treated in the last century. That extremely shrewd geometer Fermat proposed the problem to the English analysts, and Wallis called Brounker the discoverer of the solution. He reported this in Chapter 98 of his "Algebra" (*Opera Mathem. Wall.*, 2, 418 ff.).[5] Ozanam claims that it was Fermat; and Euler, who treated of it in *Comm. acad. Petrop.*, 6 [1732–33], 1738, 175,[6] *Novi comm. acad. Petrop.*, 11, [1765], 1767, 28,[u] *Algebra*, 2, 226,[7] *Opuscula Analytica*, 1, 310[8] claims that Pellius was the discoverer, and for that reason the problem is called *Pellian* by some authors. All these solutions are essentially the same as the one we would have obtained if in article 198 we had used the reduced form with $a = 1$; but no one before Lagrange was able to show that the prescribed operation necessarily comes to an end,

---

[5] Cf. p. 79. The treatise in the *Opera* is entitled "De Algebra Tractatus, Historicus et Practicus" and is in Latin. The original was entitled, "A Treatise of Algebra both Historical and Practical" and was published in London in 1685. Chapter 98 (pp. 363–71) is entitled, "A Method of Approaches for Numerical Questions; occasioned by a problem of Mons. Fermat."

[6] "De Solutione Problematum Diophantaeorum per Numeros Integros."

[u] In this commentary the algorithm we considered in article 27 is used with similar notation. We neglected to note it at that time. [The article is entitled, "De usu novi algorithmi in problemate Pelliano solvendo." Ed. Note.]

[7] *Vollständige Anleitung zur Algebra*, St. Petersburg, 1770.

[8] "Nova subsidia pro resolutione formulae $axx + 1 = yy$."

that is, that the problem is *really solvable*.[v] Consult *Mélanges de Turin*,[9] 4, 19; and for a more elegant presentation, *Hist. Acad. Berlin*, 1767, p. 237.[10] There is also an investigation of this question in the appendix to Euler's *Algebra*[7] which we have frequently commended. But our method (starting from totally different principles and not being restricted to the case $m = 1$) gives many ways of arriving at a solution because in article 198 we can begin from any reduced form $(a, b, -a')$.

▶ 203. PROBLEM. *If the forms Φ, φ are equivalent, to exhibit all the transformations of one into the other.*

Solution. When these forms are equivalent in only one way (i.e. either properly only or improperly only), by article 196 we can find one transformation $\alpha, \beta, \gamma, \delta$ of the form ϕ into Φ, and it is clear that all others are similar to this. But when ϕ, Φ are properly and improperly equivalent we will look for two dissimilar (i.e. one proper, the other improper) transformations $\alpha, \beta, \gamma, \delta$ and $\alpha', \beta', \gamma', \delta'$, and any other transformation will be similar to one of these. Suppose therefore that the form ϕ is (a, b, c), that its determinant $= D$, that m is the greatest common divisor of the numbers $a, 2b, c$ (as was always the case above), and that t, u stand in general for all numbers satisfying the equation $t^2 - Du^2 = m^2$. In the former case all transformations of the form ϕ into Φ will be contained in formula I following, and in the latter case either in the I or II.

I. $\qquad \frac{1}{m}[\alpha t - (\alpha b + \gamma c)u], \quad \frac{1}{m}[\beta t - (\beta b + \delta c)u],$

$\qquad \frac{1}{m}[\gamma t + (\alpha a + \gamma b)u], \quad \frac{1}{m}[\delta t + (\beta a + \delta b)u],$

---

[v] What Wallis, pp. 427–28, proposes for this purpose carries no weight. The paralogism consists in that fact that on p. 428, line 4, he presupposes that, given a quantity p, integers a, z can be found such that z/a is less than p and that the difference is less than an *assigned* number. This is true when the assigned difference is a *given quantity* but not when, as in the present case, it depends on a and z and thus is variable.

[9] *Mélanges de Philosphie et de Mathématique de la Société Royale de Turin pour les années 1766–69*. The article is entitled, "Solution d'un Problème d'Arithmétique" and Lagrange is the author. These proceedings are also referred to as *Miscellanea Taurinensia*.

[10] Cf. p. 11.

II. $\quad \frac{1}{m}[\alpha't - (\alpha'b + \gamma'c)u], \quad \frac{1}{m}[\beta't - (\beta'b + \delta'c)u],$

$\quad \frac{1}{m}[\gamma't + (\alpha'a + \gamma'b)u], \quad \frac{1}{m}[\delta't + (\beta'a + \delta'b)u],$

Example. We want all the transformations of the form (129, 92, 65) into the form (42, 59, 81). We found in article 195 that these are only improperly equivalent and in the following article that the improper transformation of the former into the latter is $-47, -56, 73, 87$. Therefore all transformations of the form (129, 92, 65) into (42, 59, 81) will be expressed by the formula

$$-(47t + 421u), \ -(56t + 503u), \ 73t + 653u, \ 87t + 780u$$

where t, u are all the numbers satisfying the equation $t^2 - 79u^2 = 1$; and these are expressed by the formulae

$$\pm t = \frac{1}{2}[(80 + 9\sqrt{79})^e + (80 - 9\sqrt{79})^e]$$

$$\pm u = \frac{1}{2\sqrt{79}}[(80 + 9\sqrt{79})^e - (80 - 9\sqrt{79})^e]$$

where e represents all non-negative integers.
▶ 204. Obviously a general formula representing all transformations will be simpler if the initial transformation from which the formula is deduced is simpler. Now since it does not matter from which transformation we start, very often the general formula can be simplified if from the first formula found we deduce a less complex transformation by giving specific values to t, u and use this to produce another formula. Thus, e.g., in the formula found in the preceding article by letting $t = 80$, $u = -9$, we get a transformation that is simpler than the one we found. This way we get the transformation $29, 47, -37, -60$ and the general formula $29t - 263u$, $47t - 424u$, $-37t + 337u$, $-60t + 543u$. When, therefore, by means of the preceding precepts the general formula is found, we can test whether the transformation obtained is simpler than the one from which the formula was deduced, by giving t, u specific values $\pm t', \pm u'$; $\pm t'', \pm u''$, etc. In this case a simpler formula can be derived from that transformation. But what constitutes simplicity still remains an arbitrary judgment. If it were

useful, we might be able to find a fixed norm and to assign *limits* in the series t', u'; t'', u'', etc. beyond which the transformation would become continually less simple. Thus there would be no need to look further and we could confine our search within these limits. However, for brevity's sake we will not continue this investigation because most often by the methods we have prescribed the simplest transformation arises either immediately or by using the values $\pm t', \pm u'$ for t, u.

▶ 205. PROBLEM. *To find all representations of a given number* M *by a given formula* $ax^2 + 2bxy + cy^2$ *in which the positive non-quadratic determinant* $= D$.

Solution. First we observe that the investigation of representations by values of x, y which are not relatively prime can be carried out as we did in article 181 for forms with negative determinant. Recall that we reduced it to the case where we could consider representations by values of the unknowns which were relatively prime. There is no need to repeat the argument here. Now in order to represent M by relatively prime values of x, y it is required that D be a quadratic residue of M, and if all the values of the expression $\sqrt{D}$ (mod. M) are $N, -N, N', -N', N'', -N''$, etc. (we can choose them so that none is $> M/2$), then any representation of the number M by the given form will belong to one of these values. First, therefore, we should find these values and then afterward investigate the representations belonging to each of them. There will be no representations belonging to the value N unless the forms (a, b, c) and $[M, N, (N^2 - D)/M]$ are properly equivalent; if they are, we will look for a proper transformation of the first into the second. Let it be $\alpha, \beta, \gamma, \delta$. Then we will have a representation of the number M by the form (a, b, c) belonging to the value N by letting $x = \alpha$, $y = \gamma$, and all representations belonging to this value will be expressed by the formula

$$x = \frac{1}{m}[\alpha t - (\alpha b + \gamma c)u], \quad y = \frac{1}{m}[\gamma t + (\alpha a + \gamma b)u]$$

where m is the greatest common divisor of the numbers $a, 2b, c$ and t, u represent in general all numbers satisfying the equation

$t^2 - Du^2 = m^2$. But manifestly this general formula will be simpler if the transformation $\alpha, \beta, \gamma, \delta$ from which it was deduced is simpler. Thus it would be very useful to find as we did in the preceding article the simplest transformation of the form (a, b, c) into $[M, N, (N^2 - D)/M]$ and deduce the formula from this. In exactly the same way we can produce general formulae for representations belonging to the remaining values $-N, N', -N'$, etc. (if indeed any exists).

Example. We will look for all representations of the number 585 by the formula $42x^2 + 62xy + 21y^2$. With regard to representations by values of x, y which are not relatively prime, it is immediately evident that there are no others of this kind except those in which the greatest common divisor of x, y is 3, because 585 is divisible by only one square, 9. When therefore we find all representations of the number 585/9, i.e. 65 by the form $42x'x' + 62x'y' + 21y'y'$ with x', y' relatively prime, we can derive all representations of the number 585 by the form $42x^2 + 62xy + 21y^2$ with x, y not relatively prime by letting $x = 3x'$, $y = 3y'$. The values of the expression $\sqrt{79}$ (mod. 65) are $\pm 12, \pm 27$. The representation of the number 65 belonging to the value -12 is found to be $x' = 2$, $y' = -1$. Therefore all representations of 65 belonging to this value will be expressed by the formula $x' = 2t - 41u$, $y' = -t + 53u$ and *from this* all representations of 585 by the formula $x = 6t - 123u$, $y = -3t + 159u$. In a similar way we find the general formula for all representations of the number 65 belonging to the value $+12$ to be $x' = 22t - 199u$, $y' = -23t + 211u$; and the formula for all representations of the number 585 derived from this will be $x = 66t - 597u$, $y = -69t + 633u$. But there is no representation of the number 65 belonging to the values $+27$ and -27. In order to find representations of the number 585 by relatively prime values of x, y we must first calculate the values of the expression $\sqrt{79}$ (mod. 585). They are $\pm 77, \pm 103, \pm 157, \pm 248$. There is no representation belonging to the values $\pm 77, \pm 103, \pm 248$. However the representation $x = 3$, $y = 1$ belongs to the value -157, and we deduce the general formula for all representations belonging to this value: $x = 3t - 114u$, $y = t + 157u$. Similarly we find the representation $x = 83$, $y = -87$ belonging to $+157$, and the formula in which all similar representations are contained is $x = 83t - 746u$, $y = -87t + 789u$. We have therefore

four general formulae in which are contained all representations
of the number 585 by the form $42x^2 + 62xy + 21y^2$.

$$x = 6t - 123u \qquad y = -3t + 159u$$

$$x = 66t - 597u \qquad y = -69t + 633u$$

$$x = 3t - 114u \qquad y = t + 157u$$

$$x = 83t - 746u \qquad y = -87t + 789u$$

where t, u represent in general all integers which satisfy the equation $t^2 - 79u^2 = 1$.

For brevity we will not dwell on special applications of the
preceding analysis for forms with nonquadratic positive determinant. Each one can have his own battle with these himself by
imitating the method of articles 176 and 182. We shall immediately
hurry on to consider forms with positive quadratic determinant,
which is the only case left.

▶ 206. PROBLEM. *Given the form (a, b, c) with quadratic determinant
h^2, h being the positive root, to find a form (A, B, C) which is properly
equivalent to it, in which A lies between the limits 0 and $2h - 1$
inclusively, $B = h$, $C = 0$.*

Solution. I. Since $h^2 = b^2 - ac$, we have $(h - b):a = c:-
(h + b)$. Let the ratio $\beta:\delta$ be equal to this ratio with β, δ relatively
prime, and determine α, γ so that $\alpha\delta - \beta\gamma = 1$. This can be done.
By the substitution $\alpha, \beta, \gamma, \delta$ the form (a, b, c) will be transformed
into (a', b', c') and the two will be properly equivalent. We will have

$$b' = a\alpha\beta + b(\alpha\delta + \beta\gamma) + c\gamma\delta$$
$$= (h - b)\alpha\delta + b(\alpha\delta + \beta\gamma) - (h + b)\beta\gamma$$
$$= h(\alpha\delta - \beta\gamma) = h$$
$$c' = a\beta^2 + 2b\beta\delta + c\delta^2$$
$$= (h - b)\beta\delta + 2b\beta\delta - (h + b)\beta\delta = 0$$

Further, if a' already lies between the limits 0 and $2h - 1$, the form
(a', b', c') will satisfy all the conditions.

II. But if a' lies outside the limits 0 and $2h - 1$, let A be the least positive residue of a' relative to the modulus $2h$. This manifestly will lie between those limits. And let $A - a' = 2hk$. Then the form (a', b', c'), i.e. $(a', h, 0)$ will be transformed into the form $(A, h, 0)$ by the substitution $1, 0, k, 1$. This will be properly equivalent to the forms (a', b', c'), (a, b, c) and will satisfy all the conditions. But obviously the form (a, b, c) will be transformed into the form $(A, h, 0)$ by the substitution $\alpha + \beta k, \beta, \gamma + \delta k, \delta$.

Example. Let us be given the form $(27, 15, 8)$ whose determinant $= 9$. Here $h = 3$ and $4 : -9$ is the ratio in lowest terms which is equal to the ratios $-12 : 27 = 8 : -18$. Therefore, letting $\beta = 4$, $\delta = -9$, $\alpha = -1$, $\gamma = 2$, the form (a', b', c') becomes $(-1, 3, 0)$, which goes into the form $(5, 3, 0)$ by the substitution $1, 0, 1, 1$. This is therefore the form we seek, and the given form is transformed into it by the proper substitution $3, 4, -7, -9$.

Such forms (A, B, C), in which $C = 0$, $B = h$, and A lies between the limits 0 and $2h - 1$, we will call *reduced forms*. They must be carefully distinguished from reduced forms having a determinant that is negative or nonquadratic positive.

▶ 207. THEOREM. *Two reduced forms $(a, h, 0)$, $(a', h, 0)$ which are not identical cannot be properly equivalent.*

Demonstration. If they were properly equivalent, the former could be transformed into the latter by a proper substitution $\alpha, \beta, \gamma, \delta$ and we would have the four equations:

$$a\alpha^2 + 2h\alpha\gamma = a' \tag{1}$$

$$a\alpha\beta + h(\alpha\delta + \beta\gamma) = h \tag{2}$$

$$a\beta^2 + 2h\beta\delta = 0 \tag{3}$$

$$\alpha\delta - \beta\gamma = 1 \tag{4}$$

Multiplying the second equation by β, the third by α and subtracting, we have $-h(\alpha\delta - \beta\gamma)\beta = \beta h$ or, from [4], $-\beta h = \beta h$; so necessarily $\beta = 0$. So again using [4], $\alpha\delta = 1$ and $\alpha = \pm 1$. Then from [1] $a \pm 2\gamma h = a'$, and this equation cannot be consistent unless $\gamma = 0$ (because both a and a' by hypothesis lie between 0 and $2h - 1$), i.e. unless $a = a'$ or the forms $(a, h, 0)$, $(a', h, 0)$ are identical, contrary to the hypothesis.

Thus the following problems, which offered much greater difficulty for nonquadratic determinants, can be solved with very little effort.

I. *Given two forms F, F' with the same quadratic determinant: to find whether or not they are properly equivalent.* We look for two reduced forms which are properly equivalent to the forms F, F' respectively. If they are identical the given forms will be equivalent; otherwise not.

II. *Given the same forms: to investigate whether or not they are improperly equivalent.* Let G be the form opposite to one of the given forms, e.g. the form F. If G is properly equivalent to the form F', F and F' are improperly equivalent; otherwise not.

▶ 208. PROBLEM. *Given two properly equivalent forms F, F' with determinant h^2: to find a proper transformation of one into the other.*

Solution. Let Φ be a reduced form properly equivalent to the form F. By hypothesis it will also be properly equivalent to the form F'. By article 206 we will seek a proper transformation of the form F into Φ. Let this be $\alpha, \beta, \gamma, \delta$, and let the proper transformation of the form F' into Φ be $\alpha', \beta', \gamma', \delta'$. Then Φ will be transformed into F' by the proper transformation $\delta', -\beta', -\gamma', \alpha'$ and thus F into F' by the proper substitution

$$\alpha\delta' - \beta\gamma', \quad \beta\alpha' - \alpha\beta', \quad \gamma\delta' - \delta\gamma', \quad \delta\alpha' - \gamma\beta'$$

It will be useful to develop another formula for the transformation of the form F into F' which does not involve knowing the reduced form Φ. Let us suppose that the form

$$F = (a, b, c), \quad F' = (a', b', c'), \quad \Phi = (A, h, 0)$$

Since $\beta : \delta$ is the ratio in lowest terms which is equal to the ratios $h - b : a$ or $c : -(h + b)$, it is easy to see that $(h - b)/\beta = \alpha/\delta$ is an *integer*, which we will call f; and that $c/\beta = (-h - b)/\delta$ will also be an integer, which we will call g. We have however

$A = a\alpha^2 + 2b\alpha\gamma + c\gamma^2$ and therefore $\beta A = a\alpha^2\beta + 2b\alpha\beta\gamma + c\beta\gamma^2$

or, substituting $\delta(h - b)$ for $a\beta$ and βg for c,

$$\beta A = \alpha^2\delta h + b(2\beta\gamma - \alpha\delta)\alpha + \beta^2\gamma^2 g$$

or (since $b = -h - \delta g$)

$$\beta A = 2\alpha(\alpha\delta - \beta\gamma)h + (\alpha\delta - \beta\gamma)^2 g = 2\alpha h + g$$

Similarly

$$\delta A = a\alpha^2\delta + 2b\alpha\gamma\delta + c\gamma^2\delta$$
$$= \alpha^2\delta^2 f + b(2\alpha\delta - \beta\gamma)\gamma - \beta\gamma^2 h$$
$$= (\alpha\delta - \beta\gamma)^2 f + 2\gamma(\alpha\delta - \beta\gamma)h = 2\gamma h + f$$

Therefore

$$\alpha = \frac{\beta A - g}{2h}, \quad \gamma = \frac{\delta A - f}{2h}$$

In exactly the same way by letting

$$\frac{h - b'}{\beta'} = \frac{\alpha'}{\delta'} = f', \quad \frac{c'}{\beta'} = \frac{-h - b'}{\delta'} = g'$$

we have

$$\alpha' = \frac{\beta' A - g'}{2h}, \quad \gamma' = \frac{\delta' A - f'}{2h}$$

If the values $\alpha, \gamma, \alpha', \gamma'$ are substituted in the formula which we have just given for the transformation of the form F into F', we get

$$\frac{\beta f' - \delta' g}{2h}, \quad \frac{\beta' g - \beta g'}{2h}, \quad \frac{\delta f' - \delta' f}{2h}, \quad \frac{\beta' f - \delta g'}{2h}$$

from which A has completely disappeared.

If we are given two improperly equivalent forms F, F' and we are looking for the improper transformation of one into the other, let G be the form opposite to the form F and let the proper transformation of the form G into F' be $\alpha, \beta, \gamma, \delta$. Then manifestly $\alpha, \beta, -\gamma, -\delta$ will be the improper transformation of the form F into F'.

Finally, if the given forms are properly and improperly equivalent, this method can give us two transformations, one proper, the other improper.

▶209. There remains now only to show how all other similar transformations can be deduced from one transformation. This depends on the solution of the indeterminate equation

$t^2 - h^2u^2 = m^2$, where m is the greatest common divisor of the numbers $a, 2b, c$ and (a, b, c) is one of the equivalent forms. But this equation can be solved in only two ways, that is by letting either $t = m, u = 0$ or $t = -m, u = 0$. For if we suppose that there is another solution $t = T, u = U$ where U does not $= 0$ then, since m^2 certainly divides $4h^2$, we get $4T^2/m^2 = 4h^2U^2/m^2 + 4$ and both $4T^2/m^2$ and $4h^2U^2/m^2$ will be quadratic integers. But clearly the number 4 cannot be the difference of two quadratic integers unless the smaller square is 0, i.e. $U = 0$, contrary to the hypothesis. Therefore if the form F is transformed into the form F' by the substitution $\alpha, \beta, \gamma, \delta$, there will be no other transformation similar to this except $-\alpha, -\beta, -\gamma, -\delta$. Therefore if two forms are only properly or only improperly equivalent, there will be only *two* transformations; but if they are properly and improperly equivalent there will be *four*, two proper and two improper.

▶ 210. THEOREM. *If two reduced forms* $(a, h, 0)$, $(a', h, 0)$ *are improperly equivalent, we will have* $aa' \equiv m^2$ *(mod. $2mh$) where m is the greatest common divisor of the numbers $a, 2h$ or $a', 2h$; and conversely if $a, 2h$; $a', 2h$ have the same greatest common divisor m and* $aa' \equiv m^2$ *(mod. $2mh$), the forms* $(a, h, 0)$, $(a', h, 0)$ *will be improperly equivalent.*

Demonstration. I. Let the form $(a, h, 0)$ be transformed into the form $(a', h, 0)$ by the improper substitution $\alpha, \beta, \gamma, \delta$ so that we have four equations

$$a\alpha^2 + 2h\alpha\gamma = a' \qquad [1]$$

$$a\alpha\beta + h(\alpha\delta + \beta\gamma) = h \qquad [2]$$

$$a\beta^2 + 2h\beta\delta = 0 \qquad [3]$$

$$\alpha\delta - \beta\gamma = -1 \qquad [4]$$

If we multiply [4] by h and subtract from [2] and write this as $[2] - h[4]$ it follows that

$$(a\alpha + 2h\gamma)\beta = 2h \qquad [5]$$

similarly from $\gamma\delta[2] - \gamma^2[3] - (a + a\beta\gamma + h\gamma d)[4]$, deleting the zero terms, we have

$$-a\alpha\delta = a + 2h\gamma\delta \qquad \text{or} \qquad -(a\alpha + 2h\gamma)\delta = a \qquad [6]$$

finally from $a[1]\ldots a\alpha(a\alpha + 2h\gamma) = aa'$ or

$$(a\alpha + 2h\gamma)^2 - aa' = 2h\gamma(a\alpha + 2h\gamma)$$

or

$$(a\alpha + 2h\gamma)^2 \equiv aa' \,[\text{mod. } 2h(a\alpha + 2h\gamma)] \qquad [7]$$

Now from [5] and [6] it follows that $a\alpha + 2h\gamma$ divides $2h$ and a and thus also m, which is the greatest common divisor of $a, 2h$; manifestly, however, m also divides $a\alpha + 2h\gamma$; therefore necessarily $a\alpha + 2h\gamma$ will either $= +m$ or $= -m$. And it follows immediately from [7] that $m^2 = aa'$ (mod. $2mh$).   Q.E.P.

II. If $a, 2h; a', 2h$ have the same greatest common divisor m and further $aa' \equiv m^2$ (mod. $2mh$), $a/m, 2h/m, a'/m, (aa' - m^2)/2mh$ will be integers. It is easy to confirm that the form $(a, h, 0)$ will be transformed into the form $(a', h, 0)$ by the substitution $-a'/m$, $-2h/m, (aa' - m^2)/2mh, a/m$ and that this transformation is improper. Therefore the two forms will be improperly equivalent. Q.E.S.

From this also we can immediately judge whether any given reduced form $(a, h, 0)$ is improperly equivalent to itself. That is, if m is the greatest common divisor of the numbers $a, 2h$, we ought to have $a^2 \equiv m^2$ (mod. $2mh$).

▶ 211. All reduced forms of a given determinant h^2 will be obtained if for A in the indefinite form $(A, h, 0)$ we substitute all numbers from 0 up to $2h - 1$ inclusive. There will be $2h$ of them. Clearly all forms of the determinant h^2 can be distributed into as many *classes* and they will have the same properties mentioned above (art. 175, 195) for classes of forms with negative and positive nonquadratic determinants. Thus forms with determinant 25 will be distributed into ten classes which can be distinguished by the reduced forms contained in each of them. The reduced forms will be: $(0, 5, 0), (1, 5, 0), (2, 5, 0), (5, 5, 0), (8, 5, 0)$, and $(9, 5, 0)$, each of which is improperly equivalent to itself; $(3, 5, 0)$ which is improperly equivalent to $(7, 5, 0)$; and $(4, 5, 0)$ which is improperly equivalent to $(6, 5, 0)$.

▶ 212. PROBLEM. *To find all representations of a given number M by a given form $ax^2 + 2bxy + cy^2$ with determinant h^2.*

The solution of this problem can be found from the principles of article 168 in exactly the same way that we showed above (art. 180, 181, 205) for forms with negative and nonquadratic determinants. It would be superfluous to repeat it here, since it offers no difficulty. On the other hand it would not be out of place to deduce the solution from another principle which is proper to the present case.

As in articles 206 and 208 letting

$$h - b : a = c : -(h + b) = \beta : \delta$$

$$\frac{h - b}{\beta} = \frac{a}{\delta} = f; \qquad \frac{c}{\beta} = \frac{-h - b}{\delta} = g$$

it can be shown without difficulty that the given form is a product of the factors $\delta x - \beta y$ and $fx - gy$. Thus any representation of the number M by the given form ought to provide a resolution of the number M into two factors. If therefore all divisors of the number M are d, d', d'', etc. (including also 1 and M and each taken *twice*, that is both positively and negatively), it is clear that all representations of the number M will be obtained by successively supposing

$$\delta x - \beta y = d, \quad fx - gy = \frac{M}{d}$$

$$\delta x - \beta y = d', \quad fx - gy = \frac{M}{d'}, \text{ etc.}$$

Values of x, y will be derived from this, and those representations that produce fractional values must be discarded. But manifestly from the two first equations we get

$$x = \frac{\beta M - g d^2}{(\beta f - \delta g)d} \quad \text{and} \quad y = \frac{\delta M - f d^2}{(\beta f - \delta g)d}$$

These values will always be *determined* because $\beta f - \delta g = 2h$ and thus the denominator will certainly not $= 0$. We could have solved the other problems by the same principle concerning the resolvability of any form with a quadratic determinant into two factors; but we preferred to use an analogue of the method presented above for forms with nonquadratic determinant.

Example. We will look for all representations of the number 12 by the form $3x^2 + 4xy - 7y^2$. This is resolved into the factors $x - y$ and $3x + 7y$. The divisors of the number 12 are $\pm 1, 2, 3, 4, 6, 12$. Letting $x - y = 1, 3x + 7y = 12$ we get $x = 19/10$, $y = 9/10$, which must be rejected because they are fractions. In the same way we get useless values from the divisors $-1, \pm 3, \pm 4, \pm 6, \pm 12$; but from the divisor $+2$ we get the values $x = 2$, $y = 0$ and from the divisor -2, $x = -2$, $y = 0$. There are therefore no other representations except these two.

This method cannot be used if $M = 0$. In this case manifestly all values of x, y must satisfy either the equation $\delta x - \beta y = 0$ or $fx - gy = 0$. All solutions of the former equation are contained in the formula $x = \beta z$, $y = \delta x$ where z is any integer (as long as β, δ are relatively prime as we presuppose); similarly, if we let m be the greatest common divisor of the numbers f, g, all solutions of the second equation are represented by the formula $x = gz/m$, $y = hz/m$. Thus these two general formulae include all representations of the number M in this case.

In the preceding discussion everything that pertains to equivalence, to the discovery of all transformations of forms, and to the representation of given numbers by given forms has been satisfactorily explained. It therefore remains only to show how to judge whether one of two given forms, which cannot be equivalent because they have *unequal determinants*, is contained in the other or not, and if it is to find the transformations of the former into the latter.

▶ 213. We showed above (art. 157, 158) that if the form f with determinant D implies the form F with determinant E and is transformed into it by the substitution $\alpha, \beta, \gamma, \delta$, then $E = (\alpha\delta - \beta\gamma)^2 D$; and that if $\alpha\delta - \beta\gamma = \pm 1$, the form f not only implies the form F but is equivalent to it. Therefore if the form f implies F but is not equivalent to it, the quotient E/D is an integer greater than 1. This is the problem to be solved therefore: *to judge whether a given form f with determinant D implies a given form F with determinant De^2 where e is presumed to be a positive number greater than 1.* To solve this we will show how to assign a finite number of forms contained in f, so chosen that if F is contained in f it is necessarily equivalent to one of these.

I. We will suppose that all the positive divisors of the number e (including 1 and e) are m, m', m'', etc. and that $e = mn = m'n' = m''n''$ etc. For brevity we will indicate by $(m;0)$ the form into which f is transformed by the proper substitution $m, 0, 0, n$; by $(m; 1)$ the form into which f is transformed by the proper substitution $m, 1, 0, n$, etc.; and in general by $(m; k)$ the form into which f is changed by the proper substitution $m, k, 0, n$. Similarly f will be transformed by the proper transformation $m', 0, 0, n'$ into $(m'; 0)$; by $m', 0, 1, n'$ into $(m'; 1)$ etc.; by $m'', 0, 0, n''$ into $(m''; 0)$ etc.; etc. All these forms will be properly contained in f and the determinant of each will be $= De^2$. We will designate by Ω the complex of all the forms $(m; 0)$, $(m; 1)$, $(m; 2), \ldots, (m; m - 1)$, $(m'; 0)$, $(m'; 1), \ldots$, $(m'; m' - 1)$, $(m''; 0)$, etc. Their number will be $m + m' + m'' +$ etc. and it is easy to see that they will all be different from one another.

If, e.g. the form f is $(2, 5, 7)$ and $e = 5, \Omega$ will include the following six forms $(1; 0)$, $(5; 0)$, $(5; 1)$, $(5; 2)$, $(5; 3)$, $(5; 4)$, and if they are expanded they will be $(2, 25, 175)$, $(50, 25, 7)$, $(50, 35, 19)$, $(50, 45, 35)$, $(50, 55, 55)$, $(50, 65, 79)$.

II. I now claim that if the form F with determinant De^2 is properly contained in the form f, it will necessarily be properly equivalent to one of the forms Ω. Let us suppose that the form f is transformed into F by the proper substitution $\alpha, \beta, \gamma, \delta$. We will have $\alpha\delta - \beta\gamma = e$. Let the greatest positive common divisor of the numbers γ, δ (which cannot both be 0 at the same time) be $= n$ and let $e/n = m$, which manifestly will be an integer. Pick g, h so that $\gamma g + \delta h = n$, and finally let k be the least positive residue of the number $\alpha g + \beta h$ relative to the modulus m. Then the form $(m; k)$ which is manifestly among the forms Ω will be properly equivalent to the form F and will be transformed into it by the proper substitution

$$\frac{\gamma}{n} \cdot \frac{\alpha g + \beta h - k}{m} + h, \quad \frac{\delta}{n} \cdot \frac{\alpha g + \beta h - k}{m} - g, \quad \frac{\gamma}{n}, \frac{\delta}{n}$$

For *first* it is clear that these four numbers are integers; *second* it is easy to confirm that the substitution is proper; *third* it is clear that the form into which $(m; k)$ transforms by this substitution is the same as that into which f^w transforms by the substitution

$$m\left(\frac{\gamma}{n} \cdot \frac{\alpha g + \beta h - k}{m} + h\right) + \frac{k\gamma}{n}, \; m\left(\frac{\delta}{n} \cdot \frac{\alpha g + \beta h - k}{m} - g\right) + \frac{k\delta}{n}, \; \gamma, \; \delta$$

or since $mn = e = \alpha\delta - \beta\gamma$ and thus $\beta\gamma + mn = \alpha\delta$, $\alpha\delta - mn = \beta\gamma$, this is the substitution

$$\frac{1}{n}(\alpha\gamma g + \alpha\delta h), \; \frac{1}{n}(\beta\gamma g + \beta\delta h), \; \gamma, \; \delta$$

But $\gamma g + \delta h = n$, so this is the substitution $\alpha, \beta, \gamma, \delta$. By the hypothesis this transforms f into F. So $(m; k)$ and F will be properly equivalent. Q.E.D.

From this therefore we can always judge whether a given form f with determinant D properly implies the form F with determinant De^2. If we want to find out whether f improperly implies F we need only investigate whether the form opposite to F is contained in f (art. 159).

▶ 214. PROBLEM. *Given two forms, f with determinant D and F with determinant De^2, with the former properly implying the latter: to exhibit all proper transformations of the form f into F.*

Solution. We will designate by Ω the same complex of forms as in the preceding article and extract from this complex all forms to which F is properly equivalent. Let them be Φ, Φ', Φ'', etc. Each of these forms will produce proper transformations of the form f into F and each of them will give different ones but all together will give all transformations (i.e. no transformation of the form f into F will be proper unless it is one of the forms Φ, Φ', etc.). Since the method is the same for all forms Φ, Φ', etc., we will speak of only one of them.

Let us suppose that Φ is $(M; K)$ and $e = MN$ so that f is transformed into Φ by the proper substitution $M, K, 0, N$. Further we will designate all proper transformations of the form Φ into f in general by $\mathfrak{a}, \mathfrak{b}, \mathfrak{c}, \mathfrak{d}$. Then manifestly f will be transformed into Φ by the proper substitution $M\mathfrak{a} + K\mathfrak{c}, M\mathfrak{b} + K\mathfrak{d}, N\mathfrak{c}, N\mathfrak{d}$. In this way any proper transformation of the form Φ into F will give a

---

* That is, by the substitution $m, k, 0, n$ it is transformed into $(m; k)$. See article 159.

proper transformation of the form f into F. The other forms Φ', Φ'', etc. should be treated in the same way, and each proper transformation of one of these into F will produce a proper transformation of the form f into F.

To show that this solution is complete in every way, we must show

I. *That all possible proper transformations of the form f into F are obtained in this way*. Let any proper transformation of the form f into F be $\alpha, \beta, \gamma, \delta$ and as in article 213. II, let n be the greatest common divisor of the numbers γ, δ; and let the numbers m, g, h, k be determined as they were there. Then the form $(m; k)$ will be among the forms Φ, Φ', etc., and

$$\frac{\gamma}{n} \cdot \frac{\alpha g + \beta h - k}{m} + h, \quad \frac{\delta}{n} \cdot \frac{\alpha g + \beta h - k}{m} - g, \quad \frac{\gamma}{n}, \quad \frac{\delta}{n}$$

one of the proper transformations of this form into F; from this by the rule which we just gave we will obtain the transformation $\alpha, \beta, \gamma, \delta$; all of this was shown in the preceding article.

II. *That all transformations obtained in this way are different from one another; that is, none of them is obtained twice.* It is easy to see that different transformations of the same form Φ or Φ' etc. into F cannot produce the same transformation of f into F. We will show in the following way that different forms, e.g. Φ and Φ', cannot produce the same transformation. Let us suppose that the proper transformation $\alpha, \beta, \gamma, \delta$ of the form f into F is obtained *both* from the proper transformation $\mathfrak{a}, \mathfrak{b}, \mathfrak{c}, \mathfrak{d}$ of the form Φ into F, *and* from the proper transformation $\mathfrak{a}', \mathfrak{b}', \mathfrak{c}', \mathfrak{d}'$ of the form Φ' into F. Let $\Phi = (M; K)$, $\Phi' = (M'; K')$, $e = MN = M'N'$. We will have these equations:

$$\alpha = M\mathfrak{a} + K\mathfrak{c} = M'\mathfrak{a}' + K'\mathfrak{c}' \qquad [1]$$

$$\beta = M\mathfrak{b} + K\mathfrak{d} = M'\mathfrak{b}' + K'\mathfrak{d}' \qquad [2]$$

$$\gamma = N\mathfrak{c} \qquad\quad = N'\mathfrak{c}' \qquad [3]$$

$$\delta = N\mathfrak{d} \qquad\quad = N'\mathfrak{d}' \qquad [4]$$

$$\mathfrak{a}\mathfrak{d} - \mathfrak{b}\mathfrak{c} = \mathfrak{a}'\mathfrak{d}' - \mathfrak{b}'\mathfrak{c}' = 1 \qquad [5]$$

From $\mathfrak{a}[4] - \mathfrak{b}[3]$ and using equation [5] it follows that $N = N'$ ($\mathfrak{a}\mathfrak{d}' - \mathfrak{b}\mathfrak{c}'$), and thus N' divides N; similarly from $\mathfrak{a}'[4] - \mathfrak{b}'[3]$ we get $N' = N(\mathfrak{a}'\mathfrak{d} - \mathfrak{b}'\mathfrak{c})$ and N divides N'. Now since both N and N' are supposed positive, we have necessarily $N = N'$ and $M = M'$ and thus from [3] and [4], $\mathfrak{c} = \mathfrak{c}', \mathfrak{d} = \mathfrak{d}'$. Further from $\mathfrak{a}[2] - \mathfrak{b}[1]$,

$$K = M'(\mathfrak{a}\mathfrak{b}' - \mathfrak{b}\mathfrak{a}') + K'(\mathfrak{a}\mathfrak{d}' - \mathfrak{b}\mathfrak{c}') = M(\mathfrak{a}\mathfrak{b}' - \mathfrak{b}\mathfrak{a}') + K'$$

thus $K \equiv K'$ (mod. M), which cannot be true unless $K = K'$, because both K and K' lie between the limits 0 and $M - 1$. Therefore the forms Φ, Φ' are not different, contrary to the hypothesis.

It is clear that if D is negative or square positive, this method will give us all proper transformations of the form f into F; and if D is nonquadratic positive, certain general formulae can be given that will contain all proper transformations (their number is infinite).

Finally if the form F is improperly contained in the form f, all improper transformations of the former into the latter can be found easily by the given method. That is, if $\alpha, \beta, \gamma, \delta$ designates in general all proper transformations of the form f into the form which is opposite to the form F, all improper transformations of the form f into F will be represented by $\alpha, -\beta, \gamma, -\delta$.

Example. We want all transformations of the form $(2, 5, 7)$ into $(275, 0, -1)$ which are contained in it both properly and improperly. In the preceding article we gave the complex of forms Ω for this case. If we examine these we find that both $(5; 1)$ and $(5; 4)$ are properly equivalent to the form $(275, 0, -1)$. All proper transformations of the form $(5; 1)$, i.e. $(50, 35, 19)$ into $(275, 0 - 1)$, can be found by our theorem to be continued in the general formula

$$16t - 275u, -t + 16u, -15t + 275u, t - 15u$$

where t, u are indefinite representations for all integers satisfying the equation $t^2 - 275u^2 = 1$; therefore all proper transformations of the form $(2, 5, 7)$ into $(275, 0, -1)$ will be contained in the general formula

$$65t - 1100u, -4t + 65u, -15t + 275u, t - 15u$$

Similarly all proper transformations of the form $(5; 4)$, i.e. $(50, 65, 79)$, into $(275, 0, -1)$ are contained in the general formula

$$14t + 275u, t + 14u, -15 - 275u, -t - 15u$$

and thus all proper transformations of the form $(2, 5, 7)$ into $(275, 0, -1)$ will be contained in

$$10t + 275u, \quad t + 10u, \quad -15t - 275u, \quad -t - 15u$$

These two formulae therefore include[x] all the proper transformations that we seek. In the same way we find that all improper transformations of the form $(2, 5, 7)$ into $(275, 0, -1)$ are contained in the following two formulae:

(I) $65t - 1100u, \quad 4t - 65u, \quad -15t + 275u, \quad -t + 15u$

(II) $10t + 275u, \quad -t - 10u, \quad -15t - 275u, \quad t + 15u$

▶ 215. Thus far we have excluded from our discussion forms with determinant 0. Now we must say something about these forms if our theory is to be complete in every way. Since we showed in general that if a form with determinant D implies a form with determinant D', D' is a multiple of D, it is immediately clear that a form whose determinant $=0$ cannot imply another form unless its determinant also $=0$. Thus only two problems remain to be solved: (1) *given two forms, f, F with F having determinant 0, to judge whether f implies F or not, and if it does to exhibit all the transformations involved;* (2) *to find all representations of a given number by a given form with determinant 0.* The first problem requires one method when the determinant of the form f is also 0, another when it is not 0. We now explain this.

I. Before anything else we observe that any form $ax^2 + 2bxy + cy^2$ whose determinant $b^2 - ac = 0$ can be expressed as $m(gx + hy)^2$ where g, h are relatively prime and m an integer. For let m be the greatest common divisor of a, c having the same sign as they do (it is easy to see they cannot have opposite signs). Then $a/m, c/m$ will be relatively prime non-negative integers, and their product will $= b^2/m^2$, i.e. a square, and thus each of them will also be a square (art. 21). Let $a/m = g^2$, $c/m = h^2$ and g, h will also be relatively prime, and we have $g^2h^2 = b^2/m^2$ and $gh = \pm b/m$. Thus it is clear that

$$m(gx \pm hy)^2 \text{ will be } = ax^2 + 2bxy + cy^2$$

---

[x] More concisely we can say that all proper transformations are included in the formula $10t + 55u, \ t + 2u, \ -15t - 55u, \ -t - 3u$ where t, u are all the integers satisfying the equation $t^2 - 11u^2 = 1$.

Now let two forms f, F be given, each with determinant 0 and with

$$f = m(gx + hy)^2, \quad F = M(GX + HY)^2$$

with g and h, and G and H relatively prime. Now I claim that if the form f implies the form F, m either is equal to M or at least divides M, and the quotient is a square; and conversely if M/m is a square integer, F is contained in f. For if f by the substitution

$$x = \alpha X + \beta Y, \quad y = \gamma X + \delta Y$$

is assumed to be transformed into F we have

$$\frac{M}{m}(GX + HY)^2 = [(\alpha g + \gamma h)X + (\beta g + \delta h)Y]^2$$

and it follows easily that M/m is a square. Let it $= e^2$ and we have

$$e(GX + HY) = \pm[(\alpha g + \gamma h)X + (\beta g + \delta h)Y]$$

i.e.

$$\pm eG = \alpha g + \gamma h, \qquad \pm eH = \beta g + \delta h$$

If therefore $\mathfrak{G}, \mathfrak{H}$ are so determined that $\mathfrak{G}^2 + \mathfrak{H}^2 = 1$ we get $\pm e = \mathfrak{G}(\alpha g + \gamma h) + \mathfrak{H}(\beta g + \delta h)$ and therefore an *integer*. Q.E.P.

If on the other hand we suppose that M/m is a square integer $= e^2$, the form f will imply the form F. That is, the integers $\alpha, \beta, \gamma, \delta$ can be so determined that

$$\alpha g + \gamma h = \pm eG, \quad \beta g + \delta h = \pm eH$$

For if we find integers $\mathfrak{g}, \mathfrak{h}$ so that $\mathfrak{g}g + \mathfrak{h}h = 1$, we can satisfy these equations by letting

$$\alpha = \pm eG\mathfrak{g} + hz, \quad \gamma = \pm eG\mathfrak{h} - gz$$
$$\beta = \pm eH\mathfrak{g} + hz' \quad \delta = \pm eH\mathfrak{h} - gz'$$

where z, z' can have any integral values. Thus F will be contained in f. Q.E.S. At the same time it is not difficult to see that these formulae give all the values that $\alpha, \beta, \gamma, \delta$ can assume, i.e. all transformations of the form f into F, provided z, z' assume all integral values.

II. Given the form $f = ax^2 + 2bxy + cy^2$ whose determinant does not $=0$, and the form $F = M(GX + HY)^2$ with determinant $=0$ (here as before G, H are relatively prime), I claim *first* that if f implies F, the number M can be represented by the form f; *second*, if M can be represented by f, F will be contained in f; *third*, if in this case all representations of the number M by the form f can be exhibited in general terms by $x = \xi, y = v$, all transformations of the form f into F can be exhibited by $G\xi$, $H\xi, Gv, Hv$. We show all this in the following manner.

1) Suppose f is transformed into F by the substitution $\alpha, \beta, \gamma, \delta$. Find the numbers $\mathfrak{G}, \mathfrak{H}$ so that $\mathfrak{G}G + \mathfrak{H}H = 1$. Then it is manifest that if we let $x = \alpha\mathfrak{G} + \beta\mathfrak{H}$, $y = \gamma\mathfrak{G} + \delta\mathfrak{H}$, the value of the form f will become M and thus M is representable by the form f.

2) If we suppose that $a\xi^2 + 2b\xi v + cv^2 = M$ it is manifest that by the substitution $G\xi, H\xi, Gv, Hv$ the form f will be transformed into F.

3) In this case the substitution $G\xi, H\xi, Gv, Hv$ will exhibit *all* transformations of the form f into F if we presume that ξ, v includes all values of x, y which make $f = M$. We show that as follows. Let $\alpha, \beta, \gamma, \delta$ be any transformations of the form f into F and let as before $\mathfrak{G}G + \mathfrak{H}H = 1$, then among the values of x, y will also be these

$$x = \alpha\mathfrak{G} + \beta\mathfrak{H}, \quad y = \gamma\mathfrak{G} + \delta\mathfrak{H}$$

from which we obtain the substitution

$$G(\alpha\mathfrak{G} + \beta\mathfrak{H}), \quad H(\alpha\mathfrak{H} + \beta\mathfrak{H}), \quad G(\gamma\mathfrak{G} + \delta\mathfrak{H}), \quad H(\gamma\mathfrak{G} + \delta\mathfrak{H})$$

or

$$\alpha + \mathfrak{H}(\beta G - \alpha H), \quad \beta + \mathfrak{G}(\alpha\mathfrak{H} - \beta G)$$
$$\gamma + \mathfrak{H}(\delta G - \gamma H), \quad \delta + \mathfrak{G}(\gamma H - \gamma G)$$

but since

$$a(\alpha X + \beta Y)^2 + 2b(\alpha X + \beta Y)(\gamma Y + \delta Y) + c(\gamma X + \delta Y)^2$$
$$= M(GX + HY)^2$$

we have

$$a(\alpha\delta - \beta\gamma)^2 = M(\delta G - \gamma H)^2$$
$$c(\beta\gamma - \alpha\delta)^2 = M(\beta G - \alpha H)^2$$

and thus [since the determinant of the form f multiplied by

$(\alpha\delta - \beta\gamma)^2$ is equal to the determinant of the form F, i.e. $= 0$, $\alpha\delta - \beta\gamma = 0$]

$$\delta G - \gamma H = 0, \quad \beta G - \alpha H = 0$$

Therefore the substitution in question reduces to $\alpha, \beta, \gamma, \delta$, and the formula we are considering produces *all* transformations of the form f into F.

III. It remains to show how we can exhibit all representations of a given number by a given form with determinant 0. Let this form be $m(gx + hy)^2$, and it is immediately clear that the number must be divisible by m and that the quotient is a square. If therefore we represent the given number by me^2, for all values of x, y we will have $m(gx + hy)^2 = me^2$, and for those same values $gx + hy$ will either $= +e$ or $= -e$. Thus we will have all representations if we find all integral solutions of the linear equations $gx + hy = e$, $gx + hy = -e$. It is clear that these are solvable (if indeed g, h are relatively prime as we presumed). That is, if $\mathfrak{g}, \mathfrak{h}$ are so determined that $\mathfrak{g}g + \mathfrak{h}h = 1$, the first equation will be satisfied by letting $x = \mathfrak{g}e + hz$, $y = \mathfrak{h}e - gz$; the second by letting $x = -\mathfrak{g}e + hz$, $y = -\mathfrak{h}e - gz$ with z any integer. At the same time these formulae will give *all* integral values of x, y, if we let z represent in general any integer.

Having successfully concluded these investigations, we proceed.

▶216. PROBLEM. *To find all solutions by integers of the general[y] indeterminate equation of the second degree with two unknowns*

$$ax^2 + 2bxy + cy^2 + 2dx + 2ey + f = 0$$

(*where* a, b, c, *etc. are any given integers*).

Solution. In place of the unknowns x, y we introduce others

$$p = (b^2 - ac)x + be - cd \quad \text{and} \quad q = (b^2 - ac)y + bd - ae$$

which manifestly will be integers when x, y are integers. Now we have the equation

$$ap^2 + 2bpq + cq^2 + f(b^2 - ac)^2 + (b^2 - ac)(ae^2 - 2bde + cd^2) = 0$$

_____

[y] If we have an equation in which the second, fourth, or fifth coefficient is not even multiplication by 2 will produce the same form that we are considering here.

or if for brevity we substitute the number

$$f(b^2 - ac)^2 + (b^2 - ac)(ae^2 - 2bde + cd^2) = -M$$

we have

$$ap^2 + 2bpq + cq^2 = M$$

We showed in the preceding sections how to find all solutions of this equation, i.e. all representations of the number M by the form (a, b, c). Now if for each value of p, q we determine the corresponding values of x, y with the help of the equations

$$x = \frac{p + cd - be}{b^2 - ac}, \quad y = \frac{q + ae - bd}{b^2 - ac}$$

it is obvious that all these values satisfy the given equation and that there are no integral values of x, y that are not included. If therefore we eliminate all fractions from the values of x, y thus obtained, all the solutions we want will remain.

With regard to this solution we must observe the following.

1) If either M cannot be represented by the form (a, b, c) or if we get no *integral* values of x, y from any representations, the equation cannot be solved by integers at all.

2) When the determinant of the form (a, b, c), i.e. the number $b^2 - ac$ is negative or a positive square and at the same time M does not $= 0$, the number of representations of the number M will be finite and so also the number of solutions for the given equation (if there is any at all) will be finite.

3) When $b^2 - ac$ is nonquadratic positive, or quadratic with $M = 0$, the number M will be represented *in infinitely many different ways* by the form (a, b, c) if it can be represented at all. But since it is impossible to find all these representations *individually* and to test whether they give integral values or fractions for x, y, it is necessary to establish a rule by which we can be certain when no representation at all will produce integral values of x, y (for no matter how many representations are tried, without such a rule we could never be certain). And when some representations give integral values of x, y and others fractions, we must determine how in general to distinguish one from the other.

4) When $b^2 - ac = 0$, values of x, y cannot be determined at all by the preceding formulae. Therefore for this case we will need recourse to a *special method*.

▶ 217. For the case in which $b^2 - ac$ is a positive nonquadratic number, we showed above that all representations of the number M by the form $ap^2 + 2bpq + cq^2$ (if there is any at all) can be exhibited by one or many formulae like the following

$$p = \frac{1}{m}(\mathfrak{A}t + \mathfrak{B}u), \quad q = \frac{1}{m}(\mathfrak{C}t + \mathfrak{D}u)$$

where $\mathfrak{A}, \mathfrak{B}, \mathfrak{C}, \mathfrak{D}$ are given integers, m is the greatest common divisor of the numbers $a, 2b, c$, and t, u are in general all integers satisfying the equation $t^2 - (b^2 - ac)u^2 = m^2$. Since all values of t, u both positive and negative can be taken, for each of these forms we can substitute four others:

$$p = \frac{1}{m}(\mathfrak{A}t + \mathfrak{B}u), \quad q = \frac{1}{m}(\mathfrak{C}t + \mathfrak{D}u)$$

$$p = \frac{1}{m}(\mathfrak{A}t - \mathfrak{B}u), \quad q = \frac{1}{m}(\mathfrak{C}t - \mathfrak{D}u)$$

$$p = \frac{1}{m}(-\mathfrak{A}t + \mathfrak{B}u), \quad q = \frac{1}{m}(-\mathfrak{C}t + \mathfrak{D}u)$$

$$p = -\frac{1}{m}(\mathfrak{A}t + \mathfrak{B}u), \quad q = -\frac{1}{m}(\mathfrak{C}t + \mathfrak{D}u)$$

so that the number of all formulae is now four times what it was before, and t, u are no longer all numbers satisfying the equation $t^2 - (b^2 - ac)u^2 = m^2$ but positive values only. Each of these forms therefore will be considered separately, and we must investigate which values of t, u give integral values for x, y.

From the formula

$$p = \frac{1}{m}(\mathfrak{A}t + \mathfrak{B}u), \quad q = \frac{1}{m}(\mathfrak{C}t + \mathfrak{D}u) \qquad [1]$$

the values of x, y will be these:

$$x = \frac{\mathfrak{A}t + \mathfrak{B}u + mcd - mbe}{m(b^2 - ac)}, \quad y = \frac{\mathfrak{C}t + \mathfrak{D}u + mae - mbd}{m(b^2 - ac)}$$

We showed before that all (positive) values of t form a recurrent series t^0, t', t'', etc. and similarly that the corresponding values of u also form a recurrent series u^0, u', u'', etc.; and that further we are able to assign a number ρ such that relative to any given modulus we have

$$t^\rho \equiv t^0,\ t^{\rho+1} \equiv t',\ t^{\rho+2} \equiv t'' \text{ etc.},\ u^\rho \equiv u^0,\ u^{\rho+1} \equiv u', \text{ etc.}$$

We will take for this modulus the number $m(b^2 - ac)$ and for brevity we will designate by x^0, y^0 the values of x, y which we get by letting $t = t^0,\ u = u^0$; similarly x', y' will indicate the values we get by letting $t = t',\ u = u'$, etc. Then it is not hard to see that if x^h, y^h are integers and ρ is suitably chosen, $x^{h+\rho}, y^{h+\rho}$ and $x^{h+2\rho}, y^{h+2\rho}$ and in general $x^{h+k\rho}, y^{h+k\rho}$ will also be integers; and conversely, if x^h or y^h is a fraction, $x^{h+k\rho}$ or $y^{h+k\rho}$ will also be a fraction. We conclude that if we construct the values of x, y corresponding to the indices $0, 1, 2, \ldots, \rho - 1$ and for none of these are *both* x and y integers, then there is no index at all for which both x and y will have integral values, and thus from formula [1] no integral values of x, y can be deduced. But if there are some indices, say μ, μ', μ'', etc., for which x, y have integral values, then all integral values of x, y that can be obtained from formula [1] will be those whose indices are contained in one of the formulae $\mu + k\rho, \mu' + k\rho, \mu'' + k\rho$, etc. where k is any positive integer including zero.

The other formulae containing the values of p, q can be treated in exactly the same manner. If it should happen that from none of these we get integral values of x, y, then the given equation cannot be solved at all by integers. But when it is solvable, all integral solutions can be exhibited by means of the preceding rules.

▶ 218. When $b^2 - ac$ is a square and $M = 0$, all values of p, q will be included in two formulae of the form $p = \mathfrak{A}z, q = \mathfrak{B}z$; $p = \mathfrak{A}'z, q = \mathfrak{B}'z$. Here z indicates indefinitely any integer, $\mathfrak{A}, \mathfrak{B}, \mathfrak{A}', \mathfrak{B}'$ are given integers, and the first and second do not have a common divisor nor do the third and fourth (art. 212). All integral values of x, y arising from the first formula will be contained in the formula [1]:

$$x = \frac{\mathfrak{A}z + cd - be}{b^2 - ac}, \quad y = \frac{\mathfrak{B}z + ae - bd}{b^2 - ac}$$

and all the others arising from the second formula will be contained in [2]:

$$x = \frac{\mathfrak{A}'z + cd - be}{b^2 - ac}, \quad y = \frac{\mathfrak{B}'z + ae - bd}{b^2 - ac}$$

But since either formula can produce fractional values (unless $b^2 - ac = 1$), it is necessary to separate from the others in each formula those values of z which make both x and y integral. However it is sufficient to consider the first formula only, since exactly the same method can be used for the other.

Since $\mathfrak{A}, \mathfrak{B}$ are relatively prime, we can find two numbers $\mathfrak{a}, \mathfrak{b}$ such that $\mathfrak{a}\mathfrak{A} + \mathfrak{b}\mathfrak{B} = 1$. From this we get

$$(\mathfrak{a}x + \mathfrak{b}y)(b^2 - ac) = z + \mathfrak{a}(cd - be) + \mathfrak{b}(ae - bd)$$

From this it is immediately clear that all values of z that can produce integral values of x, y must be congruent to the number $\mathfrak{a}(be - cd) + \mathfrak{b}(bd - ae)$ relative to the modulus $b^2 - ac$ or must be contained in the formula $(b^2 - ac)z' + \mathfrak{a}(be - cd) + \mathfrak{b}(bd - ae)$ where z' designates any integer. Then in place of formula [1] we easily obtain the following

$$x = (\mathfrak{A}z' + \mathfrak{b})\frac{\mathfrak{A}(bd - ae) - \mathfrak{B}(be - cd)}{b^2 - ac}$$

$$y = (\mathfrak{B}z' - \mathfrak{a})\frac{\mathfrak{A}(bd - ae) - \mathfrak{B}(be - cd)}{b^2 - ac}$$

Manifestly this gives integral values for x, y either for all values of z' or for none. The former will be true when $\mathfrak{A}(bd - ae)$ and $\mathfrak{B}(be - cd)$ are congruent relative to the modulus $b^2 - ac$, the latter when they are not congruent. We can treat formula [2] in exactly the same way and separate the integral solutions (if there is any) from the rest.

▶ 219. When $b^2 - ac = 0$, the form $ax^2 + 2bxy + cy^2$ can be expressed as $m(\alpha x + \beta y)^2$ where m, α, β are integers (art. 215). If we let $\alpha x + \beta y = z$, the given equation will be changed into

$$mz^2 + 2dx + 2ey + f = 0$$

From this and the fact that $z = \alpha x + \beta y$ we deduce that

$$x = \frac{\beta m z^2 + 2ez + \beta f}{2\alpha e - 2\beta d}, \quad y = \frac{\alpha m z^2 + 2dz + \alpha f}{2\beta d - 2\alpha e}$$

It is clear that unless $\alpha e = \beta d$ (we will consider this case separately immediately) values of x, y obtained by letting z have any value in these formulae will satisfy the given equation; therefore it remains only to show how to determine the values of z that will give integral values of x, y.

Since $\alpha x + \beta y = z$ we necessarily choose only *integers* for z. Further it is clear that if any value of z gives integral values to both x and y, all values congruent to z relative to the modulus $2\alpha e - 2\beta d$ will likewise produce integral values. Therefore if we substitute for z all integers inclusively from 0 to $2\alpha e - 2\beta d - 1$ (when $\alpha e - \beta d$ is positive) or to $2\beta d - 2\alpha e - 1$ (when $\alpha e - \beta d$ is negative), and if for none of these values are both x and y integers, then no value of z will produce integral values for x, y, and the given equation cannot be solved by integers. But if x, y have integral values for some of the values of z, say for ζ, ζ', ζ'', etc. (they can also be found by solving the second-degree congruence according to the principles of Section IV), we will find *all* solutions by letting $z = (2\alpha e - 2\beta d)v + \zeta$, $z = (2\alpha e - 2\beta d)v + \zeta'$, etc. with v taking all integral values.

▶ 220. We must find a special method now for the case which we excluded, when $\alpha e = \beta d$. Let us suppose that α, β are relatively prime, which is permissible by article 215.I. We will have $d/\alpha = e/\beta$ and it will be an integer (art. 19). Let us call it h. Then the given equation will take this form:

$$(m\alpha x + m\beta y + h)^2 - h^2 + mf = 0$$

and manifestly this cannot be solved rationally unless $h^2 - mf$ is a square. Let $h^2 - mf = k^2$, and the given equation will be equivalent to the following two:

$$m\alpha x + m\beta y + h + k = 0, \quad m\alpha x + m\beta y + h - k = 0$$

i.e. any solution of the given equation will satisfy one or the other of these equations and vice versa. Obviously the first equation cannot be solved by integers unless $h + k$ is divisible by m, and similarly the second equation will not admit solution by integers unless $h - k$ is divisible by m. These conditions are sufficient for solving each of the equations (because we presume that α, β are relatively prime) and we can find all solutions by using well-known rules.

▶ 221. We will illustrate by example the case in article 217 (because it is the most difficult). Let the given equation be $x^2 + 8xy + y^2 + 2x - 4y + 1 = 0$. By introducing other unknowns $p = 15x - 9$, $q = 15y + 6$, we derive the equation $p^2 + 8py + q^2 = -540$. All solutions by integers of this equation are then contained in the following four formulae:

$$p = 6t, \quad q = -24t - 90u$$
$$p = 6t, \quad q = -24t + 90u$$
$$p = -6t, \quad q = 24t - 90u$$
$$p = -6t, \quad q = 24t + 90u$$

where t, u denote all positive integers satisfying the equation $t^2 - 15u^2 = 1$, and they are expressed by the formula:

$$t = \tfrac{1}{2}[(4 + \sqrt{15})^n + (4 - \sqrt{15})^n]$$
$$u = \frac{1}{2\sqrt{15}}[(4 + \sqrt{15})^n - (4 - \sqrt{15})^n]$$

with n designating all positive integers (including zero). Therefore all values of x, y will be contained in these formulae:

$$x = \frac{1}{5}(2t + 3), \qquad y = -\frac{1}{5}(8t + 30u + 2)$$

$$x = \frac{1}{5}(2t + 3), \qquad y = -\frac{1}{5}(8t - 30u + 2)$$

$$x = \frac{1}{5}(-2t + 3), \quad y = \frac{1}{5}(8t - 30u - 2)$$

$$x = \frac{1}{5}(-2t + 3), \quad y = \frac{1}{5}(8t + 30u - 2)$$

If we properly apply what we have said above, we find that to produce integers we must use in the first and second formulae values of t, u which come from taking n *even*; in the third and fourth from taking n *odd*. The simplest solutions are: $x = 1, -1, -1$; $y = -2, 0, 12$ respectively.

We observe that the solution of the problem in the preceding articles can often be shortened by various means especially devised for excluding useless solutions, i.e. fractions; but we must omit this discussion in order not to prolong our discussion beyond bounds.

▶ 222. Since much of what we have explained has also been treated by other geometers, we cannot pass over their work in silence. The illustrious Lagrange undertook a general discussion concerning the *equivalence of forms*, in *Nouv. mém. Acad. Berlin*, 1773, p. 263, *and* 1775, p. 323 ff. In particular he showed that for a given determinant we can find a finite number of forms so arranged that each form of that determinant is equivalent to one of these, and thus that all forms of a given determinant can be distributed into classes. Later the distinguished Legendre discovered, for the most part by induction, many elegant properties of this classification. We present them below with demonstrations. Thus far no one has used the distinction between proper and improper equivalence, but it is a very effective instrument for more subtle investigations.

Lagrange was the first to completely solve the famous problem of article 216 et sq. (*Hist. Acad. Berlin*, 1767, p. 165, and 1768, p. 181 ff.). There is a solution also (but less complete) in the supplement to Euler's *Algebra*, which we have often praised. Euler himself attacked the same problem (*Comm. acad. Petrop.*, 7, 175; *Novi comm. acad. Petrop.*, 9, 3; ibid. *18*, 185 ff.), but he always restricted his investigation to deriving other solutions from one that he presumed already known; further his methods can give *all* solutions in only a few cases (see Lagrange, *Hist. Acad. Berlin*, 1767, p. 237). Since the last of these three commentaries is of more recent date than the solution of Lagrange, which treats the problem in all generality and leaves nothing to be desired in this respect, it seems that Euler did not then know of that solution (Vol. 18 of the *Commentarii* pertains to the year 1773 and was published in 1774). However, our solution (as well as everything else discussed in this section) is built on completely different principles.

What other authors like Diophantus and Fermat have done in relation to this subject pertains only to special cases; therefore, since we have alluded above to these comments that were worthy of note, we will not discuss them separately.

What has been said thus far concerning forms of the second degree must be regarded as only the first elements of this theory. The area left for further investigation is extremely vast and in what follows we will note anything that seems especially worthy of attention; but this line of argument is so fertile that for the sake of brevity we must pass over many other results we have discovered. And without doubt many more remain hidden, awaiting further investigation. We note here only that forms with determinant 0 are excluded from the limits of our investigations unless we specifically mention otherwise.

▶ 223. We have already showed (art. 175, 195, 211) that given any integer D (positive or negative) we can assign a finite number of forms F, F', F'', etc. with determinant D so arranged that each form with determinant D is properly equivalent to one, and only one, of these. Thus all forms with determinant D (their number is infinite) can be *classified* according to these forms by forming a first class of the complex of all forms properly equivalent to the form F; a second class of the forms which are properly equivalent to the form F', etc.

One form can be selected from each of the classes of forms with given determinant D, and this can be considered as the *representing form* of the whole class. Per se it is entirely arbitrary which form is taken from a given class, but we will always prefer the one that seems to be the *simplest*. The simplicity of a form (a, b, c) certainly ought to be judged by the size of the numbers a, b, c, and thus the form (a', b', c') will be called less simple than (a, b, c) if $a' > a$, $b' > b$, $c' > c$. But this does not give us a complete determination because it would be still undecided whether, e.g., we should choose $(17, 0, -45)$ or $(5, 0, -153)$ as the simpler form. Very often, however, it will be advantageous to observe the following norm.

I. When the determinant D is negative, we will take the reduced forms in each class as the representing forms; when two forms in the same class are reduced forms (they will be opposite; art. 172) we will take the one that has the middle term positive.

II. When the determinant D is positive nonquadratic, we will calculate the period of any reduced form contained in the class. There will be either two ambiguous forms or none (art. 187).

1) In the former case let the ambiguous forms be (A, B, C), (A', B', C'); and let M, M' be the least residues of the numbers B, B' relative to the moduli A, A' respectively (they are to be taken positively unless they $= 0$); finally, let $(D - M^2)/A = N$, $(D - M'M')/A = N'$. Having done this, from the forms $(A, M, -N), (A', M', -N')$ take as representing form the one that seems to be simpler. In judging this, the form whose middle term $= 0$ is to be preferred; when the middle term is 0 in neither or both, the form that has the smaller first term is to be preferred to the other, and when the first terms are equal in size but with different signs, the one with positive sign is to be preferred.

2) When there is no ambiguous form in the period, select that form having the smallest first term without respect to sign. If two forms occur in the same period, one having a positive sign and the other having the same term with a negative sign, the one with the positive sign should be taken. Let the chosen form be (A, B, C). Just as in the previous case deduce from it another form $(A, M, -N)$ [that is, by letting M be the absolutely least residue of B relative to the modulus A, and by letting $N = (D - M^2)/A$]. This form will be the representing form.

If it should happen that the same smallest first term A is common to several forms of the period, treat all these forms in the way we have just outlined and from the resulting forms choose as the representing form that which has the smallest middle term.

Thus, e.g., for $D = 305$ one of the periods is: $(17, 4, -17)$, $(-17, 13, 8)$, $(8, 11, -23)$, $(-23, 12, 7)$, $(7, 16, -7)$, $(-7, 12, 23)$, $(23, 11, -8)$, $(-8, 13, 17)$. First we choose the form $(7, 16, -7)$ and then deduce the representing form $(7, 2, -43)$.

III. When the determinant is a positive square $= k^2$, we look for a reduced form $(A, k, 0)$ in the class under consideration and, if $A < k$ or $= k$, this is to be taken as the representing form. But if $A > k$, take in its place the form $(A - 2k, k, 0)$. The first term will be negative but less than k.

Example. In this way all forms of the determinant -235 will be distributed into sixteen classes with the following representatives: $(1, 0, 235)$, $(2, 1, 118)$, $(4, 1, 59)$, $(4, -1, 59)$, $(5, 0, 47)$, $(10, 5, 26)$,

(13, 5, 20), (13, −5, 20) and eight others which are different from
the preceding only in having outer terms with opposite signs:
(−1, 0, −235), (−2, 1, −118), etc.

All forms with determinant 79 fall into six classes with the
following representations: (1, 0, −79), (3, 1, −26), (3, −1, −26),
(−1, 0, 79), (−3, 1, 26), (−3, −1, 26).

▶ 224. By this classification, forms that are properly equivalent
can be completely separated from all others. Two forms with the
same determinant will be properly equivalent if they are of the
same class; any number which is representable by one of them
will also be representable by the other; and if a number M can
be represented by the first form in such a way that the unknown
values are relatively prime, the same number can be represented
by the other form in the same way and, indeed, so that each
representation belongs to the same value of the expression
$\sqrt{D}$ (mod. M). If however two forms belong to different classes,
they will not be properly equivalent; and if a given number is
representable by one of the forms, nothing can be said about its
being representable by the other. On the other hand, if the
number M can be represented by one of these in such a way that
the values of the unknowns are relatively prime, we are immediately certain that there is no similar representation of the same
number by another form that belongs to the same value of the
expression $\sqrt{D}$ (mod. M) (see art. 167, 168).

It can happen however that two forms F, F' which come from
different classes K, K' are improperly equivalent. In this case
every form from one class will be improperly equivalent to *every*
form from the other class. Every form from K will have an opposite form in K' and the classes will be called *opposite*. Thus in
the first example of the preceding article the third class of forms
with determinant −235 is opposite to the fourth, the seventh to
the eighth; in the second example the second class is opposite to
the third, the fifth to the sixth. Therefore, given any two forms
from opposite classes, any number M that can be represented by
one can also be represented by the other. If in one this occurs by
relatively prime values of the unknowns, it can happen in the
other as well but in such a way that these two representations
belong to opposite values of the expression $\sqrt{D}$ (mod. M). But
the rules given above for the selection of representing forms are so

set up that opposite classes always give rise to opposite representing forms.

Finally, there are classes which are *opposite to themselves*.
That is, if any form and its opposite form are contained in the
same class, it is easy to see that all forms of this class are both
properly and improperly equivalent to one another and that they
will all have their opposites in the class. Any class will have this
property if it contains an ambiguous form and, conversely, an
ambiguous form will be found in any class which is opposite to
itself (art. 163, 165). Therefore we will call it an *ambiguous class*.
So among the classes of forms with determinant -235 there will
be eight ambiguous classes. Their representing forms are $(1, 0, 235)$,
$(2, 1, 118)$, $(5, 0, 47)$, $(10, 5, 26)$, $(-1, 0, -235)$, $(-2, 1, -118)$,
$(-5, 0, -47), (-10, 5, -26)$; among the classes of forms with
determinant 79 there will be two with representing forms:
$(1, 0, -79), (-1, 0, 79)$. But if the representing forms have been
determined according to our rules, the ambiguous classes can be
determined from them without any trouble. That is, for a nonquadratic positive determinant an ambiguous class certainly
corresponds to an ambiguous representing form (art. 194); for a
negative determinant the representing form of an ambiguous
class will either be itself ambiguous or its outer terms will be
equal (art. 172); finally, for a positive square determinant, by
article 210 it is easy to judge whether the representing form is
improperly equivalent to itself and thus whether the class which
it represents is ambiguous.

▶ 225. We showed above (art. 175) that for a form (a, b, c) with
negative determinant the outer terms must have the same sign
and that it will be the same as the sign of the outer terms of any
other form equivalent to this one. If a, c are positive we will call
the form (a, b, c) *positive*, and we will say that the whole class in
which (a, b, c) is contained, and which is made up of only positive
forms, is a *positive class*. Conversely (a, b, c) will be a *negative
form* contained in a *negative class* if a, c are negative. A negative
number cannot be represented by a positive form, nor a positive
number by a negative form. If (a, b, c) is the representing form of a
positive class, $(-a, b, -c)$ will be the representing form of a
negative class. Thus it follows that the number of positive classes
is equal to the number of negative classes, and as soon as we

know one we will know the other. Therefore in investigating forms with a negative determinant it is very often sufficient to consider positive classes, since their properties can be easily transferred to negative classes.

But this distinction holds only in forms with negative determinant; positive and negative numbers can be represented equally by forms with positive determinant, so it is not rare to find in this case the two forms (a, b, c), $(-a, b, -c)$ in the same class.

▶ 226. We call a form (a, b, c) *primitive* if the numbers a, b, c do not have a common divisor; otherwise we will call it *derived* and, indeed, if the greatest common divisor of $a, b, c = m$, the form (a, b, c) will be *derived from the primitive form* $(a/m, b/m, c/m)$. From this definition it is obvious that any form whose determinant is divisible by no square (except 1) is necessarily primitive. Further, from article 161, if we have a primitive form in any given class of forms with determinant D, all forms of this class will be primitive. In this case we will say that the class itself is *primitive*. And it is manifest that, if any form F with determinant D is derived from a primitive form f with determinant D/m^2, and if the classes in which the forms F, f respectively are contained are K, k, all forms of the class K will be derived from the primitive class k; in this case we will say that the class K itself is *derived from the primitive class k*.

If (a, b, c) is a primitive form and a, c are not both even (i.e. either one or both are odd), then clearly not only a, b, c but also $a, 2b, c$ have no common divisor. In this case the form (a, b, c) will be said to be *properly primitive* or simply a *proper form*. But if (a, b, c) is a primitive form and the numbers a, c are both even, obviously the numbers $a, 2b, c$ will have the common divisor 2 (it will also be the greatest divisor) and we will call (a, b, c) an *improperly primitive form* or simply an *improper form*.[z] In this case b will necessarily be odd [for otherwise (a, b, c) would not be a primitive form]; therefore we will have $b^2 \equiv 1$ (mod. 4) and, since ac is divisible by 4, the determinant $b^2 - ac \equiv 1$ (mod. 4). Therefore improper forms will apply only for determinants of

[z] We select here the terms *properly* and *improperly* because there are no others more suitable. We want to warn the reader not to look for any connection between this usage and that of article 157 because there is none. But there should certainly be no fear of ambiguity.

the form $4n + 1$ if they are positive or of the form $-(4n + 3)$ if they are negative. From article 161 it is obvious that if we find a properly primitive form in a given class, all forms of this class will be properly primitive: and that a class implying an improperly primitive form will be made up of only improperly primitive forms. Therefore in the former case we will call the class *properly primitive* or simply *proper*; in the latter case *improperly primitive* or *improper*. Thus, e.g., among the positive classes of forms with determinant -235 there are six proper with representing forms $(1, 0, 235), (4, 1, 59), (4, -1, 59), (5, 0, 47), (13, 5, 20), (13, -5, 20)$ and the same number of negative; and there will be two improper classes in each. All classes of forms with determinant 79 (since they are of the form $4n + 3$) are proper.

If the form (a, b, c) is derived from the primitive form $(a/m, b/m, c/m)$ this last can be either properly or improperly primitive. In the former case m will also be the greatest common divisor of the numbers $a, 2b, c$; in the latter case the greatest common divisor will be $2m$. From this we can make a clear distinction between a *form derived from a properly primitive form* and a *form derived from an improperly primitive form*; and further (since by art. 161 all forms of the same class are the same in this respect) between a *class derived from a properly primitive class* and a *class derived from an improperly primitive class*.

By means of these distinctions we have obtained the first fundamental principle on which we can construct the notion of the distribution of all classes of forms with a given determinant into various *orders*. Given two representations $(a, b, c), (a', b', c')$ we will group them in the *same order* provided the numbers a, b, c have the same greatest common divisor as a', b', c', and $a, 2b, c$ have the same greatest common divisor as $a', 2b', c'$; if one or another of these conditions is lacking, the classes will be assigned to *different orders*. It is immediately clear that all properly primitive classes will constitute one order; and all improperly primitive classes another. If m^2 is a square which divides the determinant D, the classes derived from the properly primitive classes of the determinant D/m^2 will form a special order, and the classes derived from improperly primitive classes of the determinant D/m^2 will form another, etc. If D is divisible by no square (except 1) there will be no orders of derived classes, and thus there will be either only one order (when $D \equiv 2$ or 3 relative to the modulus 4) which is an order of properly primitive classes, or two orders (when $D \equiv 1$ (mod. 4)), that is, an order of properly primitive classes and an order of improperly primitive classes. It is not difficult to establish the following general rule with the help of the principles of the calculus of combinations. We suppose that $D = D'2^{2\mu}a^{2\alpha}b^{2\beta}c^{2\gamma}\ldots$ where D' implies no quadratic factor and a, b, c, etc. are different odd prime numbers (any number can be reduced to this form by letting $\mu = 2$ when D is not divisible by 4; and when D is not divisible by an odd square we let α, β, γ, etc. $= 0$ or, what is the same thing, we omit the factors $a^{2\alpha}, b^{2\beta}, c^{2\gamma}$, etc.); thus we will have either

$$(\mu + 1)(\alpha + 1)(\beta + 1)(\gamma + 1)\ldots$$

orders when $D' \equiv 2$ or 3 (mod. 4); or

$$(\mu + 2)(\alpha + 1)(\beta + 1)(\gamma + 1)\ldots$$

orders when $D' \equiv 1$ (mod. 4). But we will not demonstrate this rule, since it is not difficult nor is it necessary here.

Example. For $D = 45 = 5 \cdot 3^2$ we have six classes with the representations $(1, 0, -45)$, $(-1, 0, 45)$, $(2, 1, -22)$, $(-2, 1, 22)$, $(3, 0, -15)$, $(6, 3, -6)$. These are distributed into four orders. Order I includes two proper classes whose representations are $(1, 0, -45)$, $(-1, 0, 45)$; order II will contain two improper classes whose representations are $(2, 1, -22)$, $(-2, 1, 22)$; order III will contain one class derived from the proper class of the determinant 5, with representation $(3, 0, -15)$; order IV will be made up of one class derived from the improper class of the determinant 5 with representation $(6, 3, -6)$.

Example 2. The positive classes of the determinant $-99 = -11 \cdot 3^2$ will be distributed into four orders: order I will include the following properly primitive classes:[a] $(1, 0, 99), (4, 1, 25)$, $(4, -1, 25), (5, 1, 20), (5, -1, 20), (9, 0, 11)$; order II will contain the improper classes $(2, 1, 50), (10, 1, 10)$; order III will contain the classes derived from proper classes of the determinant -11, namely $(3, 0, 33), (9, 3, 12), (9, -3, 12)$; order IV the single class derived from the improper class of the determinant -11, i.e.

---

[a] For brevity we use the representing forms in place of the classes whose place they take.

(6, 3, 18). Negative classes of this determinant can be distributed into orders in exactly the same way.

We observe that *opposite classes are always assigned to the same order.*

▶ 227. Of all these different orders the order of properly primitive classes merits special attention. For each derived class gets its origin from certain primitive classes (with a smaller determinant), and by considering these the properties of the classes will be made immediately clear. We will show later on that any improperly primitive class is associated with either one properly primitive class or with three (of the same determinant). Further, for negative determinants, we can omit consideration of negative classes, since they will always be linked with certain positive classes. In order to understand more fully the nature of properly primitive classes, we must first explain a certain essential difference according to which the whole order of proper classes can be subdivided into various *genera*. Since we have not yet touched on this very important subject, we will treat it from the beginning.

▶ 228. THEOREM. *There is an infinity of numbers not divisible by a given prime number p which can be represented by a properly primitive form F.*

Demonstration. If the form $F = ax^2 + 2bxy + cy^2$, manifestly p cannot divide all three numbers $a, 2b, c$. Now when a is not divisible by p, it is clear that if we choose a number for x which is not divisible by p, and for y a number which is divisible by p, the value of the form F will not be divisible by p; when c is not divisible by p the same thing will happen by giving to x a value divisible by p and to y a value which is not divisible by p; finally, when both a and c are divisible by p, and $2b$ not divisible, the form F will have a value not divisible by p if we give both x and y values which are not divisible by p. Q.E.D.

It is manifest that the theorem also holds for forms that are *improperly primitive* as long as we do not have $p = 2$.

Since many conditions of this kind can exist at the same time so that the same number is divisible by certain prime numbers but not by others (see art. 32), it is easy to see that the numbers x, y can be determined in infinitely many ways, with the result that the primitive form $ax^2 + 2bxy + cy^2$ acquires a value that is not divisible by any number of given prime numbers excluding, however, the number 2 when the form is improperly primitive. Thus we can present the theorem more generally: *We can always represent by some primitive form an infinity of numbers that are relatively prime to a given number (which is odd when the form is improperly primitive).*

▶ 229. THEOREM. *Let F be a primitive form with determinant D and p a prime number dividing D: then the numbers not divisible by p which can be represented by the form F agree in that they are either all quadratic residues of p, or they are all nonresidues.*

Demonstration. Let $F = (a, b, c)$, and m, m' be any two numbers not divisible by p which can be represented by the form F; that is

$$m = ag^2 + 2bgh + ch^2, \quad m' = ag'g' + 2bg'h' + ch'h'$$

Then we will have

$$mm' = [agg' + b(gh' + hg') + chh']^2 - D[gh' - hg']^2$$

and mm' will be congruent to a square relative to the modulus D and thus also relative to p; i.e. mm' will be a quadratic residue of p. It follows therefore that both m, m' are quadratic residues of p, or they are both nonresidues. Q.E.D.

In a similar way we can prove that when the determinant D is divisible by 4, all odd numbers representable by F are either $\equiv 1$, or $\equiv 3$ (mod. 4). That is, the product of two such numbers will always be a quadratic residue of 4 and therefore $\equiv 1$ (mod. 4); thus they will each be $\equiv 1$ or each $\equiv 3$.

Finally, when D is divisible by 8, the product of any two odd numbers which can be represented by F will be a quadratic residue of 8 and therefore $\equiv 1$ (mod. 8). So in this case all odd numbers representable by F will be $\equiv 1$, or all $\equiv 3$, or all $\equiv 5$, or all $\equiv 7$ (mod. 8).

Thus, e.g., since the number 10 which is a nonresidue of 7 can be represented by the form (10, 3, 17), all numbers not divisible by 7 which can be represented by that form will be nonresidues of 7. Since -3 is representable by the form $(-3, 1, 49)$ and is $\equiv 1$ (mod. 4), all odd numbers representable by this form will be $\equiv 1$ (mod. 4).

If it were necessary for our purposes we could easily show that numbers representable by the form F have no such fixed relationship to a prime number that does not divide D. Both residues and nonresidues of a prime number that does not divide D can be represented equally by the form F. On the contrary, with respect to the numbers 4 and 8, there exists a certain analogy in other cases also, which we cannot overlook.

I. *When the determinant D of the primitive form F is $\equiv 3$ (mod. 4). all odd numbers representable by the form F will be $\equiv 1$, or all $\equiv 3$ (mod. 4).* For if m, m' are two numbers representable by F, the product mm' can be reduced to the form $p^2 - Dq^2$ just as we did above. When each of the numbers m, m' is odd, one of the numbers p, q is necessarily even, the other odd, and therefore one of the squares p^2, q^2 will be $\equiv 0$, the other $\equiv 1$ (mod. 4). Thus $p^2 - Dq^2$ must certainly be $\equiv 1$ (mod. 4), and both m, m' must be $\equiv 1$, or both $\equiv 3$ (mod. 4). So, e.g., no odd numbers other than those of the form $4n + 1$ can be represented by the form (10, 3, 17).

II. *When the determinant D of the primitive form F is $\equiv 2$ (mod. 8): all odd numbers representable by the form F will be either partly $\equiv 1$ and partly $\equiv 7$, or partly $\equiv 3$ and partly $\equiv 5$ (mod. 8).* For let us suppose that m, m' are two odd numbers representable by F whose product mm' can be reduced to the form $p^2 - Dq^2$. When therefore both m, m' are odd, p must be odd (because D is even) and so $p^2 \equiv 1$ (mod. 8); q^2 therefore will be $\equiv 0$ or $\equiv 1$ or $\equiv 4$ and Dq^2 will be either $\equiv 0$ or $\equiv 2$. Thus $mm' = p^2 - Dq^2$ will be either $\equiv 1$ or $\equiv 7$ (mod. 8); if therefore m is either $\equiv 1$ or $\equiv 7$, m' will also be either $\equiv 1$ or $\equiv 7$; and if m is either $\equiv 3$ or $\equiv 5$, m' will also be either $\equiv 3$ or $\equiv 5$. For example, all odd numbers representable by the form (3, 1, 5) are either $\equiv 3$ or $\equiv 5$ (mod. 8), and no numbers of the form $8n + 1$ or $8n + 7$ can be represented by this form.

III. *When the determinant D of a primitive form F is $\equiv 6$ (mod. 8): odd numbers which can be represented by this form are either those that are only $\equiv 1$ and $\equiv 3$ (mod. 8) or those that are only $\equiv 5$ and $\equiv 7$ (mod. 8).* The reader can develop the argument without any trouble. It is exactly like the argument of the preceding (II). Thus, e.g., for the form (5, 1, 7), only those odd numbers can be represented which are either $\equiv 5$ or $\equiv 7$ (mod. 8).

▶ 230.   Therefore all numbers that can be represented by a given primitive form F with determinant D will have a fixed relationship to the individual prime divisors of D (by which they are not divisible). And odd numbers that can be represented by F will have also a fixed relationship to the numbers 4 and 8 in certain cases: to 4 whenever D is either $\equiv 0$ or $\equiv 3$ (mod. 4) and to 8 whenever D is $\equiv 0$ or $\equiv 2$ or $\equiv 6$ (mod. 8)[b]. We will call this type of relationship to each of these numbers the *character* or the *particular character* of the form F, and we will express it in the following manner. When only quadratic residues of a prime number p can be represented by the form F we will assign to it the character Rp, in the opposite case the character Np; similarly we will write 1, 4 when no other numbers can be represented by the form F except those that are $\equiv 1$ (mod. 4). It is immediately clear what characters are meant by 3, 4; 1, 8; 3, 8; 5, 8; 7, 8. Finally, if we have forms by which only those odd numbers that are either $\equiv 1$ or $\equiv 7$ (mod. 8) can be represented, we will assign to them the character 1 *and* 7, 8. It is immediately obvious what we mean by the characters 3 *and* 5, 8; 1 *and* 3, 8; 5 *and* 7, 8.

The different characters of a given primitive form (a, b, c) with determinant D can always be known from one at least of the numbers a, c (manifestly both are representable by that form). For whenever p is a prime divisor of D, certainly one of the numbers a, c will not be divisible by p; for if both were divisible by p, p would also divide b^2 ($= D + ac$) and therefore also b; i.e. the form (a, b, c) would not be primitive. Similarly, in those cases where the form (a, b, c) has a fixed relationship to the number 4 or 8, at least one of the numbers a, c will be odd, and we can find the relationship from that number. Thus, e.g. the character of the form $(7, 0, 23)$ with respect to the number 23 can be inferred from the number 7 to be $N23$, and the character of the same form with respect to the number 7 can be inferred from the number 23 to be $R7$; finally, the character of this form with respect to the number 4 can be found either from the number 7 or from the number 23 to be 3, 4.

Since all numbers that can be represented by a form F contained in a class K are also representable by any other form of this class,

[b] If the determinant is divisible by 8 its relationship to the number 4 can be ignored, since in this case it is already contained in the relationship to 8.

manifestly the different characters of the form F will apply to all the other forms in this class and therefore we can consider these characters as representative of the whole class. The individual characters of a given primitive class can then be known from its representing form. Opposite classes will always have the same characters.

▶ 231. The complex of *all* particular characters of a given form or class constitutes the complete character of this form or class. Thus, e.g., the complete character of the form $(10, 3, 17)$ or of the whole class which it represents will be $1, 4;$ $N7;$ $N23$. Similarly the complete character of the form $(7, 1, -17)$ will be $7, 8;$ $R3;$ $N5$. We omit the particular character $3, 4$ in this case because it is already contained in the character $7, 8$. From these results we will derive a subdivision of the whole order of properly primitive classes (positive when the determinant is negative) of a given determinant into many different *genera* by putting all classes which have the same complete character into the same genus; and into different genera those which have different complete characters. We will assign to each genus those complete characters that are possessed by the classes contained in them. Thus, e.g., for the determinant -161 we have 16 properly primitive positive classes which are distributed into 4 genera in the following way:

Character	*Representing forms of the classes*
$1, 4;$ $R7;$ $R23$	$(1, 0, 161),$ $(2, 1, 81),$ $(9, 1, 18),$ $(9, -1, 18)$
$1, 4;$ $N7;$ $N23$	$(5, 2, 33), (5, -2, 33),$ $(10, 3, 17), (10, -3, 17)$
$3, 4;$ $R7;$ $N23$	$(7, 0, 23),$ $(11, 2, 15), (11, -2, 15),$ $(14, 7, 15)$
$3, 4;$ $N7;$ $R23$	$(3, 1, 54), (3, -1, 54),$ $(6, 1, 27),$ $(6, -1, 27)$

We want to say a few words about the number of different complete characters that are possible a priori.

I. When the determinant D is divisible by 8, with respect to the number 8 four different particular characters are possible; the number 4 will supply no special character (see the preceding article). Further, with respect to each odd prime divisor of D there will be two characters; so if there are m of these divisors, there will be in all 2^{m+2} different complete characters (letting $m = 0$ as often as D is a power of 2).

II. When the determinant D is not divisible by 8 but is divisible by 4 and by m odd prime numbers, we will have in all 2^{m+1} different complete characters.

III. When the determinant is even and not divisible by 4, it will be either $\equiv 2$ or $\equiv 6$ (mod. 8). In the former case there will be two particular characters with respect to the number 8, namely 1 *and* 7, 8 and 3 *and* 5, 8; and the same number in the latter case. Letting therefore the number of odd prime divisors of $D = m$: we will have in all 2^{m+1} different complete characters.

IV. When D is odd, it will be either $\equiv 1$ or $\equiv 3$ (mod. 4). In the latter case there will be two different characters with respect to the number 4, but in the former case this relationship will not enter into the complete character. Thus if we define m as before, in the first case there will be 2^m different complete characters, in the latter case 2^{m+1}.

We want to emphasize that it does not at all follow a priori that there will always be as many genera as there are different possible characters. In our example the number of classes or genera is only half the possible number. There are no positive classes for the characters 1, 4; $R7$; $N23$ or 1, 4; $N7$; $R23$ or 3, 4; $R7$; $R23$ or 3, 4; $N7$; $N23$. We will treat this important subject more fully below.

From now on we will call the form $(1, 0, -D)$, which is un-doubtedly the simplest of all forms with determinant D, the *principal form*; and we will call the whole class in which it is found the *principal* class; and finally the whole genus in which the principal class is contained will be called the *principal genus*. We must therefore clearly distinguish between the principal form, a form of the principal class, and a form of the principal genus; and between a principal class and a class of the principal genus. We will always use this terminology even though perhaps for a particular determinant there are no other classes except the principal class or no other genera except the principal genus. This happens very often, e.g. when D is a positive prime number of the form $4n + 1$.

▶232. Although all that we have said about the characters of forms was for the purpose of finding a subdivision for the whole order of *positive properly primitive classes*, yet nothing prevents us from going farther. We can apply the same rules to negative or to improperly primitive forms and classes, and by the same principle we can subdivide into genera an improperly primitive positive order, a properly primitive negative order, and an improperly primitive negative order. Thus, e.g., after the properly primitive order of forms of the determinant 145 have been subdivided into the two following genera:

R5, R29	(1, 0, −145), (5, 0, −29)
N5, N29	(3, 1, −48), (3, −1, −48)

the improperly primitive order can then also be subdivided into two genera:

R5, R29	(4, 1, −36), (4, −1, −36)
N5, N29	(2, 1, −72), (10, 5, −12)

or, just as the positive classes of the forms of the determinant −129 are distributed into four genera:

1,4; R3; R43	(1, 0, 129), (10, 1, 13), (10, −1, 13)
1, 4; N3; N43	(2, 1, 65), (5, 1, 26), (5, −1, 26)
3, 4; R3; N43	(3, 0, 43), (7, 2, 19), (7, −2, 19)
3, 4; N3; R43	(6, 3, 23), (11, 5, 14), (11, −5, 14)

the negative classes also can be distributed into four orders:

3, 4; N3; N43	(−1, 0, −129), (−10, 1, −13), (−10, −1, −13)
3, 4; R3; R43	(−2, 1, −65), (−5, 1, −26), (−5, −1, −26)
1, 4; N3; R43	(−3, 0, −43), (−7, 2, −19), (−7, −2, −19)
1, 4; R3; N43	(−6, 3, −23), (−11, 5, −14), (−11, −5, −14)

Nevertheless, since the system of negative classes is always very similar to the system of positive classes, it seems to be superfluous to construct it separately. We will show later on how to reduce an improperly primitive order to one that is properly primitive.

Finally, with regard to the subdivision of derived orders no new rules are necessary. For since any derived order has its origin in some primitive order (with a smaller determinant), and the classes of one can be naturally related to the classes of the other, manifestly the subdivision of a derived order can be found from the subdivision of a primitive order.

▶ 233. If the (primitive) form $F = (a, b, c)$ is such that we can find two numbers g, h for which we have $g^2 \equiv a, gh \equiv b, h^2 \equiv c$

relative to a given modulus m, we will say that the form is a quadratic residue of the number m and that $gx + hy$ is a value of the expression $\sqrt{(ax^2 + 2bxy + cy^2)}$ (mod. m), or briefly that (g, h) is the value of the expression $\sqrt{(a, b, c)}$ or $\sqrt{F}$ (mod. m). More generally, if the multiplier M relatively prime to the modulus m is such that we have

$$g^2 \equiv aM, \; gh \equiv bM, \; h^2 \equiv cM \text{ (mod. } m)$$

we will say that $M \cdot (a, b, c)$ or MF is a quadratic residue of m and (g, h) the value of the expression $\sqrt{M(a, b, c)}$ or $\sqrt{MF}$ (mod. m). Thus, e.g., the form $(3, 1, 54)$ is a quadratic residue of 23 and $(7, 10)$ is a value of the expression $\sqrt{(3, 1, 54)}$ (mod. 23); similarly $(2, -4)$ is a value of the expression $\sqrt{5(10, 3, 17)}$ (mod. 23). The use of these definitions will be demonstrated below. Right now we will take note of the following propositions.

I. If $M(a, b, c)$ is a quadratic residue of the number m, m will divide the determinant of the form (a, b, c). For if (g, h) is a value of the expression $\sqrt{M(a, b, c)}$ (mod. m) that is, if

$$g^2 \equiv aM, \; gh \equiv bM, \; h^2 \equiv cM \text{ (mod. } m)$$

we will have $b^2M^2 - acM^2 \equiv 0$ which means that $(b^2 - ac)M^2$ is divisible by m. But since we supposed that M and m are relatively prime, $b^2 - ac$ will be divisible by m.

II. If $M(a, b, c)$ is a quadratic residue of m, and m is either a prime number or the power of a prime number, say $= p^\mu$, the particular character of the form (a, b, c) with respect to the number p will be Rp or Np according as M is a residue or nonresidue of p. This follows immediately, since both aM and cM are residues of m or p, and at least one of the numbers a, c is not divisible by p (art. 230).

Similarly, if (other things being equal) $m = 4$, either $1, 4$ or $3, 4$ will be a particular character of the form (a, b, c) according as $M \equiv 1$ or $M \equiv 3$; and if $m = 8$ or a higher power of the number 2, then, $1, 8$; $3, 8$; $5, 8$; $7, 8$ will be particular characters of the form (a, b, c) according as $M \equiv 1$; 3; 5; 7 (mod. 8) respectively.

III. Conversely, suppose m is a prime number or a power of an odd prime number $= p^\mu$ and that it divides the determinant $b^2 - ac$. Then if M is a residue or nonresidue of p according as the character of the form (a, b, c) with respect to p is Rp or Np, respectively, $M(a, b, c)$ will be a quadratic residue of m. For when a is not divisible by p, aM will be a residue of p and so also of m; if therefore g is a value of the expression $\sqrt{aM}$ (mod. m), h a value of the expression bg/a (mod. m) we will have $g^2 \equiv aM$, $ah \equiv bg$. Thus

$$agh \equiv bg^2 \equiv abM \qquad \text{and} \qquad gh \equiv bM$$

and finally

$$ah^2 \equiv bgh \equiv b^2 M \equiv b^2 M - (b^2 - ac)M \equiv acM$$

Thus $h^2 \equiv cM$; i.e. (g, h) is a value of the expression $\sqrt{M(a, b, c)}$. When a is divisible by m, certainly c will not be. Thus obviously we will get the same result if h assumes a value of the expression $\sqrt{cM}$ (mod. m) and g a value of the expression bh/c (mod. m).

In a similar way it can be shown that if $m = 4$ and divides $b^2 - ac$, and if the number M is taken either $\equiv 1$ or $\equiv 3$ according as 1, 4 or 3, 4 is a particular character of the form (a, b, c), then $M(a, b, c)$ will be a quadratic residue of m. Further if $m = 8$ or a higher power of 2 by which $b^2 - ac$ is divisible, and if M is taken $\equiv 1; 3; 5; 7$ (mod. 8) according as the particular character of the form (a, b, c) with respect to the number 8 demands: then $M(a, b, c)$ will be a quadratic residue of m.

IV. If the determinant of the form $(a, b, c) = D$ and $M(a, b, c)$ is a quadratic residue of D, from the number M we can immediately find all particular characters of the form (a, b, c) both with respect to each of the odd prime divisors of D, and with respect to the number 4 or 8 (if they divide D). Thus, e.g., since $3(20, 10, 27)$ is a quadratic residue of 440, that is to say that $(150, 9)$ is a value of the expression $\sqrt{3(20, 10, 27)}$ relative to the modulus 440, and $3N5, 3R11$, the characters of the form $(20, 10, 27)$ are $3, 8$; $N5$; $R11$. The particular characters with respect to the numbers 4 and 8, as long as they do not divide the determinant, are the only ones that do not have a necessary connection with the number M.

V. Conversely, if the number M is relatively prime to D and contains all particular characters of the form (a, b, c) (except for those characters with respect to the numbers 4, 8 when they do not divide D), then $M(a, b, c)$ will be a quadratic residue of D. For from III it is clear that if D is reduced to the form $\pm A^\alpha B^\beta C^\gamma \ldots$ where A, B, C, etc. are distinct prime numbers, $M(a, b, c)$ will be a quadratic residue of each of the $A^\alpha, B^\beta, C^\gamma$, etc. Now suppose the value of the expression $\sqrt{M(a, b, c)}$ relative to the modulus A^α is $(\mathfrak{A}, \mathfrak{A}')$; relative to the modulus B^β it is $(\mathfrak{B}, \mathfrak{B}')$; relative to the modulus C^γ it is $(\mathfrak{C}, \mathfrak{C}')$, etc. If the numbers g, h are so determined that $g \equiv \mathfrak{A}, \mathfrak{B}, \mathfrak{C}$ etc.; $h \equiv \mathfrak{A}', \mathfrak{B}', \mathfrak{C}'$, etc. relative to the moduli $A^\alpha, B^\beta, C^\gamma$ etc. respectively (art. 32): it is easy to see that we will have $g^2 \equiv aM$, $gh \equiv bM$, $h^2 \equiv cM$ relative to all the moduli $A^\alpha, B^\beta, C^\gamma$, etc. and thus also relative to the modulus D which is their product.

VI. For this reason such numbers as M will be called *characteristic numbers* of the form (a, b, c). Most of these numbers can be found easily by the methods of V as soon as all the particular characters of the form are known. The simplest of them will be found by trial and error. Manifestly if M is a characteristic number of the primitive form of a given determinant D, all numbers that are congruent to M relative to the modulus D will be characteristic numbers of the same form. It is also clear that forms of the same class or of different classes of the same genus have the same characteristic numbers. Consequently any characteristic number of a given form can also be ascribed to the whole class and genus. Finally, 1 is always a characteristic number of any form, class, or principal genus; that is to say, every form of a principal genus is a residue of its determinant.

VII. If (g, h) is a value of the expression $\sqrt{M(a, b, c)}$ (mod. m) and $g' \equiv g$, $h' \equiv h$ (mod. m), then (g', h') will also be a value of the same expression. Such values will be called *equivalent*. On the other hand if (g, h), (g', h') are values of the same expression $\sqrt{M(a, b, c)}$, but it is not true that $g' \equiv g$, $h' \equiv h$ (mod. m), they will be called *different*. Manifestly, whenever (g, h) is a value of such an expression, $(-g, -h)$ will also be a value, and these values will always be different unless $m = 2$. It is also easy to show that the expression $\sqrt{M(a, b, c)}$ (mod. m) cannot have more than two such (opposite) different values when m is either an odd prime number or the power of an odd prime number or $= 4$; when however $m = 8$ or a higher power of the number 2, there will be four in all. Thus from VI we see easily that if the determinant D of the form (a, b, c) is $= \pm 2^\mu A^\alpha B^\beta \ldots$ where A, B, etc. are different odd prime numbers n in number, and M is the characteristic number of the form: then there will be in all either 2^n or 2^{n+1} or 2^{n+2}

different values of the expression $\sqrt{M(a, b, c)}$ (mod. D) according as μ is either < 2 or $= 2$ or > 2. Thus, e.g., there are 16 values of the expression $\sqrt{7(12, 6, -17)}$ (mod. 240), namely $(\pm 18, \mp 11)$, $(\pm 18, \pm 29)$, $(\pm 18, \mp 91)$, $(\pm 18, \pm 109)$, $(\pm 78, \pm 19)$, $(\pm 78, \pm 59)$, $(\pm 78, \mp 61)$, $(\pm 78, \mp 101)$. Since it is not particularly necessary for what follows, we will omit a more detailed demonstration for the sake of brevity.

VIII. Finally we observe that if the determinant of two equivalent forms (a, b, c), (a', b', c') is D, the characteristic number is M, and the former can be transformed into the latter by the substitution $\alpha, \beta, \gamma, \delta$: then from any value of the expression $\sqrt{M(a, b, c)}$ such as (g, h) there follows a value of the expression $\sqrt{M(a', b', c')}$, namely $(\alpha g + \gamma h, \ \beta g + \delta h)$. The reader can demonstrate this without any trouble.

▶ 234. Now that we have explained the distribution of forms into classes, genera, and orders and the general properties that result from these distinctions, we will go on to another very important subject, the *composition* of forms. Thus far no one has considered this point. Before beginning the discussion we will insert the following lemma so as not to interrupt the continuity of our demonstration later on.

LEMMA. *Suppose we have four series of integers*

$$a, a', a'', \ldots a^n; \ b, b', b'' \ldots b^n; \ c, c', c'', \ldots c^n; \ d, d', d'', \ldots d^n$$

Each series has the same number $(n + 1)$ *of terms and they are so arranged that*

$$cd' - dc', \ cd'' - dc'' \text{ etc.}, \ c'd'' - d'c'' \text{ etc., etc.}$$

are respectively

$$= k(ab' - ba'), \ k(ab'' - ba'') \text{ etc.}, \ k(a'b'' - b'a'') \text{ etc., etc.}$$

or in general

$$c^\lambda d^\mu - d^\lambda c^\mu = k(a^\lambda b^\mu - b^\lambda a^\mu)$$

Here k is a given integer; λ, μ are any two unequal integers between 0 and n inclusively with μ being the greater of the two.[c] And there is to be no common divisor for all the $a^\lambda b^\mu - b^\lambda a^\mu$. Given these

---

[c] Taking a as a^0, b as b^0, etc. But manifestly the same equation will hold also when $\lambda = \mu$ or $\lambda > \mu$.

conditions, four integers α, β, γ, δ *can be found so that*

$$\alpha a + \beta b = c, \quad \alpha a' + \beta b' = c', \quad \alpha a'' + \beta b'' = c'', \quad \text{etc.}$$
$$\gamma a + \delta b = d, \quad \gamma a' + \delta b' = d', \quad \gamma a'' + \delta b'' = d'', \quad \text{etc.}$$

or in general

$$\alpha a^v + \beta b^v = c^v, \quad \gamma a^v + \delta b^v = d^v$$

and we have

$$\alpha\delta - \beta\gamma = k$$

Since by hypothesis the numbers $ab' - ba'$, $ab'' - ba''$, etc., $a'b'' - b'a''$ etc. [their number will $= (n + 1)n/2$] do not have a common divisor, we can find the same number of other integers so that if we multiply the first set by the second respectively the sum of the products will $= 1$ (art. 40). We will designate these multipliers by $(0, 1)$, $(0, 2)$ etc., $(1, 2)$ etc., or in general the multiplier of $a^\lambda b^\mu - b^\lambda a^\mu$ by (λ, μ) and

$$\sum(\lambda, \mu)(a^\lambda b^\mu - b^\lambda a^\mu) = 1$$

(By the letter Σ we indicate the sum of all values of the expression which it precedes when we give successively to λ, μ all unequal values between 0 and n, and such that $\mu > \lambda$). Now if we let

$$\sum(\lambda, \mu)(c^\lambda b^\mu - b^\lambda c^\mu) = \alpha, \quad \sum(\lambda, \mu)(a^\lambda c^\mu - c^\lambda a^\mu) = \beta$$
$$\sum(\lambda, \mu)(d^\lambda b^\mu - b^\lambda d^\mu) = \gamma, \quad \sum(\lambda, \mu)(a^\lambda d^\mu - d^\lambda a^\mu) = \delta$$

these numbers α, β, γ, δ will have the desired properties.

Demonstration I. If v is any integer between 0 and n, we have

$$\alpha a^v + \beta b^v = \sum(\lambda, \mu)(c^\lambda b^\mu a^v - b^\lambda c^\mu a^v + a^\lambda c^\mu b^v - c^\lambda a^\mu b^v)$$

$$= \frac{1}{k}\sum(\lambda, \mu)(c^\lambda d^\mu c^v - d^\lambda c^\mu c^v)$$

$$= \frac{1}{k}c^v\sum(\lambda, \mu)(c^\lambda d^\mu - d^\lambda c^\mu)$$

$$= c^v\sum(\lambda, \mu)(a^\lambda b^\mu - b^\lambda a^\mu) = c^v$$

And by a similar calculation we have

$$\gamma a^v + \delta b^v = d^v \qquad \text{Q.E.P.}$$

II. Since therefore

$$c^\lambda = \alpha a^\lambda + \beta b^\lambda, \; c^\mu = \alpha a^\mu + \beta b^\mu$$

we get

$$c^\lambda b^\mu - b^\lambda c^\mu = \alpha(a^\lambda b^\mu - b^\lambda a^\mu)$$

and similarly

$$a^\lambda c^\mu - c^\lambda a^\mu = \beta(a^\lambda b^\mu - b^\lambda a^\mu)$$
$$d^\lambda b^\mu - b^\lambda d^\mu = \gamma(a^\lambda b^\mu - b^\lambda a^\mu)$$
$$a^\lambda d^\mu - d^\lambda a^\mu = \delta(a^\lambda b^\mu - b^\lambda a^\mu)$$

From these formulae we can get the values of $\alpha, \beta, \gamma, \delta$ much more easily as long as λ, μ are chosen so that $a^\lambda b^\mu - b^\lambda a^\mu$ does not $= 0$. This can certainly be done because by hypothesis there is no common divisor for all the $a^\lambda b^\mu - b^\lambda a^\mu$ and so they cannot *all* $= 0$. From these same equations, if we multiply the first by the fourth, the second by the third, and subtract, we have

$$(\alpha\delta - \beta\gamma)(a^\lambda b^\mu - b^\lambda a^\mu)^2 = (a^\lambda b^\mu - b^\lambda a^\mu)(c^\lambda d^\mu - d^\lambda c^\mu)$$
$$= k(a^\lambda b^\mu - b^\lambda a^\mu)^2$$

and therefore necessarily

$$\alpha\delta - \beta\gamma = k \qquad\qquad \text{Q.E.S.}$$

▶ 235. If the form $AX^2 + 2BXY + CY^2 \ldots F$ is transformed into the product of two forms

$$ax^2 + 2bxy + cy^2 \ldots f \text{ and } a'x'x' + 2b'x'y' + c'y'y' \ldots f'$$

by the substitution

$$X = pxx' + p'xy' + p''yx' + p'''yy'$$
$$Y = qxx' + q'xy' + q''yx' + q'''yy'$$

(For the sake of brevity in what follows we shall express this operation thus: If F is transformed into ff' by the substitution p, p', p'', p'''; q, q', q'', q'''[d], we shall say simply that the form F is

---

[d] In this designation we must pay careful attention to the order of the coefficients p, p', etc. and of the forms f, f'. It is easy to see that if the order of the forms f, f' is changed so that the former becomes the latter, the coefficients p', q' must be interchanged with p'', q'' and the others left unchanged.

tranformable into *ff'*. If, further, this transformation is so constructed that the six numbers

$$pq' - qp', \ pq'' - qp'', \ pq''' - qp''', \ p'q'' - q'p'', \ p'q''' - q'p''',$$
$$p''q''' - q''p'''$$

do not have a common divisor, we will call the form *F* a *composite* of the forms *f*, *f'*.

We will begin this discussion with the most general supposition that the form *F* is transformed into *ff'* by the substitution p, p', p'', p'''; q, q', q'', q''' and find out what follows from this. Manifestly the following nine equations are completely equivalent to this supposition (i.e. whenever these equations are true, *F* will be transformed by the given substitution into *ff'*, and vice versa):

$$Ap^2 + 2Bpq + Cq^2 = aa' \quad [1]$$

$$Ap'p' + 2Bp'q' + Cq'q' = ac' \quad [2]$$

$$Ap''p'' + 2Bp''q'' + Cq''q'' = ca' \quad [3]$$

$$Ap'''p''' + 2Bp'''q''' + Cq'''q''' = cc' \quad [4]$$

$$App' + B(pq' + qp') + Cqq' = ab' \quad [5]$$

$$App'' + B(pq'' + qp'') + Cqq'' = ba' \quad [6]$$

$$Ap'p''' + B(p'q''' + q'p''') + Cq'q''' = bc' \quad [7]$$

$$Ap''p''' + B(p''q''' + q''p''') + Cq''q''' = cb' \quad [8]$$

$$A(pp''' + p'p'') + B(pq''' + qp''' + p'q'' + q'p'') + \\ C(qq''' + q'q'') = 2bb' \quad [9]$$

Let the determinants of the forms F, f, f' be D, d, d' respectively; and let the greatest common divisors of the numbers $A, 2B, C$; $a, 2b, c$; $a', 2b', c'$, be M, m, m' respectively (we suppose that all of these numbers are taken positively). Further let the six integers $\mathfrak{A}, \mathfrak{B}, \mathfrak{C}, \mathfrak{A}', \mathfrak{B}', \mathfrak{C}'$ be determined so that

$$\mathfrak{A}a + 2\mathfrak{B}b + \mathfrak{C}c = m, \ \mathfrak{A}'a' + 2\mathfrak{B}'b' + \mathfrak{C}'c' = m'$$

Finally let us designate the numbers

$$pq' - qp', \ pq'' - qp'', \ pq''' - qp''', \ p'q'' - q'p'',$$
$$p'q''' - q'p''', \ p''q''' - q''p'''$$

by P, Q, R, S, T, U respectively and let their greatest common divisor taken positively $= k$. Now, letting

$$App''' + B(pq''' + qp''') + Cqq''' = bb' + \Delta \qquad [10]$$

from equation [9] we get

$$Ap'p'' + B(p'q'' + q'p'') + Cq'q'' = bb' - \Delta \qquad [11]$$

From these eleven equations we educe the following[e]:

$$DP^2 = d'a^2 \qquad [12]$$

$$DP(R - S) = 2d'ab \qquad [13]$$

$$DPU = d'ac - (\Delta^2 - dd') \qquad [14]$$

$$D(R - S)^2 = 4d'b^2 + 2(\Delta^2 - dd') \qquad [15]$$

$$D(R - S)U = 2d'bc \qquad [16]$$

$$DU^2 = d'c^2 \qquad [17]$$

$$DQ^2 = da'a' \qquad [18]$$

$$DQ(R + S) = 2da'b' \qquad [19]$$

$$DQT = da'c' - (\Delta^2 - dd') \qquad [20]$$

$$D(R + S)^2 = 4db'b' + 2(\Delta^2 - dd') \qquad [21]$$

$$D(R + S)T = 2db'c' \qquad [22]$$

$$DT^2 = dc'c' \qquad [23]$$

And from these we deduce the following two:

$$0 = 2d'a^2(\Delta^2 - dd')$$
$$0 = (\Delta^2 - dd')^2 - 2d'ac(\Delta^2 - dd')$$

the former from equations $12 \cdot 15 - 13 \cdot 13$, the latter from equations $14 \cdot 14 - 12 \cdot 17$; and it is easy to see that $\Delta^2 - dd' = 0$ whether or

[e] The origin of these equations is as follows: [12] from $5 \cdot 5 - 1 \cdot 2$; [13] from $5 \cdot 9 - 1 \cdot 7 - 2 \cdot 6$; [14] from $10 \cdot 11 - 6 \cdot 7$; [15] from $5 \cdot 8 + 5 \cdot 8 + 10 \cdot 10 + 11 \cdot 11 - 1 \cdot 4 - 2 \cdot 3 - 6 \cdot 7 - 6 \cdot 7$; [16] from $8 \cdot 9 - 3 \cdot 7 - 4 \cdot 6$; [17] from $8 \cdot 8 - 3 \cdot 4$. We can deduce the six remaining equations by exactly the same schemes if we replace equations [2], [5], [7] by equations [3], [6], [8] respectively and leave [1], [4], [9], [10], [11] where they appear. E.g. equation [18] come from $6 \cdot 6 - 1 \cdot 3$ etc.

not $a = 0$.[f] Let us suppose therefore that we delete $\Delta^2 - dd'$ from equations [14, 15, 20, 21].

Now let

$$\mathfrak{A}P + \mathfrak{B}(R - S) + \mathfrak{C}U = mn'$$

$$\mathfrak{A}'Q + \mathfrak{B}'(R + S) + C'T = m'n$$

(where n, n' can be fractions as long as $mn', m'n$ are integers). From equations [12–17] then we deduce

$$Dm^2n'n' = d'(\mathfrak{A}a + 2\mathfrak{B}b + \mathfrak{C}c)^2 = d'm^2$$

and from equations [18–23]

$$Dm'm'n^2 = d(\mathfrak{A}'a' + 2\mathfrak{B}'b' + \mathfrak{C}'c')^2 = dm'm'$$

We have therefore $d = Dn^2$, $d' = Dn'n'$ and from this we get a FIRST CONCLUSION: *Determinants of the forms F, f, f' are related to one another as squares;* and a SECOND: *D always divides the numbers dm'm', d'm².* It is clear therefore that D, d, d' have the same sign and that no form can be transformable into the product if its determinant is greater than the greatest common divisor of the numbers $dm'm', d'm^2$.

Multiply equations [12, 13, 14] respectively by $\mathfrak{A}, \mathfrak{B}, \mathfrak{C}$ and similarly equations [13, 15, 16] and [14, 16, 17] by the same numbers. Add the three products. Divide the sum by Dmn', writing $Dn'n'$ for d'. The result gives us

$$P = an', R - S = 2bn', U = cn'$$

Similarly by multiplying equations [18, 19, 20] and [19, 21, 22] and [20, 22, 23] respectively by $\mathfrak{A}', \mathfrak{B}', \mathfrak{C}'$ we obtain

$$Q = a'n, R + S = 2b'n, T = c'n$$

From this we have a THIRD CONCLUSION: *The numbers a, 2b, c are proportional to the numbers P, R − S, U. If the ratio of the first to the second is taken as 1 to n', n' will be the square root of d'/D; similarly the numbers a', 2b', c' are proportional to the numbers Q, R + S, T and if we take the ratio as 1 to n, n will be the square root of d/D.*

[f] This manner of deriving the equation $\Delta^2 = dd'$ suffices for our present purposes. We could have deduced directly from equations [1] to [11] that $0 = (\Delta^2 - dd')^2$. It would have been a more elegant analysis but much too long at this point.

Now the quantities n, n' can be either positive or negative roots of d/D, d'/D', so we will make a distinction that seems sterile at first sight, but its use will be clear in what follows. We will say that in the transformation of the form F into ff' the form f is taken *directly* when n is positive, *inversely* when n is negative; similarly f' is taken directly or inversely according as n' is positive or negative. Given the condition that $k = 1$, the form F will be said to be composed directly of each of the forms f, f' directly *or* of each of them inversely *or* of f directly and f' inversely *or* of f inversely and f' directly, according as n, n' are both positive or both negative or the former positive, the latter negative or the former negative, the latter positive. And it is easy to see that these relations do not depend on the order in which the forms are taken (see the first note of this article).

We note further that k, the greatest common divisor of the numbers P, Q, R, S, T, U, divides the numbers $mn', m'n$ (as is clear from the values which we established above). Therefore the square k^2 divides $m^2n'n'$, $m'm'n^2$, and Dk^2 divides $d'm^2, dm'm'$. But conversely every common divisor of $mn', m'n$ divides k. Let e be such a divisor: evidently it will divide $an', 2bn', cn', a'n, 2b'n, c'n$; i.e. the numbers $P, R - S, U, Q, R + S, T$ and so also $2R$ and $2S$. Now if $2R/e$ is an odd number, $2S/e$ must also be odd (because the sum and difference are even) and the product must also be odd. This product $= 4(b'b'n^2 - b^2n'n')/e^2 = 4(d'n^2 + a'c'n^2 - dn'n' - acn'n')/e^2 = 4(a'c'n^2 - acn'n')/e^2$ and therefore even, because e divides $a'n, c'n, an', cn'$. Thus $2R/e$ is necessarily even and both R and S are divisible by e. Since therefore e divides all six P, Q, R, S, T, U, it will also divide k, their greatest common divisor. Q E.D. We conclude that k is the greatest common divisor of the numbers $mn', m'n$, and Dk^2 *will be the greatest common divisor of the numbers $dm'm', d'm^2$*. This is our FOURTH CONCLUSION. Now it is clear that whenever F is composed of f and f', D will be the greatest common divisor of the numbers $dm'm', d'm^2$ and vice versa. These properties could also be used as the definition of a composite form. The form, therefore, which is composed of the forms f, f', has the greatest possible determinant of all forms that are transformable into the product ff'.

Before we proceed farther we must first define the value of Δ more accurately. We showed that $\Delta = \sqrt{dd'} = \sqrt{D^2n^2n'n'}$, but we have not as yet determined its *sign*. For this purpose we deduce from the fundamental equations [1] to [11] that $DPQ = \Delta aa'$ (we obtain this from equations $5 \cdot 6 - 1 \cdot 11$). Thus $Daa'nn' = \Delta aa'$ and unless one of the numbers $a, a' = 0$, we have $\Delta = Dnn'$. In exactly the same way, from the fundamental equations we can deduce eight others in which we have Dnn' on the left and Δ on the right multiplied by $2ab', ac', 2ba', 4bb', 2bc', ca', 2cb', cc'$.[8] Now since not all $a, 2b, c$ nor all $a', 2b', c'$ can $= 0$, in all cases $\Delta = Dnn'$ and Δ has the same sign as D, d, d' or the opposite, according as n, n' have the same sign or different ones.

We observe that the numbers $aa', 2ab', ac', 2ba', 4bb', 2bc', ca', 2cb', cc', 2bb' + 2\Delta, 2bb' - 2\Delta$ are all divisible by mm'. This is obvious for the first nine numbers. For the other two we can show, as we did earlier, that R and S are divisible by e. It is clear that $4bb' + 4\Delta$ and $4bb' - 4\Delta$ are divisible by mm' (since $4\Delta = \sqrt{16dd'}$ and $4d$ is divisible by m^2, $4d'$ by $m'm'$, and thus $16dd'$ by $m^2m'm'$ and 4Δ by mm') and that the difference of the quotients is even. It is easy to show that the product of the quotients is even, and so each quotient is even and $2bb' + 2\Delta, 2bb' - 2\Delta$ are divisible by mm'.

Now from the eleven fundamental equations we derive the following six:

$$AP^2 = aa'q'q' - 2ab'qq' + ac'q^2$$

$$AQ^2 = aa'q''q'' - 2ba'qq'' + ca'q^2$$

$$AR^2 = aa'q'''q''' - 2(bb' + \Delta)qq''' + cc'q^2$$

$$AS^2 = ac'q''q'' - 2(bb' - \Delta)q'q'' + ca'q'q'$$

$$AT^2 = ac'q'''q''' - 2bc'q'q''' + cc'q'q'$$

$$AU^2 = ca'q'''q''' - 2cb'q''q''' + cc'q''q''$$

It follows therefore that all AP^2, AQ^2, etc. are divisible by mm', and since k^2 is the greatest common divisor of the numbers P^2, Q^2, R^2, etc. Ak^2 will also be divisible by mm'. If we substitute for $a, 2b, c, a', 2b', c'$ their values P/n' etc. or $(pq' - qp')/n'$ etc. they will be changed into six other equations in which on the right-hand side we will have products of the quantity $(q'q'' - qq''')/nn'$ into P^2, Q^2, R^2, etc. We will leave this very easy calculation to

---

[8] The reader can verify this analysis easily. For the sake of brevity we omit it.

the reader. It follows finally (since not all P^2, Q^2, etc. can $= 0$) that $Ann' = q'q'' - qq'''$.

Similarly, from the fundamental equations we can derive six other equations which differ from the preceding in that everywhere they appear, A and q, q', q'', q''' are replaced by C and p, p', p'', p''' respectively. For brevity we omit the details. And finally in the same way as above it follows that Ck^2 is divisible by mm' and $Cnn' = p'p'' - pp'''$.

And again we can deduce six other equations from the same data:

$$BP^2 = -aa'p'q' + ab'(pq' + qp') - ac'pq$$

$$BQ^2 = -aa'p''q'' + ba'(pq'' + qp'') - ca'pq$$

$$BR^2 = -aa'p'''q''' + (bb' + \Delta)(pq''' + qp''') - cc'pq$$

$$BS^2 = -ac'p''q'' + (bb' - \Delta)(p'q'' + q'p'') - ca'p'q'$$

$$BT^2 = -ac'p'''q''' + bc'(p'q''' + q'p''') - cc'p'q'$$

$$BU^2 = -ca'p'''q''' + cb'(p''q''' + q''p''') - cc'p''q''$$

and from this as before we conclude that $2Bk^2$ is divisible by mm' and $2Bnn' = pq''' + qp''' - p'q'' - q'p''$.

Now since Ak^2, $2Bk^2$, Ck^2 are divisible by mm', it is easy to see that Mk^2 must also be divisible by mm'. From the fundamental equations we know that M divides aa', $2ab'$, ac', $2ba'$, $4bb'$, $2bc'$, ca', $2cb'$, cc' and thus also am', $2bm'$, cm' (which are the greatest common divisors of the first three, the middle three, and the last three respectively); and finally that it also divides mm', which is the greatest common divisor of all of these. In this case therefore where the form F is composed of the forms f, f', that is to say $k = 1$, M necessarily $= mm'$. This is our FIFTH CONCLUSION.

If we designate the greatest common divisor of the numbers A, B, C by $\mathfrak{M}$, it will $= M$ (when the form F is properly primitive or derived from a properly primitive form) or $= M/2$ (when F is improperly primitive or derived from an improperly primitive form); similarly if we designate the greatest common divisors of the numbers a, b, c; a', b', c' by $\mathfrak{m}, \mathfrak{m}'$ respectively, $\mathfrak{m}$ will either $= m$ or $= m/2$ and $\mathfrak{m}'$ will $= m'$ or $= m'/2$. Now it is clear that $\mathfrak{m}^2$ divides d, $\mathfrak{m}'\mathfrak{m}'$ divides d'. Therefore $\mathfrak{m}^2\mathfrak{m}'\mathfrak{m}'$ divides dd' or

Δ^2 and $\mathfrak{m}\mathfrak{m}'$ divides Δ. Thus from the six last equations for BP^2 etc. it follows that $\mathfrak{m}\mathfrak{m}'$ divides Bk^2 and (since it also divides Ak^2, Ck^2) $\mathfrak{M}k^2$. Whenever therefore F is composed of f, f', $\mathfrak{m}\mathfrak{m}'$ will divide $\mathfrak{M}$. And when in this case each f, f' is properly primitive or derived from a properly primitive form or $\mathfrak{m}\mathfrak{m}' = mm' = M$, then $\mathfrak{M} = M$ or F is a similar form. But when under the same conditions one or both of f, f' are improperly primitive or derived from an improperly primitive form, then (if the form f, for example, is such) from the fundamental equations it follows that aa', $2ab'$, ac', ba', $2bb'$, bc', ca', $2cb'$, cc' are divisible by $\mathfrak{M}$ and so also am', bm', cm' and $\mathfrak{m}\mathfrak{m}' = mm'/2 = M/2$; in this case $\mathfrak{M} = M/2$ and the form F is either improperly primitive or derived from an improperly primitive form. This is our SIXTH CONCLUSION.

Finally we observe that, if we presume the following nine equations to be true,

$$an' = P, \ 2bn' = R - S, \ cn' = U$$

$$a'n = Q, \ 2b'n = R + S, \ c'n = T$$

$$Ann' = q'q'' - qq''', \ 2Bnn' = pq''' + qp''' - p'q'' - q'p'',$$
$$Cnn' = p'p'' - pp'''$$

(in what follows we will designate these conditions by Ω, since we will return to them very often) then, treating n, n' as unknown but neither $= 0$, we find by an easy substitution that the fundamental equations [1] to [9] are necessarily true, that is that the form (A, B, C) will be transformed into the product of the forms $(a, b, c)(a', b', c')$ by the substitution p, p', p'', p'''; q, q', q'', q'''. And we will also have

$$b^2 - ac = n^2(B^2 - AC), \quad b'b' - a'c' = n'n'(B^2 - AC)$$

The calculation, which would be too long to include here, we leave to the reader.

▶ 236. PROBLEM. *Given two forms whose determinants are equal or at least related to one another as squares: to find a form composed of these two.*

Solution. Let $(a, b, c) \ldots f, (a', b', c') \ldots f'$ be the forms to be composed; d, d' their determinants; m, m' the greatest common divisors of the numbers $a, 2b, c$; $a', 2b', c'$ respectively; D the

greatest common divisor of the numbers $dm'm', d'm^2$ taken with
the same sign as d, d'. Then $dm'm'/D$, $d'm^2/D$ will be relatively
prime positive numbers and their product a square; therefore
each of them will be a square (art. 21). Thus $\sqrt{d/D}, \sqrt{d'/D}$ will be
rational quantities which we will let $= n, n'$, and we will choose for
n a positive or negative value according as the form f must enter
into the composition directly or inversely. In a similar way we will
determine the sign of n' from the manner in which f' must enter
the composition; $mn', m'n$ will therefore be relatively prime
integers; n and n' can be fractions. Now we observe that $an', cn',$
$a'n, c'n, bn' + b'n, bn' - b'n$ are integers. This is obvious for the
the first four (since $an' = amn'/m$ etc.); for the last two we prove
it as we proved in the preceding article that R and S are divisible
by e.

Let us take now ad libitum four integers $\mathfrak{Q}, \mathfrak{Q}', \mathfrak{Q}'', \mathfrak{Q}'''$ with
only one condition, that the four quantities on the left of the
following equation (I) do not all $= 0$. Now form the equations:

$$\mathfrak{Q}'an' + \mathfrak{Q}''a'n + \mathfrak{Q}'''(bn' + b'n) = \mu q \qquad \text{(I)}$$

$$-\mathfrak{Q}an' + \mathfrak{Q}'''c'n - \mathfrak{Q}''(bn' - b'n) = \mu q'$$

$$\mathfrak{Q}'''cn' - \mathfrak{Q}a'n + \mathfrak{Q}'(bn' - b'n) = \mu q''$$

$$-\mathfrak{Q}''cn' - \mathfrak{Q}'c'n - \mathfrak{Q}(bn' + b'n) = \mu q'''$$

so that q, q', q'', q''' are integers not having a common divisor.
This can be done by taking for μ the greatest common divisor
of the four numbers which are on the left of the equations. Now
by article 40 we can find four integers $\mathfrak{P}, \mathfrak{P}', \mathfrak{P}'', \mathfrak{P}'''$ such that

$$\mathfrak{P}q + \mathfrak{P}'q' + \mathfrak{P}''q'' + \mathfrak{P}'''q''' = 1$$

Having done this we determine the numbers p, p', p'', p''' by the
following equations:

$$\mathfrak{P}'an' + \mathfrak{P}''a'n + \mathfrak{P}'''(bn' + b'n) = p \qquad \text{(II)}$$

$$-\mathfrak{P}an' + \mathfrak{P}'''c'n - \mathfrak{P}''(bn' - b'n) = p'$$

$$\mathfrak{P}'''cn' - \mathfrak{P}a'n + \mathfrak{P}'(bn' - b'n) = p''$$

$$-\mathfrak{P}''cn' - \mathfrak{P}'c'n - \mathfrak{P}(bn' + b'n) = p'''$$

Now make the following substitutions:

$$q'q'' - qq''' = Ann', \quad pq''' + qp''' - p'q'' - q'p'' = 2Bnn',$$
$$p'p'' - pp''' = Cnn'$$

Then A, B, C will be integers and the form $(A, B, C)\ldots F$ will be composed of the forms f, f'.

Demonstration. I. From (I) we derive the following four equations:

$$0 = q'cn' - q''c'n - q'''(bn' - b'n) \qquad (III)$$
$$0 = qcn' + q'''a'n - q''(bn' + b'n)$$
$$0 = q'''an' + qc'n - q'(bn' + b'n)$$
$$0 = q''an' - q'a'n - q(bn' - b'n)$$

II. Now let us suppose that the integers $\mathfrak{A}$, $\mathfrak{B}$, $\mathfrak{C}$, $\mathfrak{A}'$, $\mathfrak{B}'$, $\mathfrak{C}'$, $\mathfrak{N}$, $\mathfrak{N}'$ are so determined that

$$\mathfrak{A}a + 2\mathfrak{B}b + \mathfrak{C}c = m$$
$$\mathfrak{A}'a' + 2\mathfrak{B}'b' + \mathfrak{C}'c' = m'$$
$$\mathfrak{N}m'n + \mathfrak{N}'mn' = 1$$

We will then have

$$\mathfrak{A}a\mathfrak{N}'n' + 2\mathfrak{B}b\mathfrak{N}'n' + \mathfrak{C}c\mathfrak{N}'n' + \mathfrak{A}'a'\mathfrak{N}n + 2\mathfrak{B}'b'\mathfrak{N}n + \mathfrak{C}'c'\mathfrak{N}n = 1$$

From this and from equations (III), if we let

$$-q'\mathfrak{A}\mathfrak{N}' - q''\mathfrak{A}'\mathfrak{N} - q'''(\mathfrak{B}\mathfrak{N}' + \mathfrak{B}'\mathfrak{N}) = \mathfrak{q}$$
$$q\mathfrak{A}\mathfrak{N}' - q'''\mathfrak{C}'\mathfrak{N} + q''(\mathfrak{B}\mathfrak{N}' - \mathfrak{B}'\mathfrak{N}) = \mathfrak{q}'$$
$$-q'''\mathfrak{C}\mathfrak{N}' + q\mathfrak{A}'\mathfrak{N} - q'(\mathfrak{B}\mathfrak{N}' - \mathfrak{B}'\mathfrak{N}) = \mathfrak{q}''$$
$$q''\mathfrak{C}\mathfrak{N}' + q'\mathfrak{C}\mathfrak{N} + q(\mathfrak{B}\mathfrak{N}' + \mathfrak{B}'\mathfrak{N}) = \mathfrak{q}'''$$

we will get

$$\mathfrak{q}'an' + \mathfrak{q}''a'n + \mathfrak{q}'''(bn' + b'n) = q \qquad (IV)$$
$$-\mathfrak{q}an' + \mathfrak{q}'''c'n - \mathfrak{q}''(bn' - b'n) = q'$$
$$\mathfrak{q}'''cn' - \mathfrak{q}a'n + \mathfrak{q}'(bn' - b'n) = q''$$
$$-\mathfrak{q}''cn' - \mathfrak{q}'c'n - \mathfrak{q}(bn' + b'n) = q'''$$

When $\mu = 1$ these equations are not necessary and in their place we can use equations (I) which are entirely analogous. Now from equations (II), (IV) we determine the values of Ann'. $2Bnn'$, Cnn' (i.e. of the numbers $q'q'' - qq'''$ etc.) and delete the values that cancel one another, and we find that the different terms are the product of integers by nn' or of integers by $dn'n'$ or of integers by $d'nn$. Further, all the terms of $2Bnn'$ contain the factor 2. We conclude from this that A, B, C are integers (because $dn'n' = d'n^2$ and therefore $dn'n'/nn' = d'n^2/nn' = \sqrt{dd'}$ are integers).   Q.E.P.

III. If we take the values of p, p', p'', p''' from (II), use equations (III) and the following

$$\mathfrak{P}q + \mathfrak{P}'q' + \mathfrak{P}''q'' + \mathfrak{P}'''q''' = 1$$

we find that

$$pq' - qp' = an', \quad pq''' - qp''' - p'q'' + q'p'' = 2bn',$$
$$p''q''' - q''p''' = cn'$$
$$pq'' - qp'' = a'n, \quad pq''' - qp''' + p'q'' - q'p'' = 2b'n,$$
$$p'q''' - q'p''' = c'n$$

These equations are identical with the first six (Ω) of the preceding article. The remaining three are part of the hypothesis. Therefore (end of the same article) the form F will be transformed into ff' by the substitution p, p', p'', p'''; q, q', q'', q'''; its determinant will $= D$, or in other words it will equal the greatest common divisor of the numbers $dm'm', d'm^2$. According to the fourth conclusion of the preceding article this means that F is composed of f, f'.   Q.E.S. And finally, it is clear that we can know from the beginning *how* to compose F from f, f' if we properly determine the signs of the quantities n, n'.

▶ 237. THEOREM. *If the form F is transformable into the product of two forms f, f', and the form f' implies the form f'', then F will also be transformable into the product of the forms f, f''.*

Demonstration. For the forms F, f, f' all the signs of article 235 are to be retained; let $f'' = (a'', b'', c'')$ and let f' be transformed into f'' by the substitution $\alpha, \beta, \gamma, \delta$. Then F will be transformed into ff'' by the substitution

$$\alpha p + \gamma p', \quad \beta p + \delta p', \quad \alpha p'' + \gamma p''', \quad \beta p'' + \delta p'''$$
$$\alpha q + \gamma q', \quad \beta q + \delta q', \quad \alpha q'' + \gamma q''', \quad \beta q'' + \delta q''' \quad \text{Q.E.D.}$$

For brevity we will designate these coefficients as follows:

$$\alpha p + \gamma p', \ \beta p + \delta p' \text{ etc.} = \mathfrak{P}, \mathfrak{P}', \mathfrak{P}'', \mathfrak{P}'''; \ \mathfrak{Q}, \mathfrak{Q}', \mathfrak{Q}'', \mathfrak{Q}'''$$

and let the number $\alpha\delta - \beta\gamma = e$. From equations Ω, article 235, it is easy to see that

$$\mathfrak{P}\mathfrak{Q}' - \mathfrak{Q}\mathfrak{P}' = an'e$$

$$\mathfrak{P}\mathfrak{Q}''' - \mathfrak{Q}\mathfrak{P}''' - \mathfrak{P}'\mathfrak{Q}'' + \mathfrak{Q}'\mathfrak{P}'' = 2bn'e$$

$$\mathfrak{P}''\mathfrak{Q}''' - \mathfrak{Q}''\mathfrak{P}''' = cn'e$$

$$\mathfrak{P}\mathfrak{Q}'' - \mathfrak{Q}\mathfrak{P}'' = \alpha^2 a'n + 2\alpha\gamma b'n + \gamma^2 c'n = a''n$$

$$\mathfrak{P}\mathfrak{Q}''' - \mathfrak{Q}\mathfrak{P}''' + \mathfrak{P}'\mathfrak{Q}'' - \mathfrak{Q}'\mathfrak{P}'' = 2b''n$$

$$\mathfrak{P}'\mathfrak{Q}''' - \mathfrak{Q}'\mathfrak{P}''' = c''n$$

$$\mathfrak{Q}'\mathfrak{Q}'' - \mathfrak{Q}\mathfrak{Q}''' = Ann'e$$

$$\mathfrak{P}\mathfrak{Q}''' + \mathfrak{Q}\mathfrak{P}''' - \mathfrak{P}'\mathfrak{Q}'' - \mathfrak{Q}'\mathfrak{P}'' = 2Bnn'e$$

$$\mathfrak{P}'\mathfrak{P}'' - \mathfrak{P}\mathfrak{P}''' = Cnn'e$$

Now if we designate the determinant of the form f'' by d'', e will be a square root of d''/d', positive or negative according as the form f' implies the form f'' properly or improperly. Thus $n'e$ will be a square root of d''/D; and the nine preceding equations will be completely analogous to the equations Ω of article 235. The form f will be taken in the transformation of the form F into ff'' in exactly the same way that it was taken in the transformation of the form F into ff'. The form f'' in the former should be taken in the same way as f' in the latter if f' properly implies f''. If f' improperly implies f'', it should be taken oppositely.

▶ 238. THEOREM. *If the form F is contained in the form F' and is transformable into the product of the forms f, f': then the form F' will be transformable into the same product.*

Demonstration. If for the forms F, f, f' we retain the same signs as above and suppose that the form F' is transformed into F by the substitution $\alpha, \beta, \gamma, \delta$, it is easy to see that F' by the substitution

$$\alpha p + \beta q, \quad \alpha p' + \beta q', \quad \alpha p'' + \beta q'', \quad \alpha p''' + \beta q'''$$

$$\gamma p + \delta q, \quad \gamma p' + \delta q', \quad \gamma p'' + \delta q'', \quad \gamma p''' + \delta q'''$$

becomes the same as F by the substitution p, p', p'', p'''; q, q', q'', q''' and thus by this transformation F' is transformed into ff'. Q.E.D.

By a calculation similar to that of the preceding article it can also be confirmed that F' is transformable into ff' in the same way as F when F' implies F properly. But when F is contained improperly in F' the transformations of the forms F and F' into ff' will be opposite with respect to each of the forms f, f'; that is, if a form appears in one of the transformations directly, it will appear in the other inversely.

If we combine this theorem with the theorem of the preceding article we will get the following generalization. *If the form F is transformable into the product ff', if the forms f, f' respectively imply the forms g, g', and if the form F is contained in the form G: then G will be transformable into the product gg'.* For by the theorem of this article G is transformable into ff' and thus by the previous theorem into fg' and then also into gg'. It is also clear that if all three forms f, f', G properly imply the forms g, g', F, G will be transformable into gg' in the same way with respect to g, g' as F into ff' with respect to the forms f, f'. The same is true if all three implications are improper. If one of the implications is different from the other two, it is just as easy to determine *how* G is transformable into gg'.

If the forms F, f, f' are equivalent to the forms G, g, g', respectively, the latter will have the same determinants as the former. Since for the forms f, f' they are the numbers m, m', they will be the same for the forms g, g' (art. 161). Thus by the fourth conclusion of article 235 we deduce that G will be *composed* of g, g' if F is composed of f, f'; and indeed the form g will enter the former composition in the same way as f does the latter provided F is equivalent to G in the same way that f is to g, and conversely. Similarly g' must be taken in the former composition in the same or opposite way as f' in the latter, according as the equivalence of the forms f', g' is similar or dissimilar to the equivalence of the forms F, G.

▶ 239. THEOREM. *If the form F is composed of the forms f, f', any other form that is transformable into the product ff' in the same way as F will imply F properly.*

Demonstration. If we keep for F, f, f' all the signs of article 235, the equations Ω will also be pertinent here. Let us suppose that the form $F' = (A', B', C')$ whose determinant $= D$ is transformed into the product ff' by the substitution $\mathfrak{p}, \mathfrak{p}', \mathfrak{p}'', \mathfrak{p}'''$; $\mathfrak{q}, \mathfrak{q}', \mathfrak{q}'', \mathfrak{q}'''$. And let us designate the numbers

$$\mathfrak{p}\mathfrak{q}' - \mathfrak{q}\mathfrak{p}', \; \mathfrak{p}\mathfrak{q}'' - \mathfrak{q}\mathfrak{p}'', \; \mathfrak{p}\mathfrak{q}''' - \mathfrak{q}\mathfrak{p}'''$$

$$\mathfrak{p}'\mathfrak{q}'' - \mathfrak{q}'\mathfrak{p}'', \; \mathfrak{p}'\mathfrak{q}''' - \mathfrak{q}'\mathfrak{p}''', \; \mathfrak{p}''\mathfrak{q}''' - \mathfrak{q}''\mathfrak{p}'''$$

respectively by

$$P', Q', R', S', T', U'$$

Then we will have nine equations which are completely similar to those of Ω, namely

$$P' = a\mathfrak{n}', \; R' - S' = 2b\mathfrak{n}', \; U' = c\mathfrak{n}'$$

$$Q' = a'\mathfrak{n}, \; R' + S' = 2b'\mathfrak{n}, \; T' = c'\mathfrak{n}$$

$$\mathfrak{q}'\mathfrak{q}'' - \mathfrak{q}\mathfrak{q}''' = A'\mathfrak{n}\mathfrak{n}', \; \mathfrak{p}\mathfrak{q}''' + \mathfrak{q}\mathfrak{p}''' - \mathfrak{p}'\mathfrak{q}'' - \mathfrak{q}'\mathfrak{p}'' = 2B'\mathfrak{n}\mathfrak{n}'$$

$$\mathfrak{p}'\mathfrak{p}'' - \mathfrak{p}\mathfrak{p}''' = C'\mathfrak{n}\mathfrak{n}'$$

We will designate these equations by Ω'. The quantities $\mathfrak{n}, \mathfrak{n}'$ here are square roots of $d/D', d'/D'$ and have the same signs respectively as n, n'; if therefore we take the positive square root of D/D' (it will be an integer) and let it $= k$, we will have $\mathfrak{n} = kn, \mathfrak{n}' = kn'$. And then from the first six equations in Ω and Ω' we get

$$P' = kP, \; Q' = kQ, \; R' = kR$$

$$S' = kS, \; T' = kT, \; U' = kU$$

By the lemma of article 234 we can find four integers $\alpha, \beta, \gamma, \delta$ such that

$$\alpha p + \beta q = \mathfrak{p}, \quad \gamma p + \delta q = \mathfrak{q}$$

$$\alpha p' + \beta q' = \mathfrak{p}', \quad \gamma p' + \delta q' = \mathfrak{q}' \text{ etc.}$$

and

$$\alpha\delta - \beta\gamma = k$$

Substituting these values of $\mathfrak{p}, \mathfrak{q}, \mathfrak{p}', \mathfrak{q}'$, etc. into the three last equations of Ω' and using the equations $\mathfrak{n} = kn, \mathfrak{n}' = kn'$ and the three last equations of Ω, we find that

$$A'\alpha^2 + 2B'\alpha\gamma + C'\gamma^2 = A$$
$$A'\alpha\beta + B'(\alpha\delta + \beta\gamma) + C'\gamma\delta = B$$
$$A'\beta^2 + 2B'\beta\delta + C'\delta^2 = C$$

Therefore by the substitution $\alpha, \beta, \gamma, \delta$ (which will be proper since $\alpha\delta - \beta\gamma = k$ is positive) F' will be transformed into F; i.e. it will imply the form F properly. Q.E.D.

If therefore F' is composed of the forms f, f' (in the same way as F), the forms F, F' will have the same determinant and will be properly equivalent. More generally, if the form G is composed of the forms g, g' in the same way as F is composed of f, f', respectively, and the forms g, g' are properly equivalent to f, f': then the forms F, G are properly equivalent.

Since this case, where both composing forms enter into the composition directly, is the simplest one and the others can be easily reduced to it, we will consider only this in what follows. Thus if any form is said simply to be composed from two others, it will always be understood that it is properly composed of each of them.[h] The same restriction will hold whenever a form is said to be transformable into a product of two others.

▶ 240. THEOREM. *If the form F is composed of the forms f, f'; the form $\mathfrak{F}$ from F and f''; the form F' from f, f''; the form $\mathfrak{F}'$ from F' and f': then the forms $\mathfrak{F}, \mathfrak{F}'$ will be properly equivalent.*

Demonstration. I. Let

$$f = ax^2 + 2bxy + cy^2$$
$$f' = a'x'x' + 2b'x'y' + c'y'y'$$
$$f'' = a''x''x'' + 2b''x''y'' + c''y''y''$$
$$F = AX^2 + 2BXY + CY^2$$
$$F' = A'X'X' + 2B'X'Y' + C'Y'Y'$$

[h] Just as in the composition of ratios (which is very similar to the composition of forms) we normally understand that the ratios are to be taken directly unless it is otherwise indicated.

$$\mathfrak{F} = \mathfrak{A}\mathfrak{X}\mathfrak{X} + 2\mathfrak{B}\mathfrak{X}\mathfrak{Y} + \mathfrak{C}\mathfrak{Y}\mathfrak{Y}$$

$$\mathfrak{F}' = \mathfrak{A}'\mathfrak{X}'\mathfrak{X}' + 2\mathfrak{B}'\mathfrak{X}'\mathfrak{Y}' + \mathfrak{C}'\mathfrak{Y}'\mathfrak{Y}'$$

and let the determinants of these seven forms be respectively $d, d', d'', D, D', \mathfrak{D}, \mathfrak{D}'$. They will all have the same signs and be related to one another as squares. Further let m be the greatest common divisor of the numbers $a, 2b, c$, and let m', m'', M have the same signification relative to the forms f', f'', F. Then, from conclusion four of article 235, D will be the greatest common divisor of the numbers $dm'm', d'm^2$ and $Dm''m''$ the greatest common divisor of the numbers $dm'm'm''m'', dm^2m''m''$; $M = mm'$; $\mathfrak{D}$ the greatest common divisor of the numbers $Dm''m'', d''M^2$ or of the numbers $Dm''m'', d''m^2m'm'$. We conclude that $\mathfrak{D}$ is the greatest common divisor of the three numbers $dm'm'm''m''$, $d'm^2m''m''$, $d''m^2m'm'$. For similar reasons $\mathfrak{D}'$ will be the greatest common divisor of the same three numbers. Therefore since $\mathfrak{D}, \mathfrak{D}'$ have the same sign, $\mathfrak{D} = \mathfrak{D}'$ or the forms $\mathfrak{F}, \mathfrak{F}'$ will have the same determinant.

II. Now let F be transformed into ff' by the substitution

$$X = pxx' + p'xy' + p''yx' + p'''yy'$$

$$Y = qxx' + q'xy' + q''yx' + q'''yy'$$

and $\mathfrak{F}$ into Ff' by the substitution

$$\mathfrak{X} = \mathfrak{p}Xx'' + \mathfrak{p}'Xy'' + \mathfrak{p}''Yx'' + \mathfrak{p}'''Yy''$$

$$\mathfrak{Y} = \mathfrak{q}Xx'' + \mathfrak{q}'Xy'' + \mathfrak{q}''Yx'' + \mathfrak{q}'''Yy''$$

and let the positive square roots of $d/D, d'/D, D/\mathfrak{D}, d''/\mathfrak{D}$ be designated by $n, n', \mathfrak{N}, \mathfrak{n}''$. Then by article 235 we will have 18 equations, half of them belonging to the transformation of the form F into ff', the other half to the transformation of the form $\mathfrak{F}$ into Ff''. The first of them will be $pq' - qp' = an'$. The rest can be formed in the same manner, but for the sake of brevity we will omit them here. Note that the quantities $n, n', \mathfrak{N}, \mathfrak{n}''$ will be rational but not necessarily integers.

III. If the values of X, Y are substituted in the values of $\mathfrak{X}, \mathfrak{Y}$, we get a result of the form:

$$\mathfrak{X} = (1)xx'x'' + (2)xx'y'' + (3)xy'x'' + (4)xy'y''$$
$$+ (5)yx'x'' + (6)yx'y'' + (7)yy'x'' + (8)yy'y''$$

$$\mathfrak{Y} = (9)xx'x'' + (10)xx'y'' + (11)xy'x'' + (12)xy'y''$$
$$+ (13)yx'x'' + (14)yx'y'' + (15)yy'x'' + (16)yy'y''$$

Obviously by this substitution $\mathfrak{F}$ will be transformed into the product $ff'f''$. Coefficient (1) will be $= p\mathfrak{p} + q\mathfrak{p}''$, and the reader can evolve the other fifteen values. We will designate the number $(1)(10) - (2)(9)$ by $(1, 2)$, the number $(1)(11) - (3)(9)$ by $(1, 3)$ and in general $(g)(8 + h) - (h)(8 + g)$ by (g, h) where g, h are unequal integers between 1 and 16 with h being the larger;[i] in this way we will have 28 signs in all. Now if we denote the positive square roots of $d/\mathfrak{D}, d'/\mathfrak{D}$ by $\mathfrak{n}, \mathfrak{n}'$ (they will be $= n\mathfrak{R}, n'\mathfrak{R}$) we will have the following 28 equations:

$$(1, 2) = aa'\mathfrak{n}''$$
$$(1, 3) = aa''\mathfrak{n}'$$
$$(1, 4) = ab'\mathfrak{n}'' + ab''\mathfrak{n}'$$
$$(1, 5) = a'a''\mathfrak{n}$$
$$(1, 6) = a'b\mathfrak{n}'' + a'b''\mathfrak{n}$$
$$(1, 7) = a''b\mathfrak{n}' + a''b'\mathfrak{n}$$
$$(1, 8) = bb'\mathfrak{n}'' + bb''\mathfrak{n}' + b'b''\mathfrak{n} + \mathfrak{D}n\mathfrak{n}'\mathfrak{n}''$$
$$(2, 3) = ab''\mathfrak{n}' - ab'\mathfrak{n}''$$
$$(2, 4) = ac''\mathfrak{n}'$$
$$(2, 5) = a'b''\mathfrak{n} - a'b\mathfrak{n}''$$
$$(2, 6) = a'c''\mathfrak{n}$$
$$(2, 7) = bb'\mathfrak{n}' + b'b''\mathfrak{n} - bb'\mathfrak{n}'' - \mathfrak{D}n\mathfrak{n}\mathfrak{n}''$$
$$(2, 8) = bc''\mathfrak{n}' + b'c''\mathfrak{n}$$
$$(3, 4) = ac'\mathfrak{n}''$$
$$(3, 5) = a''b'\mathfrak{n} - a''b\mathfrak{n}'$$
$$(3, 6) = bb'\mathfrak{n}'' + b'b''\mathfrak{n} - bb''\mathfrak{n}' - \mathfrak{D}n\mathfrak{n}'\mathfrak{n}''$$
$$(3, 7) = a''c'\mathfrak{n}$$
$$(3, 8) = bc'\mathfrak{n}'' + b''c'\mathfrak{n}$$
$$(4, 5) = b'b''\mathfrak{n} - bb'\mathfrak{n}'' - bb''\mathfrak{n}' + \mathfrak{D}n\mathfrak{n}'\mathfrak{n}''$$
$$(4, 6) = b'c''\mathfrak{n} - bc''\mathfrak{n}'$$
$$(4, 7) = b''c'\mathfrak{n} - bc'\mathfrak{n}''$$
$$(4, 8) = c'c''\mathfrak{n}$$
$$(5, 6) = ca'\mathfrak{n}''$$

---

[i] The present meaning of these signs must not be confused with those of article 234; for the numbers expressed by these signs *here* correspond rather to those in article 234 which are multiplied by numbers denoted by similar symbols.

$$(5, 7) = ca''\mathfrak{n}'$$
$$(5, 8) = b'c\mathfrak{n}'' + b''c\mathfrak{n}'$$
$$(6, 7) = b''c\mathfrak{n}' - b'c\mathfrak{n}''$$
$$(6, 8) = cc''\mathfrak{n}'$$
$$(7, 8) = cc'\mathfrak{n}''$$

We will designate these equations by Φ, and 9 others:

$$(10)(11) - (9)(12) = a\mathfrak{n}'\mathfrak{n}''\mathfrak{A}''$$
$$(1)(12) - (2)(11) - (3)(10) + (4)(9) = 2a\mathfrak{n}'\mathfrak{n}''\mathfrak{B}$$
$$(2)(3) - (1)(4) = a\mathfrak{n}'\mathfrak{n}''\mathfrak{C},$$
$$-(9)(16) + (10)(15) + (11)(14) - (12)(13) = 2b\mathfrak{n}'\mathfrak{n}''\mathfrak{A}$$
$$\left.\begin{array}{l}(1)(16) - (2)(15) - (3)(14) + (4)(13) \\ +(5)(12) - (6)(11) - (7)(10) + (8)(9)\end{array}\right\} = 4b\mathfrak{n}'\mathfrak{n}''\mathfrak{B}$$
$$-(1)(8) + (2)(7) + (3)(6) - (4)(5) = 2b\mathfrak{n}'\mathfrak{n}''\mathfrak{C}$$
$$(14)(15) - (13)(16) = c\mathfrak{n}'\mathfrak{n}''\mathfrak{A}$$
$$(5)(16) - (6)(15) - (7)(14) + (8)(13) = 2c\mathfrak{n}'\mathfrak{n}''\mathfrak{B}$$
$$(6)(7) - (5)(8) = c\mathfrak{n}'\mathfrak{n}''\mathfrak{C}$$

we will designate by Ψ.[j]

IV. It would take too much time to derive all 37 of these equations. We will be satisfied with establishing some of them as a pattern for all the rest.

1) We have

$$(1, 2) = (1)(10) - (2)(9)$$
$$= (\mathfrak{p}\mathfrak{q}' - \mathfrak{q}\mathfrak{p}')p^2 + (\mathfrak{p}\mathfrak{q}''' - \mathfrak{q}\mathfrak{p}''' - \mathfrak{p}'\mathfrak{q}'' + \mathfrak{q}'\mathfrak{p}'')pq$$
$$+ (\mathfrak{p}''\mathfrak{q}''' - \mathfrak{q}''\mathfrak{p}''')q^2$$
$$= \mathfrak{n}''(Ap^2 + 2Bpq + Cq^2) = \mathfrak{n}''aa'$$

which is the first equation.

2) We have

$$(1, 3) = (1)(11) - (3)(9) = (\mathfrak{p}\mathfrak{q}'' - \mathfrak{q}\mathfrak{p}'')(\mathfrak{p}\mathfrak{q}' - \mathfrak{q}\mathfrak{p}') = a''\mathfrak{N}a\mathfrak{n}'$$
$$= aa''\mathfrak{n}'$$

the second equation.

[j] Observe that we could deduce 18 other equations similar to Ψ by replacing the factors a, $2b$, c by a', $2b'$, c'; a'', $2b''$, c''; but since they are not necessary to our purpose we omit them.

3) And we have

$$(1,8) = (1)(16) - (8)(9)$$

$$= (\mathfrak{p}q' - q\mathfrak{p}')pp''' + (\mathfrak{p}q''' - q\mathfrak{p}''')pq''' - (\mathfrak{p}'q'' - q'\mathfrak{p}'')qp'''$$
$$\quad + (\mathfrak{p}''q''' - q''\mathfrak{p}''')qq'''$$

$$= \mathfrak{n}''(App''' + B(pq''' + qp'') + Cqq''') + b''\mathfrak{R}(pq''' - qp''')$$

$$= \mathfrak{n}''(bb' + \sqrt{dd'}) + b''\mathfrak{R}(b'n + bn')^{\mathrm{k}}$$

$$= \mathfrak{n}''bb' + \mathfrak{n}'bb'' + \mathfrak{n}b'b'' + \mathfrak{D}\mathfrak{n}\mathfrak{n}'\mathfrak{n}''$$

the eighth equation in Φ. We leave it to the reader to confirm the remaining equations.

V. We can show as follows that by equations Φ the 28 numbers $(1, 2)$, $(1, 3)$, etc. have no common divisor. First we observe that we can form 27 products of three factors such that the first is $\mathfrak{n}$, the second is one of the numbers $a', 2b', c'$, and the third is one of the numbers $a'', 2b'', c''$; or the first is $\mathfrak{n}'$, the second one of the numbers a, $2b$, c, and the third one of the numbers a'', $2b''$, c''; or finally the first is $\mathfrak{n}''$, the second one of the numbers a, $2b$, c, and the third one of the numbers a', $2b'$, c'. And each of these 27 products, because of equations Φ, will be equal to one of the 28 numbers $(1, 2)$, $(1, 3)$, etc. or to the sum or difference of several of them [e.g. $\mathfrak{n}a'a'' = (1, 5)$, $2\mathfrak{n}a'b'' = (1, 6) + (2, 5)$, $4\mathfrak{n}b'b'' = (1, 8) + (2, 7) + (3, 6) + (4, 5)$ and so on]. Therefore if these numbers had a common divisor, it would necessarily divide all these products. So by article 40 and by the method used very frequently above, the same divisor must also divide the numbers $\mathfrak{n}m'm''$, $\mathfrak{n}'mm''$, $\mathfrak{n}''mm'$, and the square of this divisor must also divide the square of these numbers, namely $dm'm'm''m''/\mathfrak{D}$, $d'm^2m''m''/\mathfrak{D}$, $d''m^2m'm'/\mathfrak{D}$. Q.E.A., since by I the greatest common divisor of the three numerators is $\mathfrak{D}$, and so these three squares cannot have a common divisor.

VI. All of this pertains to the transformation of the form $\mathfrak{F}$ into $ff'f''$; and it can be deduced from the transformations of the form F into ff' and of the form $\mathfrak{F}$ into Ff''. In a completely similar way we can derive the transformation of the form $\mathfrak{F}$ into $ff'f''$

---

^{k} This follows from equation 10 article 235 ff. The quantity $\sqrt{dd'}$ becomes $= Dnn' = \mathfrak{D}\mathfrak{n}\mathfrak{n}'\mathfrak{R}^2 = \mathfrak{D}\mathfrak{n}\mathfrak{n}'$.

from transformations of the form F' into $f\!f''$ and of the form $\mathfrak{F}'$ into $F'f'$:

$$\mathfrak{X}' = (1)'xx'x'' + (2)'xx'y'' + (3)'xy'x'' + \text{etc.}$$
$$\mathfrak{Y}' = (9)'xx'x'' + (10)'xx'y'' + (11)'xy'x'' + \text{etc.}$$

(the coefficients here are designated in the same way as in the transformation of the form $\mathfrak{F}$ into $f\!f'f''$, but they are primed to distinguish them). From this transformation we deduce as before 28 equations analogous to the equations Φ which we will call Φ' and 9 others analogous to equations Ψ which we will call Ψ'. Thus if we denote

$(1)'(10)' - (2)'(9)'$ by $(1, 2)'$, $(1)'(11)' - (3)'(9)'$ by $(1, 3)'$, etc.

equations Φ' will be

$$(1, 2)' = aa'\mathfrak{n}'',   (1, 3)' = aa''\mathfrak{n}', \text{etc.}$$

and equations Ψ' will be

$$(10)'(11)' - (9)'(12)' = a\mathfrak{n}'\mathfrak{n}''\mathfrak{A}' \text{ etc.}$$

(For the sake of brevity we leave a more detailed derivation to the reader; the expert will find no need for a new calculation because the first analysis can be carried over by analogy.) Now from Φ and Φ' it follows immediately that

$$(1, 2) = (1, 2)',   (1, 3) = (1, 3)',   (1, 4) = (1, 4)',   (2, 3) = (2, 3)', \text{etc.}$$

And since all the $(1, 2), (1, 3), (2, 3)$, etc. have no common divisor (by V), with the help of the lemma in article 234 we can determine four integers $\alpha, \beta, \gamma, \delta$ such that

$$\alpha(1)' + \beta(9)' = (1),   \alpha(2)' + \beta(10)' = (2),   \alpha(3)' + \beta(11)' = (3),$$
$$\text{etc.}$$
$$\gamma(1)' + \delta(9)' = (9),   \gamma(2)' + \delta(10)' = (10),   \gamma(3)' + \delta(11)' = (11),$$
$$\text{etc.}$$

and $\alpha\delta - \beta\gamma = 1$.

VII. Now if from the first three equations of Ψ we substitute the values for $a\mathfrak{A}, a\mathfrak{B}, a\mathfrak{C}$, and from the first three equations of Ψ'

the values of $a\mathfrak{A}'$, $a\mathfrak{B}'$, $a\mathfrak{C}'$ we find easily that:

$$a(\mathfrak{A}\alpha^2 + 2\mathfrak{B}\alpha\gamma + \mathfrak{C}\gamma^2) = a\mathfrak{A}'$$

$$a[\mathfrak{A}\alpha\beta + \mathfrak{B}(\alpha\delta + \beta\gamma) + \mathfrak{C}\gamma\delta] = a\mathfrak{B}'$$

$$a(\mathfrak{A}\beta^2 + 2\mathfrak{B}\beta\delta + \mathfrak{C}\delta^2) = a\mathfrak{C}'$$

and unless $a = 0$ it follows that the form $\mathfrak{F}$ is transformed into $\mathfrak{F}'$ by the proper substitution $\alpha, \beta, \gamma, \delta$. If, in place of the first three equations in Ψ and Ψ' we use the next three, then we will get three equations just like the above but with the factor a replaced by b; and the same conclusion holds as long as we do not have $b = 0$. Finally if we use the last three equations in Ψ, Ψ', the same conclusion will be true unless $c = 0$. And since certainly not all a, b, c can be $= 0$ at the same time, the form $\mathfrak{F}$ will necessarily be transformed into $\mathfrak{F}'$ by the substitution $\alpha, \beta, \gamma, \delta$, and the forms will be properly equivalent.   Q.E.D.

▶ 241. If we have a form like $\mathfrak{F}$ or $\mathfrak{F}'$ which results from the composition of one of three given forms with another which is a composition of the two remaining forms, we will say that it is *composed of these three forms*. It is clear from the preceding article that it does not matter what the order of composition is for the three forms. Similarly if we have any number of forms f, f', f'', f''', etc. (and their determinants are related to one another as squares) and if we compose the form f with f', the resultant form with f'', the resultant of that with f''', etc.: we will say that the final form arising from this operation is composed *of all the forms f, f', f'', f'''*, etc. And it is easy to show that here too the order of composition is arbitrary; i.e. no matter in which order these forms are composed, the forms arising from the composition will be properly equivalent. And manifestly if the forms g, g', g'', etc. are properly equivalent to the forms f, f', f'', etc., respectively, the form composed of the former will be properly equivalent to the form composed of the latter.

▶ 242. The preceding propositions refer to the composition of forms in all its universality. We will now pass on to more particular applications for which we did not wish to interrupt the order of the argument earlier. First we will resume the problem of article 236 and limit it by the following conditions: *first* the forms to be composed should have the same determinant, i.e. $d = d'$; *second*, *m* and *m'* are to be relatively prime; *third*, the form we are seeking is to be composed directly of both f, f'. Thus also $m^2, m'm'$ will be relatively prime; and so the greatest common divisor of the numbers $dm'm', d'm^2$ i.e. D will be $= d = d'$ and $n = n' = 1$. Since we are free to choose we will let the four quantities $\mathfrak{Q}, \mathfrak{Q}', \mathfrak{Q}'', \mathfrak{Q}'''$ $= -1, 0, 0, 0$ respectively. This is permissible except when a, a', $b + b'$ are all $= 0$ at the same time, so we will omit this case. Manifestly it cannot occur except in forms with a positive square determinant. Now if μ is the greatest common divisor of the numbers $a, a', b + b'$ the numbers $\mathfrak{P}', \mathfrak{P}'', \mathfrak{P}'''$ can be chosen so that

$$\mathfrak{P}'a + \mathfrak{P}''a' + \mathfrak{P}'''(b + B) = \mu$$

As for $\mathfrak{P}$ it can be chosen arbitrarily. As a result if we substitute for p, q, p', q', etc. their values we have:

$$A = \frac{aa'}{\mu^2}, \quad B = \frac{1}{\mu}[\mathfrak{P}aa' + \mathfrak{P}'ab' + \mathfrak{P}''a'b + \mathfrak{P}'''(bb' + D)]$$

and C can be determined from the equation $AC = B^2 - D$ as long as a and a' are not both $= 0$.

Now in this solution the value of A does not depend on the values of $\mathfrak{P}, \mathfrak{P}', \mathfrak{P}'', \mathfrak{P}'''$ (which can be determined in infinitely many different ways); but B will have different values by giving various values to these numbers and so it is worthwhile to investigate how all these values of B are interconnected. For this purpose we observe

I. No matter how we determine $\mathfrak{P}, \mathfrak{P}', \mathfrak{P}'', \mathfrak{P}'''$ the resulting values of B are all congruent relative to the modulus A. Let us suppose that if

$$\mathfrak{P} = \mathfrak{p}, \quad \mathfrak{P}' = \mathfrak{p}', \quad \mathfrak{P}'' = \mathfrak{p}'', \quad \mathfrak{P}''' = \mathfrak{p}''' \quad \text{we have } B = \mathfrak{B}$$

but if we take

$$\mathfrak{P} = \mathfrak{p} + \mathfrak{d}, \quad \mathfrak{P}' = \mathfrak{p}' + \mathfrak{d}', \quad \mathfrak{P}'' = \mathfrak{p}'' + \mathfrak{d}'', \quad \mathfrak{P}''' + \mathfrak{d}'''$$

$$\text{we have } B = \mathfrak{B} + \mathfrak{D}$$

Then we will have

$$a\mathfrak{d}' + a'\mathfrak{d}'' + (b + b')\mathfrak{d}''' = 0,$$

$$aa'\mathfrak{d} + ab'\mathfrak{d}' + a'b\mathfrak{d}'' + (bb' + D)\mathfrak{d}''' = \mu\mathfrak{D}$$

Now if we multiply the first member of the second equation by $a\mathfrak{p}' + a'\mathfrak{p}'' + (b + b')\mathfrak{p}'''$, the second member by μ, and subtract from the first product the quantity

$$[ab'\mathfrak{p}' + a'b\mathfrak{p}'' + (bb' + D)\mathfrak{p}'''][a\mathfrak{d}' + a'\mathfrak{d}'' + (b + b')\mathfrak{d}''']$$

which by the first equation above manifestly $= 0$, then carry out the calculations and cancel zero terms we get

$$aa'(\mu\mathfrak{d} + [(b' - b)\mathfrak{p}'' + c'\mathfrak{p}''']\mathfrak{d}' + [(b - b')\mathfrak{p}' + c\mathfrak{p}''']\mathfrak{d}''$$

$$-[c'\mathfrak{p}' + c\mathfrak{p}'']\mathfrak{d}''') = \mu^2\mathfrak{D}$$

Manifestly $\mu^2\mathfrak{D}$ will be divisible by aa' and $\mathfrak{D}$ by aa'/μ^2 i.e. by A and

$$\mathfrak{B} \equiv \mathfrak{B} + \mathfrak{D} \ (\text{mod. } A)$$

II. If the values $\mathfrak{p}, \mathfrak{p}', \mathfrak{p}'', \mathfrak{p}'''$ of $\mathfrak{P}, \mathfrak{P}', \mathfrak{P}'', \mathfrak{P}'''$ make $B = \mathfrak{B}$, then other values of these numbers can be found which will make B equal to any given number which is congruent to $\mathfrak{B}$ relative to the modulus A, namely $\mathfrak{B} + kA$. First we observe that the four numbers $\mu, c, c', b - b'$ cannot have a common divisor; for if there was one it would divide the six numbers $a, a', b + b', c, c', b - b'$ and so also $a, 2b, c$ and $a', 2b', c'$ and therefore also m, m' which by hypothesis are relatively prime. So four integers h, h', h'', h''', can be found such that

$$h\mu + h'c + h''c' + h'''(b - b') = 1$$

And if we let

$$kh = \mathfrak{d}, \qquad k[h''(b + b') - h'''a'] = \mu\mathfrak{d}'$$

$$k[h'(b + b') + h'''a] = \mu\mathfrak{d}'', \qquad - k(h'a' + h''a) = \mu\mathfrak{d}'''$$

it is clear that $\mathfrak{d}, \mathfrak{d}', \mathfrak{d}'', \mathfrak{d}'''$ are integers and

$$a\mathfrak{d}' + a'\mathfrak{d}'' + (b + b')\mathfrak{d}''' = 0$$

$$aa'\mathfrak{d} + ab'\mathfrak{d}' + a'b\mathfrak{d}'' + (bb' + D)\mathfrak{d}'''$$

$$= \frac{aa'k}{\mu}[\mu h + ch' + c'h'' + (b - b')h'''] = \mu kA$$

From the former equation it is clear that $\mathfrak{p} + \mathfrak{d}, \mathfrak{p}' + \mathfrak{d}', \mathfrak{p}'' + \mathfrak{d}''$, $\mathfrak{p}''' + \mathfrak{d}'''$ are also values of $\mathfrak{P}, \mathfrak{P}', \mathfrak{P}'', \mathfrak{P}'''$; from the latter that these values give us $B = \mathfrak{B} + kA$. Q.E.D. Clearly then B can always be chosen so that it lies between 0 and $A - 1$ inclusively for A positive; or between 0 and $-A - 1$ for A negative.

▶ 243. From the equations

$$\mathfrak{P}'a + \mathfrak{P}''a' + \mathfrak{P}'''(b + b') = \mu,$$

$$B = \frac{1}{\mu}[\mathfrak{P}aa' + \mathfrak{P}'ab' + \mathfrak{P}''a'b + \mathfrak{P}'''(bb' + D)]$$

we deduce

$$B = b + \frac{a}{\mu}[\mathfrak{P}a' + \mathfrak{P}'(b' - b) - \mathfrak{P}'''c]$$

$$= b' + \frac{a'}{\mu}[\mathfrak{P}a + \mathfrak{P}''(b - b') - \mathfrak{P}'''c']$$

and therefore

$$B \equiv b \left(\text{mod. } \frac{a}{\mu}\right) \quad \text{and} \quad B \equiv b' \left(\text{mod. } \frac{a'}{\mu}\right)$$

Now whenever a/μ, a'/μ are relatively prime, there will be between 0 and $A - 1$ (or between 0 and $-A - 1$ when A is negative) only one number which is $\equiv b$ (mod. a/μ) and $\equiv b'$ (mod. a'/μ). If we let it $= B$ and $(B^2 - D)/A = C$, it is clear that (A, B, C) will be composed of the forms $(a, b, c), (a', b', c')$. So in this case it is not necessary to consider the numbers $\mathfrak{P}, \mathfrak{P}', \mathfrak{P}'', \mathfrak{P}'''$ in order to find the composite form.[1] Thus, e.g., if we want the form which is composed of the forms $(10, 3, 11), (15, 2, 7)$ we will have $a, a', b + b'$ respectively $= 10, 15, 5$; $\mu = 5$; so $A = 6$; $B \equiv 3$ (mod. 2) and $\equiv 2$ (mod. 3). Therefore $B = 5$ and $(6, 5, 21)$ is the form we seek. But the condition that a/μ, a'/μ be relatively prime is the same as requiring that the two numbers a, a' have no common divisor greater than the three numbers $a, a', b + b'$ or, what comes to the same thing, that the greatest common divisor of the numbers a, a' also divides the number $b + b'$. We note the following particular cases.

1) Suppose we have two forms $(a, b, c), (a', b', c')$ with the same

---

[1] We can always accomplish it by using the congruences $aB/\mu \equiv ab'/\mu$, $a'B/\mu \equiv a'b/\mu$, $(b + b')B/\mu \equiv (bb' + D)/\mu$ (mod. A).

determinant D and so related that the greatest common divisor of the numbers $a, 2b, c$ is relatively prime to the greatest common divisor of the numbers $a', 2b', c'$ and a is relatively prime to a': then the form (A, B, C) which is the composition of these two is found by letting $A = aa'$, $B \equiv b$ (mod. a) and $\equiv b'$ (mod. a'), $C = (B^2 - D)/A$. This case will always occur when one of the forms to be composed is a principal form; that is, $a = 1$, $b = 0$, $c = -D$. Then $A = a'$, B can be taken $= b'$ and we will have $C = c'$; thus *a form of this kind is composed of a principal form and any other form of the same determinant.*

2) If we are to compose two properly primitive *opposite* forms, that is (a, b, c) and $(a, -b, c)$, we will have $\mu = a$. It is easy to see that the principal form $(1, 0, -D)$ is composed of these two.

3) Suppose we are given any number of properly primitive forms $(a, b, c), (a', b', c'), (a'', b'', c'')$, etc. with the same determinant and with the first terms a, a', a'', etc. relatively prime to each other. Then we can find the form (A, B, C) which will be composed of all of them by letting A equal the product of all the a, a', a'', etc.; by taking B congruent to b, b', b'', etc. relative to the moduli a, a', a'', etc. respectively; and letting $C = (B^2 - D)/A$. Obviously the form $(aa', B, (B^2 - D)/aa')$ will be composed of the two forms (a, b, c), (a', b', c'); the form $(aa'a'', B, (B^2 - D)/aa'a'')$ will be the composition of this form and (a'', b'', c'') etc. Conversely,

4) Suppose we are given the properly primitive form (A, B, C) with determinant D. If we resolve the term A into any number of relatively prime factors a, a', a'', etc.; if we take the numbers b, b', b'', etc. all equal to B or at least choose numbers which are congruent to B relative to the moduli a, a', a'', etc. respectively; and if we choose c, c', c'', etc. so that $ac = b^2 - D$, $a'c' = b'b' - D$, $a''c'' = b''b'' - D$, etc.: then the form (A, B, C) will be composed of the forms $(a, b, c), (a', b', c'), (a'', b'', c'')$, etc. or we will say that it is *resolvable into these forms*. It is easy to show that this same proposition holds when the form (A, B, C) is improperly primitive or derived. Thus in this way any form can be resolved into others with the same determinant in which all first terms are either prime numbers or powers of prime numbers. Such a resolution can often be very useful if we want to compose one form from several given forms. Thus, e.g., if we want a composite form from the forms

(3, 1, 134), (10, 3, 41), (15, 2, 27) we resolve the second into (2, 1, 201), (5, $-$ 2, 81), the third into (3, $-$ 1, 134), (5, 2, 81). And it is clear that the form composed of the five forms (3, 1, 134), (2, 1, 201), (5, $-$2, 81), (3, $-$ 1, 134), (5, 2, 81), regardless of the order in which they are taken, will also be a composite of the three given forms. Now the composition of the first with the fourth gives the principal form (1, 0, 401); and the same results from the composition of the third with the fifth; so from the composition of all five we get the form (2, 1, 201).

5) Because of its usefulness it is worthwhile to explain this method more fully. From the preceding observation it is clear that, as long as the given forms are properly primitive with the same determinant, the problem can be reduced to the composition of forms whose initial terms are powers of prime numbers (for a prime number can be considered as its own first power). For this reason it is appropriate to consider the special case where two properly primitive forms (a, b, c), (a', b', c') are to be composed and a and a' are powers of the *same* prime number. Let therefore $a = h^\varkappa$, $a' = h^\lambda$ where h is a prime number and we shall suppose (it is legitimate) that $\varkappa$ is not less than λ. Now h^λ will be the greatest common divisor of the numbers a, a'. If it also divides $b + b'$ we have the case which we considered in the beginning of this article and (A, B, C) will be the composite form if $A = h^{\varkappa - \lambda}$, $B \equiv b$ (mod. $h^{\varkappa - \lambda}$) and $\equiv b'$ (mod. 1) (this latter condition can manifestly be omitted); and $C = (B^2 - D)/A$. If h^λ does not divide $b + b'$, the greatest common divisor of these two will necessarily be a power of h also. Let it be $= h^\nu$ and $\nu < \lambda$ ($\nu = 0$ if h^λ and $b + b'$ are relatively prime). Now if $\mathfrak{P}'$, $\mathfrak{P}''$, $\mathfrak{P}'''$ are determined so that

$$\mathfrak{P}'h^\varkappa + \mathfrak{P}''h^\lambda + \mathfrak{P}'''(b + b') \eqqcolon h^\nu$$

and $\mathfrak{P}$ is chosen arbitrarily, (A, B, C) will be a composite of the given forms if we set

$$A = h^{\varkappa + \lambda - 2\nu}, \quad B = b + h^{\varkappa - \nu}(\mathfrak{P}h^\lambda - \mathfrak{P}'(b - b') - \mathfrak{P}'''c),$$

$$C = \frac{B^2 - D}{A}$$

But it is easy to see that in this case also $\mathfrak{P}'$ can be chosen arbitrarily, so letting $\mathfrak{P} = \mathfrak{P}' = 0$ we have

$$B = b - \mathfrak{P}'''ch^{\varkappa - \nu}$$

or in general

$$B = kA + b - \mathfrak{P}'''ch^{\varkappa - \nu}$$

Here k is an arbitrary number (preceding article). Only $\mathfrak{P}'''$ enters into this very simple formula, and it is the value of the expression $h^\nu/(b + b')$ (mod. h^λ).[m] If, e.g., we want the form which is composed of (16, 3, 19) and (8, 1, 37) we have $h = 2$, $\varkappa = 4$, $\lambda = 3$, $\nu = 2$. Thus $A = 8$, $\mathfrak{P}'''$ a value of the expression 4/4 (mod. 8), say 1, and then $B = 8k - 73$. If we let $k = 9$, $B = 1$, and $C = 37$, $(8, -1, 37)$ is the form we want.

Thus if we are given several forms whose initial terms are all powers of prime numbers, we should examine whether the first terms of some of them are powers of the *same* prime number and if there are any, these forms should be composed in the manner we have just shown. In this manner we will have a set of forms whose first terms will be powers of completely different prime numbers. For example, given the forms (3, 1, 47), (4, 0, 35), (5, 0, 28), (16, 2, 9), (9, 7, 21), (16, 6, 11), from the first and fifth we get the form (27, 7, 7); from the second and fourth $(16, -6, 11)$; and from this and the sixth (1, 0, 140) which can be neglected. There remain therefore (5, 0, 28), (27, 7, 7). From these we get $(135, -20, 4)$ and in its place we take (4, 0, 35) which is properly equivalent to it. And this is the result of the composition of the six given forms.

In a similar way we could develop many more devices which are useful in practice, but in order not to extend the investigation we will omit a more lengthy treatment and go on to more difficult considerations.

▶ 244. If the number a can be represented by some form f, the number a' by the form f', and if the form F is transformable into ff': it is not difficult to see that the product aa' will be representable by the form F. It follows immediately that when the determinants of these forms are negative, the form F will be positive if

---

[m] Or of the expression $h^\nu/(b + b')$ (mod. $h^{\lambda - \nu}$) and $B \equiv b - ch^{n - \nu}[h^\nu/(b + b')] \equiv [(D + bb') : h^\nu/[(b + b') : h^\nu]$ (mod. A).

both f, f' are positive or both negative; conversely F will be negative if one of the forms f, f' is positive and the other negative. Let us examine particularly the case considered in the preceding article where F is composed of f, f' and f, f', F have the same determinant D. Let us suppose further that we have representations of the numbers a, a' by the forms f, f' by means of undetermined relatively prime values. We will also suppose that the former belongs to the value b of the expression $\sqrt{D}$ (mod. a), the latter to the value b' of the expression $\sqrt{D}$ (mod. a') and that $b^2 - D = ac$, $b'b' - D = a'c'$. Then by article 168 the forms $(a, b, c), (a', b', c')$ will be properly equivalent to the forms f, f' and so F will be composed of these two forms. But the form (A, B, C) will be composed of the same forms if the greatest common divisor of the numbers $a, a', b + b'$ is μ, and if we set $A = aa'/\mu^2$, $B \equiv b$, and $\equiv b'$ relative to the moduli $a/\mu, a'/\mu$ respectively, $AC = B^2 - D$; and this form will be properly equivalent to the form F. Now the number aa' is represented by the form $Ax^2 + 2Bxy + Cy^2$ if we let $x = \mu, y = 0$ (their greatest common divisor is μ); thus aa' can also be represented by the form F in such a way that the values of the unknowns have μ as greatest common divisor (art. 166). Whenever therefore $\mu = 1$, aa' can be represented by the form F by assigning relatively prime values to the unknowns, and this representation will belong to the value B of the expression $\sqrt{D}$ (mod. aa'), which is congruent to b, b' relative to the moduli a, a' respectively. The condition $\mu = 1$ always has place when a, a' are relatively prime; or more generally when the greatest common divisor of a, a' is relatively prime to $b + b'$.

▶ 245. THEOREM. *If the form f belongs to the same order as g, and f' is of the same order as g': then the form F composed of f, f' will have the same determinant and will be of the same order as the form G composed of g, g'.*

Demonstration. Let the forms f, f', F be $= (a, b, c), (a', b', c), (A, B, C)$, respectively, and let their determinants $= d, d', D$. Further let the greatest common divisor of the numbers $a, 2b, c$ be $= m$ and the greatest common divisor of the numbers $a, b, c = \mathfrak{m}$; and let $m', \mathfrak{m}'$ with respect to the form f' and $M, \mathfrak{M}$ with respect to the form F have similar significations. Then the order of the form f will be determined by the numbers $d, m, \mathfrak{m}$ and so these same numbers will be valid for the form g; for the same reason the numbers $d', m', \mathfrak{m}'$ will play the same role for the forms g' and f'. Now by article 235 the numbers $D, M, \mathfrak{M}$ are determined by d, d', m, m', $\mathfrak{m}, \mathfrak{m}'$; that is D will be the greatest common divisor of $dm'm'$, $d'm^2$; $M = mm'$; $\mathfrak{M} = \mathfrak{m}\mathfrak{m}'$ (if $m = \mathfrak{m}$ and $m' = \mathfrak{m}'$) or $= 2\mathfrak{m}\mathfrak{m}'$ (if $m = 2\mathfrak{m}$ or $m' = 2\mathfrak{m}'$). Since these properties of $D, M, \mathfrak{M}$ follow from the fact that F is composed of f, f', it is easy to see that $D, M, \mathfrak{M}$ perform the same function for the form G, and so G is of the same order as F.   Q.E.D.

For this reason we will say that the order of the form F is composed of the orders of the forms f, f'. Thus, e.g., if we have two properly primitive orders, their composition will be properly primitive; if one is properly primitive and the other improperly primitive, the composition will be improperly primitive. If we say that an order is composed of several other orders, it is to be understood in a similar way.

▶ 246. PROBLEM. *Given any two primitive forms f, f' and the form F composed of these two: to determine the genus to which F should be referred from the genera to which f, f' belong.*

Solution. I. Let us consider first the case where at least one of the forms f, f' (e.g. the former) is properly primitive, and let us designate the determinants of the forms f, f', F by d, d', D. D will be the greatest common divisor of the numbers $dm'm', d'$ where m' is either 1 or 2 according as the form f' is properly or improperly primitive. In the former case F will belong to a properly primitive order, in the latter to an improperly primitive order. Now the genus of the form F will be defined by its particular characters, that is with respect to the individual odd prime divisors of D and also for certain cases with respect to the numbers 4 and 8. We will consider these cases separately.

1. If p is an odd prime divisor of D, it will also necessarily divide d, d' and so also among the characters of the forms f, f' will occur their relations to p. Now if the number a can be represented by f, and the number a' by f', the product aa' can be represented by F. So if the quadratic residues of p (not divisible by p) can be represented both by f and by f', they can also be represented by F; i.e. if both f and f' have the character Rp, the form F will have the same character. For a similar reason F will have the character

Rp if both f, f' have the character Np; conversely F will have the character Np if one of the forms f, f' has the character Rp and the other Np.

2. If a relation to the number 4 enters into the total character of the form F, such a relation will also enter into the characters of the forms f, f', for this can happen only when D is $\equiv 0$ or $\equiv 3$ (mod. 4). When D is divisible by f, $dm'm'$ and d' will also be divisible by 4, and it is immediately clear that f' cannot be improperly primitive and so $m' = 1$. Then both d and d' will be divisible by 4 and a relation to 4 will enter into the character of each. When $D \equiv 3$ (mod. 4) D will divide d, d', the quotients will be squares and so d, d' will necessarily $\equiv 0$ or $\equiv 3$ (mod. 4) and a relation to the number 4 will be included among the characters of f, f'. Thus just as in (1) it will follow that the character of the form F will be 1, 4 if both f, f' have the character 1, 4 or 3, 4; conversely the character of the form F will be 3, 4 if one of the forms f, f' has the character 1, 4 and the other 3, 4.

3. When D is divisible by 8, d' will be also. Thus f' will certainly be properly primitive, $m' = 1$ and d will also be divisible by 8. And therefore one of the characters 1, 8; 3, 8; 5, 8; 7, 8 will appear among the characters of the form F only if such a relation to 8 appears also in the character of both the form f and f'. In the same way as before it is easy to see that 1, 8 will be a character of the form F if f and f' have the same character with respect to 8; that 3, 8 will be a character of the form F if one of the forms f, f' has the character 1, 8, the other 3, 8; or one of them has the character 5, 8 the other 7, 8; F will have the character 5, 8 if f, f' have 1, 8 and 5, 8 or 3, 8 and 7, 8; and F will have the character 7, 8 if f and f' have either 1, 8 and 7, 8 or 3, 8 and 5, 8 as characters.

4. When $D \equiv 2$ (mod. 8), d' will be either $\equiv 0$ or $\equiv 2$ (mod. 8) so $m' = 1$ and d will also $\equiv 0$ or $\equiv 2$ (mod. 8); but since D is the *greatest* common divisor of d, d' they cannot both be divisible by 8. Then in this case the character of the form F can only be 1 *and* 7, 8, or 3 *and* 5, 8 when both of the forms f, f' have one of these characters, or when one of them has one of these characters and the other has one of the following: 1, 8; 3, 8; 5, 8; 7, 8. The following table will determine the character of the form F. The character in the margin pertains to one of the forms f, f', and the character

at the head of the columns pertains to the other.

	1 *and* 7, 8 or 1, 8 or 7, 8	3 *and* 5, 8 or 3, 8 or 5, 8
1 *and* 7, 8	1 *and* 7, 8	3 *and* 5, 8
3 *and* 5, 8	3 *and* 5, 8	1 *and* 7, 8

5. In the same way it can be proven that F cannot have the character 1 *and* 3, 8 or 5 *and* 7, 8 unless at least one of the forms f, f' has one of these characters. The other can have one of them also or one of these: 1, 8; 3, 8; 5, 8; 7, 8. The character of the form F is determined by the following table. The characters of the forms f, f' again appear in the margin and at the head of the columns.

	1 *and* 3, 8 or 1, 8 or 3, 8	5 *and* 7, 8 or 5, 8 or 7, 8
1 *and* 3, 8	1 *and* 3, 8	5 *and* 7, 8
5 *and* 7, 8	5 *and* 7, 8	1 *and* 3, 8

II. If each of the forms f, f' is improperly primitive, D will be the greatest common divisor of the numbers $4d, 4d'$ or $D/4$ the greatest common divisor of the numbers d, d'. It follows that d and d' and $D/4$ will all be $\equiv 1 \pmod{4}$. If we let $F = (A, B, C)$ the greatest common divisor of the numbers A, B, C will $= 2$, and the greatest common divisor of the numbers $A, 2B, C$ will be 4. Thus F will be a form derived from the improperly primitive form $(A/2, B/2, C/2)$. The determinant of the latter will be $D/4$ and its genus will determine the genus of the form F. But since it is improperly primitive its character will not imply any relations to 4 or 8 but only to the individual odd prime divisors of $D/4$. Now all these divisors manifestly also divide d, d'. And if the two factors of a product are representable one by f, the other by f', then half the product is representable by the form $(A/2, B/2, C/2)$. It follows that the character of this form with respect to any odd prime number p which divides $D/4$ will be Rp when $2Rp$ and the forms f, f' have the same character with respect to p and when $2Np$ and the characters of f, f' with respect to p are opposite. Conversely the character of the form will be Np when f, f' have equal characters with respect to p and $2Np$ and when f, f' have opposite characters and $2Rp$.

▶ 247. From the solution of the preceding problem it is manifest that if g is a primitive form of the same order and genus as f, and g' is a primitive form of the same order and genus as f': then the form composed of g and g' will be of the same genus as the form composed of f and f'. From this we can easily understand what we mean when we speak of a *genus composed* of two (or even several) other genera. And further, if f, f' have the same determinant, f is a form of a principal genus, and F is composed of f and f': then F will be of the same genus as f'; and therefore the principal genus can always be omitted in the composition with other genera of the same determinant. Therefore, other things being equal, if f is not in a principal genus and f' is a primitive form: F will certainly be in a genus different from f'. Finally if f, f' are properly primitive forms of the same genus, F will be in a principal genus; if however f, f' are both properly primitive with the same determinant but in different genera, F cannot belong to a principal genus. And if a properly primitive form is composed *with itself*, the resulting form, which will also be properly primitive with the same determinant, necessarily belongs to a principal genus.

▶ 248. PROBLEM. *Given any two forms, f, f' from which F is composed: to determine the genus of the form F from those of the forms f, f'.*

Solution. Let $f = (a, b, c)$, $f' = (a', b', c')$, $F = (A, B, C)$; further, designate by $\mathfrak{m}$ the greatest common divisor of the numbers a, b, c and by $\mathfrak{m}'$ the greatest common divisor of the numbers a', b', c' so that f, f' are derived from the primitive forms $(a/\mathfrak{m}, b/\mathfrak{m}, c/\mathfrak{m}), (a'/\mathfrak{m}', b'/\mathfrak{m}', c'/\mathfrak{m}')$ which we will designate by $\mathfrak{f}, \mathfrak{f}'$ respectively. Now if at least one of the forms $\mathfrak{f}, \mathfrak{f}'$ is properly primitive, the greatest common divisor of the numbers A, B, C will be mm', and thus F will be derived from the primitive form $(A/mm', B/mm', C/mm') \ldots \mathfrak{F}$ and it is clear that the genus of the form F will depend on that of the form $\mathfrak{F}$. It is easily seen that if $\mathfrak{F}$ is transformed into $\mathfrak{f}\mathfrak{f}'$ by the same substitution that transforms F into ff' and thus that $\mathfrak{F}$ is composed of $\mathfrak{f}, \mathfrak{f}'$, its genus can be determined by the problem of article 246. But if both f, f' are improperly primitive, the greatest common divisor of the numbers A, B, C will be $2mm'$, and the form $\mathfrak{F}$ which is still composed of $\mathfrak{f}, \mathfrak{f}'$ will manifestly be derived from the properly primitive form $(A/2mm', B/2mm', C/2mm')$. The genus of this form can be determined by article 246 and since F is derived from the same form, its genus will be known as well.

From this solution it is manifest that the theorem in the preceding article which was restricted to primitive forms is valid for any forms whatsoever: *if f', g' are of the same genera respectively as f, g, the form composed of f', g' will be of the same genus as the form composed of f, g.*

▶ 249. THEOREM. *If the forms f, f' are of the same orders, genera, and classes as g, g' respectively, then the form composed of f and f' will be of the same class as the form composed of g and g'.*

From this theorem (whose truth follows immediately from art. 239) it is evident what we mean when we speak of a *class composed of two (or more) given classes.*

If any class K is composed with a principal class, the result will be the class K itself; that is, in composition with other classes of the same determinant a principal class can be ignored. From the composition of two properly primitive opposite classes we will always get a principal class of the same determinant (see art. 243). Since therefore any ambiguous class is opposite to itself, we will always get a principal class of the same determinant if we compose any properly primitive ambiguous class with itself.

The last proposition can also be converted; that is *if from the composition of a properly primitive class K with itself we get a principal class H with the same determinant, K will necessarily be an ambiguous class.* For if K' is a class opposite to K, the same class will arise from the composition of H and K' as from the three classes K, K, K'; from the latter we get K (since K and K' produce H, and H and K produce K), from the former we get K'; therefore K and K' coincide and the class is ambiguous.

We note further: *If the classes K, L are opposite to the classes K', L' respectively, the class composed of K and L will be opposite to the class composed of K' and L'.* Let the forms of the classes K, L, K', L' be respectively f, g, f', g', and let F be composed of f, g and F' composed of f', g'. Since f' is improperly equivalent to f, and g' improperly equivalent to g, while F is composed of both f and g directly: F will also be composed of f' and g' but of each of them indirectly. Thus any form which is improperly equivalent to F will be composed of f', g' directly and so will be properly equivalent to F' (art. 238, 239). Therefore F, F' will be improperly equivalent and the classes to which they belong will be opposite.

It follows from this that if an ambiguous class K is composed with an ambiguous class L we will always get an ambiguous class. For it will be opposite to the class which is composed of classes opposite to K, L; that is, to itself, since these classes are opposite to themselves.

Finally we observe that if we are given any two classes K, L of the same determinant and the former is properly primitive, we can always find a class M with the same determinant such that L is composed of M and K. Manifestly this can be done by taking for M the class which is composed of L and the class opposite to K; it is very easy to see that this class is the only one that enjoys this property; that is to say, if we compose different classes of the same determinant with the same properly primitive class, we get different classes.

It is convenient to denote composition of classes by the addition sign, $+$, and identity of classes by the equality sign. Using these signs the proposition just considered can be stated as follows: If the class K' is opposite to K, $K + K'$ will be a principal class of the same determinant, so $K + K' + L = L$; if we set $K' + L = M$ we have $K + M = L$, as was desired. Now if besides M we have another class M' with the same property, that is $K + M' = L$, we will have $K + K' + M' = L + K' = M$ and so $M' = M$. If many identical classes are composed, we can indicate this (as in multiplication) by prefixing their number, so that $2K$ means the same as $K + K$, $3K$ the same as $K + K + K$,

etc. We could also transfer the same signs to forms so that
$(a, b, c) + (a', b', c')$ would indicate the form composed of (a, b, c),
(a', b', c'); but to avoid ambiguity we prefer not to use this abbreviation, especially since we have already assigned a special
meaning to the symbol $\sqrt{M(a, b, c)}$. We will say that the class $2K$
arises from the *duplication* of the class K, the class $3K$ from
triplication etc.

▶ 250. If D is a number divisible by m^2 (we presume m positive)
there will be an order of forms of determinant D derived from the
properly primitive order of the determinant D/m^2 (when D is
negative there will be *two* of them, one positive and one negative);
manifestly the form $(m, 0, -D/m)$ will belong to that order (the
positive one) and can rightly be considered the *simplest form* in
the order (just as $(-m, 0, D/m)$ will be the simplest in the negative
order when D is negative). If we also have $D/m^2 \equiv 1$ (mod. 4)
there will also be an order of forms of determinant D derived
from the improperly primitive determinant D/m^2. The form
$[2m, m, (m^2 - D)/2m]$ will belong to this and it will be the simplest
form in the order. (When D is negative there will again be two
orders and in the negative order $[-2m, -m, (D - m^2)/2m]$ will be
the simplest form.) Thus, e.g., if we apply this to the case where
$m = 1$, the following will be the simplest in the four orders of
forms with determinant 45: $(1, 0, -45)$, $(2, 1, -22)$, $(3, 0, -15)$,
$(6, 3, -6)$.

All these considerations give rise to the following.

PROBLEM. *Given any form F of the order O, to find a properly
primitive (positive) form of the same determinant which will give
us F when it is composed with the simplest form in O.*

Solution. Let the form $F = (ma, mb, mc)$ be derived from the
primitive form $f = (a, b, c)$ of determinant d and we will suppose
first that f is properly primitive. We observe that should a and
$2dm$ not be relatively prime, there are certainly other forms
properly equivalent to (a, b, c) whose first terms have this property.
For by article 228 there are numbers relatively prime to $2dm$
representable by that form. Let such a number be $a' = a\alpha^2 +
2b\alpha\gamma + c\gamma^2$ and we shall suppose (it is legitimate) that α, γ are
relatively prime. Now if we choose β, δ so that $\alpha\delta - \beta\gamma = 1$, f will
be transformed by the substitution $\alpha, \beta, \gamma, \delta$ into the form

(a', b', c') which is properly equivalent to it and has the prescribed property. Now since F and $(a'm, b'm, c'm)$ are properly equivalent it is sufficient to consider the case where a and $2dm$ are relatively prime. Now (a, bm, cm^2) will be a properly primitive form of the same determinant as F (for if $a, 2bm, cm^2$ had a common divisor, it would also mean that $2dm = 2b^2m - 2acm$). It is easy to confirm that F will be transformed into the product of the form $(m, 0, -dm)$ into (a, bm, cm^2) by the substitution $1, 0, -b, -cm;\ 0, m, a, bm$. Note that unless F is a negative form $(m, 0, -dm)$ will be the simplest form of order O. By using the criterion of the fourth observation in article 235 we conclude that F is composed of $(m, 0, -dm)$ and (a, bm, cm^2). When however F is a negative form, it will be transformed by the substitution $1, 0, b, -cm;\ 0, -m, -a, bm$ into the product of $(-m, 0, dm)$, the simplest form of the same order, into the positive form $(-a, bm, -cm^2)$ and so it will be composed of these two.

Second, if f is an improperly primitive form we can suppose that $a/2$ and $2dm$ are relatively prime; for if this property is not yet true of the form f we can find a form properly equivalent to f which has the property. From this it follows easily that $(a/2, bm, 2cm^2)$ is a properly primitive form of the same determinant as F; and it is just as easy to confirm that F will be transformed into the product of the forms

$$[\pm 2m, \pm m, \pm\tfrac{1}{2}(m - dm)], \quad [\pm\tfrac{1}{2}a, bm, \pm 2cm^2]$$

by the substitution

$$1, 0, \tfrac{1}{2}(1 \mp b), -cm; \quad 0, \pm 2m, \pm\tfrac{1}{2}a, (b + 1)m$$

where the lower signs are to be taken when F is a negative form, the upper signs otherwise. We conclude that F is composed of these two forms and that the former is the simplest of order O, the latter a properly primitive (positive) form.

▶251. PROBLEM. *Given two forms F, f of the same determinant D and belonging to the same order O: to find a properly primitive form of determinant D which produces F when it is composed with f.*

Solution. Let ϕ be the simplest form of order O; $\mathfrak{F}$, $\mathfrak{f}$ properly primitive forms of determinant D which produce F, f respectively when composed with ϕ; and let f' be the properly primitive form which produces $\mathfrak{F}$ when composed with $\mathfrak{f}$. Then the form F will be composed of the three forms $\phi, \mathfrak{f}, f'$ or of the two forms f, f'. Q.E.I.

Therefore any class of a given order can be considered as composed of any given class of the same order and another properly primitive class of the same determinant.

▶ 252. THEOREM. *For a given determinant there are the same number of classes in every genus of the same order.*

Demonstration. Suppose the genera G and H belong to the same order, that G is composed of n classes $K, K', K'', \ldots K^{n-1}$ and that L is any class of the genus H. By the preceding article we find a properly primitive class M of the same determinant whose composition with K produces L, and we designate by $L', L'', \ldots L^{n-1}$ the classes which arise from composing the class M with $K', K'', \ldots K^{n-1}$ respectively. Then from the last observation of article 249 it follows that all the classes $L, L', L'', \ldots L^{n-1}$ are different, and by article 248 that they all belong to the same genus, i.e. to the genus H. Finally it is easy to see that H cannot contain any other classes than these since each class of genus H can be considered as composed of M and another class of the same determinant, and it will necessarily always be of genus G. Therefore H, like G, will contain n different classes. Q.E.D.

▶ 253. The preceding theorem supposes identity of order and cannot be extended to different orders. Thus for example for the determinant -171 there are 20 positive classes which are reduced to four orders: in the properly primitive order there are two genera each containing six classes; in the improperly primitive order two genera have four classes, two in each; in the order derived from the properly primitive order of determinant -19 there is only one genus containing three classes; finally, the order derived from the improperly primitive order of determinant -19 has one genus with one class. The same is true of the negative classes. It is useful therefore to inquire into the general principle which governs the relationship between the number of classes in different orders. Suppose that K, L are two classes of the same (positive) order O of determinant D, and M is a properly primitive class of the same determinant which produces L when composed with K. By article 251 such a class can always be found. Now in some cases it can happen that M is the *only* properly primitive class with this property; in other cases there can exist several different properly primitive classes with this property. Let us suppose in general that there are r properly primitive classes of this kind $M, M', M'', \ldots M^{r-1}$ and that each of them produces the same class L when composed with K. We will designate this complex by the letter W. Now let L' be another class of order O (different from class L), and N' a properly primitive class of determinant D which gives L' when composed with L. We will use W' to designate the complex of classes $N' + M, N' + M', N' + M''$, $\ldots N' + M^{r-1}$ (they will all be properly primitive and different from one another). It is easy to see that K will produce L' if it is composed with any class of W', and so we conclude that W and W' have no class in common; and every properly primitive class which produces L' when it is composed with K is contained in W'. In the same way, if L'' is another class of order O different from L, L', then there will be r properly primitive forms all different from each other and from the forms in W, W', each of which will produce L'' when composed with K. And the same thing is true for all other classes of order O. Now since any properly primitive (positive) class of determinant D produces a class of order O when composed with K, it is clear that if the number of all the classes of order O is n, the number of all properly primitive (positive) classes of the same determinant will be rn. We thus have a general rule: If we denote by K, L any two classes of order O and by r the number of different properly primitive classes of the same determinant, each of which produces L when composed with K, then the number of all the classes in the (positive) properly primitive order will be r times greater than the number of classes of order O.

Since in order O the classes K, L can be chosen arbitrarily, it is permissible to take identical classes, and it will be particularly advantageous to choose that class containing the simplest form of the order. If we therefore choose that class for K and L the operation will be reduced to assigning all properly primitive classes which duplicate K when composed with K. We will develop this method in the following.

▶ 254. THEOREM. *If $F = (A, B, C)$ is the simplest form of order O of determinant D, and $f = (a, b, c)$ is a properly primitive form of the same determinant: then the number A^2 can be represented by this form f provided F results from the composition of the forms f, F; and conversely F will be composed of itself and f if A^2 can be represented by f.*

Demonstration. I. If F is transformed into the product fF by the substitution $p, p', p'', p'''; q, q', q'', q'''$ then by article 235 we will have

$$A(aq''q'' - 2bqq'' + cq^2) = A^3$$

and therefore

$$A^2 = aq''q'' - 2bqq'' + cq^2 \quad \text{Q.E.P.}$$

II. We will *presume* that A^2 can be represented by f and designate the unknown values by which this is done as q'', $-q$; that is, $A^2 = aq''q'' - 2bqq'' + cq^2$. Further we will let

$$q''a - q(b + B) = Ap, \quad -qC = Ap', \quad q''(b - B) - qc = Ap''$$

$$-q''C = Ap''', \quad q''a - q(b - B) = Aq',$$

$$q''(b + B) - qc = Aq'''$$

It is easy to confirm that F is transformed into the product fF by the substitution $p, p', p'', p'''; q, q', q'', q'''$. If the numbers p, p', etc. are integers then F will be composed of f and F. Now from the description of the simplest form, B is either 0 or $A/2$ and so $2B/A$ is an integer; in the same way it is clear that C/A is also always an integer. Thus $q' - p, p', q''' - p'', p'''$ will be integers and it remains only to prove that p and p'' are integers. Now we have

$$p^2 + \frac{2pqB}{A} = a - \frac{q^2C}{A}, \quad p''p'' + \frac{2p''q''B}{A} = c - \frac{q''q''C}{A}$$

If $B = 0$ we get

$$p^2 = a - \frac{q^2C}{A}, \quad p''p'' = c - \frac{q''q''C}{A}$$

and so p, p'' are integers; but if $B = A/2$ we have

$$p^2 + pq = a - \frac{q^2 C}{A}, \quad p''p'' + p''q'' = c - \frac{q''q''C}{A}$$

and in this case also p and p'' are integers. Therefore F is composed of f and F.   Q.E.S.

▶ 255. The problem therefore is reduced to finding all the properly primitive classes of determinant D whose forms can represent A^2. Manifestly A^2 can be represented by any form whose first term is either A^2 or the square of a factor of A; conversely if A^2 can be represented by the form f, f will be transformed into a form whose first term is A^2/e^2 by the substitution $\alpha, \beta, \gamma, \delta$ provided we assign $\alpha e, \gamma e$ (their greatest common divisor must be e) as the values of the unknowns. This form will be properly equivalent to the form f if β, δ are chosen so that $\alpha\delta - \beta\gamma = 1$. Thus it is clear that in any class having forms that can represent A^2, we can find forms whose first term is A^2 or the square of a factor of A. The whole process then depends on finding all properly primitive classes of determinant D containing forms of this kind. We do this in the following way. Let a, a', a'' etc. be all the (positive) divisors of A; now find all values of the expression $\sqrt{D}$ (mod. a^2) between 0 and $a^2 - 1$ inclusively and call them b, b', b'', etc. We set

$$b^2 - D = a^2 c, \quad b'b' - D = a^2 c', \quad b''b'' - D = a^2 c'', \text{ etc.}$$

and designate the complex of forms $(a^2, b, c), (a^2, b', c')$, etc. by the letter V. Obviously every class of determinant D which has a form with first term a^2 also must contain some form from V. In a similar way we determine all forms of determinant D with first term $a'a'$ and second term lying between 0 and $a'a' - 1$ inclusive and designate the complex by the letter V'; by a similar construction we will let V'' be the complex of similar forms whose first term is $a''a''$ etc. Now eliminate from V, V', V'', etc. all forms which are not properly primitive and reduce the rest to classes. If there are many forms belonging to the same class, retain only one of them. In this way we will have all the classes we are looking for, and the ratio of this number with respect to unity will be the same as the ratio of the number of all properly primitive (positive) classes to the number of all classes in order O.

Example. Let $D = -531$ and O the positive order derived from the improperly primitive order of determinant -59; its simplest form is $(6, 3, 90)$, so $A = 6$. Here a, a', a'', a''' will be $1, 2, 3, 6$; V will contain the form $(1, 0, 531)$; V' will contain $(4, 1, 133)$, $(4, 3, 135)$; V'' $(9, 0, 59)$, $(9, 3, 60)$, $(9, 6, 63)$; and V''' $(36, 3, 15)$, $(36, 9, 17)$, $(36, 15, 21)$, $(36, 21, 27)$, $(36, 27, 35)$, $(36, 33, 45)$. But of these twelve forms six must be rejected, the second and third from V'', the first, third, fourth, and sixth from V'''. All of these are derived forms. All the six remaining belong to different classes. As a matter of fact the number of properly primitive (positive) classes of determinant -531 is 18; the number of improperly primitive (positive) classes of determinant -59 (or the number of classes of determinant -531 derived from these) is 3, and thus the ratio is 6 to 1.

▶ 256. This solution will be made clearer by the following general observations.

I. If the order O is derived from a properly primitive order, A^2 will divide D; but if O is improperly primitive or derived from an improperly primitive order, A will be even, D will be divisible by $A^2/4$ and the quotient will be $\equiv 1$ (mod. 4). Thus the square of any divisor of A will divide D or at least $4D$ and in the latter case the quotient will always be $\equiv 1$ (mod. 4).

II. If a^2 divides D, the values of the expression $\sqrt{D}$ (mod. a^2) which lie between 0 and $a^2 - 1$ will be $0, a, 2a, \ldots a^2 - a$ and so a will be the number of forms in V; but among them there will be only as many properly primitive forms as there are numbers among

$$\frac{D}{a^2}, \frac{D}{a^2} - 1, \frac{D}{a^2} - 4, \ldots \frac{D}{a^2} - (a - 1)^2$$

which have a common divisor with a. When $a = 1$, V will consist of only one form $(1, 0, -D)$ and it will always be properly primitive. When a is 2 or a power of 2, half the a numbers will be even, half odd; thus there will be $a/2$ properly primitive forms in V. When a is any other prime number p or a power of the prime number p, three cases must be distinguished: if D/a^2 is not divisible by p and is not a quadratic residue of p, all these a numbers will be relatively prime to a and so all forms in V will be properly primitive; but if p divides D/a^2 there will be $(p - 1)a/p$

properly primitive forms in V; finally if D/a^2 is a quadratic residue of p not divisible by p, there will be $(p-2)a/p$ properly primitive forms. All of this can be shown with no difficulty. In general, if $a = 2^\nu p^\pi q^\chi r^\rho \dots$ where p, q, r etc. are different odd prime numbers, the number of properly primitive forms in V will be $NPQR\dots$, where

$N = 1$ (if $\nu = 0$)   or $N = 2^{\nu-1}$ (if $\nu > 0$)

$P = p^\pi$ (if D/a^2 is a quadratic nonresidue of p)   or

$P = (p-1)p^{\pi-1}$ (if D/a^2 is divisible by p)   or

$P = (p-2)p^{\pi-1}$ (if D/a^2 is a quadratic residue of p not divisible by p)

and Q, R, etc. are to be defined in the same way by q, r, etc. as P by p.

III. If a^2 does not divide D, $4D/a^2$ will be an integer and $\equiv 1$ (mod. 4) and the values of the expression $\sqrt{D}$ (mod. a^2) will be $a/2, 3a/2, 5a/2, \dots a^2 - (a/2)$. Therefore the number of forms in V will be a and there will be as many properly primitive among them as there are numbers among

$$\frac{D}{a^2} - \frac{1}{4}, \frac{D}{a^2} - \frac{9}{4}, \frac{D}{a^2} - \frac{25}{4}, \dots \frac{D}{a^2} - \left(a - \frac{1}{2}\right)^2$$

which are relatively prime to a. Whenever $4D/a^2 \equiv 1$ (mod. 8) all these numbers will be even and thus there will be no properly primitive form in V; but when $4D/a^2 \equiv 5$ (mod. 8) all these numbers will be odd and so all the forms in V will be properly primitive if a is 2 or a power of 2. In this case as a general norm there will be as many properly primitive forms in V as there are numbers not divisible by any odd prime divisor of a. There will be $NPQR\dots$ of them if $a = 2^\nu p^\pi q^\chi r^\rho \dots$. Here $N = 2^\nu$ and P, Q, R, etc. are to be derived from p, q, r, etc. in the same way as in the preceding case.

IV. We have thus shown how to determine the number of properly primitive forms in V, V', V'', etc. We can find the total number by the following general rule. If $A = 2^\nu \mathfrak{A}^\alpha \mathfrak{B}^\beta \mathfrak{C}^\gamma \dots$ where $\mathfrak{A}, \mathfrak{B}, \mathfrak{C}$, etc. are different odd prime numbers, the total number of all properly primitive forms in V, V', V'', etc. will be

$$= Anabc.../2\mathfrak{A}\mathfrak{B}\mathfrak{C}... \text{ where}$$

$\mathfrak{n} = 1$ (if $4D/A^2 \equiv 1$ (mod. 8)),   or

$\mathfrak{n} = 2$ (if D/A^2 is an integer),   or

$\mathfrak{n} = 3$ (if $4D/A^2 \equiv 5$ (mod. 8));   and

$\mathfrak{a} = \mathfrak{A}$ (if $\mathfrak{A}$ divides $4D/A^2$),   or

$\mathfrak{a} = \mathfrak{A} \pm 1$ (if $\mathfrak{A}$ does not divide $4D/A^2$; the upper or lower sign is to be taken according as $4D/A^2$ is a non-residue or a residue of $\mathfrak{A}$)

Finally, b, c, etc. are to be derived from $\mathfrak{B}, \mathfrak{C}$, etc. in the same way as $\mathfrak{a}$ from $\mathfrak{A}$. Brevity does not permit us to demonstrate this more fully.

V. Now with regard to the number of classes which result from the properly primitive forms in V, V', V'', etc., we must distinguish the three following cases.

First, when D is a negative number, each of the properly primitive forms in V, V', etc. constitutes a separate class. Thus the number of classes will be expressed by the formula given in the preceding observation except for two cases, namely when $4D/A^2$ either $= -4$ or $= -3$; that is, when D either $= -A^2$ or $= -3A^2/4$. To prove this theorem we must only show that it is impossible for two different forms of V, V', V'', etc. to be properly equivalent. Let us suppose therefore that $(h^2, i, k), (h'h', i', k')$ are two different properly primitive forms of V, V', V'', etc., both belonging to the same class. And let us suppose that the former is transformed into the latter by the proper substitution $\alpha, \beta, \gamma, \delta$; we will have the equations

$$\alpha\delta - \beta\gamma = 1, \quad h^2\alpha^2 + 2i\alpha\gamma + k\gamma^2 = h'h',$$
$$h^2\alpha\beta + i(\alpha\delta + \beta\gamma) + k\gamma\delta = i'$$

From this it is easy to conclude first that γ certainly does not $= 0$ [and it follows that $\alpha = \pm 1$, $h^2 = h'h'$, $i' \equiv i$ (mod. h^2), and the proposed forms are identical, contrary to the hypothesis]; second, that γ is divisible by the greatest common divisor of the numbers h, h' (for if we let this divisor $= r$, it manifestly also divides $2i, 2i'$

and is relatively prime to k; besides, r^2 divides $h^2k - h'h'k' = i^2 - i'i'$; obviously therefore r must also divide $i - i'$; but $\alpha i' - \beta h'h' = \alpha i + \gamma k$ so γk and γ also will be divisible by r); third, $(\alpha h^2 + \gamma i)^2 - D\gamma^2 = h^2h'h'$. If therefore we let $\alpha h^2 + \gamma i = rp$, $\gamma = rq$, p and q will be integers and q will not $= 0$ and we will have $p^2 - Dq^2 = h^2h'h'/r^2$. But $h^2h'h'/r^2$ is the smallest number divisible by both h^2 and $h'h'$ and thus it will divide A^2 and $4D$. As a result $4Dr^2/h^2h'h'$ will be a (negative) integer. If we let it $= -e$ we have $p^2 - Dq^2 = -4D/e$ or $4 = (2rp/hh')^2 + eq^2$. In this equation $(2rp/hh')^2$ is necessarily a square less than 4 and so will be either 0 or 1. In the former case $eq^2 = 4$ and $D = -(hh'/rq)^2$ and it follows that $4D/A^2$ is a square with a negative sign and so certainly not $\equiv 1$ (mod. 4) and therefore O is not an improperly primitive order nor derived from an improperly primitive order. So D/A^2 will be an integer, and clearly e will be divisible by 4, $q^2 = 1$, $D = -(hh'/r)^2$ and A^2/D is also an integer. Thus $D = -A^2$ or $D/A^2 = -1$, which is the first exception. In the latter case $eq^2 = 3$ so $e = 3$ and $4D = -3(hh'/r)^2$. Thus $3(hh'/rA)^2$ will be an integer, and it cannot be anything but 3, since when we multiply it by the square integer $(rA/hh')^2$ we get 3. Therefore $4D = -3A^2$ or $D = -3A^2/4$ which is the second exception. In all remaining cases all properly primitive forms in V, V', V'', etc. will belong to different classes. For the excepted cases it is sufficient to give the result. It can be found without difficulty, but it is too long to present here. In the former case there will always be a pair of properly primitive forms in V, V', V'', etc. which belong to the same class; in the latter case a triple. Thus in the former case the number of classes will be half the value given above, in the latter case a third.

Second, when D is a positive square number, each properly primitive form in V, V', V'', etc. constitutes a separate class without exception. For let us suppose that $(h^2, i, k), (h'h', i', k')$ are two such different properly equivalent forms and that the former is transformed into the latter by the proper substitution $\alpha, \beta, \gamma, \delta$. Obviously all the arguments we used in the preceding case when we did not suppose D negative, have value here. Therefore if we determine p, q, r as above, $4Dr^2/h^2h'h'$ will be an integer here also, but positive rather than negative. Further it will be a square. If we let it $= g^2$ we will have $(2rp/hh')^2 - g^2q^2 = 4$. Q.E.A. Because the difference of two squares cannot be 4 unless the smaller is 0; thus our supposition is inconsistent.

Third, where D is positive but not a square we have as yet no general rule for comparing the number of properly primitive forms in V, V', V'', etc. with the number of different classes resulting from them. We can only assert that the latter is either equal to the former or is a factor of it. We have also discovered a connection between the quotient of these numbers and the least values of t, u satisfying the equation $t^2 - Du^2 = A^2$, but it would take too long to explain it here. And we cannot say with certitude whether it is possible to know this quotient in all cases by merely inspecting the numbers D, A (as in the preceding cases). We give a few examples and the reader can easily add some of his own. For $D = 13, A = 2$ the number of properly primitive forms in V etc. is 3, all of which are equivalent and thus make up a single class; for $D = 37, A = 2$ there will also be three properly primitive forms in V etc., but they will belong to three different classes; for $D = 588, A = 7$ we have eight properly primitive forms in V etc., and they make up four classes; for $D = 867, A = 17$ there will be 18 properly primitive forms, and the same number for $D = 1445$, $A = 17$, but for the first determinant they will divide into two classes while in the second there will be six.

VI. From the application of this general theory to the case where O is an improperly primitive order we find that the number of classes contained in this order bears the same ratio to the number of all classes in the properly primitive order as 1 does to the number of different properly primitive classes produced by the three forms $(1, 0, -D), (4, 1, (1 - D)/4), (4, 3, (9 - D)/4)$. Now when $D \equiv 1 \pmod 8$, there will be only one class because in this case the second and third forms are improperly primitive; but when $D \equiv 5 \pmod 8$ these three forms will all be properly primitive and will produce the same number of different classes if D is negative except when $D = -3$, in which case there will be only one; finally, when D is positive (of the form $8n + 5$) we have one of the cases for which there is no general rule. But we can assert that in this case the three forms will belong to three different or to only one class, never to two; for it is easily seen that if the forms $(1, 0, -D), (4, 1, (1 - D)/4), (4, 3, (9 - D)/4)$ belong respectively to the classes K, K', K'', we will have $K + K' = K'$,

$K' + K' = K''$ and thus if K and K' are identical, K' and K'' will also be identical; similarly if K and K'' are identical, K' and K'' will also be; finally, since we will have $K' + K'' = K$ if we suppose K' and K'' are identical, it follows that K and K'' will coincide. Thus the three classes K, K', K'' will all be different or all identical. For example, there are 75 numbers of the form $8n + 5$ less than the number 600. Among them there are 16 determinants for which the former case applies; that is, the number of classes in the properly primitive order is three times the number in the improperly primitive order, namely, 37, 101, 141, 189, 197, 269, 325, 333, 349, 373, 381, 389, 405, 485, 557, 573; for the other 59 cases the number of classes is the same in both orders.

VII. It is scarcely necessary to observe that the preceding method applies not only to the numbers of classes in different orders of the same determinant but also to different determinants, provided they are related to each other as squares. Thus if O is an order of determinant dm^2, and O' an order of determinant $dm'm'$, O can be compared with a properly primitive order of determinant dm^2, and this with an order derived from a properly primitive order of determinant d; or, what comes to the same thing as regards the number of classes, with this last order itself; and in a similar way the order O' can be compared with this same order.

▶ 257. Among all classes in a given order with a given determinant, ambiguous classes especially demand a further treatment, and the determination of the number of these classes opens the way to many other interesting results. It is sufficient to consider the number of classes in the properly primitive order only, since the other cases can easily be reduced to this one. We will do this in the following way. First we will determine all the properly primitive ambiguous forms (A, B, C) of determinant D in which either $B = 0$ or $B = A/2$ and then from the number of these we can find the number of all the properly primitive ambiguous classes with determinant D.

I. We will find all properly primitive forms $(A, 0, C)$ of determinant D by taking for A every divisor of D (both positive and negative) for which $C = -D/A$ is relatively prime to A. Thus when $D = -1$ there will be two of these forms: $(1, 0, 1)$, $(-1, 0, -1)$; and the same number when $D = 1$, namely $(1, 0, -1)$,

$(-1, 0, 1)$; when D is a prime number or the power of a prime number (whether the sign be positive or negative) there will be four $(1, 0, -D), (-1, 0, D), (D, 0, -1), (-D, 0, 1)$. In general when D is divisible by n different prime numbers (here we count the number 2 among them), we will have in all 2^{n+1} forms of this kind; that is to say if $D = \pm PQR \ldots$ where P, Q, R, etc. are different prime numbers or powers of primes and if their number $= n$, the values of A will be 1, P, Q, R, etc. and the products of all combinations of these numbers. By the theory of combinations the number of these values is 2^n but it must be doubled since each value is to be taken with a positive sign and a negative sign.

II. Similarly it is clear that all properly primitive forms $(2B, B, C)$ of determinant D will be obtained if for B we take all (positive and negative) divisors of D for which $C = [B - (D/B)]/2$ is an integer and relatively prime to $2B$. Since therefore C is necessarily odd and $C^2 \equiv 1$ (mod. 8), from the equation $D = B^2 - 2BC = (B - C)^2 - C^2$ it follows that D is either $\equiv 3$ (mod. 4) when B is odd, or $\equiv 0$ (mod. 8) when B is even; whenever therefore D is congruent (mod. 8) to any of the numbers 1, 2, 4, 5, 6 there will be no forms of this kind. When $D \equiv 3$ (mod. 4), C will be an integer and odd, no matter what divisor of D we take for B; but in order that C should not have a common divisor with $2B$ we must choose B so that D/B and B are relatively prime; so for $D = -1$ we have two forms $(2, 1, 1), (-2, -1, -1)$, and in general if the number of all prime divisors of D is n, there will be 2^{n+1} forms in all. When D is divisible by 8, C will be an integer if we take for B any even divisor of $D/2$; as for the other condition that $C = (B/2) - (D/2B)$ be relatively prime to $2B$, it will be satisfied *first* by taking for B all oddly even divisors of D for which D/B and B do not have a common divisor. The number of these (counting both signs) will be 2^{n+1} if D is divisible by n different odd prime numbers; *second* by taking for B all evenly even divisors of $D/2$ for which $D/2B$ and B are relatively prime. Their number will also be 2^{n+1} so that in this case we will have 2^{n+2} forms in all. Thus, if $D = \pm 2^\mu PQR \ldots$ where μ is an exponent greater than 2, P, Q, R, etc. are different odd prime numbers or powers of prime numbers, and if the number of these is n: then for *both $B/2$ and* for $D/2B$ we take all the values 1, P, Q, R, etc. and the products of any number of these numbers each with a negative and then a positive sign.

From all this we see that if D is divisible by n different odd prime numbers (letting $n = 0$ when $D = \pm 1$ or ± 2 or a power of 2) the number of all properly primitive forms (A, B, C) in which B is either 0 or $A/2$ will be 2^{n+1} when $D \equiv 1$ or $\equiv 5$ (mod. 8); 2^{n+2} when $D \equiv 2, 3, 4, 5, 6$ or 7 (mod. 8); finally 2^{n+3} when $D \equiv 0$ (mod. 8). If we compare this result with what we found in article 231 regarding the number of all possible characters of primitive forms with determinant D, we observe that the former number is precisely double this in all cases. But it is clear that when D is negative there will be as many positive as negative forms among them.

▶ 258. All the forms considered in the preceding article manifestly belong to ambiguous classes. On the other hand at least one of these forms must be contained in every properly primitive ambiguous class of determinant D; for certainly there are ambiguous forms in such a class and one of the forms of the preceding article is equivalent to any properly primitive ambiguous form (a, b, c) of determinant D, namely either

$$\left(a, 0, -\frac{D}{a}\right) \quad \text{or} \quad \left(a, \frac{1}{2}a, \frac{1}{4}a - \frac{D}{a}\right)$$

according as b is either $\equiv 0$ or $\equiv a/2$ (mod. a). Thus the problem is reduced to finding out how many classes these forms determine.

If the form $(a, 0, c)$ appears among the forms of the preceding article, the form $(c, 0, a)$ will also appear and they will be different except when $a = c = \pm 1$ and thus $D = -1$. We will set this aside for the time being. Now since these forms manifestly belong to the same class, it is sufficient to retain one, and we will reject the one whose first term is greater than the third; we will also set aside the case where $a = -c = \pm 1$ and $D = 1$. In this way we can reduce all forms $(A, 0, C)$ to half by retaining only one of every pair; and in those which remain we will always have $A < \sqrt{\pm D}$.

Similarly, if the form $(2b, b, c)$ occurs among the forms of the preceding article the following will also appear

$$(4c - 2b, 2c - b, c) = \left(-\frac{2D}{b}, -\frac{D}{b}, c\right)$$

These two will be properly equivalent but different from one another except in the case which we have omitted, where $c = b = \pm 1$ or $D = -1$. It is sufficient to retain the one of these two forms whose first term is less than the first term of the other (in this case they cannot be equal in magnitude or different in sign); thus all the forms $(2B, B, C)$ can be reduced by half by rejecting one from each pair; and in those which remain we will always have $B < D/B$ or $B < \sqrt{\pm D}$. In this way we will have only half of all the forms of the preceding article remaining. We will designate the complex by the letter W and it remains only to show how many different classes arise from these forms. Manifestly, in the case when D is negative there will be as many positive forms in W as negative.

I. When D is negative, each of the forms in W will belong to different classes. For all the forms $(A, 0, C)$ will be reduced; and all the forms $(2B, B, C)$ will be reduced except those in which $C < 2B$; for in such a form $2C < 2B + C$; therefore (since $B < D/B$, i.e. $B < 2C - B$ and $2B < 2C$ or $B < C$), $2C - 2B < C$ and $C - B < C/2$ and the reduced form is $(C, C - B, C)$ which obviously is equivalent to it. In this way we will have as many reduced forms as there are forms in W, and since no two of them will be identical or opposite (except for the case where $C - B = 0$ in which $B = C = \pm 1$ and so $D = -1$, the case we have already set aside) all will belong to different classes. Thus the number of all properly primitive ambiguous classes of determinant D will be equal to the number of forms in W or to half the number of forms in the preceding article. With regard to the excepted case when $D = -1$ the same happens by compensation; that is, there are two classes: one to which the forms $(1, 0, 1), (2, 1, 1)$ belong, the other to which $(-1, 0, -1), (-2, -1, -1)$ belong. In general, therefore, for a negative determinant the number of all properly primitive ambiguous classes is equal to the number of all assignable characters of the primitive forms of this determinant; the number of properly primitive ambiguous classes which are positive will be half of this.

II. When D is a positive square $= h^2$, it is not hard to show that each form in W belongs to a different class; but this problem can be solved more simply in the following way. By article 210 there must be one reduced form $(a, h, 0)$ contained in every properly primitive ambiguous class of determinant h^2 where a is the value of the expression $\sqrt{1}$ (mod. $2h$) lying between 0 and $2h - 1$ inclusive. Since this is so, it is clear that there are as many properly primitive ambiguous classes of determinant h^2 as there are values of this expression. From article 105 the number of these values is 2^n or 2^{n+1} or 2^{n+2} according as h is odd or oddly even or evenly even; that is, according as $D \equiv 1$ or $\equiv 4$ or $\equiv 0$ (mod. 8) and n designates the number of odd prime divisors of h or of D. Thus the number of properly primitive ambiguous classes will always be one half the number of all the forms considered in the preceding article and equal to the number of forms in W or of all possible characters.

III. When D is positive nonquadratic we will deduce from each of the forms (A, B, C) in W other forms (A', B', C') by taking $B' \equiv B$ (mod. A) and between the limits $\sqrt{D}$ and $\sqrt{D} \mp A$ (the upper or lower sign is to be used according as A is positive or negative) and $C' = (B'B' - D)/A$; we will designate this complex by the letter W'. Manifestly these forms will be properly primitive and ambiguous of determinant D, and all different. And, further, all will be reduced forms. For when $A < \sqrt{D}$, B' will be $< \sqrt{D}$ and positive; besides $B' > \sqrt{D} \mp A$ and $A > \sqrt{D} - B'$, so A taken positively will certainly lie between $\sqrt{D} + B'$ and $\sqrt{D} - B'$. When $A > \sqrt{D}$ we cannot have $B = 0$ (we rejected these forms) but B must $= A/2$. Therefore B' will be equal in magnitude to $A/2$ and positive in sign (for since $A < 2\sqrt{D}$, $\pm A/2$ will lie between the limits assigned to B' and will be congruent to B relative to the modulus A; and so $B' = \pm A/2$). As a result $B' < \sqrt{D}$ and $2B' < \sqrt{D} + B'$ or $A < \sqrt{D} + B'$ so that $\pm A$ will necessarily lie between the limits $\sqrt{D} + B'$ and $\sqrt{D} - B'$. Finally W' will contain all the properly primitive ambiguous reduced forms of determinant D; for if (a, b, c) is of this form we will have either $b \equiv 0$ or $b \equiv a/2$ (mod. a). In the former case we cannot have $b < a$ nor therefore $a > \sqrt{D}$, and so the form $(a, 0, -D/a)$ will certainly be contained in W and the corresponding form (a, b, c) in W'; in the latter case certainly $a < 2\sqrt{D}$ and so $(a, a/2, a/4 - D/a)$ will be contained in W and the corresponding form (a, b, c) in W'. Thus the number of forms in W is equal to the number of all properly primitive ambiguous reduced forms of determinant D; for, since each ambiguous class contains a *pair* of ambiguous reduced forms (art. 187, 194), the number of all properly primitive

ambiguous classes of determinant D will be half the number of
forms in W or half the number of all assignable characters.

▶ 259. The number of improperly primitive ambiguous classes of
a given determinant D is equal to the number of properly primitive
forms of the same determinant. Let K be a principal class and
K', K'', etc. the remaining properly primitive ambiguous classes of
the same determinant; L an improperly primitive ambiguous class
of the same determinant, e.g., the one containing the form $(2, 1,
1/2 - D/2)$. If we compose the class L with K we get the class L
itself; let us suppose that the composition of the class L with
K', K'', etc. produces the classes L', L'', etc. respectively. Manifestly
they will all belong to the same determinant and will be improperly
primitive and ambiguous. It is clear that the theorem will be
proven as soon as we prove that all the classes L, L', L'', etc. are
different and that there are no other improperly primitive ambiguous classes of determinant D besides these. For this purpose we
distinguish the following cases.

I. When the number of improperly primitive classes is equal to
the number of properly primitive classes, each of the former will
result from the composition of the class L with a determined
properly primitive class, and so all the L, L', L'', etc. will be
different. If we designate by $\mathfrak{L}$ any improperly primitive ambiguous
class of determinant D, there will exist a properly primitive class
$\mathfrak{R}$ such that $\mathfrak{R} + L = \mathfrak{L}$; if $\mathfrak{R}'$ is the opposite class to $\mathfrak{R}$ we will
also have (since the classes L, $\mathfrak{L}$ are their own opposites) $\mathfrak{R}' + L =
\mathfrak{L}$ and $\mathfrak{R}$ and $\mathfrak{R}'$ will be identical and so an ambiguous class. As a
result, $\mathfrak{R}$ will be found among the classes K, K', K'', etc. and $\mathfrak{L}$
among the classes L, L', L'', etc.

II. When the number of improperly primitive classes is one
third the number of properly primitive classes, let H be the class
in which the form $(4, 1, (1 - D)/4)$ appears, and H' the one in which
$(4, 3, (9 - D)/4)$ appears. H, H' will be properly primitive and
different from each other and from the principal class K, and
$H + H' = K$, $2H = H'$, $2H' = H$; and if $\mathfrak{L}$ is any improperly
primitive class of determinant D which arises from the composition of L with the properly primitive class $\mathfrak{R}$ we will also have
$\mathfrak{L} = L + \mathfrak{R} + H$ and $\mathfrak{L} = L + \mathfrak{R} + H'$. Besides the three (properly
primitive and different) classes $\mathfrak{R}, \mathfrak{R} + H$, $\mathfrak{R} + H'$ there are no
others which produce $\mathfrak{L}$ when composed with L. Since therefore if

$\mathfrak{L}$ is ambiguous and $\mathfrak{R}'$ is opposite to $\mathfrak{R}$, we will also have $L + \mathfrak{R}' = \mathfrak{L}$, $\mathfrak{R}'$ will necessarily be identical with one of the three classes. If $\mathfrak{R}' = \mathfrak{R}$, $\mathfrak{R}$ will be ambiguous; if $\mathfrak{R}' = \mathfrak{R} + H$ we will have $K = \mathfrak{R} + \mathfrak{R}' = 2\mathfrak{R} + H = 2(\mathfrak{R} + H')$ and so $\mathfrak{R} + H'$ is ambiguous; similarly if $\mathfrak{R}' = \mathfrak{R} + H'$, $\mathfrak{R} + H$ will be ambiguous and we conclude that $\mathfrak{L}$ must be found among the classes L, L', L'', etc. It is easy to see that there cannot be more than one ambiguous class among the three classes $\mathfrak{R}, \mathfrak{R} + H, \mathfrak{R} + H'$; for if both $\mathfrak{R}$ and $\mathfrak{R} + H$ were ambiguous and therefore identical with their opposites $\mathfrak{R}', \mathfrak{R}' + H'$, we would have $\mathfrak{R} + H = \mathfrak{R} + H'$; the same conclusion results from the supposition that $\mathfrak{R}$ and $\mathfrak{R} + H'$ are ambiguous; finally if $\mathfrak{R} + H, \mathfrak{R} + H'$ are ambiguous and identical with their opposites $\mathfrak{R}' + H', \mathfrak{R}' + H$, we would have $\mathfrak{R} + H + \mathfrak{R}' + H = \mathfrak{R}' + H' + \mathfrak{R} + H'$ and so $2H = 2H'$ or $H' = H$. Therefore there is only one properly primitive ambiguous class that produces $\mathfrak{L}$ when composed with L, and thus all the L, L', L'', etc. will be different.

The number of ambiguous classes in the *derived* order is obviously equal to the number of ambiguous classes in the primitive order from which it is derived, and so this number can always be determined.

▶ 260. PROBLEM. *The properly primitive class K of determinant D arises from the duplication of a properly primitive class k of the same determinant. We want all similar classes whose duplication produces K.*

Solution. Let H be the principal class of determinant D and H', H'', H''', etc. the remaining properly primitive ambiguous classes of the same determinant; $k + H', k + H'', k + H'''$, etc. are the classes which arise from the composition of these with k. We will designate them as k', k'', k''', etc. Now all the classes k, k', k'', etc. will be properly primitive of determinant D and different from one another; and the class K will result from the duplication of any one of them. If we denote by $\mathfrak{R}$ any properly primitive class of determinant D which produces the class K when it is duplicated, it will necessarily be contained among the classes k, k', k'', etc. For suppose $\mathfrak{R} = k + \mathfrak{H}$ so that $\mathfrak{H}$ is a properly primitive class of determinant D (art. 249) then $2k + 2\mathfrak{H} = 2\mathfrak{R} = K = 2k$ and thus $2\mathfrak{H}$ coincides with the principal class, $\mathfrak{H}$ is ambiguous and so contained among H, H', H'', etc., and $\Re$ among k, k', k'', etc.; therefore these classes give a complete solution of the problem.

It is manifest that when D is negative, half the classes k, k', k'', etc. will be positive, half negative.

Since therefore any properly primitive class of determinant D that can arise from the duplication of any similar class, can arise from the duplication of as many similar classes as there are properly primitive ambiguous classes of determinant D: it is clear that, if the number of all properly primitive classes of determinant D is r, and the number of all properly primitive ambiguous classes of this determinant is n, then the number of all properly primitive classes of the same determinant that can be produced by the duplication of a similar class will be r/n. The same formula results if, for a negative determinant, the characters r, n designate the number of *positive* classes. Thus, e.g., for $D = -161$, the number of all positive properly primitive classes is 16, the number of ambiguous classes 4, so the number of classes that can arise from the duplication of any class must be 4. As a matter of fact we find that all classes contained in a principal genus are endowed with this property; thus the principal class $(1, 0, 161)$ results from the duplication of the four ambiguous classes; $(2, 1, 81)$ from duplicating the classes $(9, 1, 18), (9, -1, 18), (11, 2, 15); (9, 1, 18)$ from duplicating the classes $(3, 1, 54), (6, 1, 27), (5, -2, 33), (10, 3, 17);$ finally $(9, -1, 18)$ from duplicating the classes $(3, -1, 54), (6, -1, 27), (5, 2, 33), (10, -3, 17)$.

▶ 261. THEOREM. *Half of all the assignable characters for a positive nonquadratic determinant can be applied to no properly primitive genus and, if the determinant is negative, to no properly primitive positive genus.*

Demonstration. Let m be the number of all properly primitive (positive) genera of determinant D; k the number of classes contained in each genus so that km is the number of all properly primitive (positive) classes; n the number of all the different characters assignable for this determinant. Then by article 258 the number of all properly primitive (positive) ambiguous classes will be $n/2$; and by the preceding article the number of all properly primitive classes that can result from the duplication of a similar class will be $2km/n$. But by article 247 all these classes belong to the principal genus which contains k classes; if therefore all classes of the principal genus result from the duplication of some class (we will show in what follows that this is always true) then $2km/n = k$ or $m = n/2$; but it is certain that we cannot have $2km/n > k$ nor consequently $m > n/2$. Since, therefore, the number of all properly primitive (positive) genera certainly cannot be greater than half of all assignable characters, at least half of them cannot correspond to such genera. Q.E.D. Note however that it does not yet follow from this that half of all the assignable characters actually correspond to properly primitive (positive) genera, but later we shall establish the truth of this profound proposition concerning the most abstract mystery of numbers.

Since for a negative determinant there are always as many negative genera as positive, manifestly not more than half of all assignable characters can belong to properly primitive negative genera. We will speak further of this and of improperly primitive genera below. Finally we observe that the theorem does not apply to positive square determinants. For these it is easy to see that each assignable character corresponds to a genus.

▶ 262. Thus in the case where only two different characters can be assigned for a given nonquadratic determinant D, only one will correspond to a properly primitive (positive) genus (this will have to be the principal genus). The other will correspond to no properly primitive (positive) form of that determinant. This happens for the determinants $-1, 2, -2, -4$, for positive prime numbers of the form $4n + 1$, for negative primes of the form $4n + 3$, for all positive odd powers of prime numbers of the form $4n + 1$, and for even positive or odd negative powers of prime numbers of the form $4n + 3$. From this principle we can develop a new method not only for the fundamental theorem but also for demonstrating the other theorems of the preceding section pertaining to the residues $-1, +2, -2$. This method will be completely different from those used in the preceding section and in no way less elegant. We will, however, omit consideration of the determinant -4 and determinants which are powers of prime numbers, since they will teach us nothing new.

For the determinant -1 there is no positive form with the character $3, 4$; for the determinant $+2$ there is none with the character 3 *and* $5, 8$; for the determinant -2 there will be no positive form with the character 5 *and* 7, 8; and for the determinant $-p$ where p is a prime number of the form $4n + 3$ no properly primitive (positive) form will have the chracter Np; while for the determinant $+p$ where p is a prime number of the form $4n + 1$ no properly primitive form at all will have the character Np. Therefore we will demonstrate the theorems of the preceding section in the following way.

I. -1 is a nonresidue of any (positive) number of the form $4n + 3$. For if -1 were a residue of such a number A, by setting $-1 = B^2 - AC$, (A, B, C) would be a positive form of determinant -1 with the character 3, 4.

II. -1 is a residue of any prime number p of the form $4n + 1$. For the character of the form $(-1, 0, p)$, as of all properly primitive forms of determinant p will be Rp and therefore $-1Rp$.

III. Both $+2$ and -2 are residues of any prime number p of the form $8n + 1$. For either the forms $(8, 1, (1 - p)/8)$, $(-8, 1, (p - 1)/8)$ or the forms $(8, 3, (9 - p)/8)$, $(-8, 3, (p - 9)/8)$ are properly primitive (according as n is odd or even) and so their character will be Rp; therefore $+8Rp$ and $-8Rp$ and also $2Rp$ and $-2Rp$.

IV. $+2$ is a nonresidue of any number of the form $8n + 3$ or $8n + 5$. For if it were a residue of such a number A, there would be a form (A, B, C) of determinant $+2$ with the character 3 *and* 5, 8.

V. Similarly -2 is a nonresidue of any number of the form $8n + 5$ or $8n + 7$, for otherwise there would be a form (A, B, C) of determinant -2 with the character 5 *and* 7, 8.

VI. -2 is a residue of any prime number p of the form $8n + 3$. We will show this proposition by a double method. *First*, since by IV we have $+2Np$ and by I $-1Np$, we necessarily have $-2Rp$. The *second* demonstration begins with a consideration of the determinant $+2p$. For this, four characters are assignable, namely $Rp, 1$ *and* 3, 8; $Rp, 5$ *and* 7, 8; $Np, 1$ *and* 3, 8; $Np, 5$ *and* 7, 8. Of these at least two will correspond to no genus. Now the form $(1, 0, -2p)$ will agree with the first character; the form $(-1, 0, 2p)$ the fourth; therefore the second and third must be rejected. And since the character of the form $(p, 0, -2)$ relative to the number 8 is 1 *and* 3, 8, its character relative to p must be Rp and so $-2Rp$.

VII. $+2$ is a residue of any prime number p of the form $8n + 7$. This can be shown by two methods. *First*, since by I and V $-1Np$,

$-2Np$ we will have $+2Rp$. *Second*, since either $(8, 1, (1 + p)/8)$ or $(8, 3, (9 + p)/8)$ is a properly primitive form of determinant $-p$ (according as n is even or odd), its character will be Rp and so $8Rp$ and $2Rp$.

VIII. Any prime number p of the form $4n + 1$ is a nonresidue of any odd number q that is a nonresidue of p. For clearly if p were a residue of q there would be a properly primitive form of determinant p with the character Np.

IX. Similarly if an odd number q is a nonresidue of a prime number p of the form $4n + 3$, $-p$ will be a nonresidue of q; otherwise there would be a properly primitive positive form of determinant $-p$ with character Np.

X. Any prime number p of the form $4n + 1$ is a residue of any other prime number q which is a residue of p. If q is also of the form $4n + 1$ this follows immediately from VIII; but if q is of the form $4n + 3$, $-q$ will also be a residue of p (by II) and so pRq (by IX).

XI. If a prime number q is a residue of another prime number p of the form $4n + 3$, $-p$ will be a residue of q. For if q is of the form $4n + 1$ it follows from VIII that pRq and so (by II) $-pRq$; this method does not apply when q is of the form $4n + 3$, but it can easily be resolved by considering the determinant $+pq$. For, since of the four characters assignable for this determinant Rp, Rq; Rp, Nq; Np, Rq; Np, Nq, two of them cannot correspond to any genus and since the characters of the forms $(1, 0, -pq)$, $(-1, 0, pq)$ are the first and fourth respectively, the second and third are the characters which correspond to no properly primitive form of determinant pq. And since by hypothesis the character of the form $(q, 0, -p)$ with respect to the number p is Rp, its character with respect to the number q must be Rq and therefore $-pRq$. Q.E.D.

If in propositions VIII and IX we suppose q is a prime number, these propositions joined with X and XI will give us the fundamental theorem of the preceding section.

▶ 263. Now that we have given a new proof of the fundamental theorem, we are going to concern ourselves with the half of the characters of a given nonquadratic determinant that cannot correspond to any properly primitive (positive) forms. We can treat of this rather briefly, since the basis of our discussion is already contained in articles 147–150. Let e^2 be the largest square that divides the given determinant D, and let $D = D'e^2$ so that D' does not involve a square factor. Further, let a, b, c, etc. be all the odd prime divisors of D'. Thus D' except perhaps for sign will be a product of these numbers or double this product. Let Ω designate the complex of particular characters Na, Nb, Nc, etc. taken alone when $D' \equiv 1$ (mod. 4); taken with the added character 3, 4 when $D' \equiv 3$ and e is odd or oddly even; taken with 3, 8 and 7, 8 when $D' \equiv 3$ and e is evenly even; taken with either the character 3 *and* 5, 8 or the two characters 3, 8 and 5, 8 when $D' \equiv 2$ (mod. 8) and e is either odd or even; finally taken with either the character 5 *and* 7, 8 or the two characters 5, 8 and 7, 8 when $D' \equiv 6$ (mod. 8) and e is either odd or even. This done, no properly primitive (positive) genera of determinant D can correspond to all the integral characters containing odd number of particular characters Ω. In every case the particular characters, which express a relation to those prime divisors of D that do not divide D', contribute nothing to the possibility or impossibility of genera. From the theory of combinations however it is very easy to see that in this way half of all the assignable integral characters are excluded.

We demonstrate this as follows. By the principles of the preceding section or by the theorems we have just demonstrated in the preceding article, it is clear that if p is an (odd positive) prime number which does not divide D and which has one of the rejected characters corresponding to it, D' will involve an odd number of factors which are nonresidues of p. Therefore D' and D also will be nonresidues of p. Further, the product of odd numbers relatively prime to D, which have none of the rejected characters as correspondents, cannot correspond to any such character. And, conversely, any odd positive number relatively prime to D which corresponds to one of the rejected characters certainly implies some prime factor of the same quality. If, therefore, we have a properly primitive (positive) form of determinant D corresponding to one of the rejected characters, D would be a nonresidue of some odd positive number relatively prime to it and representable by such a form. But this is manifestly inconsistent with the theorem of article 154.

The classifications in article 231, 232 give good examples of this, and the reader can augment their number at will.

▶ 264. Thus if we are given a nonquadratic determinant, all assignable characters will be equally distributed into two kinds, P, Q, in such a way that no properly primitive (positive) form can correspond to any of the characters Q. As for the characters P, from what we know so far, there is nothing to hinder them from belonging to forms of this sort. We take special note of the following proposition concerning these types of characters, which can be easily deduced from the criterion concerning them. If we compose a character of P with a character of Q (according to the form of article 246 just as if that genus applied here also) we will get a character of Q; but if we compose two characters of P or two of Q, the resulting character will belong to P. With the help of this theorem we can also exclude half of all assignable characters for negative and improperly primitive genera in the following way.

I. For a negative determinant D, negative genera will be contrary to positive genera in this respect, that none of the characters of P will belong to a properly primitive negative genus. All such genera will have characters of Q. For when $D' \equiv 1$ (mod. 4), $-D'$ will be a positive number of the form $4n + 3$, and thus among the numbers a, b, c, etc. there will be an odd number of the form $4n + 3$ and -1 will be a nonresidue of each of them. It follows in this case that the complete character of the form $(-1, 0, D)$ will include an odd number of particular characters of Ω and thus it will belong to Q; when $D' \equiv 3$ (mod. 4), among the numbers a, b, c, etc. there will be either no number of the form $4n + 3$ or two or four, etc. And, since in this case $3, 4$ or $3, 8$ or $7, 8$ will occur among the particular characters of the form $(-1, 0, D)$, it is clear that the complete character of this form will also belong to Q. We will obtain the same conclusion just as easily for the remaining cases so that the negative form $(-1, 0, D)$ will always have a character of Q. But since this form composed with any other properly primitive negative form of the same determinant will produce a similar positive form, it is clear that no properly primitive negative form can have a character of P.

II. We can prove in the same way that improperly primitive (positive) genera have either the same or the opposite property as for properly primitive genera according as $D \equiv 1$ or $\equiv 5$ (mod. 8). For in the former case we will also have $D' \equiv 1$ (mod. 8), and we can conclude that among the numbers a, b, c, etc. there will be either no numbers of the form $8n + 3$ and $8n + 5$ or two or four, etc. (that is, the product of any number of odd numbers which include an odd number of numbers of the form $8n + 3$ and $8n + 5$ will always be $\equiv 3$ or $\equiv 5$ (mod. 8), and the product of all the numbers a, b, c, etc. will be equal either to D' or $-D'$); thus the complete character of the form $(2, 1, (1 - D)/2)$ will involve either no particular character of Ω, or two or four etc., and so will belong to P. Now since any improperly primitive (positive) form of determinant D can be considered as composed of $(2, 1, (1 - D)/2)$ and a properly primitive (positive) form of the same determinant, it is obvious that no improperly primitive (positive) form can have one of the characters of Q in this case. In the other case, when $D \equiv 5$ (mod. 8), everything is quite different, that is D' which will also $\equiv 5$, must certainly imply an odd number of factors of the form $8n + 3$ and $8n + 5$. Thus the character of the form $(2, 1, (1 - D)/2)$ and also the character of any improperly primitive (positive) form of determinant D will belong to Q and no improperly primitive positive genus can have a character in P.

III. Finally, for a negative determinant the improperly primitive negative genera are again contrary to improperly primitive positive genera. They cannot have a character belonging to P or Q according as $D \equiv 1$ or $\equiv 5$ (mod. 8) or according as $-D$ is of the form $8n + 7$ or $8n + 3$. We deduce this from the fact that if we compose the form $(-1, 0, D)$ whose character is in Q with improperly primitive negative forms of the same determinant, we get improperly primitive positive forms. Thus when characters of Q are excluded from these, characters of P must also be excluded, and conversely.

▶ 265. All the above is based on the considerations of articles 257, 258 concerning the number of ambiguous classes. There are many other conclusions very worthy of consideration, which, for the sake of brevity we will omit, but we cannot pass over the following one which is remarkable for its elegance. For a positive determinant p which is a prime number of the form $4n + 1$ we have shown that there is only one properly primitive ambiguous class. Thus all properly primitive ambiguous forms will be properly equivalent. If therefore b is the positive integer immediately less than $\sqrt{p}$ and $p - b^2 = a'$, the forms $(1, b, -a')$, $(-1, b, a')$ will be properly equivalent and, since they are both reduced forms, one will be contained in the period of the other. If we give the index 0 to the former form in its period, the index of the latter will necessarily be odd (since the first terms of these two forms have opposite signs); let us suppose therefore that this index $= 2m + 1$. It is easy to see that if the forms of indices 1, 2, 3, etc. are respectively

$$(-a', b', a''), \quad (a'', b'', -a'''), \quad (-a''', b''', a''''), \quad \text{etc.}$$

the following forms will correspond to the indices $2m, 2m - 1$, $2m - 2, 2m - 3$, etc. respectively

$$(a', b', -1), \quad (-a'', b', a'), \quad (a''', b'', -a''), \quad (-a'''', b''', a'''), \quad \text{etc.}$$

Thus if the form of index m is (A, B, C), $(-C, B, -A)$ will be the same and so $C = -A$ and $p = B^2 + A^2$. Therefore any prime number of the form $4n + 1$ can be decomposed into two squares (we deduced this proposition from entirely different principles in art. 182). And we can find this decomposition by a very simple and completely uniform method; that is, by evolving the period of the reduced form whose determinant is that prime number and whose first term is 1 into the form whose outer terms are equal in magnitude but opposite in sign. Thus, e.g., for $p = 233$ we have $(1, 15, -8)$, $(-8, 9, 19)$, $(19, 10, -7)$, $(-7, 11, 16)$, $(16, 5, -13)$, $(-13, 8, 13)$ and $233 = 64 + 169$. It is clear that A is necessarily odd (since $(A, B, -A)$ must be a properly primitive form) and B even. Since for the positive determinant p which is a prime number of the form $4n + 1$ only one ambiguous class is contained in the improperly primitive order, it is clear that if g is an odd number immediately less than $\sqrt{p}$ and $p - g^2 = 4h$, the improperly primitive reduced forms $(2, g, -2h), (-2, g, 2h)$ will be properly equivalent and thus one will be contained in the period of the other. So by a similar reasoning we conclude that we can find a form in the period of the form $(2, g, -2h)$ which has outer terms of equal magnitude and opposite sign. Thus we can decompose the number p into two squares in this case also. The outer terms of this form will be even, the middle one odd; and since it is clear that a prime number can be decomposed into two squares in only one way, the form which we find by this method will be either $(B, \pm A, -B)$ or $(-B, \pm A, B)$. Thus in our example for $p = 233$ we will have $(2, 15, -4)$, $(-4, 13, 16)$, $(16, 3, -14)$, $(-14, 11, 8)$, $(8, 13, -8)$ and $233 = 169 + 64$ as above.

▶ 266. Thus far we have restricted our discussion to functions of the second degree with *two* unknowns and there was no need to give them a special name. But manifestly this argument is only one section of the general treatise concerning *rational algebraic functions which are integral and homogeneous in many unknowns and many dimensions.* Such functions relative to the number of dimensions can properly be divided into *forms of the second, third, fourth degree, etc.* and relative to the number of unknowns into *binary, ternary, quaternary, etc. forms.* Thus the forms we have been considering can be called simply *binary forms of the second degree.* But functions like

$$Ax^2 + 2Bxy + Cy^2 + 2Dxz + 2Eyz + Fz^2$$

where $(A, B, C, D, E, F$ are integers) are called *ternary forms of the second degree* and so forth. We have devoted the present section to the treatment of binary forms of the second degree. But there are many beautiful truths concerning these forms which are properly considered in the theory of ternary forms of the second degree. We will therefore make a brief digression into this theory and will especially treat of those elements which are necessary to complete the theory of binary forms, hoping thereby to please geometers who would be disappointed if we ignore them or treated them in a less natural manner. We must, however, reserve a more exact treatment of this important subject for another occasion because its usefulness far exceeds the limits of this work and because, hopefully, we will be able to enrich the discussion by more profound insights later on. At this time we will completely exclude from the discussion quaternary, quinary, etc. forms and all forms of higher degrees.[n] It is sufficient to draw this broad field to the attention of geometers. There is ample material for the exercise of their genius, and transcendental Arithmetic will surely benefit by their efforts.

▶ 267. It will be of great advantage for our understanding to establish a fixed order for the unknown values of the ternary form just as we did for binary forms so that we can distinguish the *first, second,* and *third unknowns* from each other. In disposing the different parts of a form we will always observe the following

[n] For this reason whenever we speak simply of binary or ternary forms we will always mean binary or ternary forms of the *second degree.*

order: we will set in the first place the term that refers to the square of the first unknown, then the term that refers to the square of the second unknown, the square of the third unknown, the double product of the second by the third, the double product of the first by the third, and then the double product of the first by the second. Finally, we will call the integers by which these squares and double products are multiplied in the same order the *first*, *second*, *third*, *fourth*, *fifth*, and *sixth coefficients*. Thus

$$ax^2 + a'x'x' + a''x''x'' + 2bx'x'' + 2b'xx'' + 2b''xx'$$

will be a ternary form in proper order. The first unknown is x, the second x', the third x''. The first coefficient is a etc., the fourth is b etc. But since it contributes very much to brevity, even though it is not always necessary, to denote the unknowns of a ternary form by special letters, we will also designate such a form by

$$\begin{pmatrix} a, & a', & a'' \\ b, & b', & b'' \end{pmatrix}$$

By letting

$$b^2 - a'a'' = A, \quad b'b' - aa'' = A, \quad b''b'' - aa' = A''$$
$$ab - b'b'' = B, \quad a'b' - bb'' = B', \quad a''b'' - bb' = B''$$

we will get another form

$$\begin{pmatrix} A, & A', & A'' \\ B, & B', & B'' \end{pmatrix} \dots F$$

which we will call the *adjoint* of the *form*

$$\begin{pmatrix} a, & a', & a'' \\ b, & b', & b'' \end{pmatrix} \dots f$$

Again if we denote for brevity the number

$$ab^2 + a'b'b' + a''b''b'' - aa'a'' - 2bb'b'' \text{ by } D$$

we will have

$$B^2 - A'A'' = aD, \quad B'B' - AA'' = a'D, \quad B''B'' - AA' = a''D$$
$$AB - B'B'' = bD, \quad A'B' - BB'' = b'D, \quad A''B'' - BB' = b''D$$

and it is obvious that the adjoint of the form F will be the form

$$\begin{pmatrix} aD, & a'D, & a''D \\ bD, & b'D, & b''D \end{pmatrix}$$

The properties of the ternary form f depend in a special way on the nature of the number D. We will call it the *determinant* of this form. In the same way the determinant of the form F will $= D^2$, that is, equal to the square of the determinant of the form f whose adjoint it is.

Thus, e.g., the adjoint of the ternary form $\begin{pmatrix} 29; & 13; & 9 \\ 7, & -1, & 14 \end{pmatrix}$ is $\begin{pmatrix} -68, & -260, & -181 \\ 217, & -111, & 133 \end{pmatrix}$ and the determinant of each $= 1$.

We will entirely exclude from our following investigation ternary forms of determinant 0. We will show at another time when we treat more fully the theory of ternary forms that these are ternary forms only in *appearance*. They are actually equivalent to binary forms.

▶ 268. A ternary form f of determinant D and with unknowns x, x', x'' (the first $= x$ etc.) is transformed into a ternary form g of determinant E and unknowns y, y', y'' by a substitution such as this

$$x = \alpha y + \beta y' + \gamma y''$$
$$x' = \alpha' y + \beta' y' + \gamma' y''$$
$$x'' = \alpha'' y + \beta'' y' + \gamma'' y''$$

where the nine coefficients α, β, etc. are all integers. For brevity we will neglect the unknowns and say simply that f is transformed into g by the substitution (S)

$$\begin{matrix} \alpha, & \beta, & \gamma \\ \alpha', & \beta', & \gamma' \\ \alpha'', & \beta'', & \gamma'' \end{matrix}$$

and that f *implies* g or that g is *contained* in f. From this supposition will follow six equations for the six coefficients in g, but it is unnecessary to transcribe them here. And from these the following conclusions result:

I. If for brevity we denote the number

$$\alpha\beta'\gamma'' + \beta\gamma'\alpha'' + \gamma\alpha'\beta'' - \gamma\beta'\alpha'' - \alpha\gamma'\beta'' - \beta\alpha'\gamma'' \quad \text{by } k$$

we find after suitable calculation that $E = k^2 D$. Thus D divides
E and the quotient is a square. It is clear that with regard to transformations of ternary forms the number k is similar to the number
$\alpha\delta - \beta\gamma$ of article 157 with respect to transformations of binary
forms, namely the square root of a quotient of determinants.
We could conjecture that in this case also a difference of the sign
of k indicates an essential difference between proper and improper
transformations and implications. But if we examine the situation
more closely we see that f is transformed into g by this substitution
also

$$
\begin{array}{ccc}
-\alpha, & -\beta, & -\gamma \\
-\alpha', & -\beta', & -\gamma' \\
-\alpha'', & -\beta'', & -\gamma''
\end{array}
$$

And if in the equation k we put $-\alpha$ for α, $-\beta$ for β, etc. we will
get $-k$. Thus this substitution would be dissimilar to the substitu-
tion S and any ternary form that implies another in one way
would also imply the same form in the other. So we will abandon
this distinction entirely, since it is of no use for ternary forms.

II. If we denote by F, G the forms that are adjoint to f, g,
respectively, the coefficients in F will be determined by the coefficients in f, the coefficients in G by the values of the coefficients
of the form g from the equations which the substitution S pro-
vides. If we express the coefficients of the form f by letters and
compare the values of the coefficients of the forms F, G, it is easy
to see that F implies G and that it is transformed into G by the
substitution (S')

$$
\begin{array}{ccc}
\beta'\gamma'' - \beta''\gamma', & \gamma'\alpha'' - \gamma''\alpha', & \alpha'\beta'' - \alpha''\beta' \\
\beta''\gamma - \beta\gamma'', & \gamma''\alpha - \gamma\alpha'', & \alpha''\beta - \alpha\beta'' \\
\beta\gamma' - \beta'\gamma, & \gamma\alpha' - \gamma'\alpha, & \alpha\beta' - \alpha'\beta
\end{array}
$$

Since the calculation presents no difficulty we will not write it
down.

III. By the substitution (S'')

$$
\begin{array}{ccc}
\beta'\gamma'' - \beta''\gamma', & \beta''\gamma - \beta\gamma'', & \beta\gamma' - \beta'\gamma \\
\gamma'\alpha'' - \gamma''\alpha', & \gamma''\alpha - \gamma\alpha'', & \gamma\alpha' - \gamma'\alpha \\
\alpha'\beta'' - \alpha''\beta', & \alpha''\beta - \alpha\beta'', & \alpha\beta' - \alpha'\beta
\end{array}
$$

g will be transformed into the same form as f by the substitution

$$k, \quad 0, \quad 0$$
$$0, \quad k, \quad 0$$
$$0, \quad 0, \quad k$$

This is the form which comes from multiplying each of the coefficients of the form f by k^2. We will designate this form by f'.

IV. In exactly the same way we prove that, by the substitution (S''')

$$\alpha, \quad \alpha', \quad \alpha''$$
$$\beta, \quad \beta', \quad \beta''$$
$$\gamma, \quad \gamma', \quad \gamma''$$

the form G will be transformed into a form that comes from F by multiplying each coefficient by k^2. We will designate this form by F'.

We will say that the substitution S''' arises from the *transposition* of the substitution S, and manifestly we will get S again from the transposition of the substitution S'''; in the same way we will get each of the substitutions S', S'' from a transposition of the other. We can call the substitution S' the *adjoint* to the substitution S, and the substitution S'' will be the adjoint to the substitution S'''.

▶ 269. If the form f implies g and g also implies f we will call f, g *equivalent* forms. In this case D divides E, but E also divides D and so $D = E$. Conversely if the form f implies a form g of the same determinant, these forms will be equivalent. For (if we use the same symbols as in the preceding article except for the case when $D = 0$) we have $k = \pm 1$ and so the form f' into which g is transformed by the substitution S'' is identical with f and f is contained in g. Further, in this case the forms F, G which are adjoint to f, g will be equivalent to each other, and the latter will be transformed into the former by the substitution S'''. Finally, as a converse, if we *presume* that the forms F, G are equivalent and the former is transformed into the latter by the substitution T, the forms f, g will also be equivalent, and f will be transformed into g by the adjoint substitution of T and g into f by the substitution which comes from the transposition of the substitution T.

For by these two substitutions, respectively, the form adjoint to F will be transformed into the form adjoint to G and vice versa. These two forms however come from f, g by multiplying all the coefficients by D; so we can conclude that f is transformed into g and g into f, respectively, by these same substitutions.

▶ 270. If the ternary form f implies the ternary form f' and f' implies the form f'', then f will also imply f''. For it is easy to see that if f is transformed into f' by the substitution

$$
\begin{array}{ccc}
\alpha, & \beta, & \gamma \\
\alpha', & \beta', & \gamma' \\
\alpha'', & \beta'', & \gamma''
\end{array}
$$

and f' into f'' by the substitution

$$
\begin{array}{ccc}
\delta, & \varepsilon, & \zeta \\
\delta', & \varepsilon', & \zeta' \\
\delta'', & \varepsilon'', & \zeta''
\end{array}
$$

then f will be transformed into f'' by the substitution

$$
\begin{array}{ccc}
\alpha\delta + \beta\delta' + \gamma\delta'', & \alpha\varepsilon + \beta\varepsilon' + \gamma\varepsilon'', & \alpha\zeta + \beta\zeta' + \gamma\zeta'' \\
\alpha'\delta + \beta'\delta' + \gamma'\delta'', & \alpha'\varepsilon + \beta'\varepsilon' + \gamma'\varepsilon'', & \alpha'\zeta + \beta'\zeta' + \gamma'\zeta'' \\
\alpha''\delta + \beta''\delta' + \gamma''\delta'', & \alpha''\varepsilon + \beta''\varepsilon' + \gamma''\varepsilon'', & \alpha''\zeta + \beta''\zeta' + \gamma''\zeta''
\end{array}
$$

And in the case where f is equivalent to f' and f' to f'', the form f will also be equivalent to the form f''. It is immediately obvious how these theorems apply to a series of several forms.

▶ 271. It is apparent from what we have seen that ternary forms, like binary forms, can be distributed into *classes* by assigning equivalent forms to the same class and nonequivalent forms to different classes. Forms with different determinants will certainly therefore belong to different classes and thus there will be infinitely many classes of ternary forms. Ternary forms of the same determinant sometimes have a large number of classes and sometimes a small number, but it is an important property of these forms that *all forms of the same given determinant constitute a finite number of classes*. Before we discuss this important theorem in detail, we must explain the following essential difference that obtains among ternary forms.

Certain ternary forms are so constructed that they can represent positive and negative numbers without distinction, e.g. the form $x^2 + y^2 + z^2$. We will call them *indefinite forms*. On the other hand there are forms that cannot represent negative numbers but (except for zero which is determined by making each unknown $= 0$) only positive numbers, e.g. $x^2 + y^2 + z^2$. We will call these *positive forms*. Finally there are others that cannot represent positive numbers, e.g. $-x^2 - y^2 - z^2$. These will be called *negative forms*. Positive and negative forms are both called *definite forms*. We will now give general criteria for determining how to distinguish these properties of forms.

If we multiply the ternary form

$$f = ax^2 + a'x'x' + a''x''x'' + 2bx'x'' + 2b'xx'' + 2b''xx'$$

of determinant D by a, and if we denote the coefficients of the form which is adjoint to f as in article 267 by A, A', A'', B, B', B'', we have

$$(ax + b''x' + b'x'')^2 - A''x'x' + 2Bx'x'' - A'x''x'' = g$$

and multiplying by A' we get

$$A'(ax + b''x' + b'x'')^2 - (A'x'' - Bx')^2 + aDx'x' = h$$

If both A' and aD are negative numbers, all values of h will be negative, and manifestly the form f can represent only numbers whose sign is opposite that of aA', i.e. identical with the sign of a or opposite to the sign of D. In this case f will be a definite form and it will be positive or negative according as a is positive or negative or according as D is negative or positive.

But if aD, A' are both positive or one positive and the other negative (neither $= 0$) h can produce either positive or negative quantities by a suitable choice of x, x', x''. Thus in this case f can produce values with either the same or opposite values as aA', and it will be an indefinite form.

For the case where $A' = 0$ and a does not $= 0$, we have

$$g = (ax + b''x' + b'x'')^2 - x'(A''x' - 2Bx'')$$

By giving x' an arbitrary value (other than 0) and by taking x'' in such a way that $(A''x'/2B) - x''$ has the same sign as Bx'

(this can be done since B cannot $=0$ for then we would have $B^2 - A'A'' = aD = 0$, and $D = 0$, the excluded case) we will make $x'(A''x' - 2Bx'')$ a positive quantity, and therefore x can be chosen so as to make g a negative quantity. Manifestly all these values can be chosen so that, if we wish, all are integers. Finally, no matter what values are given to x', x'', x can be taken so large as to make g positive. Therefore in this case f will be an indefinite form.

Finally, if $a = 0$ we have

$$f = a'x'x' + 2bx'x'' + a''x''x'' + 2x(b''x' + b'x'')$$

Now if we take x', x'' arbitrarily but in such a way that $b''x' + b'x''$ does not $=0$ (obviously this can be done unless both b' and b'' $=0$; but then we would have $D = 0$) it is easy to see that x can be so chosen that f will have both positive and negative values. And in this case also f will be an indefinite form.

In the same way that we were able to determine the property of the form f from the numbers aD, A', we could also have used aD and A'' so that the form f would be definite if both aD and A'' were negative; indefinite in all other cases. We could for the same purpose consider the numbers $a'D$ and A, or $a'D$ and A'', or $a''D$ and A, or finally $a''D$ and A'.

From all this it follows that in the definite form the six numbers $A, A', A'', aD, a'D, a''D$ are all negative. For the positive form a, a', a'' will be positive and D negative; for the negative form a, a', a'' will be negative, D positive. Therefore all ternary forms with a given positive determinant can be distributed into negative and indefinite forms; all with a negative determinant into positive and indefinite forms; and there are no positive forms with a positive determinant, no negative forms with a negative determinant. And it is easy to see that the adjoint of a definite form is always definite and *negative*, and the adjoint of an indefinite form is always indefinite.

Since all numbers which are representable by a given ternary form can also be represented by all forms that are equivalent to it, ternary forms of the same class are either all indefinite, all positive, or all negative. Thus it is legitimate to transfer these designations to whole classes also.

▶ 272. We will consider the theorem of the preceding article, which says that all ternary forms of a given determinant can be distributed into a finite number of classes, and treat it as we did for binary forms. First we will show how each ternary form can be reduced to a simpler form and then show that the number of simplest forms (which result from such reductions) is finite for any given determinant. Let us suppose in general that the given form is the ternary form $f = \binom{a,\ a',\ a''}{b,\ b',\ b''}$ of determinant D (different from zero) and that it is transformed into the equivalent form $g = \binom{m,\ m',\ m''}{n,\ n',\ n''}$ by the substitution (S):

$$\alpha, \quad \beta, \quad \gamma$$
$$\alpha', \quad \beta', \quad \gamma'$$
$$\alpha'', \quad \beta'', \quad \gamma''$$

It remains for us to so determine α, β, γ, etc. that g will be simpler than f. Let the forms which are adjoint to f, g respectively be $\binom{A,\ A',\ A''}{B,\ B',\ B''}$, $\binom{M,\ M',\ M''}{N,\ N',\ N''}$ and designate them by F, G. Then by article 269 F will be transformed into G by a substitution which is adjoint to S, and G will be transformed into F by a substitution derived from a transposition of S. The number

$$\alpha\beta'\gamma'' + \alpha'\beta''\gamma + \alpha''\beta'\gamma' - \alpha''\beta'\gamma - \alpha\beta''\gamma' - \alpha'\beta\gamma''$$

must either $= +1$ or $= -1$. We will denote it by k. We observe the following:

I. If we have $\gamma = 0$, $\gamma' = 0$, $\alpha'' = 0$, $\beta'' = 0$, $\gamma'' = 1$ then

$$m = a\alpha^2 + 2b''\alpha\alpha' + a'\alpha'\alpha', \quad m' = a\beta^2 + 2b''\beta\beta' + a'\beta'\beta',$$
$$m'' = a''$$

$$n = b\beta' + b'\beta, \quad n' = b\alpha' + b'\alpha, \quad n'' = a\alpha\beta + b''(\alpha\beta' + \beta\alpha') + a'\alpha'\beta'$$

Further, $\alpha\beta' - \beta\alpha'$ must either $= +1$ or $= -1$. Thus it is manifest that the binary form (a, b'', a') whose determinant is A' will be transformed by the substitution $\alpha, \beta, \alpha', \beta'$ into the binary form (m, n'', m') of determinant M'' and, since $\alpha\beta' - \beta\alpha' = \pm 1$, they will be equivalent and hence $M'' = A''$. This can also be confirmed directly. Unless therefore (a, b'', a') is already the simplest form in its class, we can so determine $\alpha, \beta, \alpha', \beta'$ that (m, n'', m') is a simpler form. From the theory of the equivalence of binary forms it is easy to conclude that this can be done so that m is not greater than $\sqrt{-4A''/3}$ if A'' is negative, or not greater than $\sqrt{A''}$ when

A'' is positive or so that $m = 0$ when $A'' = 0$. Thus in all cases the (absolute) value of m can be made either less than or at least equal to $\sqrt{\pm 4A''/3}$. In this way the form f is reduced to another with a smaller first coefficient, if this is possible. And the form which is adjoint to this has the same third coefficient as the form F which is adjoint to f. This is the *first reduction*.

II. But if $\alpha = 1$, $\beta = 0$, $\gamma = 0$, $\alpha' = 0$, $\alpha'' = 0$ we will have $k = \beta'\gamma'' - \beta''\gamma' = \pm 1$; thus the substitution which is adjoint to S will be

$$
\begin{array}{ccc}
\pm 1, & 0, & 0 \\
0, & \gamma'', & -\beta'' \\
0, & -\gamma', & \beta'
\end{array}
$$

and by this substitution F will be transformed into G and we will have

$$m = a, \quad n' = \beta'\gamma'' + b''\gamma', \quad n'' = b'\beta'' + b''\beta'$$
$$m' = \alpha'\beta'\beta' + 2b\beta'\beta'' + a''\beta''\beta''$$
$$m'' = a'\gamma'\gamma' + 2b\gamma'\gamma'' + a''\gamma''\gamma''$$
$$n = a'\beta'\gamma' + b(\beta'\gamma'' + \gamma'\beta'') + a''\beta''\gamma''$$
$$M' = A'\gamma''\gamma'' - 2B\gamma'\gamma'' + A''\gamma'\gamma'$$
$$N = -A'\beta''\gamma'' + B(\beta'\gamma'' + \gamma'\beta'') - A''B\gamma'$$
$$M'' = A'\beta''\beta'' - 2B\beta'\beta'' + A''\beta'\beta'$$

Thus it is clear that the binary form (A'', B, A') whose determinant is Da will be transformed by the substitution $\beta', -\gamma', -\beta'', \gamma''$ into the form (M'', N, M') of determinant Dm and thus (since $\beta'\gamma'' - \gamma'\beta'' = \pm 1$ or since $Da = Dn$) equivalent to it. Unless therefore (A'', B, A') is already the simplest form of its class, the coefficients $\beta', \gamma', \beta'', \gamma''$ can be so determined that (M'', N, M') is simpler. And this can always be done so that without respect to sign, M'' is not greater than $\sqrt{\pm 4Da/3}$. In this way the form f is reduced to another with the same first coefficient. But the form which is adjoint to this will have, if possible, a smaller third coefficient than the form F which is adjoint to f. This is the *second reduction*.

III. Now if neither the first nor the second reduction is applicable to the ternary form f, i.e. if f cannot be transformed by either one into a simpler form: then necessarily a^2 will be either $<$ or $=4A''/3$ and $A''A''$ will be either $<$ or $=4aD/3$ without respect to sign. Thus a^4 will be $<$ or $=16A''A''/9$ and so a^4 will be $<$ or $=64aD/27$, a^3 $<$ or $=64D/27$, and a $<$ or $=4\sqrt[3]{D}/3$; and again $A''A''$ will be $<$ or $=16\sqrt[3]{D^4}/9$ and A'' $<$ or $=4\sqrt[3]{D^2}/3$. Therefore whenever a or A'' exceeds these limits, one or another of the preceding reductions necessarily applies to the form f. But this conclusion cannot be converted, since it often happens that the first coefficient and the third coefficient of the adjoint form of a ternary form are already below these limits; nevertheless it can be made simpler by one or another of the reductions.

IV. Now if we apply alternately the first and second reduction to a given ternary form of determinant D, i.e. we apply the first or the second, then to the result we apply the second or first, and to the result of this again the first or second, etc., it is clear that eventually we will arrive at a form to which neither can any longer be applied. For the absolute magnitude of the first coefficients of the forms themselves and of the third coefficients of the adjoint forms will alternately remain the same and then decrease and thus the progression will eventually stop; otherwise we would have two infinite series of continually decreasing numbers. We have therefore this remarkable theorem: *Any ternary form of determinant D can be reduced to an equivalent form with the property that its first coefficient is not greater than $4\sqrt[3]{D}/3$ and the third coefficient of the adjoint form is not greater than $4\sqrt[3]{D^2}/3$ disregarding sign, provided the proposed form does not already have these properties.* In place of the first coefficient of the form f and of the third coefficient of the adjoint form we could have considered in exactly the same way either the first coefficient of the form itself and the second of the adjoint; or the second of the form itself and the first or third of the adjoint; or the third of the form and the first or second of the adjoint. Eventually we would arrive at the same conclusion; but it is more advantageous to use one method consistently so that the operations involved can be reduced to a fixed algorithm. We observe finally that if we had separated the forms into definite and indefinite we could have fixed lower limits for the two coefficients that we have been treating; but this is not necessary for our present purposes.

▶ 273. These examples illustrate the preceding principles.

Example 1. Let $f = \left(\begin{smallmatrix} 19, & 21, & 50 \\ 15, & 28, & 1 \end{smallmatrix}\right)$, then $F = \left(\begin{smallmatrix} -825, & -166, & -398 \\ 257, & 573, & -370 \end{smallmatrix}\right)$ and $D = -1$. Since $(19, 1, 21)$ is a reduced binary form and there is no other equivalent to it which has its first term less than 19, the first reduction is not applicable here; the binary form $(A'', B, A') = (-398, 257, -166)$, by the theory of the equivalence of binary forms, can be transformed into a simpler equivalent $(-2, 1, -10)$ by the substitution 2, 7, 3, 11. Therefore, if we let $\beta' = 2$, $\gamma' = -7$, $\beta'' = -3$, $\gamma'' = 11$ and if we apply the substitution

$$\begin{Bmatrix} 1, & 0, & 0 \\ 0, & 2, & -7 \\ 0, & -3, & 11 \end{Bmatrix}$$

to the form f, it will be transformed into $\left(\begin{smallmatrix} 19, & 354, & 4769 \\ -1299, & 301, & -82 \end{smallmatrix}\right)\ldots f'$. The third coefficient of the adjoint form is -2 and in this respect f' is simpler than f.

The first reduction can be applied to the form f'. That is, since the binary form $(19, -82, 354)$ is transformed into $(1, 0, 2)$ by the substitution 13, 4, 3, 1 the substitution

$$\begin{Bmatrix} 13, & 4, & 0 \\ 3, & 1, & 0 \\ 0, & 0, & 1 \end{Bmatrix}$$

can be applied to the form f' and it will be transformed into $\left(\begin{smallmatrix} 1, & 2, & 4769 \\ -95, & 16, & 0 \end{smallmatrix}\right)\ldots f''$.

We can again apply the second reduction to the form f'' whose adjoint is $\left(\begin{smallmatrix} -513, & -4513, & -2 \\ -95, & 32, & 1520 \end{smallmatrix}\right)$. That is $(-2, -95, -4513)$ will be transformed into $(-1, 1, -2)$ by the substitution 47, 1, -1, 0; so the substitution

$$\begin{Bmatrix} 1, & 0, & 0 \\ 0, & 47, & -1 \\ 0, & 1, & 0 \end{Bmatrix}$$

can be applied to f'' and it will be transformed into $\left(\begin{smallmatrix}1; & 257; & 2\\1; & 0; & 16\end{smallmatrix}\right)\ldots f'''$. The first coefficient of this form cannot be any further reduced by the first reduction, nor can the third coefficient of the adjoint be further reduced by the second reduction.

Example 2. Let the given form be $\left(\begin{smallmatrix}10; & 26; & 2\\7; & 0; & 4\end{smallmatrix}\right)\ldots f$. Its adjoint is $\left(\begin{smallmatrix}-3; & -20; & -244\\70; & -28; & 8\end{smallmatrix}\right)$ and its determinant $=2$. Applying alternately the second and first reduction

by the substitutions	we transform	into
$\begin{pmatrix}1, & 0, & 0\\0, & -1, & 0\\0, & 4, & -1\end{pmatrix}$	f	$\begin{pmatrix}10, & 2, & 2\\-1, & 0, & -4\end{pmatrix}=f'$
$\begin{pmatrix}0, & -1, & 0\\1, & -2, & 0\\0, & 0, & 1\end{pmatrix}$	f'	$\begin{pmatrix}2, & 2, & 2\\2, & -1, & 0\end{pmatrix}=f''$
$\begin{pmatrix}1, & 0, & 0\\0, & -1, & 0\\0, & 2, & -1\end{pmatrix}$	f''	$\begin{pmatrix}2, & 2, & 2\\-2, & 1, & -2\end{pmatrix}=f'''$
$\begin{pmatrix}1, & 0, & 0\\1, & 1, & 0\\0, & 0, & 1\end{pmatrix}$	f'''	$\begin{pmatrix}0, & 2, & 2\\-2, & -1, & 0\end{pmatrix}=f''''$

The form f'''' cannot be any further reduced by the first or second reduction.

▶ 274. When we have a ternary form, and the first coefficient of the form itself and the third coefficient of the adjoint form have been reduced as far as possible by the preceding methods, the following method will supply a further reduction.

Using the same signs as in article 272 and letting $\alpha = 1$, $\alpha' = 0$, $\beta' = 1$, $\alpha'' = 0$, $\beta'' = 0$, $\gamma'' = 1$, i.e. using the substitution

$$\begin{matrix}1, & \beta, & \gamma\\0, & 1, & \gamma'\\0, & 0, & 1\end{matrix}$$

we will have

$$m = a, \quad m' = a' + 2b''\beta + a\beta^2,$$

$$m'' = a'' + 2b\gamma' + 2b'\gamma + a\gamma^2 + 2b''\gamma\gamma' + a'\gamma'\gamma'$$

$$n = b + a'\gamma' + b'\beta + b''(\gamma + \beta\gamma') + a\beta\gamma, \quad n' = b' + a\gamma + b''\gamma',$$

$$n'' = b'' + a\beta$$

and further

$$M'' = A'', \quad N = B - A''\gamma', \quad N' = B' - N\beta - A''\gamma$$

Such a transformation does not change the coefficients a, A'' which were decreased by the preceding reductions. It remains therefore to find a suitable determination of β, γ, γ' so that the remaining coefficients may be lessened. We observe first that if $A'' = 0$ we can suppose that $a = 0$ also; for if a did not $= 0$, the first reduction would be applicable once more, since a form like $(0, 0, h)$ is equivalent to any binary form of determinant 0 and its first term $= 0$ (see art. 215). For a completely similar reason it is legitimate to suppose that A'' would also $= 0$ if $a = 0$, and thus either both or neither of the numbers a, A'' will be 0.

If neither is 0, β, γ, γ' can be so determined that, disregarding sign, n'', N, N' respectively are not greater than $a/2, A''/2, A''/2$. Thus in the first example of the preceding article the last form $\binom{1;\ 257;\ 2}{1;\ \ 0;\ 16}$ whose adjoint is $\binom{-513;\ -2;\ -1}{1;\ -16;\ 32}$ will be transformed by the substitution

$$\begin{pmatrix} 1, & -16, & 16 \\ 0, & 1, & -1 \\ 0, & 0, & 1 \end{pmatrix}$$

into the form $\binom{1;\ 1;\ 1}{1;\ 0;\ 0}\ldots f''''$, whose adjoint is $\binom{-1;\ -1;\ -1}{0;\ 0;\ 0}$.

In the case where $a = A'' = 0$ and thus also $b'' = 0$ we will have

$$m = 0, \quad m' = a', \quad m'' = a'' + 2b\gamma' + 2b'\gamma + a'\gamma'\gamma'$$

$$n = b + a'\gamma' + b'\beta, \quad n' = b', \quad n'' = 0$$

and so

$$D = a'b'b' = m'n'n'$$

It is easy to see that β and γ' can be determined in such a way that n will be equal to the absolutely least residue of b relative to the modulus which is the greatest common divisor of a', b'; i.e. so that

n will not be greater than half this divisor, disregarding sign, and we will have $n = 0$ whenever a', b' are relatively prime. If β, γ' are determined in this way, the value of γ can be taken so that m'' is not greater than b' disregarding sign. This of course would be impossible if $b' = 0$, but then we would have $D = 0$ which is the case we excluded. Thus for the last form in the second example of the preceding article $n = -2 - \beta + 2\gamma'$, and if we set $\beta = -2$, $\gamma' = 0$ we will have $n = 0$; further $m'' = 2 - 2\gamma$, and if we set $\gamma = 1$ then $m'' = 0$. Thus we would have the substitution

$$\left\{ \begin{array}{ccc} 1, & -2, & 1 \\ 0, & 1, & 0 \\ 0, & 0, & 1 \end{array} \right\}$$

by which that form would be transformed into $\left(\begin{smallmatrix} 0; & 2; & 0 \\ 0; & -1; & 0 \end{smallmatrix} \right) \ldots f''''$.

▶ 275. If we have a series of equivalent ternary forms f, f', f'', f''', etc. and the transformations of each of these forms into its successor: then from the transformation of the form f into f' and of the form f' into f'', by article 270 we can deduce a transformation of the form f into f''; from this and from the transformation of the form f'' into f''' there would result a transformation of the form f into f''' etc., and by this process we could find the transformation of the form f into any other form of the series. And since from the transformation of the form f into any other equivalent form g we can deduce a transformation of the form g into f (S'' from S, art. 268, 269) we can in this way produce a transformation of any one of the series f', f'', etc. into the first form f. Thus for the forms of the first example of the preceding article we find the substitutions

13,	4,	0	13,	188,	−4	13,	−20,	16
6,	2,	−7	6,	87,	−2	6,	−9,	7
−9,	−3,	11	−9,	−130,	3	−9,	14,	−11

by which f will be transformed into f'', f''', f'''' respectively and from the last substitution we can derive

$$\left\{ \begin{array}{ccc} 1, & 4, & 4 \\ 3, & 1, & 5 \\ 3, & -2, & 3 \end{array} \right\}$$

by which f'''' will be transformed into f. Similarly for example 2 of the preceding article we have the substitutions

$$\begin{array}{rrr|rrr}
1, & -1, & 1 & 2, & -3, & -1 \\
-3, & 4, & -3 & 3, & 1, & 0 \\
10, & -14, & 11 & 2, & 4, & 1
\end{array}$$

by which the form $\left(\begin{smallmatrix}10, & 26, & 2 \\ 7, & 0, & 4\end{smallmatrix}\right)$ is transformed into $\left(\begin{smallmatrix}0, & 2, & 0 \\ 0, & -1, & 0\end{smallmatrix}\right)$ and conversely.

▶ 276. THEOREM. *The number of classes into which all ternary forms of a given determinant are distributed is always finite.*

Demonstration. I. The number of all forms $\left(\begin{smallmatrix}a, & a', & a'' \\ b, & b', & b''\end{smallmatrix}\right)$ of a given determinant D in which $a = 0, b'' = 0$, b not greater than half the greatest common divisor of the numbers a', b', and a'' not greater than b', is obviously finite. For since we must have $a'b'b' = D$, the only values that can be taken for b' are $+1$, -1, and the roots of squares that divide D (if there are others besides 1) taken positively and negatively. The number of these is finite. For each of the values of b', however, the value of a' is determined, and obviously the number of values of b, a'' is finite.

II. Suppose that a is not $= 0$ nor greater than $(4/3)\sqrt[3]{\pm D}$; that $b''b'' - aa' = A''$ and is not $= 0$ nor greater than $(4/3)\sqrt[3]{D^2}$; that b'' is not greater than $a/2$; that $ab - b'b'' = B$ and $a'b' - bb''$ $= B'$ and neither is greater than $A''/2$. In this case an argument similar to the one used above shows that the number of all forms $\left(\begin{smallmatrix}a, & a', & a'' \\ b, & b', & b''\end{smallmatrix}\right)$ of determinant D is finite. For the number of all combinations of values of a, b'', A'', B, B' will be finite, and when these have been determined, the remaining coefficients of the form, namely, a', b, b', a'' and the coefficients of the adjoint form

$$b^2 - a'a'' = A, \quad b'b' - aa'' = A', \quad a''b'' - bb' = B''$$

will be determined by the following equations:

$$a' = \frac{b''b - A''}{a}, \quad A' = \frac{B^2 - aD}{A''}, \quad A = \frac{B'B' - a'D}{A''},$$

$$B'' = \frac{BB' + b''D}{A''}$$

$$b = \frac{AB - B'B''}{D} = -\frac{Ba' + B'b''}{A''}, \quad b' = \frac{A'B' - BB''}{D}$$

$$= -\frac{Bb'' + B'a}{A''}$$

$$a'' = \frac{b'b' - A'}{a} = \frac{b^2 - A}{a'} = \frac{bb' + B''}{b''}$$

Now when all these forms have been obtained, if we choose from all the combinations of the values of a, b'', A'', B, B' those that make a', a'', b, b' integral, there will be a finite number of them.

III. Thus all the forms in I and II constitute a finite number of classes, and if any forms are equivalent, there will be fewer classes than forms. Since, by what we have said above, any ternary form of determinant D is necessarily equivalent to one of these forms, i.e. it belongs to one of the classes which these forms determine, these classes will include all forms of determinant D; i.e. all ternary forms of determinant D will be distributed into a finite number of classes. Q.E.D.

▶ 277. The rules for forming all the forms in I and II of the preceding article follow naturally from their explanation; therefore it is sufficient to give some examples. For $D = 1$, the forms·I produce the following six (taking one of the double signs at a time):

$$\begin{pmatrix} 0, & 1, & 0 \\ 0, & \pm 1, & 0 \end{pmatrix}, \begin{pmatrix} 0, & 1, & \pm 1 \\ 0, & \pm 1, & 0 \end{pmatrix}$$

For the forms II, a and A'' can have no other values but $+1$ and -1, and thus for each of the four resulting combinations b'', B, and B' must $=0$ and we get the four forms:

$$\begin{pmatrix} 1, & -1, & 1 \\ 0, & 0, & 0 \end{pmatrix}, \begin{pmatrix} -1, & 1, & 1 \\ 0, & 0, & 0 \end{pmatrix} \begin{pmatrix} 1, & 1, & -1 \\ 0, & 0, & 0 \end{pmatrix}, \begin{pmatrix} -1, & -1, & -1 \\ 0, & 0, & 0 \end{pmatrix},$$

Similarly for $D = -1$ we get the six forms I and the four forms II:

$$\begin{pmatrix} 0, & -1, & 0 \\ 0, & \pm 1, & 0 \end{pmatrix}, \begin{pmatrix} 0, & -1, & \pm 1 \\ 0, & \pm 1, & 0 \end{pmatrix}, \begin{pmatrix} 1, & -1, & -1 \\ 0, & 0, & 0 \end{pmatrix}, \begin{pmatrix} -1, & 1, & -1 \\ 0, & 0, & 0 \end{pmatrix},$$

$$\begin{pmatrix} -1, & -1, & 1 \\ 0, & 0, & 0 \end{pmatrix}, \begin{pmatrix} 1, & 1, & 1 \\ 0, & 0, & 0 \end{pmatrix}$$

For $D = 2$ we have the six forms I:

$$\begin{pmatrix} 0, & 2, & 0 \\ 0, & \pm 1, & 0 \end{pmatrix}, \begin{pmatrix} 0, & 2, & \pm 1 \\ 0, & \pm 1, & 0 \end{pmatrix}$$

and the eight forms II:

$$\begin{pmatrix} 1, & -1, & 2 \\ 0, & 0, & 0 \end{pmatrix}, \begin{pmatrix} -1, & 1, & 2 \\ 0, & 0, & 0 \end{pmatrix}, \begin{pmatrix} 1, & 1, & -2 \\ 0, & 0, & 0 \end{pmatrix}, \begin{pmatrix} -1, & -1, & -2 \\ 0, & 0, & 0 \end{pmatrix},$$

$$\begin{pmatrix} 1, & -2, & 1 \\ 0, & 0, & 0 \end{pmatrix}, \begin{pmatrix} -1, & 2, & 1 \\ 0, & 0, & 0 \end{pmatrix}, \begin{pmatrix} 1, & 2, & -1 \\ 0, & 0, & 0 \end{pmatrix}, \begin{pmatrix} -1, & -2, & -1 \\ 0, & 0, & 0 \end{pmatrix}$$

But the number of classes of forms in these three cases is much less than the number of forms. It is easy to confirm that

I. The form $\begin{pmatrix} 0; & 1; & 0 \\ 0; & 1; & 0 \end{pmatrix}$ is transformed into

$$\begin{pmatrix} 0, & 1, & 0 \\ 0, & -1, & 0 \end{pmatrix}, \begin{pmatrix} 0, & 1, & 1 \\ 0, & \pm 1, & 0 \end{pmatrix}, \begin{pmatrix} 0, & 1, & -1 \\ 0, & \pm 1, & 0 \end{pmatrix}, \begin{pmatrix} 1, & 1, & -1 \\ 0, & 0, & 0 \end{pmatrix}$$

respectively by the substitutions

1,	0,	0	0,	0,	1	0,	0,	1	1,	0,	-1
0,	1,	0	0,	1,	-1	0,	1,	1	1,	1,	-1
0,	0,	-1	± 1,	1,	0	± 1,	-1,	-1	0,	-1,	1

and that the form $\begin{pmatrix} 1; & 1; & -1 \\ 0; & 0; & 0 \end{pmatrix}$ is transformed into $\begin{pmatrix} 1; & -1; & 1 \\ 0; & 0; & 0 \end{pmatrix}$ and $\begin{pmatrix} -1; & 1; & 1 \\ 0; & 0; & 0 \end{pmatrix}$ merely by changing the unknowns. Thus the ten ternary forms of determinant 1 are reduced to these two: $\begin{pmatrix} 0; & 1; & 0 \\ 0; & 1; & 0 \end{pmatrix}$, $\begin{pmatrix} -1; & -1; & -1 \\ 0; & 0; & 0 \end{pmatrix}$; for the former, if you prefer, we can take $\begin{pmatrix} 1; & 0; & 0 \\ 0; & 0; & 0 \end{pmatrix}$. And since the first form is indefinite and the second definite, it is manifest that any indefinite ternary form of determinant 1 is equivalent to the form $x^2 + 2yz$ and any definite form to $-x^2 - y^2 - z^2$.

II. In a similar way we find that any indefinite ternary form of determinant -1 is equivalent to the form $-x^2 + 2yz$ and any definite form to $x^2 + y^2 + z^2$.

III. For the determinant 2 the second, sixth, and seventh of the eight forms (II) can be immediately rejected because they can be derived from the first by merely changing the unknowns. Similarly, the fifth can be derived from the third and the eighth from the fourth. The three remaining forms together with the six forms I will make up three classes; that is $\left(\begin{smallmatrix}0; & 2; & 0\\0; & 1; & 0\end{smallmatrix}\right)$ will be transformed into $\left(\begin{smallmatrix}0; & 2; & 0\\0; & -1; & 0\end{smallmatrix}\right)$ by the substitution

$$\begin{Bmatrix} 1, & 0, & 0 \\ 0, & 1, & 0 \\ 0, & 0, & -1 \end{Bmatrix}$$

and the form $\left(\begin{smallmatrix}1; & 1; & -2\\0; & 0; & 0\end{smallmatrix}\right)$ is transformed into

$$\begin{pmatrix} 0, & 2, & 1 \\ 0, & 1, & 0 \end{pmatrix}, \begin{pmatrix} 0, & 2, & 1 \\ 0, & -1, & 0 \end{pmatrix}, \begin{pmatrix} 0, & 2, & -1 \\ 0, & 1, & 0 \end{pmatrix}, \begin{pmatrix} 0, & 2, & -1 \\ 0, & -1, & 0 \end{pmatrix},$$

$$\begin{pmatrix} 1, & -1, & 2 \\ 0, & 0, & 0 \end{pmatrix}$$

respectively by the substitutions

1,	0,	1	1,	0,	−1	1,	0,	0	1,	0,	0	1,	0,	0
1,	2,	0	1,	2,	0	1,	2,	−1	1,	2,	1	0,	1,	2
1,	1,	0	1,	1,	0	1,	1,	−1	1,	1,	1	0,	1,	1

Therefore any ternary form of determinant 2 is reducible to one of the three forms

$$\begin{pmatrix} 0, & 2, & 0 \\ 0, & 1, & 0 \end{pmatrix}, \begin{pmatrix} 1, & 1, & -2 \\ 0, & 0, & 0 \end{pmatrix}, \begin{pmatrix} -1, & -1, & -2 \\ 0, & 0, & 0 \end{pmatrix}$$

and, if you prefer, $\left(\begin{smallmatrix}2; & 0; & 0\\1; & 0; & 0\end{smallmatrix}\right)$ can be put in place of the first. Manifestly any definite ternary form will necessarily be equivalent to the third $-x^2 - y^2 - 2z^2$, since the first two are indefinite. And an indefinite form will be equivalent to the first or second; to the first, $2x^2 + 2yz$, if its first, second, and third coefficients are all even (obviously such a form will be transformed into a similar form by any substitution and so it cannot be equivalent to the second form); to the second form, $x^2 + y^2 - 2z^2$, if its first, second, and third coefficients are not all even but one, two, or all are odd (for the first form, $2x^2 + 2yz$, is not transformable by any substitution into such a form).

According to this argument we could have foreseen a priori in the examples of article 273, 274 that the definite form $\left(\begin{smallmatrix}19; & 21; & 50\\15; & 28; & 1\end{smallmatrix}\right)$ of determinant -1 would reduce to $x^2 + y^2 + z^2$ and that the indefinite form $\left(\begin{smallmatrix} 19; & 26; & 2 \\ 7; & 0; & 4 \end{smallmatrix}\right)$ of determinant 2 to $2x^2 - 2yz$ or (what comes to the same thing) to $2x^2 + 2yz$.

▶ 278. If the unknowns of a ternary form are x, x', x'', the form will *represent* numbers by giving determined values to x, x', x'' and will represent binary forms by the substitutions

$$x = mt + nu, \qquad x' = m't + n'u, \qquad x'' = m''t + n''u$$

where m, n, m', etc. are determined numbers and t, u the unknowns of the binary form. Now to complete the theory of ternary forms we require a solution of the following problem. I, To find all representations of a given number by a given ternary form. II, To find all representations of a given binary form by a given ternary form. III, To judge whether or not two given ternary forms of the same determinant are equivalent and, if they are, to find all transformations of one into the other. IV, To judge whether or not a given ternary form implies another given form of a greater determinant and, if it does, to assign all transformations of the first into the second. Since these problems are much more difficult than the analogous problems in binary forms we will treat them more in detail at another time.

For the present we will restrict our investigation to showing how the first problem can be reduced to the second and the second to the third, we will show how to solve the third for very simple cases which especially illustrate the theory of binary forms, and we will exclude the fourth altogether.

▶ 279. LEMMA. *Given any three integers a, a', a'' (not all $= 0$): to find six others B, B', B'', C, C', C'' so disposed that*

$$B'C'' - B''C' = a, \qquad B''C - BC'' = a', \qquad BC' - B'C = a''$$

Solution. Let α be the greatest common divisor of a, a', a'' and take the integers A, A', A'' so that

$$Aa + A'a' + A''a'' = \alpha$$

Now take arbitrarily three integers $\mathfrak{C}$, $\mathfrak{C}'$, $\mathfrak{C}''$ with the sole condition that the three numbers $\mathfrak{C}'A'' - \mathfrak{C}''A'$, $\mathfrak{C}''A - \mathfrak{C}A''$, $\mathfrak{C}A' - \mathfrak{C}'A$ are not all $= 0$. We will designate these numbers respectively by b, b', b'' and their greatest common divisor by β. Then if we let

$$a'b'' - a''b' = \alpha\beta C, \qquad a''b - ab'' = \alpha\beta C', \qquad ab' - a'b = \alpha\beta C''$$

it is clear that C, C', C'' are integers. Finally if we choose integers $\mathfrak{B}$, $\mathfrak{B}'$, $\mathfrak{B}''$ so that

$$\mathfrak{B}b + \mathfrak{B}'b' + \mathfrak{B}''b'' = \beta$$

and let

$$\mathfrak{B}a + \mathfrak{B}'a' + \mathfrak{B}''a'' = h$$

and set

$$B = \alpha\mathfrak{B} - hA, \qquad B' = \alpha\mathfrak{B}' - hA', \qquad B'' = \alpha\mathfrak{B}'' - hA''$$

the values B, B', B''. C, C', C'' will satisfy the given equations.

For we find that

$$aB \quad + \quad a'B' + a''B'' = 0$$

$$bA + b'A' + b''A'' = 0 \quad \text{and therefore} \quad bB + b'B' + b''B'' = \alpha\beta$$

Now from the values of C', C'' we have

$$\begin{aligned} \alpha\beta(B'C'' - B''C') &= ab'B' - a'bB' - a''bB'' + ab''B'' \\ &= a(bB + b'B' + b''B'') - b(aB + a'B' + a''B'') \\ &= \alpha\beta a \end{aligned}$$

and so $B'C'' - B''C' = a$; similarly we find that $B''C - BC'' = a'$ and $BC' - B'C = a''$. Q.E.F. But we have to omit here the analysis by which we found this solution and the method of finding all of them from one solution.

▶ 280. Let us suppose that the binary form

$$at^2 + abtu + cu^2 \ldots \phi$$

whose determinant $= D$ is represented by the ternary form f with unknowns x, x', x'' by letting

$$x = mt + nu, \qquad x' = m't + n'u, \qquad x'' = m''t + n''u$$

and that the adjoint of f is the form F with unknowns X, X', X''. Then it is easy to confirm by calculating (letting the coefficients of the forms f, F be designated by letters) or by deduction from article 268. II, that the number D can be represented by F if we let

$$X = m'n'' = m''n', \qquad X' = m''n - mn'', \qquad X'' = mn' - m'n$$

We will say that this representation of the number D is the *adjoint* of the representation of the form ϕ by f. If the values of X, X', X'' do not have a common divisor, for brevity we will call this representation of D *proper*, otherwise, *improper*, and we will give these same designations also to the representation of the form ϕ by f to which that representation of D is adjoint. Now the discovery of all proper representations of the number D by the form F is based on the following considerations:

I. There is no representation of D by F that cannot be deduced from some representation of a form of determinant D by the form f, i.e. which is adjoint to such a representation.

For let some representation of D by F be such that $X = L$, $X' = L'$, $X'' = L''$; and by the lemma of the preceding article choose m, m', m'', n, n', n'' so that

$$m'n'' - m''n' = L, \quad m''n - mn'' = L', \quad mn' - m'n = L''$$

and let f be transformed into the binary form $\phi = at^2 + 2btu + cu^2$ by the substitution

$$x = mt + nu, \qquad x' = m't + n'u, \qquad x'' = m''t + n''u$$

It is easy to see that D will be the determinant of the form ϕ and that the representation of D by F will be adjoint to the representation of ϕ by f.

Example. Let $f = x^2 + x'x' + x''x''$ and $F = -X^2 - X'X' - X''X''$; $D = -209$; its representation by F will be $X = 1$, $X' = 8$, $X'' = 12$; and we find the values of m, m', m'', n, n', n'' to be $-20, 1, 1, -12, 0, 1$ respectively and $\phi = 402t^2 + 482tu + 145u^2$.

II. If ϕ, χ are properly equivalent binary forms, any representation of D by F which is adjoint to a representation of the form ϕ by f will also be adjoint to a representation of the form χ by f.

Let p, q be the unknowns of the indeterminate form χ; let ϕ be transformed into χ by the proper substitution $t = \alpha p + \beta q$, $u = \gamma p + \delta q$ and let a representation of the form ϕ by f be

$$x = mt + nu, \quad x' = m't + n'u, \quad x'' = m''t + n''u \qquad (R)$$

Then if we let

$$\alpha m + \gamma n = g, \qquad \alpha m' + \gamma n' = g', \qquad \alpha m'' + \gamma n'' = g''$$
$$\beta m + \delta n = h, \qquad \beta m' + \delta n' = h', \qquad \beta m'' + \delta n'' = h''$$

the form χ will be represented by f if we set

$$x = gp + hq, \qquad x' = g'p + h'q, \qquad x'' = g''p + h''q \quad (R')$$

and by calculation (since $\alpha\delta - \beta\gamma = 1$) we find

$$g'h'' - g''h' = m'n'' - m''n', \qquad g''h - gh'' = m''n - mn'',$$
$$gh' - g'h = mn' - m'n$$

i.e. the same representation of D by F is adjoint to the representations R, R'.

Thus in the preceding example the form ϕ is equivalent to $\chi = 13p^2 - 10pq + 18q^2$ and it is transformed into it by the proper substitution $t = -3p + q, u = 5p - 2q$; and the representation of the form χ by f is: $x = 4q, x' = -3p + q, x'' = 2p - q$. From this we deduce the same representation of the number -209 that we had before.

III. Finally, if two binary forms ϕ, χ of determinant D with unknowns $t, u; p, q$ can be represented by f, and if the same proper representation of D by F is adjoint to a representation of each of these, the two forms must be properly equivalent. Let us suppose that ϕ is represented by f by letting

$$x = mt + nu, \qquad x' = m't + n'u, \qquad x'' = m''t + n''u$$

and that χ is represented by f by setting

$$x = gp + hq, \qquad x' = g'p + h'q, \qquad x'' = g''p + h''q$$

and that

$$m'n'' - m''n' = g'h'' - g''h' = L$$
$$m''n - mn'' = g''h - gh'' = L'$$
$$mn' - m'n = gh' - g'h = L''$$

Now choose the integers l, l', l'' so that $Ll + L'l' + L''l'' = 1$ and let

$$n'l'' - n''l' = M, \qquad n''l - nl'' = M', \qquad nl' - n'l = M''$$
$$l'm'' - l''m' = N, \qquad l''m - lm'' = N', \qquad lm' - l'm = N''$$

and finally let

$$gM + g'M' + g''M'' = \alpha, \qquad hM + h'M' + h''M'' = \beta$$
$$gN + g'N' + g''N'' = \gamma, \qquad hN + h'N' + h''N'' = \delta$$

From this it is easy to deduce

$$\alpha m + \gamma n = g - l(gL + g'L' + g''L'') = g$$
$$\beta m + \delta n = h - l(hL + h'L' + h''L'') = h$$

and similarly

$$\alpha m' + \gamma n' = g', \qquad \beta m' + \delta n' = h', \qquad \alpha m'' + \gamma n'' = g'',$$
$$\beta m'' + \delta n'' = h''$$

From this it is clear that $mt + nu$, $m't + n'u$, $m''t + n''u$ will be transformed into $gp + hq$, $g'p + h'q$, $g''p + h''q$, respectively, by the substitution

$$t = \alpha p + \beta q, \qquad u = \gamma p + \delta q \ldots (S)$$

and by the substitution S, ϕ will be transformed into the same form as f when we let

$$x = gp + hq, \qquad x' = g'p + h'q, \qquad x'' = g''p + h''q$$

that is to say into χ to which it must therefore be equivalent. Finally, by proper substitutions we find that

$$\alpha\delta - \beta\gamma = (Ll + L'l' + L''l'')^2 = 1$$

Therefore the substitution S is proper and the forms ϕ, χ are properly equivalent.

As a result of these observations we derive the following rules for finding all proper representations of D by F: Find all classes of binary forms with determinant D and from each of them select one form arbitrarily; find all proper representations of each of these forms by f (rejecting any that cannot be represented by f) and from each of these representations deduce representations of the number D by F. By I and II it is manifest that in this way we obtain all possible proper representations and that thus the solution is complete; by III it is clear that transformations of forms from different classes certainly produce different representations.

▶ 281. The investigation of *improper* representations of a given number D by the form F can easily be reduced to the preceding case. It is evident that if D is divisible by no square (except 1) there will be no representations of this kind at all; but if λ^2, μ^2, ν^2, etc. are the square divisors of D, all improper representations of

D by F can be found if we find all proper representations of the numbers $D/\lambda^2, D/\mu^2, D/\nu^2$, etc. by this same form and multiply the values of the unknowns by λ, μ, ν, etc. respectively.

Therefore, finding all representations of a given number by a given ternary form *which is adjoint to another ternary form* depends on the second problem. And even though at first sight this seems to be a very particular case, all other cases can be reduced to it as follows: let D be the number which is to be represented by the form $(\begin{smallmatrix} g, & g', & g'' \\ h, & h', & h'' \end{smallmatrix})$ of determinant Δ whose adjoint is the form $(\begin{smallmatrix} G, & G', & G'' \\ H, & H', & H'' \end{smallmatrix}) = f$. Then the adjoint of f will be $(\begin{smallmatrix} \Delta g, & \Delta g', & \Delta g'' \\ \Delta h, & \Delta h', & \Delta h'' \end{smallmatrix}) = F$, and it is clear that the representations of the number ΔD by F (this investigation depends on the preceding) will be identical with the representations of the number D by the given form. But when all coefficients of the form f have a common divisor μ, it is evident that all coefficients of the form F will be divisible by μ^2 and so ΔD must also be divisible by μ^2 (otherwise there would be no representations); and representations of the number D by the proposed form will coincide with representations of the number $\Delta D/\mu^2$ by the form that results from F by dividing each of its coefficients by μ^2, and this form will be adjoint to the one that results from f by dividing each of the coefficients by μ.

We observe, finally, that this solution of the first problem is not applicable when $D = 0$; for in this case the binary forms of determinant D are not distributed into a finite number of classes; we will solve this case later by using different principles.

▶ 282. The investigation of the representations of a given binary form with determinant not $= 0°$ by a given ternary form depends on the following observations.

I. From any proper representation of a binary form $(p, q, r) = \phi$ of determinant D by a ternary form f of determinant Δ we can find integers B, B' so that we have

$$B^2 \equiv \Delta p, \qquad BB' \equiv -\Delta q, \qquad B'B' \equiv \Delta r \text{ (mod. } D)$$

i.e. the value of the expression $\sqrt{\Delta(p, -q, r)}$ (mod. D). Suppose we have the following proper representation of the form ϕ by f

$$x = \alpha t + \beta u, \qquad x' = \alpha' t + \beta' u, \qquad x'' = \alpha'' t + \beta'' u$$

_____
° For brevity we will omit here the treatment of the zero case because it requires a somewhat different method.

(where x, x', x''; t, u designate the unknowns of the forms f, ϕ); choose integers $\gamma, \gamma', \gamma''$ so that

$$(\alpha'\beta'' - \alpha''\beta')\gamma + (\alpha''\beta - \alpha\beta'')\gamma' + (\alpha\beta' - \alpha'\beta)\gamma'' = k$$

with k being either $= +1$ or $= -1$, and let f be transformed by the substitution

$$\alpha, \quad \beta, \quad \gamma$$
$$\alpha', \quad \beta', \quad \gamma'$$
$$\alpha'', \quad \beta'', \quad \gamma''$$

into the form $\binom{a, \ a', \ a''}{b, \ b', \ b''} = g$ whose adjoint is $\binom{A, \ A', \ A''}{B, \ B', \ B''} = G$. Then manifestly we will have $a = p$, $b'' = q$, $a' = r$, $A'' = D$, and Δ the determinant of the form g; therefore

$$B^2 = \Delta p + A'D, \qquad BB' = -\Delta q + B''D, \qquad B'B' = \Delta r + AD$$

So, e.g., the form $19t^2 + 6tu + 41u^2$ is represented by $x^2 + x'x' + x''x''$ by letting $x = 3t + 5u$, $x' = 3t - 4u$, $x'' = t$; and if we let $\gamma = -1$, $\gamma' = 1$, $\gamma'' = 0$ we will have $B = -171$, $B' = 27$ or $(-171, 27)$ as a value of the expression $\sqrt{-1}(19, -3, 41)$ (mod. 770).

It follows from this that if $\Delta(p, -q, r)$ is not a quadratic residue of D, ϕ will not be properly representable by any ternary form of determinant Δ; in the case therefore where Δ, D are relatively prime, Δ will have to be the characteristic number of the form ϕ.

II. Since $\gamma, \gamma', \gamma''$ can be determined in infinitely many different ways, different values of B, B' will result. Let us see what connection they will have with one another. Suppose we have also chosen $\delta, \delta', \delta''$ so that $(\alpha'\beta'' - \alpha''\beta')\delta + (\alpha''\beta - \alpha\beta'')\delta' + (\alpha\beta' - \alpha'\beta)\delta'' = \mathfrak{k}$ becomes either $= +1$ or $= -1$ and that the form f is transformed by the substitution.

$$\alpha, \quad \beta, \quad \delta$$
$$\alpha', \quad \beta', \quad \delta'$$
$$\alpha'', \quad \beta'', \quad \delta''$$

into $\binom{a, \ a', \ a''}{b, \ b', \ b''} = \mathfrak{g}$ with adjoint $\binom{\mathfrak{A}, \ \mathfrak{A}', \ \mathfrak{A}''}{\mathfrak{B}, \ \mathfrak{B}', \ \mathfrak{B}''} = \mathfrak{G}$. Then $g, \mathfrak{g}$ will be equivalent and so also G and $\mathfrak{G}$ and by applying the principles[p]

---

[p] We derive the transformation of the form g into f from the transformation of the form f into g; from this and from the transformation of the form f into $\mathfrak{g}$ we get the transformation of the form g into $\mathfrak{g}$; and from this by transposition the transformation of the form $\mathfrak{G}$ into G.

given in articles 269, 270 we will find that if we let

$$(\beta'\gamma'' - \beta''\gamma')\delta + (\beta''\gamma - \beta\gamma'')\delta' + (\beta\gamma' - \beta'\gamma)\delta'' = \zeta$$

$$(\gamma'\alpha'' - \gamma''\alpha')\delta + (\gamma''\alpha - \gamma\alpha'')\delta' + (\gamma\alpha' - \gamma'\alpha)\delta'' = \eta$$

the form $\mathfrak{G}$ will be transformed into G by the substitution

$$k, \quad 0, \quad 0$$
$$0, \quad k, \quad 0$$
$$\zeta, \quad \eta, \quad \mathfrak{k}$$

Thus we will have

$$B = \eta\,\mathfrak{k}D + \mathfrak{k}k\mathfrak{B}, \qquad B' = \zeta\,\mathfrak{k}D + \mathfrak{k}k\mathfrak{B}'$$

and so, since $\mathfrak{k}k = \pm 1$, either $B \equiv \mathfrak{B}$, $B' \equiv \mathfrak{B}'$ or $B \equiv -\mathfrak{B}$, $B' \equiv -\mathfrak{B}'$ (mod. D). In the first case we will say that the values $(B, B'), (\mathfrak{B}, \mathfrak{B}')$ are equivalent, in the second case that they are opposite; and we will say that the representation of the form ϕ *belongs* to any value of the expression $\sqrt{\Delta(p, -q, r)}$ (mod. D) which can be deduced from it by the method of I. Thus all values to which the same representation belongs will be either equivalent or opposite.

III. Conversely, if as in I, $x = \alpha t + \beta u$ etc. is a representation of the form ϕ by f, and if this representation belongs to the value (B, B') which is deduced from it by the transformation

$$\alpha, \quad \beta, \quad \gamma$$
$$\alpha', \quad \beta', \quad \gamma'$$
$$\alpha'', \quad \beta'', \quad \gamma''$$

the same representation will also belong to any other value $(\mathfrak{B}, \mathfrak{B}')$ which is either equivalent or opposite to it; i.e. in place of $\gamma, \gamma', \gamma''$ we can take other integers $\delta, \delta', \delta''$ for which the equation

$$(\alpha'\beta'' - \alpha''\beta')\delta + (\alpha''\beta - \alpha\beta'')\delta' + (\alpha\beta' - \alpha'\beta)\delta'' = +1\ldots(\Omega)$$

applies and which are so chosen that if f is transformed into its adjoint form by the substitution (S):

$$\alpha, \quad \beta, \quad \delta$$
$$\alpha', \quad \beta', \quad \delta'$$
$$\alpha'', \quad \beta'', \quad \delta''$$

the fourth and fifth coefficients of the adjoint form will respectively $= \mathfrak{B}, \mathfrak{B}'$. For suppose we let

$$\pm B = \mathfrak{B} + \eta D, \qquad \pm B' = \mathfrak{B}' + \zeta D$$

[here and later we take the upper or lower sign according as the values $(B, B'), (\mathfrak{B}, \mathfrak{B}')$ are equivalent or opposite]; ζ, η will be integers and by the substitution

$$1, \quad 0, \quad \zeta$$
$$0, \quad 1, \quad \eta$$
$$0, \quad 0, \quad \pm 1$$

$\mathfrak{g}$ will be transformed into the form $\mathfrak{g}$ with determinant Δ. It is easy to see that coefficients 4 and 5 of the adjoint form will $= \mathfrak{B}, \mathfrak{B}'$ respectively. If however we set

$$\alpha\zeta + \beta\eta \pm \gamma = \delta, \qquad \alpha'\zeta + \beta'\eta \pm \gamma' = \delta', \qquad \alpha''\zeta + \beta''\eta \pm \gamma = \delta''$$

it is not hard to see that f will be transformed by the substitution (S) into $\mathfrak{g}$ and that the equation (Ω) will be satisfied. Q.E.D.

▶ 283. From these principles we can deduce the following method of finding all proper representations of the binary form

$$\phi = pt^2 + 2qtu + ru^2$$

of determinant D by the ternary form f of determinant Δ.

I. Find all the different (i.e. nonequivalent) values of the expression $\sqrt{\Delta}(p, -q, r)$ (mod. D). For the case where ϕ is a primitive form and Δ and D are relatively prime, this problem was solved before (art. 233), and the remaining cases can be reduced to this very easily; but brevity does not permit a fuller explanation. We observe only that as long as Δ and D are relatively prime, the expression $\Delta(p, -q, r)$ cannot be a quadratic residue of D unless ϕ is a primitive form. For let us suppose that

$$\Delta p = B^2 - DA', \qquad -\Delta q = BB' - DB'', \qquad \Delta r = B'B' - DA$$

then

$$(DB'' - \Delta q)^2 = (DA' + \Delta p)(DA + \Delta r)$$

and by manipulating and substituting $q^2 - pr$ for D we have

$$(q^2 - pr)(B''B'' - AA') - \Delta(Ap + 2B''q + A'r) + \Delta^2 = 0$$

And it is easy to conclude that if p, q, r have a common divisor, it will also divide Δ^2; consequently Δ and D cannot be relatively prime. Therefore p, q, r cannot have a common divisor and ϕ is a primitive form.

II. Let us designate the number of these values by m and suppose that among them there are n values opposite to themselves (we will let $n = 0$ when there is none). Then manifestly the remaining $m - n$ values will always be composed of pairs that are opposite to one another (since we have supposed that all values are included); now from every pair of opposite values reject one arbitrarily and there will be left $(m + n)/2$ values in all. Thus, e.g., we have eight values of the expression $\sqrt{} - 1(19, -3, 41)$ (mod. 770), namely, $(39, 237)$, $(171, -27)$, $(269, -83)$, $(291, -127)$, $(-39, -237)$, $(-171, 27)$, $(-269, 83)$, $(-291, 127)$. We reject the last four as opposites of the first four. But it is evident that if (B, B') is a value that is opposite to itself, $2B, 2B'$ and also $2\Delta p, 2\Delta q, 2\Delta r$ will be divisible by D; and if therefore Δ, D are relatively prime, $2p, 2q, 2r$ will also be divisible by D. By I in this case p, q, r cannot have a common divisor, so 2 must be divisible by D. This cannot happen unless D either $= \pm 1$ or $= \pm 2$. Thus for all values of D greater than 2, we will always have $n = 0$ if Δ and D are relatively prime.

III. Realizing this, it is evident that any proper representation of the form ϕ by f must belong to one of the remaining values and only to one. We should therefore run through all of these values in succession to find the representations belonging to each one. In order to find the representations belonging to a *given* value (B, B') we must first determine the ternary form $g = \left(\begin{smallmatrix} a, & a', & a'' \\ b, & b', & b'' \end{smallmatrix}\right)$ whose determinant $= \Delta$ and in which $a = p, b'' = q, a' = r$, $ab - b'b'' = B, a'b' - bb'' = B'$; the values a'', b, b' can be found with the help of the equations in article 276. II. And from these it is easy to see that in the case where Δ, D are relatively prime, b, b', a'' must be integers (because these three numbers give integral values when multiplied by D and by Δ). Now if one of the coefficients b, b', b'' is a fraction or the forms f, g are not equivalent, there will be no representations of the form ϕ by f belonging to (B, B'); but if b, b', b'' are integers and the forms f, g are equivalent, then any transformation of f into g, say

$$\alpha, \quad \beta, \quad \gamma$$
$$\alpha', \quad \beta', \quad \gamma'$$
$$\alpha'', \quad \beta'', \quad \gamma''$$

will supply such a representation, namely,

$$x = \alpha t + \beta u, \quad x' = \alpha' t + \beta' u, \quad x'' = \alpha'' t + \beta'' u$$

Manifestly there is no representation of this kind that cannot be deduced from some transformation. And thus that part of the second problem which is concerned with *proper* representations is reduced to the third problem.

IV. Now different transformations of the form f into g always produce different representations, the only exception being the case where the value (B, B') is opposite to itself. In this case two transformations give only one representation. For let us suppose that f is also transformed into g by the substitution

$$
\begin{array}{ccc}
\alpha, & \beta, & \delta \\
\alpha', & \beta', & \delta' \\
\alpha'', & \beta'', & \delta''
\end{array}
$$

(which gives the same representation as the preceding one) and let us denote by $k, \mathfrak{t}, \zeta, \eta$ the same numbers as in II of the preceding article. We will have

$$B = k\,\mathfrak{t}B + \eta\,\mathfrak{t}D, \qquad B' = k\,\mathfrak{t}B' + \zeta\,\mathfrak{t}D$$

If we suppose that both $k, \mathfrak{t} = +1$ or both $= -1$, we find (since we have excluded the case where $D = 0$) that $\zeta = 0$, $\eta = 0$ and it follows that $\delta = \gamma, \delta' = \gamma', \delta'' = \gamma''$; these two transformations can be different only when one of the numbers $k, \mathfrak{t}$ is $+1$ and the other -1; then we have $B \equiv -B, B' \equiv -B'$ (mod. D) or the value of (B, B') is opposite to itself.

V. From what we said above (art. 271) about the distinctions between definite and indefinite forms it follows easily that if Δ is positive, D negative, and ϕ a negative form, g will be a definite negative form; but if Δ is positive and either D positive, or D negative and ϕ a positive form, g will be an indefinite form. Now since f, g certainly cannot be equivalent unless they are similar in this respect, it is clear that binary forms with positive determinant and positive forms cannot be properly represented by a negative ternary form and that negative binary forms cannot be represented by an indefinite ternary form with positive determinant; a ternary

form of the former type can represent a binary form of the latter
type only, and a ternary form of the latter type can represent a
binary form of the former type only. Similarly, we conclude that
a definite (i.e. positive) ternary form with negative determinant
can represent positive binary forms only, and that an indefinite
ternary form can represent only negative binary forms and forms
with a positive determinant.

▶ 284. Now *improper* representations of the binary form ϕ with
determinant D by the ternary form f whose adjoint is F are the
ones from which we deduce improper representations of the
number D by the form F. Manifestly, therefore, ϕ cannot be
improperly represented by f unless D implies square factors. Let
us suppose that all the squares (except 1) that divide D are e^2, $e'e'$,
$e''e''$, etc. (their number is finite since we presume that we do not
have $D = 0$). Every improper representation of the form ϕ by f
will give a representation of the number D by F in which the values
of the unknowns will have one of the numbers e, e', e'', etc.
as the greatest common divisor. For this reason we will say
briefly that an improper representation of the form ϕ belongs to
the quadratic divisor e^2 or $e'e'$ or $e''e''$ etc. Now we can use the
following rules to find all representations of the form ϕ belonging
to the same *given* divisor e^2 (we will suppose that its root e is
taken positively). For brevity we will give a synthetic demonstration, but it will be easy to reconstruct the analysis that leads to
the result.

First, find all binary forms of determinant D/e^2 which are
transformed into the form ϕ by a proper substitution such as
$T = \varkappa t + \lambda u$, $U = \mu u$ where T, U are the unknowns of the form;
t, u the unknowns of the form ϕ; $\varkappa, \mu$ positive integers (whose
product therefore $= e$); λ a positive integer less than μ (it can be
zero). These forms with the corresponding transformations are
found as follows.

Let $\varkappa$ equal successively each of the divisors of e taken positively
(including 1 and e) and let $\mu = e/\varkappa$; for each of the values of $\varkappa, \mu$
assign to λ all integral values from 0 to $\mu - 1$ and we will assuredly
have all transformations. Now we can find the form that is trans-
formed into ϕ by a substitution $T = \varkappa t + \lambda u$, $U = \mu u$ by investigating the form into which ϕ is transformed by $t = (T/\varkappa) - (\lambda U/e)$,
$u = U/\mu$; thus we obtain the forms corresponding to each of the

transformations; but only those forms are to be retained in which all three coefficients are integers.[q]

Second, suppose that Φ is one of the forms which is transformed into ϕ by the substitution $T = \varkappa t + \lambda u$, $U = \mu u$; let us investigate all *proper* representations of the form Φ by f (if any exists) and express them in general by the formula

$$x = \mathfrak{A}T + \mathfrak{B}U, \qquad x' = \mathfrak{A}'T + \mathfrak{B}'U,$$

$$x'' = \mathfrak{A}''T + \mathfrak{B}''U \qquad (\mathfrak{R})$$

From each of the $(\mathfrak{R})$ we can deduce the representation

$$x = \alpha t + \beta u, \qquad x' = \alpha' t + \beta' u, \qquad x'' = \alpha'' t + \beta'' u \qquad (\rho)$$

by the equations

$$\alpha = \varkappa \mathfrak{A}, \qquad \alpha' = \varkappa \mathfrak{A}', \qquad \alpha'' = \varkappa \mathfrak{A}'' \qquad (R)$$

$$\beta = \lambda \mathfrak{A} + \mu \mathfrak{B}, \qquad \beta' = \lambda \mathfrak{A}' + \mu \mathfrak{B}', \qquad \beta'' = \lambda \mathfrak{A}'' + \mu \mathfrak{B}''$$

Let all the other forms which we found by the first rule (if there are more) be treated in the same way and thus other representations will be derived from each proper representation of each form. In this way we will get all representations of the form ϕ belonging to the divisor e^2 and each of them only once.

Demonstration. I. It is so obvious that the ternary form f is transformed into ϕ by each substitution (ρ) that it needs no further explanation; that each representation (ρ) is improper and belongs to the divisor e^2 is clear from the fact that the numbers $\alpha'\beta'' - \alpha''\beta'$, $\alpha''\beta - \alpha\beta''$, $\alpha\beta' - \alpha'\beta$ respectively are $= e(\mathfrak{A}'\mathfrak{B}'' - \mathfrak{A}''\mathfrak{B}')$, $e(\mathfrak{A}''\mathfrak{B} - \mathfrak{A}\mathfrak{B}'')$, $e(\mathfrak{A}\mathfrak{B}' - \mathfrak{A}'\mathfrak{B})$ and their greatest common divisor will be e (since $(\mathfrak{R})$ is a proper representation).

II. We show that from any given representation (ρ) of the form ϕ we can find a proper representation of a form of determinant D/e^2 contained among the forms found by the first rule; that is, from the given values of $\alpha, \alpha', \alpha'', \beta, \beta', \beta''$ we can deduce integral values of $\varkappa, \lambda, \mu$ with the prescribed conditions as well as values of $\mathfrak{A}, \mathfrak{A}', \mathfrak{A}'', \mathfrak{B}, \mathfrak{B}', \mathfrak{B}''$ satisfying equations $(\mathfrak{R})$ uniquely. It is

[q] If we could treat this problem more fully, we would be able to abbreviate the solution a great deal. It is immediately obvious that for $\varkappa$ we need consider only those divisors of e whose squares divide the first coefficient of the form ϕ. We will reserve for a more suitable occasion a further consideration of this problem. We note that we can deduce from it also simpler solutions of the problems of articles 213, 214.

immediately clear from the first three equations in ($\mathfrak{R}$) that for $\varkappa$ we should take the greatest common divisor of $\alpha, \alpha', \alpha''$ with positive sign (for since $\mathfrak{A}'\mathfrak{B}'' - \mathfrak{A}''\mathfrak{B}', \mathfrak{A}''\mathfrak{B} - \mathfrak{A}\mathfrak{B}'', \mathfrak{A}\mathfrak{B}' - \mathfrak{A}'\mathfrak{B}$ will not have a common divisor, neither will $\mathfrak{A}, \mathfrak{A}', \mathfrak{A}''$); as a result we can also determine $\mathfrak{A}, \mathfrak{A}', \mathfrak{A}''$ and $\mu = e/x$ (it is easy to see that it will necessarily be an integer). Let us suppose that we take the three integers $\mathfrak{a}, \mathfrak{a}', \mathfrak{a}''$ in such a way that $\mathfrak{a}\mathfrak{A} + \mathfrak{a}'\mathfrak{A}' + \mathfrak{a}''\mathfrak{A}'' = 1$ and let us for brevity write k for $\mathfrak{a}\mathfrak{B} + \mathfrak{a}'\mathfrak{B}' + \mathfrak{a}''\mathfrak{B}''$. Then from the last three equations (R) it follows that $\mathfrak{a}\beta + \mathfrak{a}'\beta' + \mathfrak{a}''\beta'' = \lambda + \mu k$ and from this it is immediately evident that there is only one value of λ between the limits 0 and $\mu - 1$. When we have done this, the values $\mathfrak{B}, \mathfrak{B}', \mathfrak{B}''$ will also be determined, so it remains only to show that they will always be integers. Now we have

$$\mathfrak{B} = \frac{1}{\mu}(\beta - \lambda\mathfrak{A}) = \frac{1}{\mu}[\beta(1 - \mathfrak{a}\mathfrak{A}) - \mathfrak{A}(\mathfrak{a}'\beta' + \mathfrak{a}''\beta'')] + \mathfrak{A}k$$

$$= \frac{1}{\mu}[\mathfrak{a}''(\mathfrak{A}''\beta - \mathfrak{A}\beta'') - \mathfrak{a}'(\mathfrak{A}\beta' - \mathfrak{A}'\beta)] + \mathfrak{A}k$$

$$= \frac{1}{e}[\mathfrak{a}''(\alpha''\beta - \alpha\beta'') - \mathfrak{a}'(\alpha\beta' - \alpha'\beta)] + \mathfrak{A}k$$

Manifestly this shows that $\mathfrak{B}$ is an integer, and in the same way we can show that $\mathfrak{B}', \mathfrak{B}''$ are integers. From these arguments we see that there cannot be any improper representation of the form ϕ by f belonging to the divisor e^2 which cannot be obtained uniquely by the method we have used.

And if we treat the remaining quadratic divisors of D in the same way and evolve the representations belonging to each of them, we will have all the improper representations of the form ϕ by f.

From this solution it is easy to deduce that the theorem enunciated at the end of the preceding article for proper representations also applies to improper representations; that is, in general no positive binary form with negative determinant can possibly be represented by a negative ternary form etc. For if ϕ were such a binary form that according to the theorem could not be represented by f properly, then also all forms with determinants D/e^2, $D/e'e'$, etc. which imply ϕ could not be properly represented by f. The reason is that all these forms have a determinant with the same sign as ϕ, and when these determinants are negative, all the forms will be positive or negative according as ϕ belongs to positive or negative forms.

▶ 285. We can give here only a few details concerning the third problem (to which we have reduced the other two); that is to say, concerning the manner of judging whether or not two given ternary forms of the same determinant are equivalent and, if they are, of finding all transformations of one into the other. The reason is that the complete solution, such as we have given for analogous problems in binary forms, would present great difficulties here. We will therefore limit our discussion to some particular cases pertinent to this digression.

I. For the determinant $+1$ we showed above that all ternary forms are distributed into two classes, one containing all indefinite forms, the other containing all (negative) definite forms. We immediately conclude that any two ternary forms of determinant 1 are equivalent if they are both definite or both indefinite; if one is definite and the other indefinite, they are not equivalent (manifestly the last part of the proposition holds generally for forms of any determinant). Similarly, any two forms with determinant -1 are certainly equivalent if they are both definite or both indefinite. Two definite forms with determinant 2 are always equivalent; two indefinite forms are not equivalent if in one the three first coefficients are all even and in the other not all are even; in the remaining cases (either all three first coefficients in each form or none are even) the forms will be equivalent. We could show many more propositions of this special nature if we had developed more examples above (art. 277).

II. For all these cases we could find a transformation of one of two equivalent ternary forms f, f into the other. For in all cases, in any class of ternary forms the number of forms assigned is small enough so that any form of the same class can be reduced by uniform methods to one of them; and we have also showed how to reduce all of them to a single form. Let F be this form in the class to which f, f' belong, and by the methods given above we can find transformations of the forms f, f' into F and of the form F into f, f'. Then by article 270 we can deduce transformations of the form f into f' and of the form f' into f.

III. It remains therefore only to show how all possible transformations can be derived from one transformation of a ternary form f into another f'. This problem depends on a simpler problem, that of finding all transformations of a ternary form f into itself. For if f is transformed into itself by various substitutions $(\tau), (\tau')$, (τ''), etc. and if it is transformed into f' by the substitution (t), it is clear that we can combine the transformation (t) with $(\tau), (\tau'), (\tau'')$, etc. according to the norm of article 270 and produce transformations, all of which will take f into f'. By further calculation it is easy to prove that any transformation of the form f into f' can be deduced in this way from the combination of a given transformation (t) of the form f into f' together with one (and *only* one) transformation of the form f into itself. Thus from the combination of a given transformation of f into f' with *all* transformations of the form f into itself, we will get *all* transformations of the form f into f' and, indeed, each of them only once.

We will restrict our investigation of all transformations of the form f into itself to the case where f is a definite form whose 4th, 5th, and 6th coefficients all $= 0$.[r] Therefore let $f = \left(\begin{smallmatrix} a, & a', & a'' \\ 0, & 0, & 0 \end{smallmatrix}\right)$ and let all substitutions by which f is transformed into itself be represented in general by

$$\begin{array}{ccc} \alpha, & \beta, & \gamma \\ \alpha', & \beta', & \gamma' \\ \alpha'', & \beta'', & \gamma'' \end{array}$$

so that the following equations are satisfied

$$a\alpha^2 + a'\alpha'\alpha' + a''\alpha''\alpha'' = a \qquad (\Omega)$$
$$a\beta^2 + a'\beta'\beta' + a''\beta''\beta'' = a'$$
$$a\gamma^2 + a'\gamma'\gamma' + a''\gamma''\gamma'' = a''$$
$$a\alpha\beta + a'\alpha'\beta' + a''\alpha''\beta'' = 0$$
$$a\alpha\gamma + a'\alpha'\gamma' + a''\alpha''\gamma'' = 0$$
$$a\beta\gamma + a'\beta'\gamma' + a''\beta''\gamma'' = 0$$

Now three cases must be distinguished:

I. When a, a', a'' (they all have the same sign) are unequal, let

---

[r] The other cases in which f is a definite form can be reduced to this one; but if f is an indefinite form a completely different method must be used and the number of transformations will be infinite.

us suppose that $a < a', a' < a''$ (if there is a different order of magnitude the same conclusions will result from a similar method). Then the first equation in (Ω) evidently requires that $\alpha' = \alpha'' = 0$ and so $\alpha = \pm 1$; then by equations 4, 5 we will have $\beta = 0, \gamma = 0$; similarly from equation 2 we have $\beta'' = 0$ and therefore $\beta' = \pm 1$; now from equation 6, $\gamma' = 0$ and by 3, $\gamma'' = \pm 1$ and thus (because of the individual ambiguity of signs) we will have in all 8 transformations.

II. When two of the numbers a, a', a'' are equal, e.g. $a' = a''$, and the third is unequal, let us suppose:

First that $a < a'$. Then in the same way as in the preceding case we will have $\alpha' = 0, \alpha'' = 0, \alpha = \pm 1, \beta = 0, \gamma = 0$; and from equations 2, 3, 6 it is easy to deduce that we must have either $\beta' = \pm 1, \gamma' = 0, \beta'' = 0, \gamma'' = \pm 1$ or $\beta' = 0, \gamma' = \pm 1, \beta'' = \pm 1, \gamma'' = 0$.

But if, *second*, $a > a'$ we will obtain the same conclusions in this way; from equations 2, 3 we have necessarily $\beta = 0, \gamma = 0$, and either $\beta' = \pm 1, \gamma' = 0, \beta'' = 0, \gamma'' = \pm 1$ or $\beta' = 0, \gamma' = \pm 1, \beta'' = \pm 1, \gamma'' = 0$; in either case from equations 4, 5 we will have $\alpha' = 0, \alpha'' = 0$ and from equation 1, $\alpha = \pm 1$. And thus for each case there will be 16 different transformations. The two remaining cases where either $a = a''$ or $a = a'$ can be resolved in an entirely similar manner. In the former case we need merely commute the the characters $\alpha, \alpha', \alpha''$ with β, β', β'' respectively; in the second case they are to be commuted with $\gamma, \gamma', \gamma''$ respectively.

III. When all the a, a', a'' are equal, equations 1, 2, 3 require that in each of the three sets of three numbers $\alpha, \alpha', \alpha'', \beta, \beta', \beta'', \gamma, \gamma', \gamma''$ two of the numbers $= 0$, the third $= \pm 1$. By equations 4, 5, 6 it is easy to see that only one of the three numbers α, β, γ can be $= \pm 1$. The same is true of the set α', β', γ' and the set $\alpha'', \beta'', \gamma''$. Therefore there are only six possible combinations:

α	α	α'	α'	α''	α''	$= \pm 1$
β'	β''	β	β''	β	β'	$= \pm 1$
γ''	γ'	γ''	γ	γ'	γ	$= \pm 1$

The remaining six coefficients will $= 0$

and thus we have in all 48 transformations. The same table also includes the preceding cases, but only the first column is to be taken when a, a', a'' are all unequal; the first and second when $a' = a''$; the first and third when $a = a'$; the first and sixth when $a = a''$.

In summary, if the form $f = ax^2 + a'x'x' + a''x''x''$ is transformed into another equivalent form f' by the substitution

$$x = \delta y + \varepsilon y' + \zeta y'', \qquad x' = \delta' y + \varepsilon' y' + \zeta' y'',$$
$$x'' = \delta'' y + \varepsilon'' y' + \zeta'' y''$$

all transformations of the form f into f' will be contained in the following scheme:

x	x	x'	x'	x''	x''	$= \pm(\delta y + \varepsilon y' + \zeta y'')$
x'	x''	x	x''	x	x'	$= \pm(\delta' y + \varepsilon' y' + \zeta' y'')$
x''	x'	x''	x	x'	x	$= \pm(\delta'' y + \varepsilon'' y' + \zeta'' y'')$

with this difference, that the first six columns will all be used when $a = a' = a''$; columns 1 and 2 when $a' = a''$; 1 and 3 when $a = a'$; 1 and 6 when $a = a''$; and the first column only when a, a', a'' are all unequal. In the first case the number of transformations will be 48, in the second, third, and fourth cases 16, in the fifth 8.

SOME APPLICATIONS TO THE THEORY OF BINARY FORMS

Since the basic elements of the theory of ternary forms have been succinctly developed, we will proceed to some special applications. Among them the following problem merits first place.

▶ 286. PROBLEM. *Given a binary form $F = (A, B, C)$ of determinant D belonging to a principal genus: to find the binary form f from whose duplication we get the form F.*

Solution. I. Let F' be opposite to the form F. We want a proper representation of $F' = AT^2 - 2BTU + CU^2$ by the ternary form $x^2 - 2yz$. Suppose it is

$$x = \alpha T + \beta U, \qquad y = \alpha' T + \beta' U, \qquad z = \alpha'' T + \beta'' U$$

It is clear that this can be done from the preceding theory on ternary forms. For, since by hypothesis F belongs to a principal genus, there is a value of the expression $\sqrt{(A, B, C)}$ (mod. D) from which can be found a ternary form ϕ of determinant 1 in which $(A, -B, C)$ will be a factor, and all its coefficients will be integers. It is equally obvious that ϕ will be an indefinite form (since by hypothesis F is certainly not a negative form); and so it will necessarily be equivalent to the form $x^2 - 2yz$. Consequently we can assign a transformation of ϕ into F which gives us a proper representation of the form F' by the form $x^2 - 2yz$. As a result we have

$$A = \alpha^2 - 2\alpha'\alpha'', \quad -B = \alpha\beta - \alpha'\beta'' - \alpha''\beta', \quad C = \beta^2 - 2\beta'\beta''$$

and further, if we designate the numbers $\alpha\beta' - \alpha'\beta$, $\alpha'\beta'' - \alpha''\beta'$, $\alpha''\beta - \alpha\beta''$ by a, b, c, respectively, they will not have a common divisor and $D = b^2 - 2ac$.

II. With the help of the last observation of article 235 it is easy to conclude that by the substitution $2\beta', \beta, \beta, \beta''$; $2\alpha', \alpha, \alpha, \alpha''$, F will be transformed into the product of the form $(2a, -b, c)$ into itself and by the substitution $\beta', \beta, \beta, 2\beta''$; $\alpha', \alpha, \alpha, 2\alpha''$ into the product of the form $(a, -b, 2c)$ into itself. Now the greatest common divisor of the numbers $2a, 2b, 2c$ is 2; therefore if c is odd, the numbers $2a, 2b, c$ will not have a common divisor, so $(2a, -b, c)$ will be a properly primitive form; similarly, if a is odd, $(a, -b, 2c)$ will be a properly primitive form. In the former case F will be derived from a duplication of the form $(2a, -b, c)$, in the latter case from a duplication of the form $(a, -b, 2c)$ (see conclusion 4, art. 235). And certainly one of these cases will always be true. For if both a, c were even, b would necessarily be odd; now it is easy to confirm that $\beta''a + \beta b + \beta'c = 0$, $\alpha''a + \alpha b + \alpha'c = 0$ and it follows that $\beta b, \alpha b$ will be even and so also α and β. From this it would follow that A and C are even, but this is contrary to the hypothesis according to which F is a form of a principal genus and thus of a properly primitive order. But it can happen that both a and c are odd. In this case we will immediately have two forms which will produce F by their duplication.

Example. Suppose we are given the form $F = (5, 2, 31)$ with determinant -151. The value of the expression $\sqrt{(5, 2, 31)}$ will be $(55, 22)$; by the norms of article 272 we find that the ternary form $\phi = \left(\begin{smallmatrix} 5, & 31; & -4 \\ 1 1; & 0; & -2 \end{smallmatrix}\right)$ is equivalent to the form $\left(\begin{smallmatrix} 1; & 1; & -1 \\ 0; & 0; & 0 \end{smallmatrix}\right)$ and this will be transformed into ϕ by the substitution

$$\begin{Bmatrix} 2, & 2, & 1 \\ 1, & -6, & -2 \\ 0, & 3, & 1 \end{Bmatrix}$$

And with the help of the transformations given in article 277 we
(find that $(_{-1}^{\ 1};\ _0^0;\ _0^0)$ is transformed into ϕ by the substitution

$$\begin{Bmatrix} 3, & -7, & -2 \\ 2, & -1, & 0 \\ 1, & -9, & -3 \end{Bmatrix}$$

Thus $a = 11$, $b = -17$, $c = 20$; therefore since a is odd, F will be
derived from the duplication of the form $(11, 17, 40)$ and will be
transformed into the product of this form into itself by the substitution $-1, -7, -7, -18; 2, 3, 3, 2$.

▶ 287. We add the following observations on the problem that
was solved in the preceding article.

I. If the form F is transformed into a product of the two forms
(h, i, k), (h', i', k') by the substitution p, p', p'', p'''; q, q', q'', q''' (we
suppose that each is always taken properly) we will have the equa-
tions easily deduced from conclusion 3 of article 235:

$$p''hn' - p'h'n - p(in' - i'n) = 0$$
$$(p'' - p')(in' + i'n) - p(kn' - k'n) + p'''(hn' - h'n) = 0$$
$$p'kn' - p''k'n - p'''(in' - i'n) = 0$$

and three others which are derived from these by interchanging
the numbers p, p', p'', p''' and q, q', q'', q'''; n, n' are the positive
square roots which result from the division of the determinants
of the forms (h, i, k), (h', i', k') by the determinant of the form F.
Thus if these forms are identical, that is, $n = n'$, $h = h'$, $i = i'$,
$k = k'$, the equations will become

$$(p'' - p')hn = 0, \qquad (p'' - p')in = 0, \qquad (p'' - p')kn = 0$$

and *necessarily* $p' = p''$ and similarly $q' = q''$. If therefore we assign
to the forms (h, i, k), (h', i', k') the *same* unknowns t, u and designate
the unknowns of the form F by T, U then F will be transformed
by the substitution

$$T = pt^2 + 2p'tu + p''u^2, \qquad U = qt^2 + 2q'tu + q'''u^2$$
$$\text{into} \qquad (ht^2 + 2itu + ku^2)^2$$

II. If the form F is derived from a duplication of the form f,
it will also be derived from a duplication of any other form

contained in the same class as f; that is, the class of the form F will be derived from a duplication of the class of the form f (see art. 238). Thus in the example of the preceding article $(5, 2, 31)$ will also be derived from a duplication of the form $(11, -5, 16)$ which is properly equivalent to the form $(11, 17, 40)$. From one class which by duplication produces the class of the form F we find all (if there are more than one) such classes with the help of problem 260; in our example there is no other positive class because there exists only one properly primitive positive ambiguous class of determinant -151 (the principal class); and since, from the composition of the single negative ambiguous class $(-1, 0, -151)$ with the class $(11, -5, 16)$, we get the class $(-11, -5, -16)$, this will be the only negative class and from its duplication we derive the class $(5, 2, 31)$.

III. Since by the solution of the problem of the preceding article it is clear that any properly primitive (positive) class of binary forms belonging to a principal genus can be derived from the duplication of any properly primitive class of the same determinant, we can expand the theorem of article 261. This theorem stated that we can be sure that *at least* half of all assignable characters for a given nonquadratic determinant D cannot correspond to properly primitive (positive) genera. Now we can say that *exactly* half of all such characters correspond to such genera and that none of the other half can correspond to such a genus (see the demonstration of the theorem). In article 264 we distributed all those characters into two equal groups P, Q. We proved that none of Q can correspond to properly primitive (positive) forms. It remained uncertain whether there were genera corresponding to each of the P. Now this doubt is removed, and we are certain that in the whole complex of characters P there is none that does not correspond to a genus. It was shown in article 264. I that for a negative determinant it is impossible for P, and possible *only* for Q, to have any members in a properly primitive *negative* order. Now we can deduce that it is possible to have *all* the members of Q. For if K is any character in Q, f an arbitrary form in the order of properly primitive negative forms of determinant D, and K' its character, then K' will be in Q; from this it is easy to see that the character composed from K, K' (according to the norm of art. 246) belongs to P and so there are properly primitive positive forms of determinant D corresponding to it. If we compose this form with f we will have a properly primitive negative form of determinant D whose character will be K. In a similar way we can prove that those characters in an improperly primitive order, which by the methods of articles 264. II, III are found to be the *only* ones possible, are actually *all* possible whether they be in P or Q. We believe that these theorems are among the most beautiful in the theory of binary forms, especially because, despite their extreme simplicity, they are so profound that a rigorous demonstration requires the help of many other investigations.

We turn now to another application of the preceding digression, the decomposition of numbers and binary forms into three squares. We begin with the following.

▶ 288. PROBLEM. *Given a positive number M, to find the conditions which negative primitive binary forms of determinant $-M$ must satisfy in order that they be quadratic residues of M, that is, in order that they have 1 as a characteristic number.*

Solution. Let us designate by Ω the complex of all particular characters that give the relations of the number 1 both to the individual prime (odd) divisors of M and to the number 8 or 4 when it divides M; manifestly these characters will be Rp, Rp', Rp'', etc. where p, p', p'', etc. are the prime divisors, and $1, 4$ when 4 divides M; $1, 8$ when 8 divides M. Furthermore we will use the letters P, Q with the same meaning as in the preceding article and in article 264. Now we distinguish the following cases.

I. When M is divisible by $4, \Omega$ will be an integral character and it is clear from article 233. V that 1 can be a characteristic number of only those forms whose character is Ω. But it is manifest that Ω is a character of the principal form $(1, 0, M)$ and so belongs to P and cannot coincide with a properly primitive negative form; therefore, since there are no improperly primitive forms for this determinant, in this case there will be no negative primitive forms which are residues of M.

II. When $M \equiv 3$ (mod. 4) the same reasoning holds with this exception that in this case an *improperly* primitive negative order exists in which the characters P will be possible or impossible according as $M \equiv 3$ or $\equiv 7$ (mod. 8) (see art. 264. III). In the former case therefore there will be a genus for this order whose character is

Ω, so 1 will be a characteristic number of all forms contained in it; in the latter case there cannot be any forms at all with this property.

III. When $M \equiv 1$ (mod. 4), Ω is not yet a complete character, but we must add to it a relation to the number 4; it is clear, however, that Ω must enter into the character of a form whose characteristic number is 1, and conversely, that any form whose character is either $\Omega; 1, 4$ or $\Omega; 3, 4$ has 1 as a characteristic number. Now $\Omega; 1, 4$ is manifestly a character of a principal genus that belongs to P and so is impossible in a negative properly primitive order; for the same reason $\Omega; 3, 4$ will belong to Q (art. 263), so there will be a corresponding genus in a properly primitive negative order all of whose forms will have 1 as a characteristic number. In this case, as in the following, there will be no improperly primitive order.

IV. When $M \equiv 2$ (mod. 4) we must add to Ω a relation to 8 in order to get a complete character. These relations will be 1 *and* 3, 8 or 5 *and* 7, 8 when $M \equiv 2$ (mod. 8); and either 1 *and* 7, 8 or 3 *and* 5, 8 when $M \equiv 6$ (mod. 8). In the former case the character $\Omega; 1$ *and* 3, 8 will obviously belong to P and so $\Omega; 5$ *and* 7, 8 to Q. As a result there will be a properly primitive negative genus corresponding to it. For a similar reason in the latter case there will be one genus in a properly primitive negative order whose form has the prescribed property; that is, its character is $\Omega; 3$ *and* 5, 8.

From all this it follows that there are negative primitive forms of determinant $-M$ with characteristic number 1 only when M is congruent to one of the numbers 1, 2, 3, 5, 6 relative to the modulus 8, and they will belong to only one genus, which is improper, when $M \equiv 3$; there are no such forms at all when $M \equiv 0, 4,$ or 7 (mod. 8). But manifestly if $(-a, -b, -c)$ is a primitive negative form with characteristic number $+1$, (a, b, c) will be a positive primitive form with characteristic -1. From this it is clear that in the five former cases (when $M \equiv 1, 2, 3, 5, 6$) there is one positive primitive genus whose forms have -1 as a characteristic number, and it is *improper* if $M \equiv 3$; in the last three cases however (when $M \equiv 0, 4, 7$) there are no such positive forms at all.

▶ 289. With regard to proper representations of binary forms by the ternary form $x^2 + y^2 + z^2 = f$ we can derive the following from the general theory of article 282.

I. The binary form ϕ cannot be properly represented by f unless it is a positive primitive form and -1 (i.e. the determinant of the form f) is its characteristic number. Thus for a positive determinant, as well as for the negative determinant $-M$ when M is either divisible by 4 or of the form $8n + 7$, there are no binary forms properly representable by f.

II. Now if $\phi = (p, q, r)$ is a positive primitive form of determinant $-M$, and -1 is the characteristic number of the form ϕ and so also of the opposite form $(p, -q, r)$, there will be proper representations of the form ϕ by f belonging to any given value of the expression $\sqrt{-(p, -q, r)}$. That is, all coefficients of the ternary form g of determinant -1 (art. 283) will necessarily be integers, the form g will be definite and so certainly equivalent to f (art. 285. I).

III. We know by article 283. III that the number of representations belonging to the same value of the expression $\sqrt{-(p, -q, r)}$ in all cases except when $M = 1$ and $M = 2$ is equal in magnitude to the number of transformations of the form f into g, and so, by article 285, $= 48$; thus if we have one representation belonging to a given value, the 47 others can be derived from it by permuting the values of x, y, z in all possible ways and by changing their signs; as a result, all 48 representations will produce *only one* decomposition of the form ϕ into three squares, if we consider the squares themselves only and not their order or the signs of their roots.

IV. If we let the number of all the different odd prime numbers that divide $M = \mu$, it is not difficult to conclude from article 233 that the number of different values of the expression $\sqrt{-(p, -q, r)}$ (mod. M) will be $= 2^{\mu}$. And according to article 283 we need consider only half of these (when $M > 2$). Therefore the number of all proper representations of the form ϕ by f will be $= 48 \cdot 2^{\mu-1} = 3 \cdot 2^{\mu+3}$; but the number of different decompositions into three squares is $= 2^{\mu-1}$.

Example. Let $\phi = 19t^2 + 6tu + 41u^2$ so that $M = 770$; here we must consider (art. 283) the following four values of the expression $\sqrt{-(19, -3, 41)}$ (mod. 770): $(39, 237)$, $(171, -27)$, $(269, -83)$, $(291, -127)$. In order to find representations belonging to the value $(39, 237)$ we must determine the ternary form $\left(\begin{smallmatrix} 19, & 41, & 2 \\ 3, & 6, & 3 \end{smallmatrix}\right) = g$. By the methods of articles 272, 275 we find that f will be transformed into this form by the substitution

$$\left\{ \begin{array}{rrr} 1, & -6, & -0 \\ -3, & -2, & -1 \\ -3, & -1, & -1 \end{array} \right\}$$

and the representation of the form ϕ by f is:

$$x = t - 6u, \qquad y = -3t - 2u, \qquad z = -3t - u$$

For the sake of brevity we will not write the 47 remaining representations belonging to the same value, which result from the permutation of these values and the conversion of signs. All 48 representations produce the same decomposition of the form ϕ into three squares

$$t^2 - 12tu + 36u^2, \qquad 9t^2 + 12tu + 4u^2, \qquad 9t^2 + 6tu + u^2$$

In a similar way the value $(171, -27)$ will give as a decomposition into squares $(3t + 5u)^2$, $(3t - 4u)^2$, t^2; the value $(269, -83)$ will give $(t + 6u)^2 + (3t + u)^2 + (3t - 2u)^2$; and the value $(291, -127)$ will give $(t + 3u)^2 + (3t + 4u)^2 + (3t - 4u)^2$; each of these decompositions is equivalent to 48 representations. Outside of these 192 representations or four decompositions there are no others, since 770 is not divisible by any square and so there cannot be any improper representations.

▶ 290. Forms of determinant -1 and -2 are subject to certain exceptions so we will say a few words about them separately. We begin with the general observation that if ϕ, ϕ' are any two equivalent binary forms, (Θ) a given transformation of the first into the second, then by a combination of any representation of the form ϕ by a ternary form f with the substitution (Θ) we get a representation of the form ϕ' by f. Further, from proper representations of ϕ we get proper representations of the form ϕ', from different representations of ϕ we get different ones for ϕ', and if we take all representations of the first, we will get all representations of the second. All of this can be proved by very easy calculation. Therefore one of the forms ϕ, ϕ' can be represented by f in the same number of ways as the other.

I. First let $\phi = t^2 + u^2$ and ϕ' any other positive binary form of determinant -1 to which, therefore, ϕ is equivalent; and let the transformation of ϕ into ϕ' be by the substitution $t = \alpha t' + \beta u'$, $u = \gamma t' + \delta u'$. The form ϕ is represented by the ternary form $f = x^2 + y^2 + z^2$ by letting $x = t$, $y = u$, $z = 0$; if we permute x, y, z we will have *six* representations and from each of these, four more by changing the signs of t, u. Thus in all we will have 24 different representations and to these there will correspond only one decomposition into three squares. It is easy to see that there will be no other representations than these. And we conclude that the form ϕ' also can be decomposed into three squares in only one way, namely, $(\alpha t' + \beta u')^2$, $(\gamma t' + \delta u')^2$, and 0. This decomposition will be equivalent to the 24 representations.

II. Let $\phi = t^2 + 2u^2$, ϕ' any other positive binary form of determinant -2, into which ϕ is transformed by the substitution $t = \alpha t' + \beta u'$, $u = \gamma t' + \delta u'$. Then in a manner similar to that of the preceding case we conclude that ϕ and also ϕ' can be decomposed into three squares in only one way, namely, ϕ into $t^2 + u^2 + u^2$ and ϕ' into $(\alpha t' + \beta u')^2 + (\gamma t' + \delta u')^2 + (\gamma t' + \delta u')^2$; it is obvious that this decomposition is equivalent to the 24 representations.

From all this it follows that binary forms of determinants -1 and -2 with respect to the number of representations by the ternary form $x^2 + y^2 + z^2$ are completely like other binary forms; for since in both cases we have $\mu = 0$, the formula given in IV of the preceding article will give 24 representations. The reason for this is that the two exceptions to which such forms are subject mutually compensate one another.

For the sake of brevity we omit applying to the form $x^2 + y^2 + z^2$ the general theorem regarding improper representations which was given in article 284.

▶ 291. The question of finding all proper representations of a given positive *number* M by the form $x^2 + y^2 + z^2$ is first reduced by article 281 to the investigation of the proper representations of the number $-M$ by the form $-x^2 - y^2 - z^2 = f$; by the methods of article 280 these can be found in the following way.

I. We find all classes of binary forms of determinant $-M$ whose forms can be properly represented by $X^2 + Y^2 + Z^2 = F$ (which has f as adjoint). When $M \equiv 0, 4$, or 7 (mod. 8) by article 288 there are no such classes, and so M cannot be decomposed into three squares that do not have a common divisor.[5] But when $M \equiv 1, 2, 5$, or 6 there will be a properly primitive positive genus, and when $M \equiv 3$ an improperly primitive one which includes all those classes. We will indicate the number of these classes by k.

II. Now select arbitrarily one form from each of these classes and call them ϕ, ϕ', ϕ'', etc.; investigate all proper representations of all of these by F. The number of them will be $3 \cdot 2^{\mu+3} k = K$ where μ is the number of (odd) prime factors of M; finally, from each of these representations in order to have

$$X = mt + nu, \qquad Y = m't + n'u, \qquad Z = m''t + n''u$$

we derive the following representation of M by $x^2 + y^2 + z^2$:

$$x = m'n'' - m''n', \qquad y = m''n - mn'', \qquad z = mn' - m'n$$

All representations of M are necessarily contained in the complex of these K representations which we will designate by Ω.

III. It remains therefore only to find whether there are any representations in Ω which are *identical*; and since from article 280. III it is already clear that those representations in Ω that are derived from different forms, e.g. from ϕ and ϕ', must be different, the only question that remains is whether different representations of the same form, e.g. of ϕ, by F can produce identical representations of the number M by $x^2 + y^2 + z^2$. Now it is immediately evident that if among the representations of ϕ we find

$$X = mt + nu, \qquad Y = m't + n'u, \qquad Z = m''t + n''u \qquad (r)$$

we will also find among the same representations

$$X = -mt - nu, \qquad Y = -m't - n'u, \qquad Z = -m''t - n''u \qquad (r')$$

and from each we can derive the same representation of M which we will call (R); let us examine therefore whether the representation (R) can result from still other representations of the form ϕ. From article 280. III if we let $\chi = \phi$ and if we exhibit all transformations of the proper form ϕ into itself by

$$t = \alpha t + \beta u, \qquad u = \gamma t + \delta u$$

we can deduce that all those representations of the form ϕ from

---

[5] This impossibility is also clear from the fact that the sum of three odd squares must be $\equiv 3$ (mod. 8); the sum of two odd and one even is either $\equiv 2$ or $\equiv 6$; the sum of one odd and two even either $\equiv 1$ or $\equiv 5$; and finally the sum of three even is either $\equiv 0$ or $\equiv 4$; but in the latter case the representation is manifestly improper.

which R is derived will be expressed by

$$x = (\alpha m + \gamma n)t + (\beta m + \delta n)u$$
$$y = (\alpha m' + \gamma n')t + (\beta m' + \delta n')u$$
$$z = (\alpha m'' + \gamma n'')t + (\beta m'' + \delta n'')u$$

But from the theory of the transformations of binary forms with a negative determinant, as explained in article 179, it follows that in all cases except when $M = 1$ and $M = 3$ there are only two proper transformations of the form ϕ into itself, namely, $\alpha, \beta, \gamma, \delta = 1, 0, 0, 1$ and $= -1, 0, 0, -1$, respectively [for since ϕ is a primitive form, the number we designated in article 179 by m will be either 1 or 2 and so, except for the excluded cases, 1) will certainly apply]. Therefore (R) can come only from r, r' and every proper representation of the number M will be found twice, and not more often, in Ω; and the number of all the different proper representations of M will be $K/2 = 3 \cdot 2^{\mu + 2}k$.

With regard to the excepted cases, the number of proper transformations of the form ϕ into itself by article 179 will be 4 for $M = 1$, 6 for $M = 3$; and it is easy to confirm that the number of proper representations of the numbers 1, 3 is $K/4$, $K/6$ respectively; that is, each number can be decomposed into three squares in only one way, 1 into $1 + 0 + 0$, 3 into $1 + 1 + 1$. The decomposition of 1 supplies six, the decomposition of 3, eight representations which are different; now for $M = 1$ we have $K = 24$ (here $\mu = 0$, $k = 1$) and for $M = 3$ we have $K = 48$ (here $\mu = 1$, $k = 1$).

Let h designate the number of classes in the principal genus. By article 252 it will be equal to the number of classes in any other properly primitive genus. We observe that $k = h$ for $M \equiv 1, 2, 5$ or 6 (mod. 8), but $k = h/3$ for $M \equiv 3$ (mod. 8) with the single case $M = 3$ excepted (where $k = h = 1$). Thus the number of representations *in general* for numbers of the form $8n + 3$ is $= 2^{\mu + 2}h$, since for the number 3 the two exceptions compensate one another.

▶ 292. We have distinguished decompositions of numbers (as well as of binary forms) into three squares by representations of the form $x^2 + y^2 + z^2$ in such a way that in the former we paid attention only to the magnitude of the squares, and in the latter we also regarded the order of the roots and their signs. Thus we consider the representations $x = a$, $y = b$, $z = c$ and $x = a'$,

$y = b'$, $z = c'$ to be different unless simultaneously $a = a'$, $b = b'$, $c = c'$; and we consider the decompositions into $a^2 + b^2 + c^2$ and into $a'a' + b'b' + c'c'$ to be the same if, without regard to order, the squares in one are equal to the squares in the other. From this it is clear:

I. That the decomposition of the number M into $a^2 + b^2 + c^2$ is equivalent to 48 representations if none of the squares $=0$ and all are unequal; but only 24 if *either* one of them $= 0$ and the others are unequal, *or* none of them $= 0$ and two of them are equal. If however in the decomposition of a given number into three squares, two of the squares $= 0$, *or* one $= 0$ and the others are equal, *or* they are all equal, the decomposition will be equivalent to 6 *or* 12 *or* 8 representations; but this cannot happen unless we have the special cases where $M = 1$ or 2 or 3, respectively, at least if the representations are to be proper. Excluding these three cases, let us suppose that the number of all decompositions of a number M into three squares (which do not have a common divisor) is E, and that among them we have e decompositions in which one square is 0, and e' in which two squares are equal; the former can be regarded as decompositions into two squares, the latter as decompositions into a square and twice a square. Then the number of all proper representations of the number M by $x^2 + y^2 + z^2$ will be

$$= 24(e + e') + 48(E - e - e') = 48E - 24(e + e')$$

But from the theory of binary forms it is easy to deduce that e will either $=0$ or $=2^{\mu-1}$, according as -1 is a nonresidue or a quadratic residue of M, and that e' will be $= 0$ or $= 2^{\mu-1}$ according as -2 is a nonresidue of residue of M. Here μ is the number of (odd) prime factors of M (see art. 182; we omit here a more complete exposition). From all this we have

$E = 2^{\mu-2}k$, if both -1 and -2 are nonresidues of M

$E = 2^{\mu-2}(k + 2)$, if both numbers are residues; and finally

$E = 2^{\mu-2}(k + 1)$, if one is a residue, the other a nonresidue

In the excluded cases where $M = 1$ and $M = 2$ this formula would make $E = 3/4$ whereas it should have $E = 1$. For $M = 3$, however, we get the correct value, $E = 1$, because the exceptions mutually compensate.

Therefore if M is a prime number, we will have $\mu = 1$ and so $E = (k + 2)/2$ when $M \equiv 1$ (mod. 8); $E = (k + 1)/2$ when $M \equiv 3$ or $\equiv 5$. These special theorems were discovered by the illustrious Legendre by induction and were published by him in that splendid commentary which we have so often praised (*Hist. Acad. Paris*, 1785, p. 530 ff.).[11] If he presented them in a little different form, it is only because he did not distinguish between proper and improper forms and so mixed opposite forms together.

II. In order to find all decompositions of a number M into three squares (with no common divisor) it is not necessary to derive all proper representations of all the forms ϕ, ϕ', ϕ''. For it is easy to confirm that all (48) representations of the form ϕ belonging to the same value of the expression $\sqrt{-(p, -q, r)}$ [where $\phi = (p, q, r)$] will give the same decomposition of the number M, and so it is sufficient if we have one of them or, what comes to the same thing, if only we know all the different decompositions' of the form f into three squares. The same thing holds for ϕ', ϕ'', etc. Now if ϕ belongs to a nonambiguous class, it is quite permissible to neglect the form which was chosen from the opposite class; that is, it is sufficient to consider only one of two opposite classes. For since it is entirely arbitrary which form we select from a class, let us suppose that from the class that is opposite to the class containing ϕ we select the form ϕ', which is opposite to the form ϕ. Then it is not hard to show that if we represent the decompositions of the proper form ϕ by the general expression.

$$(gt + hu)^2 + (g't + h'u)^2 + (g''t + h''u)^2$$

all decompositions of the form ϕ' will be expressed by

$$(gt - hu)^2 + (g't - h'u)^2 + (g''t - h''u)^2$$

and the same decompositions of the number M will be derived from both. Finally, for the case where ϕ is a form of an ambiguous class but not of a principal class nor equivalent to the form $(2, 0, M/2)$ or $(2, 1, (M + 1)/2)$ it is permissible to omit half the values

---

' We must always understand the word "proper" if we want to transfer this expression from representations to decompositions.

[11] Cf. pp. 27, 105.

of the expression $\sqrt{-(p, -q, r)}$; but for brevity we will not give the details of this simplification. We can also use these simplifications when we want all the proper representations of M by $x^2 + y^2 + z^2$, since the latter can be derived very easily from the decompositions.

As an example we will investigate all decompositions of the number 770 into three squares. Here $\mu = 3$, $e = e' = 0$, and so $E = 2k$. Since it is easy to apply the norms of article 231 to classify the positive binary forms of determinant -770, we will omit this operation for brevity's sake. We find that the number of positive classes $= 32$. All of them are properly primitive and are distributed into eight genera so that $k = 4$ and $E = 8$. The genus whose characteristic number is -1 manifestly has the particular characters $R5$; $N7$; $N11$ with respect to the numbers $5, 7, 11$, and by article 263 we conclude that its character with respect to the number 8 must be 1 *and* 3, 8. Now in the genus with the character 1 *and* 3, 8; $R5$; $N7$; $N11$ we find four classes. From them we select the following as representatives, $(6, 2, 129)$, $(6, -2, 129)$, $(19, 3, 41)$, $(19, -3, 41)$ and reject the second and fourth, since they are opposite to the first and third. In article 289 we gave four decompositions of the form $(19, 3, 41)$. From these we get decompositions of the number 770 into $9 + 361 + 400$, $16 + 25 + 729$, $81 + 400 + 289$, $576 + 169 + 25$. Similarly we can find four decompositions of the form $6t^2 + 4tu + 129u^2$ into

$$(t - 8u)^2 + (2t + u)^2 + (t + 8u)^2,$$
$$(t - 10u)^2 + (2t + 5u)^2 + (t + 2u)^2$$
$$(2t - 5u)^2 + (t + 10u)^2 + (t + 2u)^2,$$
$$(2t + 7u)^2 + (t - 8u)^2 + (t - 4u)^2$$

These come respectively from the values $(48, 369)$, $(62, -149)$, $(92, -159)$, $(202, 61)$ of the expression $\sqrt{-(6, -2, 129)}$. As a result we have the decompositions of the number 770 into $225 + 256 + 289$, $1 + 144 + 625$, $64 + 81 + 625$, $16 + 225 + 529$. And there are no decompositions besides these eight.

With regard to the decomposition of numbers into three squares which have common divisors, it follows so easily from the general theory of article 281 that there is no need to recall it here.

▶ 293. The preceding arguments also provide a demonstration of that famous theorem: *any positive integer can be decomposed into three trigonal numbers.* It was discovered by Fermat but until now there has been no rigorous proof for it. It is manifest that any decomposition of the number M into the three trigonals

$$\tfrac{1}{2}x(x + 1) + \tfrac{1}{2}y(y + 1) + \tfrac{1}{2}z(z + 1)$$

will produce a decomposition of the number $8M + 3$ into three odd squares

$$(2x + 1)^2 + (2y + 1)^2 + (2z + 1)^2$$

and vice versa. By the preceding theory any positive integer $8M + 3$ is resolvable into three squares which will necessarily be odd (see the note to art. 291); and the number of resolutions depends both on the number of prime factors of $8M + 3$ and on the number of classes into which the binary forms of determinant $-(8M + 3)$ are distributed. There will be the same number of decompositions of the number M into three trigonal numbers. We have supposed however that for any integral value of x, the number $x(x + 1)/2$ is looked at as a trigonal number; and if we prefer to exclude zero, the theorem should be changed as follows: any positive integer is either trigonal or resolvable into two or three trigonal numbers. A similar change would have to be made in the following theorem if we prefer to exclude zero as a square.

From these same principles we can demonstrate another theorem of Fermat, which says that *any positive integer can be decomposed into four squares.* If we subtract from a number of the form $4n + 2$ any square whatsoever (less than the number), from a number of the form $4n + 1$ an even square, from a number of the form $4n + 3$ an odd square, the residue in all these cases will be resolvable into three squares and the given number therefore into four. Finally, a number of the form $4n$ can be represented as $4^\mu N$ in such a way that N belongs to one of the three preceding forms; and when N is resolved into four squares, $4^\mu N$ will also be resolvable. We could also remove from a number of the form $8n + 3$ the square of an evenly even root, from a number of the form $8n + 7$ the square of an oddly even root, from a number of the form $8n + 4$ an odd square, and the residue will be resolvable into three squares. But this theorem has already been proven by the

illustrious Lagrange, *Nouv. mém. Acad. Berlin*, 1770, p. 123.[12] And
the illustrious Euler explained it more fully (in a manner quite
different from ours) in *Acta. acad. Petrop. 2* [1778], 1780, 48.[13]
There are other theorems of Fermat which are a kind of continuation of the preceding ones. They say that any integer is resolvable
into five pentagonal numbers, six hexagonal numbers, seven
heptagonal numbers, etc. But they still lack proof and they seem
to require different principles for their solution.

▶ 294. THEOREM. *If the numbers a, b, c are relatively prime and
none of them $=0$ nor is divisible by a square, the equation*

$$ax^2 + by^2 + cz^2 = 0 \ldots (\Omega)$$

*cannot be solved by integers (except when $x = y = z = 0$, which
we do not consider), unless $-bc$, $-ac$, $-ab$ respectively are quadratic residues of a, b, c and these numbers do not have different signs;
but when these four conditions hold, (Ω) will be solvable by integers.*

Demonstration. If (Ω) is completely solvable by integers, it will
also be solvable by values of x, y, z which do not have a common
divisor; for any values that satisfy the equation (Ω) will also satisfy
it if they are divided by their greatest common divisor. Now if we
suppose that $ap^2 + bq^2 + cr^2 = 0$ and that p, q, r are free from a
common divisor, they will also be relatively prime to one another;
for if q, r had a common divisor μ, it would be relatively prime to
p, but μ^2 would divide ap^2 and thus also a, contrary to the hypothesis; as a result p, r; p, q must be relatively prime. Now let $-ap^2$ be
represented by the binary form $by^2 + cz^2$ by assigning to y, z the
relatively prime values q, r; thus its determinant $-bc$ will be a
quadratic residue of ap^2 and so also of a (art. 154); in the same
way we will have $-acRb$, $-abRc$. As for the condition that (Ω)
does not admit a resolution if a, b, c have the same sign, it is so
obvious that it needs no explication.

To demonstrate the inverse proposition that constitutes the
second part of the theorem, we will *first* show how to find a
ternary form which is equivalent to $\left(\begin{smallmatrix} a, & b, & c \\ 0, & 0, & 0 \end{smallmatrix}\right) \ldots f$ and so chosen that

---

[12] "Démonstration d'un Théoreme d'Arithmétique."

[13] "Dilucidationes super methodo elegantissima qua illustris de La Grange usus est in
integranda aequatione differentiali $dx/\sqrt{X} = dy/\sqrt{Y}$."

the second, third and fourth coefficients are divisible by abc; and *second*, we will deduce a solution of the equation (Ω) from this.

I. We want three integers A, B, C which do not have a common divisor and so chosen that A is relatively prime to b and c; B is relatively prime to a and c; C is relatively prime to a and b; and $aA^2 + bB^2 + cC^2$ is divisible by abc. We do this in the following way. Let $\mathfrak{A}, \mathfrak{B}, \mathfrak{C}$ respectively be values of the expressions $\sqrt{-bc}$ (mod. a), $\sqrt{-ac}$ (mod. b), $\sqrt{-ab}$ (mod. c). They will necessarily be relatively prime to a, b, c respectively. Now take any three integers $\mathfrak{a}, \mathfrak{b}, \mathfrak{c}$ with the only condition that they be relatively prime to a, b, c respectively (e.g. let all of them $= 1$), and determine A, B, C so that

$$A \equiv \mathfrak{b}c \text{ (mod. } b) \qquad \text{and} \qquad \equiv \mathfrak{c}\mathfrak{C} \text{ (mod. } c)$$

$$B \equiv \mathfrak{c}a \text{ (mod. } c) \qquad \text{and} \qquad \equiv \mathfrak{a}\mathfrak{A} \text{ (mod. } a)$$

$$C \equiv \mathfrak{a}b \text{ (mod. } a) \qquad \text{and} \qquad \equiv \mathfrak{b}\mathfrak{B} \text{ (mod. } b)$$

Then we will have

$$aA^2 + bB^2 + cC^2 \equiv \mathfrak{a}^2(b\mathfrak{A}^2 + cb^2) \equiv \mathfrak{a}^2(b\mathfrak{A}^2 - \mathfrak{A}^2 b) \equiv 0 \text{ (mod. } a)$$

Thus it will be divisible by a and consequently by b, c and so also by abc. It is further evident that A is necessarily relatively prime to b and c; B to a and c; C to a and b. Now if the values A, B, C prove to have a (greatest) common divisor μ, it will necessarily be relatively prime to a, b, c and so also to abc; therefore if we divide those values by μ we will get new ones that do not have a common divisor and which will produce a value of $aA^2 + bB^2 + cC^2$ which is still divisible by abc, and thus satisfies all conditions.

II. If we determine the numbers A, B, C in this way, the numbers Aa, Bb, Cc will not have a common divisor either. For if they did have a common divisor μ, it would necessarily be relatively prime to a (which is of course relatively prime to both Bb and Cc) and similarly to b and c; therefore μ would also have to divide A, B, C, contrary to the hypothesis. We can therefore find integers α, β, γ so that $\alpha Aa + \beta Bb + \gamma Cc = 1$. We also want six integers $\alpha', \beta', \gamma', \alpha'', \beta'', \gamma''$ so that

$$\beta'\gamma'' - \gamma'\beta'' = Aa, \qquad \gamma'\alpha'' - \alpha'\gamma'' = Bb, \qquad \alpha'\beta'' - \beta'\alpha'' = Cc$$

Now let f be transformed by the substitution

$$\begin{array}{ccc} \alpha, & \alpha', & \alpha'' \\ \beta, & \beta', & \beta'' \\ \gamma, & \gamma', & \gamma'' \end{array}$$

into $\left(\begin{smallmatrix} m, & m', & m'' \\ n, & n', & n'' \end{smallmatrix}\right) = g$ (which will be equivalent to f) and I say that m', m'', n will be divisible by abc. For let

$$\beta''\gamma - \gamma''\beta = A', \qquad \gamma''\alpha - \alpha''\gamma = B', \qquad \alpha''\beta - \beta''\alpha = C'$$
$$\beta\gamma' - \gamma\beta' = A'', \qquad \gamma\alpha' - \alpha\gamma' = B'', \qquad \alpha\beta' - \beta\alpha' = C''$$

and we will have

$$\alpha' = B''Cc - C''Bb, \qquad \beta' = C''Aa - A''Cc, \qquad \gamma' = A''Bb - B''Aa$$
$$\alpha'' = C'Bb - B'Cc, \qquad \beta'' = A'Cc - C'Aa, \qquad \gamma'' = B'Aa - A'Bb$$

If we substitute these values in the equations

$$m' = a\alpha'\alpha' + b\beta'\beta' + c\gamma'\gamma'$$
$$m'' = a\alpha''\alpha'' + b\beta''\beta'' + c\gamma''\gamma''$$
$$n = a\alpha'\alpha'' + b\beta'\beta'' + c\gamma'\gamma''$$

we have relative to the modulus a

$$m' \equiv bcA''A''(B^2b + C^2c) \equiv 0$$
$$m'' \equiv bcA'A'(B^2b + C^2c) \equiv 0$$
$$n \equiv bcA'A''(B^2b + C^2c) \equiv 0$$

i.e. m', m'', n will be divisible by a; in a similar way we can show that the same numbers are divisible by b, c and thus find that they are divisible by abc. Q.E.P.

III. Let us suppose for the sake of elegance that the determinant of the forms f, g, i.e. the number $-abc, = d$, then

$$md = M, \qquad m' = M'd, \qquad m'' = M''d, \qquad n = Nd,$$
$$n' = N', \qquad n'' = N''$$

It is clear that f is transformed by the substitution (S)

$$\alpha d, \quad \alpha', \quad \alpha''$$

$$\beta d, \quad \beta', \quad \beta''$$

$$\gamma d, \quad \gamma', \quad \gamma''$$

into the ternary form $\left(\begin{smallmatrix} Md, & M'd, & M''d \\ Nd, & N'd, & N''d \end{smallmatrix}\right) = g'$ of determinant d^3 which therefore will be contained in f. Now I claim that the form $\left(\begin{smallmatrix} d, & 0, & 0 \\ d, & 0, & 0 \end{smallmatrix}\right) = g''$ is necessarily equivalent to g'. For it is clear that $\left(\begin{smallmatrix} M, & M', & M'' \\ N, & N', & N'' \end{smallmatrix}\right) = g'''$ is a ternary form of determinant 1; further, since by hypothesis a, b, c cannot have the same signs, f will be an indefinite form, and we easily conclude that g' and g''' must also be indefinite; therefore g''' will be equivalent to the form $\left(\begin{smallmatrix} 1, & 0, & 0 \\ 1, & 0, & 0 \end{smallmatrix}\right)$ (art. 277), and we can find a transformation (S') from g''' into it; manifestly however (S') will give us a transformation of g' into g''. Therefore g'' will also be contained in f and from a combination of the substitutions $(S), (S')$ we can deduce a transformation of f into g''. If this transformation is

$$\delta, \quad \delta', \quad \delta''$$

$$\varepsilon, \quad \varepsilon', \quad \varepsilon''$$

$$\zeta, \quad \zeta', \quad \zeta''$$

manifestly we have a double solution of the equation (Ω), namely $x = \delta', y = \varepsilon', z = \zeta'$ and $x = \delta'', y = \varepsilon'', z = \zeta''$; it is likewise clear that not all these values can $= 0$ at the same time, since we must have

$$\delta\varepsilon'\zeta'' + \delta'\varepsilon''\zeta + \delta''\varepsilon\zeta' - \delta\varepsilon''\zeta' - \delta'\varepsilon\zeta'' - \delta''\varepsilon'\zeta = d \qquad \text{Q.E.S.}$$

Example. Let the given equation be $7x^2 - 15y^2 + 23z^2 = 0$. It is solvable because $345R7$, $-161R15$, $105R23$. Here the values of $\mathfrak{A}, \mathfrak{B}, \mathfrak{C}$ will be $3, 7, 6$; by letting $\mathfrak{a} = \mathfrak{b} = \mathfrak{c} = 1$ we find that $A = 98$, $B = -39$, $C = -8$. From this we get the substitution

$$\begin{Bmatrix} 3, & 5, & 22 \\ -1, & 2, & -28 \\ 8, & 25, & -7 \end{Bmatrix}$$

by which f is transformed into $\left(\begin{smallmatrix} 1520, & 14490, & -7245 \\ -2415, & -1246, & 4735 \end{smallmatrix}\right) = g$. And as a result we have

$$(S) = \begin{Bmatrix} 7245, & 5, & 22 \\ -2415, & 2, & -28 \\ 19320, & 25, & -7 \end{Bmatrix}, \quad g''' = \begin{pmatrix} 3670800, & 6, & -3 \\ -1, & -1246, & 4735 \end{pmatrix}$$

The form g''' is found to be transformed into $\left(\begin{smallmatrix} 1, & 0, & 0 \\ 1, & 0, & 0 \end{smallmatrix}\right)$ by the substitution

$$\begin{Bmatrix} 3, & 5, & 1 \\ -2440, & -4066, & -813 \\ -433, & -722, & -144 \end{Bmatrix} \dots (S')$$

If we combine this with (S) we get:

$$\begin{Bmatrix} 9, & 11, & 12 \\ -1, & 9, & -9 \\ -9, & 4, & 3 \end{Bmatrix}$$

which will transform f into g''. We have therefore a double solution of the proposed equation $x = 11$, $y = 9$, $z = 4$ and $x = 12$, $y = -9$, $z = 3$; the second solution is made simpler by dividing the values by their common divisor 3, and we have $x = 4$, $y = -3$, $z = 1$.

▶ 295. The second part of the theorem in the preceding article can also be solved as follows. We will look for an integer h such that $ah \equiv \mathfrak{C} \pmod{c}$ (we give the characters $\mathfrak{A}$, $\mathfrak{B}$, $\mathfrak{C}$ the same meaning as in the preceding article) and we will have $ah^2 + b = ci$. It is easy to see that i is an integer and that $-ab$ is the determinant of the binary form $(ac, ah, i) \dots \phi$. This form will certainly not be positive (for, since by hypothesis a, b, c do not have the same signs, ab and ac cannot be positive at the same time); further, it will have the characteristic number -1. We can show this synthetically as follows. Determine the integers e, e' so that

$$e \equiv 0 \pmod{a} \quad \text{and} \quad \equiv \mathfrak{B} \pmod{b}, \quad ce' \equiv \mathfrak{A} \pmod{a}$$

$$\text{and} \quad \equiv h\mathfrak{B} \pmod{b}$$

and (e, e') will be a value of the expression $\sqrt{-(ac, ah, i)}$ (mod. $-ab$). For relative to the modulus a we have

$$e^2 \equiv 0 \equiv -ac, \qquad ee' \equiv 0 \equiv -ah$$

$$c^2 e'e' \equiv \mathfrak{A}^2 \equiv -bc \equiv -c^2 i \quad \text{and so} \quad e'e' \equiv -i$$

but relative to the modulus b we have

$$e^2 \equiv \mathfrak{B}^2 \equiv -ac, \qquad cee' \equiv h\mathfrak{B}^2 \equiv -ach \quad \text{and so} \quad ee' \equiv -ah$$

$$c^2 e'e' \equiv h^2 \mathfrak{B}^2 \equiv -ach^2 \equiv c^2 i \quad \text{and so} \quad e'e' \equiv -i$$

and the same three congruences that hold relative to each of the moduli a, b separately will also hold relative to the modulus ab. Then by the theorem of ternary forms it is easy to conclude that ϕ is representable by the form $(^{-1;\,0;\,0}_{\ 1;\,0;\,0})$. Suppose then that

$$act^2 + 2ahtu + iu^2 = -(\alpha t + \beta u)^2 + 2(\gamma t + \delta u)(\varepsilon t + \zeta u)$$

Multiplying by c we get

$$a(ct + hu)^2 + bu^2 = -c(\alpha t + \beta u)^2 + 2c(\gamma t + \delta u)(\varepsilon t + \zeta u)$$

Now if we give to t, u such values that either $\gamma t + \delta u$ or $\varepsilon t + \zeta u$ will $= 0$, we will have a solution of the equation (Ω) which will be satisfied both by

$$x = \delta c - \gamma h, \qquad y = \gamma, \qquad z = \alpha\delta - \beta\gamma$$

and by

$$x = \zeta c - \varepsilon h, \qquad y = \varepsilon, \qquad z = \alpha\zeta - \beta\varepsilon$$

Manifestly not all of the values in either set can $= 0$ at the same time; for if $\delta c - \gamma h = 0$, $\gamma = 0$ we would also have $\delta = 0$ and $\phi = -(\alpha t + \beta u)^2$ and as a result $ab = 0$, contrary to the hypothesis. Similarly for the other equations. In our example we found that the form ϕ is $(161, -63, 24)$, that the value of the expression $\sqrt{-\phi}$ (mod. 105) $= (7, -51)$, and that the representation of the form ϕ by $(^{-1;\,0;\,0}_{\ 1;\,0;\,0})$ is

$$\phi = -(13t - 4u)^2 + 2(11t - 4u)(15t - 5u)$$

This gives us the solutions $x = 7$, $y = 11$, $z = -8$; $x = 20$, $y = 15$, $z = -5$ or dividing by 5 and neglecting the sign of z, $x = 4$, $y = 3$, $z = 1$.

Of the two methods of solving the equation (Ω), the second is preferable because most often it uses smaller numbers; the former, however, which can be shortened by various devices which we will here omit, seems to be the more elegant, especially because the numbers a, b, c are treated in the same manner and the calculation is not changed by permuting them. It is otherwise in the second method where we have the most convenient calculation if we let a be the smallest and c the largest of the three given numbers, as we have done in our example.

▶ 296. The elegant theorem we have been explaining in the preceding articles was first discovered by the illustrious Legendre (*Hist. Acad. Paris*, 1785, p. 507[14]) and he justified it with a beautiful demonstration (entirely different from our two). At the same time this outstanding geometer tried to derive from it a demonstration of propositions that harmonize with the fundamental theorem of the preceding section, but we have already said in article 151 that it does not seem to be suitable for this purpose. This, therefore, is the place to explain this demonstration (extremely elegant in itself) briefly and to give the reasons for our judgment. We begin with the following observation: *If the numbers a, b, c are all $\equiv 1$ (mod. 4), the equation $ax^2 + by^2 + cz^2 = 0 \dots (\Omega)$ is not solvable.* For it is very easy to see that in this case the value of $ax^2 + by^2 + cz^2$ will necessarily be either $\equiv 1$, or $\equiv 2$, or $\equiv 3$ (mod. 4) unless all the x, y, z are even at the same time; if therefore (Ω) were solvable, this could not happen except by even values of x, y, z. Q.E.A., since any values satisfying the equation (Ω) would still satisfy it if they were divided by their greatest common divisor, so at least one of the values must be odd. Now the different cases of the theorem that must be demonstrated involve the following changes.

I. If p, q are (unequal, positive) prime numbers of the form $4n + 3$ we cannot have pRq, qRp at the same time. For if it were possible, manifestly by letting $1 = a$, $-p = b$, $-q = c$ all the conditions for solving the equation $ax^2 + by^2 + cz^2 = 0$ would be fulfilled (art. 294); but by the preceding observation this equation has no solution; therefore our supposition is inconsistent. From this follows immediately proposition 7 of article 131.

---

[14] Cf. pp. 27, 105, 340.

II. If p is a prime number of the form $4n + 1$, q a prime number of the form $4n + 3$, we cannot have at the same time qRp, pNq. For otherwise we would have $-pRq$ and equation $x^2 + py^2 - qz^2 = 0$ is solvable. But according to our preceding observation it has no solution. From this we derive cases 4 and 5 of article 131.

III. If p, q are prime numbers of the form $4n + 1$, we cannot have at the same time pRq, qNp. Take another prime number r of the form $4n + 3$ which is a residue of q and of which p is a nonresidue. Then by the cases already (II) demonstrated we will have qRr, rNp. If therefore we have pRq, qNp, we would have $qrRp$, $prRq$, $pqNr$ and eventually $-pqRr$. This would make the equation $px^2 + qy^2 - rz^2 = 0$ solvable, contrary to the preceding observation; and the supposition would be inconsistent. From this follow cases 1 and 2 of article 131.

This case can be treated more elegantly in the following way. Let r be a prime number of the form $4n + 3$ of which p is a nonresidue. Then we will also have rNp and therefore (supposing pRq, qNp) $qrRp$; besides, we have $-pRq$, $-pRr$, and so also $-pRqr$ and the equation $x^2 + py^2 - qrz^2 = 0$ would be solvable, contrary to the preceding observation. As a result etc.

IV. If p is a prime number of the form $4n + 1$, q a prime of the form $4n + 3$, we cannot have pRq, qNp at the same time. Take an auxiliary prime number r for the form $4n + 1$ which is a nonresidue of both p, q. Then we will have (by II) qNr and (by III) pNr; therefore $pqRr$; if therefore pRq, qNp we would also have $prNq$, $-prRq$, $qrRp$; thus the equation $px^2 - qy^2 + rz^2 = 0$ would be solvable. Q.E.A. From this we derive cases 3 and 6 of article 131.

V. If p, q are prime numbers of the form $4n + 3$ we cannot have pNq, qNp at the same time. For if we suppose this were possible, and we take the auxiliary prime number r of the form $4n + 1$ which is a nonresidue of both p, q, we will have $qrRp$, $prRq$; further (by II), pNr, qNr and therefore $pqRr$ and $-pqRr$; so the equation $-px^2 - qy^2 + rz^2 = 0$ is possible, contrary to the preceding observation. From this we derive case 8 of article 131.

▶ 297. By examining the preceding demonstration carefully anyone can easily see that cases I and II are absolute, so that there is no room for objection. But the demonstrations of the remaining cases depend on the existence of auxiliary numbers, and since their existence has not yet been proven, the method manifestly loses all its force. Even though these suppositions are so plausible that there seems to be no need of demonstration, and even though they certainly give to the theorem we are trying to prove the highest degree of *probability*, nevertheless, if we want geometric rigor we cannot just accept them gratuitously. As for the supposition in IV and V that there exists a prime number r of the form $4n + 1$ which is a nonresidue of two other given primes p, q, it is easy to conclude from Section IV that all numbers less than $4pq$ and relatively prime to them [their number is $2(p - 1)(q - 1)$] can be equally distributed into four classes. One of them will contain the nonresidues of both p, q, the three remaining will contain the residues of p which are nonresidues of q, the nonresidues of p which are residues of q, and those which are residues of both p, q; and in each class half will be numbers of the form $4n + 1$, half of the form $4n + 3$. Among them therefore there will be $(p - 1)(q - 1)/4$ which are nonresidues of both p, q of the form $4n + 1$. We will designate them by g, g', g'', etc. and the remaining $7(p - 1)(q - 1)/4$ numbers by h, h', h'', etc. Manifestly all numbers contained in the forms $4pqt + g$, $4pqt + g'$, $4pqt + g''$, etc. ... (G) will also be nonresidues of p, q of the form $4n + 1$. Now it is clear that to establish our supposition it is only necessary to show that the forms (G) certainly contain *prime numbers*. This already seems very plausible, since these forms along with the forms $4pqt + h, 4pqt + h'$, etc. ... (H) contain all numbers which are relatively prime to $4pq$ and therefore all absolutely prime numbers (except $2, p, q$); and there is no reason why this series of prime numbers is not distributed equally among these forms so that one-eighth belongs to (G), the rest to (H). But obviously such reasoning is far from geometric rigor. The illustrious Legendre himself confessed that the demonstration of the theorem which asserts that prime numbers are certainly contained in such a form $kt + l$ (where k, l are given relatively prime numbers, t indefinite) is quite difficult, and he suggests in passing a method that may prove useful. It seems to us that many preliminary investigations are necessary before we can arrive at a rigorous demonstration in this way. With regard to the other supposition (III, second method) that there exists a prime number r of the form $4n + 3$ of which another given prime number p of the form $4n + 1$ is a nonresidue, Legendre adds nothing at all. We have shown above (art. 129) that there certainly are prime numbers for which p is a nonresidue, but our method hardly seems convenient for showing that there exist such prime numbers *which are also of the form* $4n + 3$ (as is required here but not in our first demonstration). However, we can easily prove the truth of this proposition as follows. By article 287 there exists a positive category of binary forms of determinant $-p$ whose character is $3, 4$; Np. Let (a, b, c) be such a form and a odd (this is permissible). Then a will be of the form $4n + 3$ and either itself prime or at least it will imply a prime factor r of the form $4n + 3$. We have, however, $-pRa$ and so also $-pRr$ and as a result pNr. But we must note carefully that the propositions of articles 263, 287 depend on the fundamental theorem, and so we would have a vicious circle if we based any part of this discussion on them. Finally the supposition in the first method of III is so much more gratuitous that there is no point in adding more about it here.

Let us add an observation about case V which is indeed not sufficiently proven by the preceding method. It will however be satisfactorily resolved by what follows. If pNq, qNp were true at the same time, we would have $-pRq$, $-qRp$ and it is easy to derive the fact that -1 is a characteristic number of the form $(p, 0, q)$ which could then (according to the theory of ternary forms) be represented by the form $x^2 + y^2 + z^2$. Let

$$pt^2 + qu^2 = (\alpha t + \beta u)^2 + (\alpha' t + \beta' u)^2 + (\alpha'' t + \beta'' u)^2$$

or

$$\alpha^2 + \alpha'\alpha' + \alpha''\alpha'' = p, \qquad \beta^2 + \beta'\beta' + \beta''\beta'' = q,$$

$$\alpha\beta + \alpha'\beta' + \alpha''\beta'' = 0$$

and we will have from equations 1 and 2 that all the numbers $\alpha, \alpha', \alpha'', \beta, \beta', \beta''$ are odd; but then manifestly the third equation cannot be consistent. The second case can be resolved in a manner which is not at all dissimilar to this.

▶ 298. PROBLEM. *Given any numbers whatsoever a, b, c all different from 0: to find the conditions for solving the equation*

$$ax^2 + by^2 + cz^2 = 0 \dots (\omega)$$

Solution. Let $\alpha^2, \beta^2, \gamma^2$ be the largest squares that divide bc, ac, ab respectively and let $\alpha a = \beta\gamma A$, $\beta b = \alpha\gamma B$, $\gamma c = \alpha\beta C$. Then A, B, C will be integers which are relatively prime to one another; the equation (ω) will be solvable or not according as

$$AX^2 + BY^2 + CZ^2 = 0 \dots (\Omega)$$

does or does not admit a solution according to the norms of article 294.

Demonstration. Set $bc = \mathfrak{A}\alpha^2$, $ac = \mathfrak{B}\beta^2$, $ab = \mathfrak{C}\gamma^2$. $\mathfrak{A}, \mathfrak{B}, \mathfrak{C}$ will be integers free from square factors and $\mathfrak{A} = BC$, $\mathfrak{B} = AC$, $\mathfrak{C} = AB$; as a result $\mathfrak{A}\mathfrak{B}\mathfrak{C} = (ABC)^2$ and so $ABC = A\mathfrak{A} = B\mathfrak{B} = C\mathfrak{C}$ is necessarily an integer. Let m be the greatest common divisor of the numbers $\mathfrak{A}, A\mathfrak{A}$. Then $\mathfrak{A} = gm$, $A\mathfrak{A} = hm$ and g will be relatively prime to h and (because $\mathfrak{A}$ is free from quadratic factors) to m. Now we have $h^2m = gA^2\mathfrak{A} = g\mathfrak{B}\mathfrak{C}$ so g divides h^2m, which is obviously impossible unless $g = \pm 1$. Thus $\mathfrak{A} = \pm m$, $A = \pm h$ and therefore integral. Consequently B, C will also be integers. Q.E.P. Since $\mathfrak{A} = BC$ has no quadratic factors, B, C must be relatively prime; and similarly A will be relatively prime to C and to B. Q.E.S. Finally, if $X = P$, $Y = Q$, $Z = R$ satisfy equation (Ω), equation (ω) will be solved by $x = \alpha P$, $y = \beta Q$, $z = \gamma r$; conversely if (ω) is satisfied by $x = p$, $y = q$, $z = r$, (Ω) will be satisfied by $X = \beta\gamma p$, $Y = \alpha\gamma q$, $Z = \alpha\beta r$ and so either both or neither will be solvable. Q.E.T.

▶ 299. PROBLEM. *Given the ternary form*

$$f = ax^2 + a'x'x' + a''x''x'' + 2bx'x'' + 2b'xx'' + 2b''xx'$$

to find whether zero can be represented by this form (with not all values of the unknowns $=0$ at the same time).

Solution. I. When $a = 0$ the values of x', x'' can be taken arbitrarily and it is clear from the equation

$$a'x'x' + 2bx'x'' + a''x''x'' = -2x(b'x'' + b''x')$$

that x will assume a determined rational value; whenever in this way we get a fraction as the value of x, we need only multiply the values of x, x', x'' by the denominator of the fraction and we will have integers. The only values of x', x'' that must be excluded are those that make $b'x'' + b''x' = 0$ unless they also make $a'x'x' + 2bx'x'' + a''x''x'' = 0$. In this case x can be chosen arbitrarily. Thus all possible solutions can be obtained. But the case where b' and $b'' = 0$ does not belong here, for then x does not enter into the determination of f; that is, f is a binary form and the representability of zero by f should be decided from the theory of such forms.

II. When we do not have $a = 0$, the equation $f = 0$ will be equivalent to

$$(ax + b''x' + b'x'')^2 - A''x'x' + 2Bx'x'' - A'x''x'' = 0$$

by letting

$$b''b'' - aa' = A'', \qquad ab - b'b'' = B, \qquad b'b' - aa'' = A'$$

Now when $A' = 0$, and B is not $= 0$, it is manifest that if we take $ax + b''x' + b'x''$ and x'' arbitrarily, x and x' will be determined to be rational numbers, and when they are not integers, they can be made such by a suitable multiplication. For one value of x'', namely, for $x'' = 0$, the value of $ax + b''x' + b'x''$ is not arbitrary but must also $= 0$; but then x' can be taken with complete freedom and it will produce a rational value of x. When A'' and $B = 0$ at the same time, it is clear that if A' is a square $= k^2$, the equation $f = 0$ is reduced to these two linear equations (one *or* the other of them must be true)

$$ax + b''x' + (b' + k)x'' = 0, \qquad ax + b''x' + (b' - k)x'' = 0$$

but if (under the same hypothesis) A' is nonquadratic, the solution of the proposed equation depends on the following (*both* of which must hold): $x'' = 0$ and $ax + b''x' = 0$.

It is scarcely necessary to observe that the method of I also applies when a' or $a'' = 0$ and the method of II when $A' = 0$.

III. When neither a nor $A'' = 0$, the equation $f = 0$ will be equivalent to

$$A''(ax + b''x' + b'x'')^2 - (A''x' - Bx'')^2 + Dax''x'' = 0$$

where D is the determinant of the form f or Da is the number $B^2 - A'A''$. When $D = 0$ we will have a solution like the one at the end of the preceding case; that is, if A'' is a square $= k^2$, the proposed equation is reduced to these

$$kax + (kb'' - A'')x' + (kb' + B)x'' = 0$$
$$kax + (kb'' + A'')x' + (kb'' - B)x'' = 0$$

but if A'' is nonquadratic, we must have

$$ax + b''x' + b'x'' = 0, \qquad A''x' - Bx'' = 0$$

When, however, D is not $=0$, we are reduced to the equation

$$A''t^2 - u^2 + Dav^2 = 0$$

whose possibility can be judged by the preceding article. And if this cannot be solved unless we let $t = 0$, $u = 0$, $v = 0$, the proposed equation cannot be solved unless we let $x = 0$, $x' = 0$, $x'' = 0$; but if it otherwise solvable, from any set of integral values of t, u, v we can, by the equations

$$ax + b''x' + b'x'' = t, \qquad A''x' - Bx'' = u, \qquad x'' = v$$

derive at least rational values for x, x', x''. If these involve fractions, we can make them integers by a suitable multiplication.

As soon as *one* solution of the equation $f = 0$ by integers is found, the problem is reduced to case I and all solutions can be found in the following way. Let the values of x, x', x'' which satisfy the equation $f = 0$ be $\alpha, \alpha', \alpha''$. We will suppose that they are free from common factors. Now (by art. 40, 279) choose integers $\beta, \beta', \beta'', \gamma, \gamma', \gamma''$ in such a way that

$$\alpha(\beta'\gamma'' - \beta''\gamma') + \alpha'(\beta''\gamma - \beta\gamma'') + \alpha''(\beta\gamma' - \beta'\gamma) = 1$$

and by the substitution

$$x = \alpha y + \beta y' + \gamma y'', \qquad x' = \alpha'y + \beta'y' + \gamma'y'',$$
$$x'' = \alpha''y + \beta''y' + \gamma''y'' \qquad (S)$$

let f be transformed into

$$g = cy^2 + c'y'y' + c''y''y'' + 2dy'y'' + 2d'yy'' + 2d''yy'$$

Then we will have $c = 0$ and g will be equivalent to f. It is easy to conclude that all solutions by integers of the equation $f = 0$ can be derived (by S) from all solutions of the equation $g = 0$.

And by I all solutions of the equation $g = 0$ are contained in the formulae

$$y = -z(c'p^2 + 2dpq + c''q^2), \qquad y' = 2z(d''p^2 + d'pq),$$
$$y'' = 2z(d''pq + d'q^2)$$

where p, q are indefinite integers, z an indefinite number which can be a fraction as long as y, y', y'' are integers. If we substitute these values of y, y', y'' into (S), we will have all solutions of the equation $f = 0$ by integers. Thus, e.g., if

$$f = x^2 + x'x' + x''x'' - 4x'x'' + 2xx'' + 8xx'$$

and one solution of the equation $f = 0$ is $x = 1, x' = -2, x'' = 1$: by setting $\beta, \beta', \beta'', \gamma, \gamma', \gamma'' = 0, 1, 0, 0, 0, 1$ we have

$$g = y'y' + y''y'' - 4y'y'' + 12yy''$$

So all solutions of the equation $g = 0$ by integers will be contained in the formula

$$y = -z(p^2 - 4pq + q^2), \qquad y' = 12zpq, \qquad y'' = 12zq^2$$

and all solutions of the equation $f = 0$ in the formula

$$x = -z(p^2 - 4pq + q^2)$$
$$x' = 2z(p^2 + 2pq + q^2)$$
$$x'' = -z(p^2 - 4pq - 11q^2)$$

▶ 300. From the problem of the preceding article we have immediately a solution of the indeterminate equation

$$ax^2 + 2bxy + cy^2 + 2dx + 2ey + f = 0$$

if we want only rational values. We have already solved it above (art. 216 et sqq.) for integral values. All rational values of x, y can be represented by $t/v, u/v$ where t, u, v are integers. Thus it is clear that the solution of this equation by rational numbers is identical with the solution by integers of the equation

$$at^2 + 2btu + cu^2 + 2dtv + 2euv + fv^2 = 0$$

And this coincides with the equation treated in the preceding article. We exclude only those solutions where $v = 0$; but there can be none of these when $b^2 - ac$ is a nonquadratic number.

Thus, e.g., all solutions by rational numbers of the equation (solved in general by integers in art. 221)

$$x^2 + 8xy + y^2 + 2x - 4y + 1 = 0$$

will be contained in the formula

$$x = \frac{p^2 - 4pq + q^2}{p^2 - 4pq - 11q^2}, \qquad y = -\frac{2p^2 + 4pq + 2q^2}{p^2 - 4pq - 11q^2}$$

where p, q are any integers. But we have treated only briefly here these two problems which are so intimately connected, and we have omitted many pertinent observations in order not to be too prolix. We have another solution of the problem of the preceding article based on general principles, which, however, must be reserved to another occasion because it demands a more penetrating examination of ternary forms.

▶ 301. We return now to the consideration of binary forms, concerning which we still have many remarkable properties to examine. First we will add some observations about the number of classes and genera in a properly primitive order (positive if the determinant is negative), and for the sake of brevity we will have to restrict our investigation to these.

The *number of genera* into which all (properly primitive positive) forms with a given positive or negative determinant $\pm D$ are distributed is always 1, 2, 4 or a higher power of the number 2. Its exponent depends on the factors of D, and it can be found a priori by the arguments that we presented above. Now, since in a series of natural numbers prime numbers are mixed with more or less composite numbers, it happens that for many successive determinants $\pm D$, $\pm(D + 1)$, $\pm(D + 2)$, etc. the number of genera now increases, now decreases, and there seems to be no order in this series. Nevertheless if we add the numbers of genera corresponding to many successive determinants

$$\pm D, \pm(D + 1), \ldots \pm(D + m)$$

and divide the sum by the number of determinants, we get the *average number of genera*. We can think of this as applying to about the middle determinant $\pm[D + (m/2)]$ and it establishes a very regular progression. We have supposed, however, that not only is m sufficiently large, but also that D is much larger, so that the ratio of the extreme determinants $D, D + m$ does not differ too much from equality. This is what we mean by the regularity of this progression: if D' is a number much greater than D, the average number of genera around the determinant $\pm D'$ will be noticeably greater than that around D; and if D, D' do not differ by much, the average numbers of genera around D and D' will be about equal. But the average number of genera around the positive determinant $+ D$ will always be about equal to the average number around the corresponding negative determinant, and the greater D is, the more exactly this will be true, since for small values the former is a little larger than the latter. These observations will be better illustrated by the following examples taken from a table which classifies more than 4000 binary forms. Among the hundred determinants from 801 to 900 there are 7 which correspond to only one genus, 32, 52, 8, 1 which correspond respectively to 2, 4, 8, 16 genera. There are in all 359 genera and the average number $= 3.59$. The hundred negative determinants from -801 to -900 produce 360 genera. The following examples all have to do with negative determinants. In the sixteenth hundred (from -1501 to -1600) the average number of genera is 3.89; in the twenty-fifth it is 4.03; in the fifty-first it is 4.24; for the 600 determinants from -9401 to -10000 it is 4.59. From these examples it is clear that the average number of genera decreases much more slowly than the determinants themselves; but we want to know the law of this progression. By a theoretical discussion so difficult that it would be too long to explain here, we have found that the average number of genera around the determinant $+D$ or $-D$ can be calculated approximately from the formula

$$\alpha \log D + \beta$$

where α, β are constant quantities and indeed

$$\alpha = \frac{4}{\pi^2} = 0.4052847346$$

(π is half the circumference of a circle of radius 1),

$$\beta = 2\alpha g + 3\alpha^2 h - \tfrac{1}{6}a \log 2 = 0.8830460462$$

where g is the sum of the series

$$1 - \log(1 + 1) + \tfrac{1}{2} - \log(1 + \tfrac{1}{2}) + \tfrac{1}{3} - \log(1 + \tfrac{1}{3}) + \text{etc.}$$
$$= 0.5772156649$$

(see Euler, *Inst. Calc. Diff.*, p. 444) and h is the sum of the series

$$\tfrac{1}{4}\log 2 + \tfrac{1}{9}\log 3 + \tfrac{1}{16}\log 4 + \text{etc.}$$

which approximately $= 0.9375482543$. From this formula it is clear that the average number of genera increases in an arithmetic progression if the determinants increase in a geometric progression. The values of this formula for $D = 850/2$, $1550/2$, $2450/2$, $5050/2$, $9700/2$ are found to be 3.617, 3.86, 4.046, 4.339, 4.604, which differ little from the average numbers given above. The greater the middle determinant and the larger the number of determinants from which the average is computed, the less the true value will differ from the formula. With the help of this formula we can also find approximately the sum of the numbers of genera for successive determinants $\pm D$, $\pm(D + 1), \ldots \pm(D + m)$ by adding together the average numbers corresponding to each, no matter how far apart are the extremes $D, D + m$. This sum will be

$$= \alpha[\log D + \log(D + 1) + \text{etc.} + \log(D + m)] + \beta(m + 1)$$

or with reasonable exactitude

$$= \alpha[(D + m)\log(D + m) - (D - 1)\log(D - 1)] + (\beta - \alpha)(m + 1)$$

In this way the sum of the number of genera for the determinants -1 to -100 is found to be $= 234.4$, whereas it is actually 233; similarly, from -1 to -2000 the formula gives us 7116.6, whereas it is actually 7112; from -9001 to -10000 it is 4595 and the formula gives us 4594.9. We could not hope for a better approximation.

▶ 302. With respect to the *number of classes* (we always presume that they are properly primitive positive) positive determinants act in an entirely different way from negative determinants; we shall therefore consider each of them separately. They do agree in that for a given determinant there is an equal number of classes in each genus, and therefore the number of all classes is equal to the product of the number of genera times the number of classes contained in each.

First, with regard to negative determinants the number of classes corresponding to several successive determinants, $-D$, $-(D + 1)$, $-(D + 2)$, etc. forms a progression that is equally as irregular as the number of genera. The average number of classes however (there is no need of a definition) increases very regularly, as will be evident from the following examples. The hundred determinants from -500 to -600 produce 1729 classes, so the average number $= 17.29$. Similarly, in the fifteenth hundred the average number of classes is 28.26; for the twenty-fourth and twenty-fifth hundreds we have 36.28; for the sixty-first, the sixty-second, and sixty-third we have 58.50; for each of the five hundreds from 91 to 95 we have 71.56; finally for the five from 96 to 100 we have 73.54. These examples show that the average number of classes increases more slowly than the determinants but much more quickly than the average number of genera; a little attention shows that it increases almost exactly in proportion to the square roots of the middle determinants. As a matter of fact we have found by a theoretical investigation that the average number of classes for the determinant $-D$ can be expressed approximately by

$$\gamma\sqrt{D} - \delta$$

where

$$\gamma = 0.7467183115 = \frac{2\pi}{7e}$$

where e is the sum of the series

$$1 + \tfrac{1}{8} + \tfrac{1}{27} + \tfrac{1}{64} + \tfrac{1}{125} + \text{etc.}$$

$$\delta = 0.2026423673 = \frac{2}{\pi^2}$$

The average values computed by this formula differ little from those which we have recorded above from a table of classifications. With the help of this formula we can also find approximately the number of all (properly primitive positive) classes corresponding to the successive determinants $-D$, $-(D + 1)$, $-(D + 2), \ldots$ $-(D + m - 1)$ no matter how far apart the extremes are, by adding together the average number corresponding to these determinants according to the formula. Thus we will have a number

$$= \gamma[\sqrt{D} + \sqrt{(D + 1)} + \text{etc.} + \sqrt{(D + m - 1)}] - \delta m$$

or approximately

$$= \tfrac{2}{3}\gamma[(D + m - \tfrac{1}{2})^{3/2} - (D - \tfrac{1}{2})^{3/2}] - \delta m$$

Thus, e.g., from the formula the sum for the hundred determinants -1 to -100 will $=481.1$ whereas it is actually 477; the thousand determinants -1 to -1000 according to the table will give 15533 classes, from the formula 15551.4; the second thousand according to the table gives 28595 classes, the formula 28585.7; similarly the third thousand actually has 37092 classes, the formula gives 37074.3; the tenth thousand has 72549 by the table, 72572 by the formula.

303. The table of negative determinants arranged according to the various classifications offers many other remarkable observations. For determinants of the form $-(8n + 3)$ the number of classes (both the total number and the number contained in each properly primitive genus) is always divisible by 3 except for the single determinant, -3. The reason for this is clear from article 256.VI. For those determinants whose forms are contained in only one class, the number of classes is always odd; for, since for such a determinant there is only one ambiguous class—the principal one—the number of other classes which are always opposed in pairs is necessarily even and thus the total number is odd; this latter property also holds for positive determinants. Further, the series of determinants corresponding to the same given classification (i.e. the given number of both genera and classes) always seems to terminate with a finite number. We illustrate this rather remarkable observation by some examples. (The Roman numeral indicates the number of properly primitive positive genera; the Arabic numeral the number of classes contained in each genus; then follows a series of determinants which correspond to this classification. For brevity we omit the negative sign.)

I.   1 ... 1, 2, 3, 4, 7
I.   3 ... 11, 19, 23, 27, 31, 43, 67, 163

I.   5 ... 47, 79, 103, 127
I.   7 ... 71, 151, 223, 343, 463, 487
II.   1 ... 5, 6, 8, 9, 10, 12, 13, 15, 16, 18, 22, 25, 28, 37, 58
II.   2 ... 14, 17, 20, 32, 34, 36, 39, 46, 49, 52, 55, 63, 64, 73, 82, 97,
            100, 142, 148, 193
IV.   1 ... 21, 24, 30, 33, 40, 42, 45, 48, 57, 60, 70, 72, 78, 85, 88, 93,
            102, 112, 130, 133, 177, 190, 232, 253
VIII. 1 ... 105, 120, 165, 168, 210, 240, 273, 280, 312, 330, 345, 357,
            385, 408, 462, 520, 760
XVI. 1 ... 840, 1320, 1365, 1848

Similarly, there are 20 determinants (the largest $= -1423$) which correspond to the classification I.9; 4 (the largest $= -1303$) which correspond to the classification I.11 etc.; the classifications II.3, II.4, II.5, IV.2 correspond to not more than 48, 31, 44, 69 determinants respectively. The greatest are $-652, -862, -1318, -1012$. Since the table from which we drew these examples has been extended[u] far beyond the largest determinants that occur here, and since it furnishes no others belonging to these classes, there seems to be no doubt that the preceding series does in fact terminate, and by analogy it is permissible to extend the same conclusion to any other classifications. For example, since in the whole tenth thousand of the determinants there is none corresponding to fewer than 24 classes, it is extremely probable that the classifications I.23, I.21, etc., II.11, II.10 etc., IV.5, IV.4, IV.3, VIII.2 are all complete before we reach the number -9000, or at least that they have very few determinants greater than -10000. However, *rigorous* proofs of these observations seem to be very difficult. It is no less worthy of notice that all determinants whose forms are distributed into 32 or more genera have at least two classes in each genus, and thus that the classifications XXXII.1, LXIV.1, etc. do not exist at all (the smallest determinant among these is -9240 and it corresponds to the classification XXXII.2); and it seems quite probable that as the number of genera increases, more classifications will disappear. In this respect the 65 determinants given above for the classifications I.1, II.1, IV.1, VIII.1,

<hr/>

[u] While this was being printed we worked out the table up to -3000, and also for the whole tenth thousand, for many separate hundreds, and for many carefully selected individual determinants.

XVI.1 are quite exceptional, and it is easy to see that they and they alone enjoy two remarkable properties: all the classes of the forms belonging to them are ambiguous, and all forms contained in the same genus are both properly and improperly equivalent. The illustrious Euler (*Nouv. mém. Acad. Berlin*, 1776, p. 338[15]) has already singled out these 65 numbers (under a slightly different aspect and with a criterion that is easy to demonstrate; we will mention it later on).

▶ 304. The number of properly primitive classes which are formed by binary forms with a positive *square* determinant, k^2, can be completely determined a priori. It is equal to the number of numbers which are relatively prime to $2k$ and less than it; from this fact and by a reasoning that is not difficult, but which we omit here, we deduce that the average number of classes around k^2 belonging to such determinants is approximately equal to $8k/\pi^2$. In this respect, however, positive nonquadratic determinants present singular phenomena. That is, there is only a small number of classes for small negative and quadratic determinants, e.g., classification I.1 or I.3 or II.1 etc., and the series ceases altogether in a very short time; for positive nonquadratic determinants, on the contrary, as long as they are not too large, by far the greatest number of them produce classifications in which only one class is contained in each genus. Thus classifications like I.3, I.5, II.2, II.3, IV.2, etc. are very rare. For example, among the 90 nonquadratic determinants below 100 we find 11, 48, 27 which correspond to the classifications I.1, II.1, IV.1 respectively; only one (37) has I.3; two (34 and 82) have II.2; one (79) has II.3. Nevertheless, as the determinants increase, larger numbers of classes appear and do so more frequently; thus among the 96 nonquadratic determinants from 101 to 200, two (101, 197) have classification I.3; four (145, 146, 178, 194) have II.2; three (141, 148, 189) have II.3. Among the 197 determinants from 801 to 1000, three have I.3; four II.2; fourteen have II.3; two have II.5; two have II.6; fifteen have IV.2; six have IV.3; two have IV.4; four have VIII.2. The remaining 145 have one class in each genus. It is a curious question and it would not be unworthy of a geometer's talent to investigate the law that

---

[15] "Extrait d'un lettre de M. Euler à M. Beguelin en Mai 1778."

governs the fact that determinants having one class in a genus
become increasingly rare. Up to the present we cannot decide
theoretically nor is it possible to confirm by observation whether
there is only a finite number of them (this hardly seems probable),
or that they occur *infinitely rarely*, or that their frequency tends
more and more to a fixed limit. The average number of classes
increases by a ratio that is hardly greater than that of the number
of genera and far more slowly than the square roots of the determinants. Between 800 and 1000 we find that it $= 5.01$. We can
add to these observations another that restores somewhat the
analogy between positive and negative determinants: we have
found that for a positive determinant D it is not so much the
number of classes but this number multiplied by the logarithm
of the quantity $t + u\sqrt{D}$ (t, u are the smallest numbers outside of
$1, 0$ which satisfy the equation $t^2 - Du^2 = 1$) that is analogous
to the number of classes for a negative determinant. We cannot
explain this more thoroughly, but the average value of that
product is approximately expressed by a formula like $m\sqrt{(D - n)}$.
However, we have not thus far been able to determine the values
of the constant quantities m, n theoretically. If it is permissible
to draw a conclusion from the comparison of some hundreds of
determinants, m seems to be very nearly $7/3$. We will reserve for
another occasion a more complete discussion of the principles
of the preceding discussion concerning average values of quantities which do not follow an analytic law but rather approximate
such a law asymptotically. We pass on now to another discussion
concerning the comparison among themselves of different properly
primitive classes of the same determinant.

▶ 305. THEOREM. *If K is a principal class of forms of a given
determinant D, and C is any other class of a principal genus of the
same determinant ; and if $2C$, $3C$, $4C$, etc. are the classes that result
(as in art. 249) from the duplication, triplication, quadruplication,
etc. of class C: then by continuing the progression C, $2C$, $3C$, etc.
long enough we will finally arrive at a class which is identical with
K ; and if we suppose that mC is the first that is identical with K
and that the number of classes in the principal genus $= n$, then we
will have either $m = n$, or m will be a factor of n.*

Demonstration. I. Since all the classes K, C, $2C$, $3C$, etc.

necessarily belong to a principal genus (art. 247), the first $n + 1$ classes of this series, $K, C, 2C, \ldots nC$, manifestly cannot all be different. Therefore either K will be identical with one of the classes $C, 2C, 3C, \ldots nC$ or at least two of these classes are identical with each other. Let $rC = sC$ and $r > s$. We will have also

$$(r - 1)C = (s - 1)C, \qquad (r - 2)C = (s - 2)C, \text{ etc.}$$

and

$$(r + 1 - s)C = C$$

therefore $(r - s)C = K$. Q.E.P.

II. It also follows from this that either $m = n$ or $m < n$, and it remains only to show that in the first case m is a factor of n. Since the classes

$$K, C, 2C, \ldots (m - 1)C$$

which complex we will designate by $\mathfrak{C}$ do not completely exhaust the principal genus, let C' be a class of this genus not contained in $\mathfrak{C}$. Now let $\mathfrak{C}'$ designate the complex of classes which result from the composition of C' with the individual classes of C, namely

$$C', C' + C, C' + 2C, \ldots C' + (m - 1)C$$

Now obviously all the classes in $\mathfrak{C}'$ will be different from one another and from all the classes in $\mathfrak{C}$ and will belong to the principal genus; if $\mathfrak{C}$ and $\mathfrak{C}'$ completely exhaust this genus, we will have $n = 2m$; otherwise $2m < n$. In the latter case let C'' be any class of the principal genus which is not contained in either $\mathfrak{C}$ or in $\mathfrak{C}'$, and designate by $\mathfrak{C}''$ the complex of classes resulting from the composition of C'' with the individual classes of $\mathfrak{C}$; i.e.

$$C'', C'' + C, C'' + 2C, \ldots C'' + (m - 1)C$$

and it is clear that all of these are different from one another and from all the classes in $\mathfrak{C}$ and $\mathfrak{C}'$ and will belong to the principal genus. Now if $\mathfrak{C}, \mathfrak{C}', \mathfrak{C}''$ exhaust this genus, we have $n = 3m$; otherwise $n > 3m$. In this case there is another class C''' contained in the principal genus and not in $\mathfrak{C}, \mathfrak{C}'$ or $\mathfrak{C}''$. In a similar way we find that $n = 4m$ or $n > 4m$ and so on. Now since n and m are finite numbers, the principal genus must eventually be exhausted, and n will be a multiple of m, or m a factor of n. Q.E.S.

Example. Let $D = -356$, $C = (5, 2, 72)$.^{v} We will have $2C = (20, 8, 21)$, $3C = (4, 0, 89)$, $4C = (20, -8, 21)$, $5C = (5, -2, 72)$, $6C = (1, 0, 356)$. Here $m = 6$ and for this determinant n is 12. If we take $(8, 2, 45)$ for the class C', the remaining five classes in $\mathfrak{C}'$ will be $(9, -2, 40)$, $(9, 2, 40)$, $(8, -2, 45)$, $(17, 1, 21)$, $(17, -1, 21)$.

306. The demonstration of the preceding theorem is quite analogous to the demonstrations in articles 45, 49 and, in fact, the theory of the multiplication of classes has a great affinity in every way with the argument given in Section III. But the limits of this work do not permit us to pursue this theory though it is worthy of greater development; we will add only a few observations here, suppress those demonstrations that require too much detail, and reserve a more complete discussion to another occasion.

I. If the series $K, C, 2C, 3C$, etc. is extended beyond $(m - 1)C$ we will get the same classes over again

$$mC = K, \quad (m + 1)C = C, \quad (m + 2)C = 2C, \text{ etc.}$$

and in general (for elegance taking K as $0C$) the classes $gC, g'C$ will be identical or different, according as g and g' are congruent or noncongruent relative to the modulus m. Therefore the class nC will always be identical with the principal class K.

II. The complex of classes $K, C, 2C, \ldots (m - 1)C$ which we designated above by $\mathfrak{C}$ will be called the *period* of the class C. This expression must not be confused with the *periods* of reduced *forms* of a nonquadratic positive determinant as treated in article 186. It is clear therefore that from a composition of any number of classes contained in the same period we will get a class that is also contained in that period

$$gC + g'C + g''C \text{ etc.} = (g + g' + g'' + \text{etc.})C$$

III. Since $C + (m - 1)C = K$, the classes C and $(m - 1)C$ will be opposite and so also $2C$ and $(m - 2)C$, $3C$ and $(m - 3)C$, etc. Therefore, if m is even, the class $mC/2$ will be opposite to itself and so *ambiguous*; conversely, if in $\mathfrak{C}$ there occurs any class other than K which is ambiguous, say gC, we will have $gC = (m - g)C$ and so $g = m - g = m/2$. It follows that if m is even, there cannot be any other ambiguous class in $\mathfrak{C}$ except K and $mC/2$; if m is odd, none except K alone.

---

^{v} We will always express the classes by the (simplest) *forms* contained in them.

IV. If we suppose that the period of any class hC contained in $\mathfrak{C}$ is

$$K, hC, 2hC, 3hC, \ldots (m' - 1)hC$$

it is manifest that $m'h$ is the smallest multiple of h divisible by m. If therefore m and h are relatively prime, we will have $m' = m$, and the two periods will contain the same classes but arranged in a different order. In general, however, if we let μ be the greatest common divisor of m, h we have $m' = m/\mu$. Thus it is clear that the number of classes contained in the period of any class of $\mathfrak{C}$ will be either m or a factor of m; and indeed there will be as many classes in $\mathfrak{C}$ having periods of m terms as there are numbers from the series $0, 1, 2, \ldots m - 1$ which are relatively prime to m, or ϕm if we use the symbol of article 39; and in general there will be as many classes in $\mathfrak{C}$ which have periods of m/μ terms as there are numbers from the series $0, 1, 2, \ldots m - 1$ having μ as the greatest common divisor of themselves and m. It is easy to see that the number of these will be $\phi(m/\mu)$. If therefore $m = n$, or the *whole* principal genus is contained in $\mathfrak{C}$, there will be in this genus ϕn classes in all whose periods include the whole genus, and ϕe classes whose periods will consist of e terms, where e is any divisor of n. This conclusion is true in general when we have any class in a principal genus whose period consists of n terms.

V. Under the same supposition, the best way to handle a system of classes of the principal genus is to take as *base* a class that has a period of n terms, and then to place the classes of the principal genus in the same order in which they appear in this period. Now if we assign the *index* 0 to the principal class, the index 1 to the class which we took as base, etc.: then by only adding indices we can tell which class will result from a composition of any classes of the principal genus. Here is an example for the determinant -356 where we take the class $(9, 2, 40)$ as base:

0	(1,	0,	356)	4	(20,	8,	21)	8	(20,	−8,	21)
1	(9,	2,	40)	5	(17,	1,	21)	9	(8,	2,	45)
2	(5,	2,	72)	6	(4,	0,	89)	10	(5,	−2,	72)
3	(8,	−2,	45)	7	(17,	−1,	21)	11	(9,	−2,	40)

VI. Although analogy with Section III and an induction on more than 200 negative determinants and many more positive nonquadratic determinants would seem to give the highest probability that the supposition is true for *all* determinants, nevertheless such a conclusion would be false, and would be disproven by a continuation of the table of classifications. For the sake of brevity we will call *regular* those determinants for which the whole principal genus can be included in one period, and *irregular* those for which this is not true. We can illustrate with only a few observations this argument which depends on the most profound mysteries of higher Arithmetic and involves the most difficult investigation. We will begin with the following general statement.

VII. If C, C' are classes of the principal genus, if their periods consist of m, m' classes, and if M is the smallest number divisible by m and m': then there will be classes in the same genus whose periods contain M terms. Resolve M into two relatively prime factors r, r' where one (r) divides m, the other (r') divides m' (see art. 73), and the class $(mC/r) + (m'C'/r') = C''$ will have the desired property. For suppose that the period of the class C'' consists of g terms, we will have

$$K = grC'' = gmC + \frac{grm'}{r'}C' = K + \frac{grm'}{r'}C' = \frac{grm'}{r'}C'$$

so grm'/r' must be divisible by m' or gr by r' and thus g by r'. Similarly we find that g will be divisible by r and therefore by $rr' = M$. But since $MC'' = K$, M will be divisible by g, and necessarily $M = g$. It follows from this that the *greatest* number of classes (for a given determinant) contained in any period is divisible by the number of classes in any other period (of a class in the same principal genus). We can also derive here a method for finding the class that has the largest period (for a regular determinant this period will include the whole principal genus). This method is completely analogous to that of articles 73, 74, but in practice we can shorten the labor by various devices. If we divide the number n by the number of classes in the largest period, we will get the number 1 for regular determinants and an integer greater than 1 for irregular determinants. This quotient is suitable for expressing the different kinds of irregularity. For this reason we will call it the *exponent of irregularity.*

VIII. Until now we have not had a general rule by which we can distinguish a priori regular determinants from irregular ones, especially since among the latter there are both prime and composite numbers. It will be sufficient therefore to append some particular observations here. When more than two ambiguous classes are found in the principal genus, the determinant is certainly irregular and the exponent of irregularity is even; but when the genus has only one or two, the determinant will be regular or at least the exponent of irregularity will be odd. All negative determinants of the form $-(216k + 27)$ with the exception of -27, are irregular and the exponent of irregularity is divisible by 3; the same is true for negative determinants of the form $-(1000k + 75)$ and $-(1000k + 675)$ with the exception of -75, and for an infinity of others. If the exponent of irregularity is a prime number p, or at least divisible by p, n will be divisible by p^2. It follows from this that if n admits of no quadratic divisor, the determinant is certainly regular. It is only for positive *square* determinants e^2 that we can determine a priori whether they are regular or irregular; they are regular when e is 1 or 2 or an odd prime number or the power of an odd prime number; in all other cases they are irregular. For negative determinants, irregulars are continually more frequent as the determinants become larger; e.g. in the whole first thousand we find 13 irregulars (omitting the negative sign) 576, 580, 820, 884, 900 whose exponent of irregularity is 2, and 243, 307, 339, 459, 675, 755, 891, 974 whose exponent of irregularity is 3; in the tenth thousand there are 31 with exponent of irregularity 2, and 32 with exponent of irregularity 3. We cannot yet decide whether determinants with exponent of regularity greater than 3 occur below -10000; beyond this limit we can find determinants with any given exponent. It is quite probable that as determinants increase in size the frequency of irregular negative determinants approaches a constant ratio with respect to the frequency of regulars. The determination of this ratio is indeed worthy of the talents of geometers. For nonquadratic positive determinants irregulars are much more rare; there are certainly infinitely many whose exponent of irregularity is even (e.g. 3026 for which it is 2); and there seems to be no doubt that there are some whose exponent of irregularity is odd, although we must confess that none has come to our attention thus far.

IX. For brevity's sake we cannot treat here of the most useful disposition of the system of classes contained in a principal genus with an irregular determinant; we observe only that, since one base is not sufficient, we must take two or more classes and from their multiplication and composition produce all the rest. Thus we will have *double or multiple indices* which will perform the same task that simple ones do for regular determinants. But we will treat this subject more fully at another time.

X. We observe finally that, since all the properties we considered in this and the preceding articles depend especially on the number n, which plays a role somewhat similar to that of $p - 1$ in Section III, this number is worthy of careful attention. It is very desirable, therefore, to discover the general connection between this number and the determinant to which it belongs. And we should not despair of finding an answer, since we have already succeeded in establishing (art. 302) the analytic formulae for the average value of the product of n into the number of genera (and these can be assigned a priori), at least for determinants that are negative.

▶ 307. The discussions of the preceding articles take into account only principal classes of a genus, and so they are sufficient for positive determinants when only one genus is given, and for negative determinants when only one positive genus is given; we do not wish to consider the negative genus. It remains only to add a few comments concerning the remaining (properly primitive) genera.

I. When G' is a genus different from the principal genus G (of the same determinant) and there is an ambiguous class in G', there will be just as many as in G. Let the ambiguous classes in G be L, M, N, etc. (including the principal class K) and in G', L', M', N', etc. and let us designate the former complex by A and the latter by A'. Since manifestly all the classes $L + L'$, $M + L'$, $N + L'$, etc. are ambiguous and different from one another and belong to G', and so must also be contained in A', the number of classes in A' cannot be less than the number in A; and similarly, since the classes $L' + L'$, $M' + L'$, $N' + L'$, etc. are different from one another and ambiguous and belong to G, and so are contained in A, the number of classes in A cannot be less than the number in A'. Thus the number of classes in A and A' are necessarily equal.

II. Since the number of all ambiguous classes is equal to the number of genera (art. 261, 287.III): it is manifest that if there is only one ambiguous class in G, there must be one ambiguous class contained in *every* genus; if there are two ambiguous classes in G there will be two in half of all the genera and none in the remaining; finally, if several ambiguous classes are contained in G, say a of them,^w the ath part of all the genera will contain a ambiguous classes, the rest none.

III. For the case where G contains two ambiguous classes, let G, G', G'', etc. be the genera which contain two, and H, H', H'', etc. be the genera which contain no ambiguous classes, and designate the first complex by $\mathfrak{G}$ and the second by $\mathfrak{H}$. Since we always get an ambiguous class from the composition of two ambiguous classes (art. 249), it is not hard to see that from the composition of two genera of $\mathfrak{G}$ we will always get a genus of $\mathfrak{G}$. And, further, from the composition of a genus of $\mathfrak{G}$ with a genus of $\mathfrak{H}$ we get a genus of $\mathfrak{H}$; for if, e.g., $G' + H$ does not belong to $\mathfrak{H}$ but to $\mathfrak{G}$, $G' + H + G'$ must also be contained in $\mathfrak{G}$. Q.E.A. Because $G' + G' = G$ and so $G' + H + G' = H$. Finally the genera $G + H$, $G' + H$, $G'' + H$, etc. and $H + H$, $H' + H$, $H'' + H$, etc. are all different and so taken together they must be identical with $\mathfrak{G}$ and $\mathfrak{H}$; but by what we have just shown, the genera $G + H$, $G' + H$, $G'' + H$, etc. all belong to $\mathfrak{H}$ and exhaust this complex. Necessarily therefore, the others, $H + H$, $H' + H$, $H'' + H$, etc., all belong to $\mathfrak{G}$; i.e. from the composition of two genera of $\mathfrak{H}$ we always get a genus of $\mathfrak{G}$.

IV. If E is a class of the genus V, different from the principal genus G, it is clear that $2E$, $4E$, $6E$, etc. all belong to G; and $3E$, $5E$, $7E$, etc. to V. If therefore the period of the class $2E$ contains m terms, manifestly in the series $E, 2E, 3E$, etc. the class $2mE$, and none before it, will be identical with K; that is, the period of the class E will consist of $2m$ terms. So the number of terms in the period of any class other than one from the principal

---

^w This can happen only for irregular determinants, and a will always be a power of 2.

genus will be $2n$ or a factor of $2n$, where n designates the number of classes in each genus.

V. Let C be a given class of the principal genus G, E a class of the genus V which gives rise to C when it is duplicated (there is always one such, art. 286), and let K, K', K'', etc. be all (properly primitive) ambiguous classes of the same determinant, and let $E(= E + K)$, $E + K'$, $E + K''$, etc. be *all* classes which produce C when they are duplicated. We will call this last complex Ω. The number of these classes will be equal to the number of ambiguous classes or to the number of genera. Manifestly there will be as many classes in Ω belonging to the genus V as there are ambiguous classes in G. If therefore we designate this number by a, in every genus there will be either a classes of Ω or none. As a result, when $a = 1$ every genus will contain one class of Ω; when $a = 2$, half of all the genera will contain two classes of Ω, the rest none. Indeed the first half will either coincide totally with 𝕲 (according to the meaning in III above) and the second half with 𝕳, or vice versa. When a is larger, the ath part of all the genera will include classes of Ω (a in each class).

VI. Let us suppose now that C is a class whose period contains n terms. It is obvious that in the case when $a = 2$ and n even, no class of Ω can belong to G [for then this class would be contained in the period of the class C; if, therefore it were $= rC$, that is $2rC = C$, we would have $2r \equiv 1$ (mod. n). Q.E.A.]. Therefore, since G belongs to 𝕲 all the classes of Ω must be distributed among the genera 𝕳. As a result, since (for a regular determinant) there are in all ϕn classes in G having periods of n terms, for the case where $a = 2$ there will be in all $2\phi n$ classes in each genus of 𝕳 with periods of $2n$ terms, and they will include both their own genus and the principal one. When $a = 1$ there will be ϕn of these classes in every genus except the principal one.

VII. Given these observations we now establish the following method of constructing the system of *all* properly primitive class for any given regular determinant (for we have set irregular determinants completely aside). Arbitrarily select a class E with a period of $2n$ terms. It will include its own genus which we will call V, and the principal genus G; the classes of these two genera are so disposed that they occur in that period. The job will already be finished when there are no other genera except these two, or when it does not seem necessary to add the rest of them (e.g. for a negative determinant which has only two positive genera). But when there are four or more genera, the remaining ones will be treated as follows. Let V' be any one of them and $V + V' = V''$. In V' and V'' there will be two ambiguous classes (either one in each, or two in one and none in the other). Select one of these, A, arbitrarily and it is clear that if A is composed with each of the classes in G and V, we will get $2n$ different classes belonging to V' and V'' which will completely exhaust these genera; thus these genera also can be ordered. If there are other genera besides these four, let V''' be one of the remaining and V'''', V''''', V'''''' be the genera resulting from the composition of V''' with V, V' and V''. These four genera $V''' \ldots V''''''$ will contain four ambiguous classes, and if one of these A' is selected and composed with each of the classes in G, V, V', V'' we will get all the classes in $V''' \ldots V''''''$. If there are still more genera remaining, continue in the same way until they are all gone. Obviously, if the number of all genera constructed is 2^μ we will need $\mu - 1$ ambiguous classes in all, and every class in these genera can be produced either by a multiplication of the class E or by composing a class that results from such a multiplication with one or more of the ambiguous classes. Two examples to illustrate these procedures follow; we will say no more about the use of such a construction or about the artifices by which the labor involved can be lightened.

I. The determinant -161

Four positive genera; four classes in each

G	V
1, 4; R7; R23	3, 4; N7; R23
$(1, \quad 0, \quad 161) = K$	$(3, \quad 1, \quad 54) = E$
$(9, \quad 1, \quad 18) = 2E$	$(6, -1, \quad 27) = 3E$
$(2, \quad 1, \quad 81) = 4E$	$(6, \quad 1, \quad 27) = 5E$
$(9, -1, \quad 18) = 6E$	$(3, -1, \quad 54) = 7E$

V'	V''
3, 4; R7; N23	1, 4; N7; N23
$(7, \quad 0, \quad 23) = A$	$(10, \quad 3, \quad 17) = A + E$
$(11, \ -2, \quad 15) = A + 2E$	$(5, \quad 2, \quad 33) = A + 3E$
$(14, \quad 7, \quad 15) = A + 4E$	$(5, \ -2, \quad 33) = A + 5E$
$(11, \quad 2, \quad 15) = A + 6E$	$(10, \ -3, \quad 17) = A + 7E$

II. The determinant -546

Eight positive categories; three classes in each

G	V
1 and 3, 8; R3; R7; R13	5 and 7, 8; N3; N7; N13
$(1, \quad 0, \quad 546) = K$	$(5, \quad 2, \quad 110) = E$
$(22, \ -2, \quad 25) = 2E$	$(21, \quad 0, \quad 26) = 3E$
$(22, \quad 2, \quad 25) = 4E$	$(5, \ -2, \quad 110) = 5E$

V'	V''
1 and 3, 8; N3; R7; N13	5 and 7, 8; R3; N7; R13
$(2, \quad 0, \quad 273) = A$	$(10, \quad 2, \quad 55) = A + E$
$(11, \ -2, \quad 50) = A + 2E$	$(13, \quad 0, \quad 42) = A + 3E$
$(11, \quad 2, \quad 50) = A + 4E$	$(10, \ -2, \quad 55) = A + 5E$

V'''	V''''
1 and 3, 8; N3; N7; R13	5 and 7, 8; R3; R7; N13
$(3, \quad 0, \quad 182) = A'$	$(15, \ -3, \quad 37) = A' + E$
$(17, \quad 7, \quad 35) = A' + 2E$	$(7, \quad 0, \quad 78) = A' + 3E$
$(17, \ -7, \quad 35) = A' + 4E$	$(15, \quad 3, \quad 37) = A' + 5E$

V''''	V'''''
1 and 3, 8; R3; N7; N13	5 and 7, 8; N3; R7; R13
$(6, \quad 0, \quad 91) = A + A'$	$(23, \quad 11, \quad 29) = A + A' + E$
$(19, \quad 9, \quad 33) = A + A' + 2E$	$(14, \quad 0, \quad 39) = A + A' + 3E$
$(19, \ -9, \quad 33) = A + A' + 4E$	$(23, \ -11, 29) = A + A' + 5E$

SECTION VI

VARIOUS APPLICATIONS OF THE PRECEDING DISCUSSIONS

▶ 308. We have often indicated how fruitful higher Arithmetic can be for truths that pertain to other branches of mathematics. It is therefore worthwhile to discuss some applications that deserve further development without, however, attempting to exhaust a subject which could easily fill several volumes. In this section we will treat first of the resolution of fractions into simpler ones and of the conversion of common fractions into decimals. We will then explain a new method of exclusion which will be useful for the solution of indeterminate equations of the second degree. Finally, we will give new shortened methods for distinguishing prime numbers from composite numbers and for finding the factors of the latter. In the following section we will establish the general theory of a special kind of functions, which has broad significance in all of analysis and is closely connected with the higher Arithmetic. We shall in particular study the extension of the theory of the sections of a circle. Until now only the first elements of this theory have been known.

▶ 309. PROBLEM. *To decompose the fraction m/n whose denominator n is the product of two relatively prime numbers a, b into two others whose denominators are a, b.*

Solution. Let the desired fractions be $x/a, y/b$ and we should have $bx + ay = m$; thus x will be a root of the congruence $bx \equiv m$ (mod. a). It can be found by the methods of Section II. And y will be $= (m - bx)/a$.

It is clear that the congruence $bx \equiv m$ has infinitely many roots, all congruent relative to a; but there is only one that is positive and less than a. It is also possible that y is negative. It is hardly necessary to point out that we can also find y by the congruence

$ay \equiv m$ (mod. b) and x by the equation $x = (m + ay)/b$. For example, given the fraction 58/77, 4 will be a value of the expression 58/11 (mod. 7), so 58/77 will be decomposed into $(4/7) + (2/11)$.

▶ 310. If we are given the fraction m/n with a denominator n which is the product of any number of mutually relatively prime factors a, b, c, d, etc., then by the preceding article it can first be resolved into two fractions whose denominators will be a and bcd, etc.; then the second of these into two fractions with denominators b and cd, etc.; the last of these into two others and so forth until at length the given fraction is reduced to this form

$$\frac{m}{n} = \frac{\alpha}{a} + \frac{\beta}{b} + \frac{\gamma}{c} + \frac{\delta}{d} + \text{etc.}$$

Manifestly we can take the numerators $\alpha, \beta, \gamma, \delta$, etc. positive and less than their denominators, except for the last, which is no longer arbitrary once the rest are determined. It can be negative or greater than its denominator (if we do not presuppose that $m < n$). In that case it will most often be advantageous to put it in the form $(\varepsilon/e) \mp k$ where ε is positive and less than e and k is an integer. And finally a, b, c, etc. can be taken to be prime numbers or the powers of prime numbers.

Example. The fraction 391/924 whose denominator $= 4 \cdot 3 \cdot 7 \cdot 11$ is resolved in this way into $(1/4) + (40/231)$; 40/231 into $(2/3) - (38/77)$; $- 38/77$ into $(1/7) - (7/11)$; and so writing $(4/11) - 1$ for $- 7/11$ we have $391/924 = (1/4) + (2/3) + (1/7) + (4/11) - 1$.

▶ 311. The fraction m/n can be decomposed *in only one way* into the form $(\alpha/a) + (\beta/b) +$ etc. $\mp k$ so that α, β, etc. are positive and less than a, b, etc.; that is, if we suppose that

$$\frac{m}{n} = \frac{\alpha}{a} + \frac{\beta}{b} + \frac{\gamma}{c} + \text{etc.} \mp k = \frac{\alpha'}{a} + \frac{\beta'}{b} + \frac{\gamma'}{c} + \text{etc.} \mp k'$$

and if α', β', etc. are also positive and less than a, b, etc., we will have necessarily $\alpha = \alpha'$, $\beta = \beta', \gamma = \gamma'$, etc., $k = k'$. For if we multiply by $n = abc$ etc. we have $m \equiv \alpha bcd$ etc. $\equiv \alpha'bcd$ etc. (mod. a) and so, since bcd etc. is relatively prime to a, necessarily $\alpha \equiv \alpha'$ and therefore $\alpha = \alpha'$ and then $\beta = \beta'$ etc., and immediately $k = k'$. Now since it is completely arbitrary which denominator is taken first, it is manifest that *all* numerators can be investigated as we did α in the preceding article, namely β by the congruence βacd etc. $\equiv m$ (mod. b), γ by γabd etc. $\equiv m$ (mod. c) etc. The sum of all fractions thus found will be equal to the fraction m/n or the difference will be the integer k. This gives us a means of checking the calculation. Thus in the preceding article the values of the expression 391/231 (mod. 4), 391/308 (mod. 3), 391/132 (mod. 7), 391/84 (mod. 11) will immediately supply the numerators 1, 2, 1, 4 corresponding to the denominators 4, 3, 7, 11, and the sum of these fractions will exceed the given fraction by unity.

▶ 312. *Definition.* If a common fraction is converted into a decimal, the series of decimal figures[a] (excluding the integral part if there is one), whether it be finite or infinite, we will call the *mantissa* of the fraction. Here we take an expression that until now has been used only for logarithms in a wider sense. Thus, e.g., the mantissa of the fraction 1/8 is 125, the mantissa of the fraction 35/16 is 1875, and of the fraction 2/37 it is 054054... repeated infinitely often.

From this definition it is immediately clear that fractions of the same denominator l/n, m/n will have the same or different mantissas according to whether the numerators l, m are congruent or noncongruent relative to n. A finite mantissa is not changed if any number of zeros is added on the right. The mantissa of the fraction $10m/n$ is obtained by dropping from the mantissa of the fraction m/n the first figure and in general the mantissa of the fraction $10^v m/n$ is found by dropping the first v figures from the mantissa of m/n. The mantissa of the fraction $1/n$ begins immediately with a significant figure (i.e. different from zero) if n is not > 10; but if $n > 10$ and equal to no power of 10, the number of figures of which it is made up is k, the first $k - 1$ figures of the mantissa of $1/n$ will be zeros and the kth one will be significant. Therefore if l/n, m/n have different mantissas (i.e. if l, m are noncongruent relative to n), they certainly cannot have the first k figures the same but must differ at least in the kth.

▶ 313. PROBLEM. *Given the denominator of the fraction m/n and the first k figures of its mantissa, to find the numerator m, presuming that it is less than n.*

[a] For brevity we will restrict the following discussion to the system which is commonly called decimal, but it can easily be extended to any other.

Solution. Let us consider the k figures as an integer. Multiply by n and divide the product by 10^k (or drop the last k figures). If the quotient is an integer (or all the dropped figures are zeros), it will manifestly be the number we seek and the given mantissa is complete; otherwise the numerator we want will be the next larger integer, or the quotient increased by unity, after the following decimal figures are dropped. The reason for this rule is so easily understood from the statement at the end of the preceding article that there is no need for a more detailed explanation.

Example. If we know that the first two figures of the mantissa of a fraction having denominator 23, is 69, we have the product $23 \cdot 69 = 1587$. Dropping the last two figures and adding unity, we have the number 16 for the numerator we want.

▶ 314. We begin with a consideration of those fractions whose denominators are prime numbers or powers of primes, and afterward we will show how to reduce the rest to this case. And we observe immediately that the mantissa of the fraction a/p^μ (we suppose that its numerator a cannot be divided by the prime number p) is finite and that it consists of μ figures, if $p = 2$ or $= 5$; in the former case this mantissa considered as an integer will be $= 5^\mu a$, in the latter case $= 2^\mu a$. This is so obvious that it needs no explanation.

But if p is another prime number, $10^r a$ will never be divisible by p^μ no matter how large we take r, and therefore the mantissa of the fraction $F = a/p^\mu$ must be infinite. Let us suppose that 10^e is the lowest power of the number 10 which is congruent to unity relative to the modulus p^μ (cf. Section III, where we showed that e is either equal to the number $(p - 1)p^{\mu - 1}$ or to an aliquot part of it). It is obvious that $10^e a$ is the first number in the series $10a$, $100a$, $1000a$, etc. which is congruent to a relative to the same modulus. Now since, according to article 312, we get the mantissas of the fractions $10a/p^\mu$, $100a/p^\mu, \ldots 10^e a/p^\mu$ by suppressing the first figure of the fraction F, then the first two figures, etc., until we have suppressed the first e figures, it is manifest that only after the first e figures, and not before, the same ones will be repeated. We can call these first e figures that form the mantissa by repeating themselves infinitely often the *period* of this mantissa or of the fraction F. The magnitude of the period, i.e. the number of figures in it (which $= e$), is completely independent of the numerator a and is determined by the denominator alone. Thus, e.g., the period of the fraction 1/11 is 09 and the period of the fraction 3/7 is 428571.[b]

▶ 315. Thus, when we know the period of any fraction, we can obtain the mantissa to as many figures as we wish. Now if $b \equiv 10^\lambda a$ (mod. p^μ), we can get the period for the fractions b/p^μ if we write the first λ figures of the period of the fraction F (we suppose that $\lambda < e$ which is permissible) after the remaining $e - \lambda$. Thus along with the period of the fraction F we will have at the same time the periods of all fractions whose numerators are congruent to the numbers $10a$, $100a$, $1000a$, etc. relative to the denominator p^μ. Thus, e.g., since $6 \equiv 3 \cdot 10^2$ (mod. 7), the period of the fraction 6/7 can be immediately deduced from the period of the fraction 3/7, and it is 857142.

Therefore, whenever 10 is a primitive root (art. 57, 89) for the modulus p^μ, from the period of the fraction $1/p^\mu$ we can immediately deduce the period of any other fraction m/p^μ (if the numerator m is not divisible by p). We do it by taking from the left and writing to the right as many figures as the index of m has unities when the number 10 is taken as base. Thus it is clear why in this case the number 10 was always taken as the base in Table 1 (see art. 72).

When 10 is not a primitive root, the only periods of fractions that can be derived from the period of the fraction $1/p^\mu$ are those whose numerators are congruent to some power of 10 relative to p^μ. Let 10^e be the lowest power of 10 which is congruent to unity relative to p^μ; let $(p - 1)p^{\mu - 1} = ef$ and take as base the primitive root r in such a way that f is the index of the number 10 (art. 71). In this system the numerators of the fractions whose periods can be derived from the period of the fraction $1/p^\mu$ will have as indices f, $2f$, $3f$, ... $ef - f$; similarly, from the period of the fraction r/p^μ we can deduce periods for fractions whose numerators $10r$, $100r$, $1000r$, etc. correspond to the indices $f + 1$, $2f + 1$, $3f + 1$, etc.; from the period of the fraction with numerator r^2 (whose index is 2) we can deduce the periods of the fractions whose numerators have indices $f + 2$, $2f + 2$, $3f + 2$, etc.; and in general from the period of the fraction with numerator r^i we can derive

[b] Robertson ("Of the Theory of Circulating Decimal Fractions," *Philosophical Transactions* [for 1768], London, 1769, p. 207) indicates the beginning and end of the period by writing a dot above the first and last figures. We do not think this is necessary here.

the periods of fractions whose numerators have indices $f + i$, $2f + i$, $3f + i$, etc. Thus if only we have the periods of the fractions whose numerators are $1, r, r^2, r^3, \ldots r^{f-1}$, we can get all the others by only one transposition with the aid of the following rule: Let i be the index of the numerator m of a given fraction m/p^μ in a system where r is taken as base (we suppose that i is less than $(p-1)p^{\mu-1}$); dividing by f we get $i = \alpha f + \beta$ where α, β are positive integers (or 0) and $\beta < f$; having done this we can get the period of the fraction m/p^μ from the period of the fraction whose numerator is r^β (it is 1 when $\beta = 0$), by putting the first α figures after the rest (when $\alpha = 0$ we keep the same period). This explains why in constructing Table 1 we followed the rule stated in article 72.

▶ 316. According to these principles we have constructed a table for all denominators of the form p^μ under 1000. Given the opportunity we will publish it in its entirety or even with further expansion. For the present we give as a specimen Table 3, which extends only to 100. It needs no explanation. For denominators that have 10 as a primitive root it gives the periods of fractions with the numerator 1 (e.g. for 7, 17, 19, 23, 29, 47, 59, 61, 97); for the rest it gives the f periods corresponding to the numerators $1, r, r^2, \ldots r^{f-1}$. They are distinguished by the numbers (0), (1), (2) etc.; for the base r we have always taken the same primitive root as in Table 1. Thus after the index of the numerator is derived from Table 1, the period of any fraction whose denominator is contained in this table can be calculated by the rules given in the preceding article. But for very small denominators we can accomplish the same thing without Table 1, if by ordinary division we compute as many initial figures of the mantissa as, according to article 313, are necessary to distinguish it from all others of the same denominator (for Table 3 not more than two are necessary). Now we examine all periods corresponding to the given denominator until we find these initial figures, and we will have the whole period. We must remember that these figures can be separated so that one (or more) appears at the end of a period, and the others at the beginning.

Example. We want the period of the fraction 12/19. For the modulus 19 by Table 1 we have ind. $12 = 2$ ind. $2 +$ ind. $3 = 39 \equiv 3 \pmod{18}$ (art. 57). Since for this case there is only one period corresponding to the numerator 1, we have to transpose its first three figures to the end and we have the period we want: 631578947368421052. It would have been just as easy to find the beginning of the period from the first two figures, 63.

Suppose we want the period of the fraction 45/53. For the modulus 53, ind. $45 = 2$ ind. $3 +$ ind. $5 = 49$. The number of periods here is $4 = f$, and $49 = 12f + 1$. Therefore from the period marked (1) we must transpose the first 12 figures to the end position and the period we want is 8490566037735. The initial figures 84 are separated in the table.

We observe here, as we promised in article 59, that with the help of Table 3 we can also find the number that corresponds to a given index for a given modulus (in the table the modulus is listed as a denominator). For it is clear from the preceding that we can find the period of a fraction to whose numerator (although unknown) the given index corresponds. It is sufficient to take as many initial figures of this period as there are figures in the denominator. From these, by article 313 we can derive the numerator or the number corresponding to the given index.

▶ 317. By the preceding method the mantissa of any fraction whose denominator is a prime number, or the power of a prime number within the limits of the table, can be calculated without computation to any number of figures. But with the help of what we said in the beginning of this section we can extend the use of this table farther and include all fractions whose denominators are the product of prime numbers or the powers of primes lying within its limits. For, since such a fraction can be decomposed into others whose denominators are these factors, and these can be converted into decimal fractions with any number of figures, we need only combine all these into a sum. There is scarcely any need to point out that the last figure of the sum can prove to be a little less than it ought to be, but manifestly the errors cannot add up to as many unities as there are individual fractions being added, so it will be appropriate to compute these to more figures than the given fraction would demand. For example, let us consider the fraction $6099380351/1271808720 = F.$[c] Its denominator is the product of the numbers 16, 9, 5, 49, 13, 47, 59. By the rules given above we find

[c] This fraction is one of those that approximates the square root of 23, and the excess is less than seven unities in the twentieth decimal figure.

that $F = 1 + (11/16) + (4/9) + (4/5) + (22/49) + (5/13) + (7/47) + (52/59)$. The individual fractions are converted into decimals as follows:

$$1 = 1$$

$$\frac{11}{16} = 0.6875$$

$$\frac{4}{5} = 0.8$$

$$\frac{4}{9} = 0.4444444444\ 4444444444\ 44$$

$$\frac{22}{49} = 0.4489795918\ 3673469387\ 75$$

$$\frac{5}{13} = 0.3846153846\ 1538461538\ 46$$

$$\frac{7}{47} = 0.1489361702\ 1276595744\ 68$$

$$\frac{52}{59} = 0.8813559322\ 0338983050\ 84$$

$$F = 4.7958315233\ 1271954166\ 17$$

The error in this sum is certainly less than five unities in the twenty-second figure and so the first twenty-one cannot be changed by it. Carrying the calculation to more figures we get in place of the last two figures 17, the number 1893936... It will be obvious to everyone that this method of converting common fractions into decimals is especially useful when we want a great many decimal figures; when we want only a few, ordinary division or logarithms can be used just as easily.

▶ 318. Therefore, since we have reduced the resolution of such fractions with denominators composed of several different prime numbers to the case where the denominator is prime or the power of a prime, we need add only a few remarks concerning their mantissas. If the denominator does not contain the factor 2 and 5, the mantissa will still be composed of periods, because in this case also in the series 10, 100, 1000, etc. we will eventually come to a term that is congruent to unity relative to the denominator. At the same time the exponent of this term, which can easily be determined by the methods of article 92, will indicate the size of the period independently of the numerator, as long as it is relatively prime to the denominator. If the denominator is of the form $2^\alpha 5^\beta N$ with N designating a number relatively prime to 10, α and β numbers of which at least one is not 0, the mantissa of the fraction will become periodic after the first α or β figures (whichever is the larger). These periods will be comparable in length to the periods of fractions which have the denominator N. This is easy to see, since the fraction is resolvable into two others with denominators $2^\alpha 5^\beta$ and N and the first of these will be cut off after the first α or β figures. We could easily add many other observations concerning this subject, especially concerning devices for constructing a tabulation like Table 3. We will omit this discussion, however, for the sake of brevity and because a great deal of it has already been published by Robertson (loc. cit.) and by Bernoulli (*Nouv. mém. Acad. Berlin*, 1771, p. 273[1]).

▶ 319. With regard to the congruence $x^2 \equiv A$ (mod. m) which is equivalent to the indeterminate equation $x^2 = A + my$, in Section IV (art. 146) we have treated its *possibility* in a way that does not seem to demand anything further. For finding the unknown itself, however, we observed above (art. 152) that indirect methods are to be preferred to direct ones. If m is a prime number (the other cases can easily be reduced to this one) we can use the table of indices I (combined with III according to the observation of art. 316) for this purpose as we showed more generally in article 60, but the method would be restricted by the limits of the table. For these reasons we hope that the following general and brief method will please the lovers of arithmetic.

First we observe that it is sufficient if we have only those values of x that are positive and not greater than $m/2$ since the others will be congruent modulo m to one of these taken either positively or negatively. For such a value of x the value of y is necessarily contained within the limits $-A/m$ and $(m/4) - (A/m)$. The obvious method therefore consists in this, that for each value of y contained within these limits (we will denote the complex by Ω) the value of $A + my$ will be computed (we will call it V) and we retain only those values for which V is a square. When m is a small number (e.g. below 40) the number of trials is so small that there is scarcely need of a shortcut; but when m is large, the labor can be shortened as much as you please by the *method of exclusion*.

▶ 320. Let E be an arbitrary integer relatively prime to m and greater than 2; and let all its different quadratic nonresidues

---

[1] "Sur les fractions decimales periodiques."

(i.e. noncongruent relative to E) be a, b, c, etc.; and let the roots of the congruences

$$A + my \equiv a, \qquad A + my \equiv b, \qquad A + my \equiv c, \text{ etc.}$$

relative to the modulus E be α, β, γ, etc., all positive and less than E. Let y have a value which is congruent relative to E to one of the numbers α, β, γ, etc. Then the resulting value of $V = A + my$ will be congruent to one of the a, b, c, etc. and so a nonresidue of E and nonquadratic. Thus we can immediately exclude as useless all values of Ω which are contained in the forms $Et + \alpha$, $Et + \beta$, $Et + \gamma$, etc.; it will be sufficient to test the rest and we will call that complex Ω'. In this operation the number E can be said to be an *excluding* number.

If we take another similar excluding number E', in the same way we can find as many numbers α', β', γ', etc. as there are different quadratic nonresidues; y cannot be congruent to them relative to the modulus E'. Now we can again remove from Ω' all numbers contained in the forms $E't + \alpha'$, $E't + \beta'$, $E't + \gamma'$, etc. In this way we can continue excluding numbers until those contained in Ω are so reduced that it will be no more difficult to test the ones that remain than to construct new exclusions.

Example. Given the equation $x^2 = 22 + 97y$, the limits of the values of y will be $-22/97$ and $(24/4) - (22/97)$. So (since the value 0 is obviously useless) Ω will include the numbers $1, 2, 3, \ldots$ 24. For $E = 3$ we have only one nonresidue, $a = 2$; so $\alpha = 1$ and we must exclude from Ω all numbers of the form $3t + 1$; the number remaining in Ω' will be 16. Similarly, for $E = 4$ we have $a = 2$, $b = 3$ and so $\alpha = 0$, $\beta = 1$; and we must reject numbers of the form $4t$ and $4t + 1$. The eight remaining numbers are 2, 3, 6, 11, 14, 15, 18, 23. Then for $E = 5$ we find that we must reject numbers of the form $5t$ and $5t + 3$ and so we have 2, 6, 11, 14 remaining. Excluding 6 would remove numbers of the forms $6t + 1$ and $6t + 4$, but these have already been removed (since they are also numbers of the form $3t + 1$). Excluding 7 rejects numbers of the forms $7t + 2$, $7t + 3$, $7t + 5$ and leaves 6, 11, 14. If we substitute these for y we get for V, 604, 1089, 1380 respectively. Only the second value is a square, so $x = \pm 33$.

▶ 321. Since the operation with excluding E rejects from the values of V and the corresponding values of y in Ω all those that are quadratic nonresidues of E but does not touch the residues of the same number, it is obvious that the use of the excluding E and $2E$ does not differ if E is odd, since in this case E and $2E$ have the same residues and nonresidues. Thus if we use successively the numbers 3, 4, 5, etc. as excluding, we can omit the oddly even numbers 6, 10, 14, etc. as superfluous. And the double operation using E, E' as excluding removes all those values of V that are nonresidues of both E, E' or of one of them, and leaves those that are residues of both. Now, since in the case where E and E' do not have a common divisor, the ejected numbers are all nonresidues and the remaining ones are residues of the product EE', it is manifest that using the excluding EE' would effect the same result as using the two E, E', and its use is therefore superfluous. Thus it is permissible to omit all those excluding numbers that can be resolved into two relatively prime factors, and it is sufficient to use those that are either primes (not dividing m) or powers of primes. And finally it is manifest that, after using the excluding p^μ which is a power of the prime number p, the excluding p or p^ν, when $\nu < \mu$, would be superfluous. For, since p^μ leaves only its residues among the values of V, there will certainly be no nonresidues of p or of a lower power p^ν. If p or p^ν was used before p^μ, the latter manifestly can eject only those values of V which are the same time residues of p (or p^ν) and nonresidues of p^μ; therefore it is sufficient to take for a, b, c, etc. only nonresidues of p^μ.

▶ 322. Calculation of the numbers α, β, γ, etc. corresponding to any given excluding E can be greatly shortened by the following observations. Let $\mathfrak{A}, \mathfrak{B}, \mathfrak{C}$, etc. be roots of the congruences $my \equiv a, my \equiv b, my \equiv c$, etc. (mod. E) and k a root of $my \equiv -A$. It is clear that $\alpha \equiv \mathfrak{A} + k$, $\beta \equiv \mathfrak{B} + k$, $\gamma \equiv \mathfrak{C} + k$, etc. Now if it were necessary to find $\mathfrak{A}, \mathfrak{B}, \mathfrak{C}$, etc. by solving these congruences, this method of finding α, β, γ, etc. would be no shorter than the one we showed above; but this is not necessary. For if E is a prime number and m a quadratic residue of E, it is clear by article 98 that $\mathfrak{A}, \mathfrak{B}, \mathfrak{C}$, etc., i.e. the values of the expressions $a/m, b/m, c/m$, etc. (mod. E), are different nonresidues of E and so are identical with α, β, γ, etc. if we pay no attention to their order, which does not matter here anyway. If in the same supposition m is a nonresidue of E, the numbers $\mathfrak{A}, \mathfrak{B}, \mathfrak{C}$, etc. are identical with all the quadratic residues, disregarding 0. If E is the square of an (odd) prime number, $= p^2$, and p has already been used as excluding, it is sufficient according to the preceding article to assume for a, b, c, etc. those nonresidues of p^2 that are residues of p, i.e. the numbers $p, 2p, 3p, \ldots p^2 - p$ (all numbers less than p^2 which are divisible by p, except 0); thus for $\mathfrak{A}, \mathfrak{B}, \mathfrak{C}$, etc. we must get exactly the same numbers but in a different order. Similarly if we set $E = p^3$ after applying the excluding numbers p and p^2, it will suffice to take for a, b, c, etc. the products of each of the nonresidues of p into p^2. As a result we will get for $\mathfrak{A}, \mathfrak{B}, \mathfrak{C}$, etc. either the same numbers or the products of p^2 into each residue of p except 0, according as m is a residue or nonresidue of p. In general if we take for E any power of a prime number, say p^μ, after applying all lower powers, we will get for $\mathfrak{A}, \mathfrak{B}, \mathfrak{C}$, etc. the products of $p^{\mu-1}$ either into all numbers less than p, 0 excepted when μ is even, or into all nonresidues of p which are less than p when μ is odd and mRp, or into all residues when mNp. If $E = 4$ and $a = 2$, $b = 3$ we have for $\mathfrak{A}, \mathfrak{B}$ either 2 and 3 or 2 and 1, according as $m \equiv 1$ or $\equiv 3$ (mod. 4). If after using the excluding 4, we set $E = 8$, we will have $\alpha = 5$, and $\mathfrak{A}$ will become $5, 7, 1, 3$ according as $m \equiv 1, 3, 5, 7$ (mod. 8). In general if E is a higher power of 2, say 2^μ, and all the lower powers have already been applied, we should set $a = 2^{\mu-1}$, $b = 3 \cdot 2^{\mu-2}$ when μ is even. This gives us $\mathfrak{A} = 2^{\mu-1}$ and $\mathfrak{B} = 3 \cdot 2^{\mu-2}$ or $= 2^{\mu-2}$ according as $m \equiv 1$ or $\equiv 3$. But when μ is odd, we must set $a = 5 \cdot 2^{\mu-3}$ and $\mathfrak{A}$ will be equal to the product of the number $2^{\mu-3}$ into 5, 7, 1, or 3 according as $m \equiv 1, 3, 5, 7$ (mod. 8).

But skilled mathematicians will easily find a method of *mechanically* ejecting useless values of y from Ω after computing the number α, β, γ, etc. for as many exclusions as seem necessary. But we have not space to discuss this or other labor-saving devices.

▶ 323. In Section V we gave a general method for finding all representations of a given A by the binary form $mx^2 + ny^2$. This is of course the same as finding the solutions of the indeterminate equation $mx^2 + ny^2 = A$. The method leaves nothing to be desired from the point of view of brevity if we already have all the values of the expression $\sqrt{-mn}$ relative to the modulus A itself or to A divided by its quadratic factors. For the case, however, wherever mn is positive we will give a solution which is much shorter than the direct one when those values have not yet been computed. Let us suppose that the numbers m, n, and A are positive and relatively prime to one another, since the other cases can easily be reduced to this. It will also suffice to derive positive values of x, y, since the others can be deduced from these by changing signs.

Clearly x must be such that $(A - mx^2)/n$, which we will designate by V, is positive, an integer, and a square. The first condition requires that x be not greater than $\sqrt{A/m}$; the second holds when $n = 1$, otherwise it requires that the value of the expression A/m (mod. n) be a quadratic residue of n. And if we designate all the different values of the expression $\sqrt{A/m}$ (mod. n) by $\pm r, \pm r'$, etc., x will have to be contained in one of the forms $nt + r, nt - r$, $nt + r'$, etc. The simplest way would be to substitute for x all numbers of these forms below the limit $\sqrt{A/m}$ (we will call this complex Ω) and keep only those for which V is a square. In the following article we will show how to reduce the number of these trials as much as we wish.

▶ 324. The method of exclusions by which we accomplish this, just as in the preceding discussion, consists in taking arbitrarily several numbers which we will call *excluding* numbers. Next we find the values of x for which the value of V becomes a nonresidue of the excluding numbers. Such x we reject from Ω. The reasoning here is completely analogous to that of article 321, and so we should use as excluding numbers only those that are primes or powers of primes, and in the latter case we should reject only those nonresidues among the values of V that are residues of all lower powers of the same prime number, presuming we have begun our exclusion with these.

Therefore let our excluding number be $E = p^\mu$ (we can have $\mu = 1$) with p a prime number which does not divide m, and suppose[d] that p^ν is the highest power of p that divides n. Let a, b, c, etc. be quadratic nonresidues of E (all of them when $\mu = 1$; the necessary ones or those that are residues of lower powers when $\mu > 1$). Compute the roots of the congruences $mz \equiv A - na$, $mz \equiv A - nb$, $mz \equiv A - nc$, etc. (mod. $Ep^\nu = p^{\mu+\nu}$). Designate these roots by α, β, γ, etc. It is easy to see that if for some value of x we have $x^2 \equiv \alpha$ (mod. Ep^ν), the corresponding value of V will be

---

[d] For brevity we will consider together the two cases in which n is divisible and not divisible by p; in the latter case we should let $\nu = 0$.

$\equiv a$ (mod. E), that is, a nonresidue of E. Similarly for the remaining numbers β, γ, etc. Conversely, it is just as easy to see that if a value of x produces $V \equiv a$ (mod. E), for the same value we will have $x^2 \equiv \alpha$ (mod. Ep^ν). Thus all values of x for which x^2 is not congruent to any of the numbers α, β, γ, etc. (mod. Ep^ν) will produce similar values of V which are not congruent to any of the numbers a, b, c, etc. (mod. E). Now select from the numbers α, β, γ, etc. all quadratic residues of Ep^ν and call them g, g', g'', etc. Compute the values of the expressions $\sqrt{g}, \sqrt{g'}, \sqrt{g''}$, etc. (mod. Ep^ν) and designate them as $\pm h, \pm h', \pm h''$, etc. Having done this, all numbers of the forms $Ep^\nu t \pm h$, $Ep^\nu t \pm h'$, $Ep^\nu t \pm h''$, etc. can safely be ejected from Ω, and no value of V contained in the forms $Eu + a, Eu + b, Eu + c$, etc. can correspond to any value of x in Ω after this exclusion. And it is manifest that no value of x in Ω can produce such values of V when none of the numbers α, β, γ, etc. is a quadratic residue of Ep^ν. In this case, therefore, the number E cannot be used as excluding. In this way we can use as many excluding numbers as we wish and consequently diminish the numbers in Ω at will.

Let us see now whether it is permissible to use prime numbers that divide m and powers of such numbers as excluding numbers. Let B be a value of the expression A/n (mod. m) and it is clear that V will always be congruent to B relative to the modulus m no matter what value we take for x. Thus for the proposed equation to be possible it is necessary that B be a quadratic residue of m. Let p be any odd prime divisor of m. By hypothesis it does not divide n or A and therefore not B. For any value of x, V will be a residue of p and thus also of any power of p; therefore neither p nor any of its powers can be taken as excluding. Similarly, when m is divisible by 8, to make the proposed equation possible it is required that $B \equiv 1$ (mod. 8) and so for any value of x, V will be $\equiv 1$ (mod. 8) and the powers of 2 will not be suitable for excluding. However, when m is divisible by 4 but not by 8, for the same reason we must have $B \equiv 1$ (mod. 4) and the value of the expression A/n (mod. 8) will be either 1 or 5. We will designate it by C. For an even value of x we will have $V \equiv C$; for an odd value $V \equiv C + 4$ (mod. 8). And so even values must be rejected when $C = 5$, odd values when $C = 1$. Finally, when m is divisible by 2 but not by 4, let C as before be a value of the expression A/n (mod. 8). It will be 1, 3, 5, or 7; and let D be a value of $m/2n$ (mod. 4) which will be 1 or 3. Now since the value of V is always $\equiv C - 2Dx^2$ (mod. 8) and so for x even, $\equiv C$, for x odd, $\equiv C - 2D$, it follows that all odd values of x should be rejected when $C = 1$, all the even values when $C = 3$ and $D = 1$ or $C = 7$ and $D = 3$. The remaining values will all produce $V \equiv 1$ (mod. 8); that is to say, V is a residue of any power of 2. In the remaining cases, namely when $C = 5$, or $C = 3$ and $D = 3$, or $C = 7$ and $D = 1$, we have $V \equiv 3, 5$ or 7 (mod. 8) no matter whether x is odd or even. It follows in these cases that the proposed equation has no solution at all.

Now in the same way that we found x by the method of exclusion we could also have found y. Thus there are always two ways of applying the method of exclusion for the solution of a given problem (unless $m = n = 1$, when the two coincide). We will always prefer the method for which the number of terms Ω is smaller. This can easily be estimated in advance. It is scarcely necessary to observe that if, after a number of exclusions, *all* the numbers in Ω are rejected, this must be considered a certain indication of the impossibility of the proposed equation.

▶ 325. *Example.* Let the given equation be $3x^2 + 455y^2 = 10857362$. We will solve it in two ways, *first* by investigating the values of x, and then the values of y. The limit in this case is $\sqrt{(10857362/3)}$ which falls between 1902 and 1903; the value of the expression $A/3$ (mod. 455) is 354 and the values of the expression $\sqrt{354}$ (mod. 455) are $\pm 82, \pm 152, \pm 173, \pm 212$. So Ω consists of the following 33 numbers: 82, 152, 173, 212, 243, 282, 303, 373, 537, 607, 628, 667, 698, 737, 758, 828, 992, 1062, 1083, 1122, 1153, 1192, 1213, 1283, 1447, 1517, 1538, 1577, 1608, 1647, 1668, 1738, 1902. The number 3 cannot be used in this case for exclusion because it divides m. For the excluding number 4, we have $a = 2$, $b = 3$ so $\alpha = 0$, $\beta = 3$; $g = 0$ and the values of the expression $\sqrt{g}$ (mod. 4) are 0 and 2; thus all numbers of the forms $4t$ and $4t + 2$, i.e. all even numbers, must be rejected from Ω; we will designate the remaining (16) by Ω'. For $E = 5$, which also divides n, the roots of the congruences $mz \equiv A - 2n$ and $mz \equiv A - 3n$ (mod. 25) are 9 and 25, both of them residues of 25. The values of the expressions $\sqrt{9}$ and $\sqrt{24}$ (mod. 25) are $\pm 3, \pm 7$. When we reject from Ω' all numbers of the forms $25t \pm 3, 25t \pm 7$ there remain these ten (Ω''): 173, 373, 537, 667, 737, 1083, 1213, 1283, 1517, 1577. For $E = 7$ the roots of the congruences $mz \equiv A - 3n$, $mz \equiv A - 5n$, $mz \equiv A - 6n$ (mod. 49) are 32, 39, 18, all of them residues of 49, and the values of the expression $\sqrt{32}$, $\sqrt{39}$, $\sqrt{18}$ (mod. 49) are ± 9, ± 23, ± 19. When we eject from Ω'' numbers of the forms $49t \pm 9$, $49t \pm 19$, $49t \pm 23$ these five (Ω''') remain: 537, 737, 1083, 1213, 1517. For $E = 8$ we have $a = 5$ so $\alpha = 5$, a nonresidue of 8; therefore excluding 8 cannot be used. The number 9 must be rejected for the same reason as 3. for $E = 11$ the numbers a, b, etc. become 2, 6, 7, 8, 10; $\nu = 0$; so the numbers α, β, etc. $= 8, 10, 5, 0, 1$. Three of them, 0, 1, 5, are residues of 11. For this reason we reject from Ω''' numbers of the form $11t$, $11t \pm 1$, $11t \pm 4$. There remain the numbers 537, 1083, 1213. Using these we get for V the values 21961, 16129, 14161 respectively. Only the second and third are squares. So the given equation admits of only two solutions by positive values of x, y: $x = 1083$, $y = 127$ and $x = 1213$, $y = 119$.

Second, if we prefer to find the other unknown of this same equation by exclusions, commute x and y and put it in the form $455x^2 + 3y^2 = 10857362$ so that we can retain the notation of articles 323, 324. The limit of the values of x falls between 154 and 155; the value of the expression A/m (mod. n) is 1; the values of $\sqrt{1}$ (mod. 3) are $+1$ and -1. Therefore Ω contains all numbers of the forms $3t + 1$ and $3t - 1$, i.e. all numbers up to 154 inclusive that are not divisible by 3. There are 103 of them. Applying the rules given above for excluding 3, 4, 9, 11, 17, 19, 23 we must reject numbers of the forms $9t \pm 4$; $4t, 4t \pm 2$ or all even numbers; $27t \pm 1, 27t \pm 10$; $11t, 11t \pm 1, 11t \pm 3$; $17t \pm 3, 17t \pm 4, 17t \pm 5$, $17t \pm 7$; $19 \pm 2, 19t \pm 3, 19t \pm 8, 19t \pm 9$; $23t, 23t \pm 1, 23t \pm 5$, $23t \pm 7, 23t \pm 9, 23t \pm 10$. After all these have been deleted we have left the numbers 119, 127, both of which give V a quadratic value and produce the same solutions that we got above.

▶ 326. The preceding method is already so consise that it leaves scarcely anything to be desired. Nevertheless there are many devices for shortening the operation. We can touch on only a few here and will therefore restrict our discussion to the case where the excluding number is an odd prime not dividing A, or a power of such a prime. The remaining cases can be treated in an analogous way or reduced to this. Let us suppose *first* that excluding $E = p$ is a prime number not dividing m, n and the values of the expressions A/m, $-na/m$, $-nb/m$, $-nc/m$, etc. (mod. p) are k, $\mathfrak{A}$, $\mathfrak{B}$, $\mathfrak{C}$, etc. respectively. Derive the numbers α, β, γ, etc. from the congruences $\alpha \equiv k + \mathfrak{A}$, $\beta \equiv k + \mathfrak{B}$, $\gamma \equiv k + \mathfrak{C}$, etc. (mod. p). The numbers $\mathfrak{A}$, $\mathfrak{B}$, $\mathfrak{C}$, etc. can be determined without calculating the congruences by a device just like the one we used in article 322. They will be identical with all nonresidues or all residues of p (except 0) according as the value of the expression $-m/n$ (mod. p), or (what is the same thing) the number $-mn$, is a residue or nonresidue of p. Thus in example II of the preceding article, for $E = 17$ we have $k = 7$; $-mn = -1365 \equiv 12$ is a nonresidue of 17; so the number $\mathfrak{A}$, $\mathfrak{B}$, etc. will be 1, 2, 4, 8, 9, 13, 15, 16 and the numbers α, β, etc. will be 8, 9, 11, 15, 16, 3, 5, 6. The residues of these are 8, 9, 11, 15, 16 so $\pm h$, h' etc. become $\pm 5, 3, 7, 4$. Those who often have to solve problems of this type will find it extremely useful if they calculate for several prime numbers p the values of h, h', etc. corresponding to individual values of k $(1, 2, 3, \ldots p - 1)$ under the double supposition (namely, that $-mn$ is a residue and a nonresidue of p). We observe that the number of numbers h, $-h$, h', etc. is always $(p - 1)/2$ when the numbers k and $-mn$ are both residues or both nonresidues of p; $(p - 3)/2$ when the former is a residue, the latter a nonresidue; $(p + 1)/2$ when the former is a nonresidue, the latter a residue; but we must omit the demonstration of this theorem lest we become too prolix.

Second, we can explain rather expeditiously the cases where E is a prime number which divides n, or the power of an (odd) prime number which divides or does not divide n. We will treat all these cases together and, retaining the notation of articles 324, we will let $n = n'p^v$ so that n' is not divisible by P. The numbers a, b, c, etc. will be the products of the number $p^{\mu-1}$ either into all numbers less than p (except 0) or into all nonresidues of p less than p, according as μ is even or odd. We will express them indefinitely by $up^{\mu-1}$. Let k be the value of the expression A/m (mod. $p^{\mu+v}$). It will not be divisible by p because A is not. All the α, β, γ, etc. will be congruent to k modulo p, and so p^μ will exclude nothing from Ω if kNp; suppose, however, that kRp and so also $kRp^{\mu+v}$. Let r be a value of the expression $\sqrt{k}$ (mod. $p^{\mu+v}$) which is not divisible by p, and let e be value of $-n'/2mr$ (mod. p).

Then we will have $\alpha \equiv r^2 + 2erap^v$ (mod. $p^{\mu+v}$) and clearly α is a residue of $p^{\mu+v}$ and the values of the expression $\sqrt{\alpha}$(mod. $p^{\mu+v}$) become $\pm(r + eap^v)$; thus all the h, h', h'', etc. are expressed by $r + uep^{\mu+v-1}$. Finally we conclude that the numbers h, h', h'', etc. come from the addition of the number r with the products of the number $p^{\mu+v-1}$ *either* into all numbers less than p (except 0) when μ is even; *or* into all the non-residues of p less than this limit when μ is odd and eRp or, what comes to the same thing when $-2mrn'Rp$; *or* into all residues (except 0) when μ is odd and $-2mrn'Np$.

But just as we found the numbers h, h', etc. for each of the excluding numbers, we can accomplish the exclusion itself by mechanical operations. If it seems useful, the reader can easily develop such devices.

Finally we ought to observe that any equation $ax^2 + 2bxy + cy^2 = M$ in which $b^2 - ac$ is negative and $= -D$, can be easily reduced to the form which we considered in the preceding articles. For if we let m be the greatest common divisor of the numbers a, b and set

$$a = ma', \quad b = mb', \quad \frac{D}{m} = a'c - mb'b' = n, \quad a'x + b'y = x'$$

the equation will obviously equal $mx'x' + ny^2 = a'M$. This can be solved by the rules we gave above. Only those solutions are to be retained in which $x' - b'y$ is divisible by a' or which gives integral values of x.

▶ 327. The direct solution of the equation $ax^2 + 2bxy + cy^2 = M$ contained in Section V presumes that we know the values of the expression $\sqrt{(b^2 - ac)}$(mod. M). So far the case where $b^2 - ac$ is negative, the indirect solution above gives a very quick method of finding those values and is to be preferred to the method of article 322 et seq. especially for a very large value of M. Suppose however that M is a prime number or, at least, if it is composite, that its factors are still unknown. For if it were clear that the prime number p divides M and if $M = p^\mu M'$ in such a way that M' does not imply the factor p, it would be far more convenient to explore the values of the expression $\sqrt{(b^2 - ac)}$ for the moduli p^μ and M' separately (getting the former from the values relative to the modulus p, art. 101) and then deduce the values relative to the modulus M from their combination (art. 105).

Let us therefore find all values of the expression $\sqrt{} - D$ (mod. M) where D and M are positive, and M is contained in a form of the divisors of $x^2 + D$ (art. 147 et seq.). Otherwise it would be a priori evident that there are no numbers which satisfy the given expression. The values we seek will always be opposite in pairs. Let them be $\pm r$, $\pm r'$, $\pm r''$, etc., and $D + r^2 = Mh$, $D + r'r' = Mh'$, $D + r''r'' = Mh''$, etc.; further designate the classes to which the forms (M, r, h), $(M, -r, h)$, (M, r', h'), $(M, -r', h')$, (M, r'', h''), $(M, -r'', h'')$, etc. belong respectively by $\mathfrak{C}$, $-\mathfrak{C}$, $\mathfrak{C}'$, $-\mathfrak{C}'$, $\mathfrak{C}''$, $-\mathfrak{C}''$, etc. and their complex by $\mathfrak{G}$. Speaking in general, these classes are to be considered as unknown. However it is clear, *first*, that they are all positive and properly primitive, *second*, that they all belong to the same genus whose *character* is easily recognizable from the nature of the number M, i.e. from its relations to each of the prime divisors of D (and to 4 or 8 when these are necessary) (cf. art. 230). Since we supposed that M is contained in a form of the divisors of $x^2 + D$, we can be a priori certain that there is a properly primitive positive genus corresponding to this character even if perhaps it cannot satisfy the expression $\sqrt{} - D$ (mod. M). Since therefore this genus is known, we can find all the classes contained in it. Let them be designated as C, C', C'', etc. and their complex by G. It is clear therefore that the individual classes $\mathfrak{C}$, $-\mathfrak{C}$, etc. must be identical with some class in $\mathfrak{G}$; it can also happen that several classes in $\mathfrak{G}$ will be identical with each other and with the same class in G; and when G contains only one class, certainly all in $\mathfrak{G}$ will coincide with it. Therefore if from the classes C, C', C'', etc. we select the (simplest) forms f, f', f'', etc. (one from each), among them will appear a form from each class in $\mathfrak{G}$. Now if $ax^2 + 2bxy + cy^2$ is one of the forms contained in $\mathfrak{C}$, there will be two representations of the number M belonging to the value r by this form, and if one is $x = m$, $y = n$ the other will be $x = -m$, $y = -n$. The only exception occurs when $D = 1$, in which case there will be four representations (see art. 180).

It follows from this that if we find all representations of the number M by the individual forms f, f', f'', etc. (using the indirect method of the preceding articles) and deduce from these the values of the expression $\sqrt{-D}$ (mod. M) to which each belongs (art. 154 et seq.), we will get *all* the values of this expression, and indeed each of them twice or, if $D = 1$, four times. Q.E.F. If we find any forms among the f, f', etc. by which M cannot be represented, this is an indication that they belong to no class in $\mathfrak{G}$ and so should be neglected. But if M can be represented by none of these forms, $-D$ is necessarily a quadratic nonresidue of M. Regarding these operations the following observations should be kept in mind.

I. The representations of the number M by the forms f, f', etc. which we use here are those in which the values of the unknowns are relatively prime; if others occur in which these values have a common divisor μ (this can happen only when μ^2 divides M, and certainly happens when $-DRM/\mu^2$) they are to be completely disregarded for our present purposes, even though they can be useful in another respect.

II. Other things being equal, it is obvious that the labor involved will be easier the smaller the number of classes f, f', f'', etc. Consequently it is the shortest possible when D is one of the 65 numbers treated in article 303 because they have only one class in each genus.

III. Since there are always two representations $x = m, y = n$ and $x = -m, y = -n$ belonging to the same value, it is obviously sufficient to consider only those representations in which y is positive. Thus different representations will always correspond to different values of the expression $\sqrt{-D}$ (mod. M), and the number of all different values will be equal to the number of representations (always excepting the case where $D = 1$; here the first number will be half the second).

IV. Since as soon as we know one of the two opposite values $+r, -r$ we immediately know the other, the operations can be abbreviated somewhat. If the value r is obtained from the representation of the number M by a form contained in the class C, i.e. if $\mathfrak{C} = C$, the opposite value $-r$ manifestly comes from the representation by a form contained in the class which is opposite to C, and this class will always be different from C unless C is ambiguous. It follows that when not all the classes in G are ambiguous, only half the remaining need be considered. We can neglect one of each pair of opposites and immediately write down both values after calculating only one. When C is ambiguous, both values r and $-r$ will emerge at the same time; that is to say, if we take the ambiguous form $ax^2 + 2bxy + cy^2$ from C and the value r is produced by the representation $x = m, y = n$, the value $-r$ will result from the representation $x = -\text{m} - (2bn/a)$, $y = n$.

V. For the case where $D = 1$, there is only one class from which we can select the form $x^2 + y^2$. And if the value r results from the representation $x = m$, $y = n$ it will also result from $x = -m$, $y = -n$; $x = n, y = -m$; $x = -n, y = m$ and the opposite, $-r$, will result from $x = m, y = -n$; $x = -m, y = n$; $x = n, y = m$; $x = -n, y = -m$. Thus of these eight representations that constitute only one decomposition, one is sufficient as long as we associate the opposite value with the one that results from our investigation.

VI. The value of the expression $\sqrt{-D}$ (mod. M) to which the representation $M = am^2 + 2bmn + cn^2$ belongs is, by article 155, $\mu(mb + nc) - \nu(ma + nb)$ or any number congruent to it relative to M. The numbers μ, ν are to be taken in such a way that $\mu m + \nu n = 1$. If we designate this value by ν, we will have

$$mv \equiv \mu m(mb + nc) - \nu(M - mnb - n^2 c) \equiv (\mu m + \nu n)(mb + nc)$$
$$\equiv mb + nc \ (\text{mod. } M)$$

Thus it is clear that v is a value of the expression $(mb + nc)/m$ (mod. M); similarly we find that it is a value of the expression $-(ma + nb)/n$ (mod. M). These formulae are very often to be preferred to the one from which they were deduced.

▶ 328. *Examples.* I. We want all values of the expression $\sqrt{-1365}$ (mod. $5428681 = M$); the number M is $\equiv 1, 1, 1, 6, 11$ (mod. $4, 3, 5, 7, 13$) and so it is contained under the form of the divisors of $x^2 + 1$, $x^2 + 3$, $x^2 - 5$ and under the form of the nondivisors of $x^2 + 7$, $x^2 - 13$ and therefore under the form of the divisors of $x^2 + 1365$; the character of the genus in which the classes $\mathfrak{G}$ will be found, are $1, 4; R3; R5; N7; N13$. There is only one class contained in this genus. From it we select the form $6x^2 + 6xy + 229y^2$; in order to find all representations of the number M by this form, we let $2x + y = x'$, and we have

$3x'x' + 455y^2 = 2M$. This equation admits of four solutions in which y is positive, $y = 127$, $x' = \pm 1083$, $y = 119$, $x' = \pm 1213$. From these we get four solutions of the equation $6x^2 + 6xy + 229y^2 = M$ in which y is positive,

x	478	-605	547	-666
y	127	127	119	119

The first solution gives for v the value of the expression $30517/478$ or $-3249/127$ (mod. M) and it is found to be 2350978; the second produces the opposite value -2350978; the third, the value 2600262; and the fourth, its opposite -2600262.

II. If we want the values of the expression $\sqrt{-268}$ (mod. $4272943 = M$), the character of the genus in which the classes $\mathfrak{G}$ are contained, will be 1 *and* 7, 8; $R11$; $R13$. It will therefore be a principal genus in which three classes are contained, represented by the forms $(1, 0, 286), (14, 6, 23), (14, -6, 23)$. We can neglect the third of these, since it is opposite to the second. By the form $x^2 + 286y^2$ we find two representations of the number M in which y is positive, $y = 103$, $x = \pm 1113$. From these we deduce the values of the given expression 1493445, -1493445. We find that M is not representable by the form $(14, 6, 23)$, and we conclude that these are the only two values.

III. Given the expression $\sqrt{-70}$ (mod. 997331), the classes $\mathfrak{G}$ must be contained in the genus whose character is 3 *and* 5, 8; $R5$; $N7$. There is only one class and its representing form is $(5, 0, 14)$. When we do the calculation we find that the number 997331 is not representable by the form $(5, 0, 14)$, and so -70 will necessarily be a quadratic nonresidue of that number.

▶ 329. The problem of distinguishing prime numbers from composite numbers and of resolving the latter into their prime factors is known to be one of the most important and useful in arithmetic. It has engaged the industry and wisdom of ancient and modern geometers to such an extent that it would be superfluous to discuss the problem at length. Nevertheless we must confess that all methods that have been proposed thus far are either restricted to very special cases or are so laborious and prolix that even for numbers that do not exceed the limits of tables constructed by estimable men, i.e. for numbers that do not yield to artificial methods, they try the patience of even the practiced calculator. And these methods do not apply at all to larger numbers. Even though the tables, which are available to everyone and which we hope will continue to be extended, are indeed sufficient for most ordinary cases, it frequently happens that the trained calculator will be sufficiently rewarded by reducing large numbers to their factors so that it will compensate for the time spent. Further, the dignity of the science itself seems to require that every possible means be explored for the solution of a problem so elegant and so celebrated. For these reasons we do not doubt that the two following methods, whose efficacy and brevity we can confirm from long experience, will prove rewarding to the lovers of arithmetic. It is in the nature of the problem that *any* method will become more prolix as the numbers get larger. Nevertheless, in the following methods the difficulties increase rather slowly, and numbers with seven, eight, or even more digits have been handled with success and speed beyond expectation, especially by the second method. The techniques that were previously known would require intolerable labor even for the most indefatigable calculator.

Before calling upon the following methods, it is always very useful to try to divide the given number by some of the smaller primes, say by 2, 3, 5, 7, etc. up to 19 or a little beyond, in order to avoid using subtle and artificial methods when division alone would be easier;[e] and also, because when no division is successful, the application of the second method uses with great benefit the *residues* derived from these divisions. Thus, e.g., if the number 314159265 is to be resolved into its factors, division by 3 is successful twice and afterward, divisions by 5 and 7. Thus we have $314159265 = 9 \cdot 5 \cdot 7 \cdot 997331$ and it is sufficient to examine by more subtle means the number 997331, which is not divisible by 11, 13, 17, 19. Similarly, given the number 43429448, we can remove the factor 8 and apply more artificial methods to the quotient 5428681.

▶ 330. The foundation of the FIRST METHOD is the theorem which states that *any positive or negative number which is a quadratic residue of another number M, is also a residue of any divisor of M.*

[e] More so, because generally speaking among any given six numbers there will scarcely be *one* which is not divisible by one of the numbers 2, 3, 5, ... 19.

We know in general that if M is divisible by no prime number below $\sqrt{M}$, M is certainly prime; but if all prime numbers below this limit that divide M are p, q, etc., the number M is composed of these *alone* (or their powers), or there is only *one* other prime factor greater than $\sqrt{M}$. It is found by dividing M by p, q, etc. as often as we can. Therefore, if we designate the complex of all prime numbers below $\sqrt{M}$ (excluding those which we already know do not divide the number) by Ω, manifestly it will be sufficient to find all the prime divisors of M contained in Ω. Now if we know in some manner that a number r (nonquadratic) is a quadratic residue of M, certainly no prime number of which r is a nonresidue can be a divisor of M; therefore we can remove from Ω all prime numbers of this type (they will usually compose about half the numbers in Ω). And if it becomes clear that another nonquadratic number r' is a residue of M, we can exclude from the remaining prime numbers in Ω those for which r' is a nonresidue. Again we will reduce these numbers by almost half, provided the residues r and r' are independent (i.e. unless one of them is necessarily a residue of all numbers of which the other is a residue; this happens when rr' is a square). If we know still other residues of M, r'', r''', etc. all of which are independent[f] of those remaining, we can institute similar exclusions with each of them. Thus the number of numbers in Ω will diminish rapidly until they are all removed, in which case M will certainly be a prime number, or so few will remain (obviously all prime divisors of M will appear among them, if there are any such) that division by them can be tried without too much difficulty. For a number that does not exceed a million or so, six or seven exclusions will usually suffice; for a number with eight or nine digits, nine or ten exclusions will certainly suffice. There remain now two things to do, *first* to find suitable residues of M and a sufficient number of them, *then* to effect the exclusion in the most convenient way. But we will invert the order of the questions, especially since the second will show us which residues are the most suitable for this purpose.

▶ 331. We have shown at length in Section IV how to distinguish

---

[f] If the product of any number of numbers r, r', r'', etc. is a square, each of them, e.g. r, will be a residue of any prime number (which does not divide any one of them) that is a residue of the others, r', r'', etc. Thus for the residues to be independent, no product of pairs, or triples, etc. of them can be a square.

prime numbers whose residue is a given number r (we can suppose that it is not divisible by a square) from those for which it is a nonresidue; that is to say, how to distinguish divisors of the expression $x^2 - r$ from nondivisors. All the former are contained under formulae like $rz + a$, $rz + b$, etc. or like $4rz + a$, $4rz + b$, etc. and the latter under similar formulae. Whenever r is a very small number, we can evolve satisfactory exclusions with the help of these formulae; e.g. when $r = -1$ all numbers of the form $4z + 3$ will be excluded; when $r = 2$ all numbers of the forms $8z + 3$ and $8z + 5$ etc. But since it is not always possible to find residues like this for a given number M, and the application of the formulae is not very convenient when the value of r is large, much will be gained and the work of exclusion will be greatly reduced if we have a table for a sufficiently large number of numbers (r) both positive and negative which are not divisible by squares. The table should distinguish prime numbers which have each (r) as residue from those for which they are nonresidues. Such a table can be arranged like the example at the end of this book which we have already described above; but in order that it be useful for our present purposes, the prime numbers (moduli) in the margin should be continued much farther, to 1000 or 10000. It would be still more convenient if composite and negative numbers also were listed at the top, although this is not absolutely necessary, as is clear from Section IV. And the maximum utility would result if the individual vertical columns were removable and could be reassembled on plates or rods (like those of Napier). Then those that are necessary in each case, i.e. those which correspond to r, r', r'', etc., the residues of the given number, could be examined separately. If these are *properly* placed next to the first column of the table which contains the moduli (i.e. in such a way that the position in each of the rods which corresponds to the same number in the first column is placed in the corresponding horizontal line) those prime numbers that remain after the exclusions from Ω corresponding to the residues r, r', r'', etc. can be recognized immediately by inspection. They are the numbers in the first column that have little dashes in *all* the adjacent rods. If any rod has an empty space in *any* column it must be rejected. An example will illustrate this sufficiently well. If somehow we know that the numbers $-6, +13$, $-14, +17, +37, -53$ are residues of 997331, then we should join together a first column (which in this case should be continued as far as the number 997, i.e. up to the prime number next smaller than $\sqrt{997331}$) and the columns which have at the top the numbers -6, $+13$, etc. Here is a section of this scheme:

	-6	$+13$	-14	$+17$	$+37$	-53
3	—	—	—		—	—
5	—		—			
7	—		—		—	
11	—				—	
13		—	—	—		—
17		—		—		—
19			—	—		—
23		—	—			—
			etc.			
113		—	—			—
127	—	—	—	—	—	—
131	—	—	—			
			etc.			

From merely inspecting the numbers *contained in this part of the scheme* we see that after all the exclusions with the residues $-6, 13$, etc. only the number 127 remains in Ω. And the whole scheme extended to the number 997 would show us that there would be no other number remaining in Ω. When we try it, we find that 127 actually divides 997331. In this way we find that this number can be resolved into the prime factors 127.7853.[g]

From this example it is abundantly clear that this method is useful if the residues are not too large or at least if they can be decomposed into prime factors that are too large. For the immediate use of the auxiliary table does not extend beyond the numbers at the head of the columns, and the mediate use includes only

[g] The author has constructed for his own use a large section of the table described here, and he would gladly have published it if the small number of those for whom it would be useful had sufficed to justify such an undertaking. If there is any devotee of arithmetic who understands the principles involved and desires to construct such a table on his own, the author would find great pleasure in communicating to him by letter all the procedures and devices that he used.

those numbers that can be resolved into factors contained in the table.

▶ 332. We will give three methods for finding the residues of a given number M, but before we explain them we want to make two observations which will help us derive the simplest residues when the ones we have are not too suitable. *First*, if the number ak^2 which is divisible by the square k^2 (we presume it is relatively prime to M) is a residue of M, a will also be a residue. For this reason residues which are divisible by large squares are just as useful as small residues, and we will presume that all residues used in the following methods are free from square factors. *Second*, if two or more numbers are residues, their product will also be a residue. Combining this observation with the preceding, we can often deduce from several residues that are not all simple enough, another which is simple, provided the residues have a great number of common factors. For this reason it is very useful to have residues that are composed of many factors which are not too large, and all such should be immediately resolved into their factors. The force of these observations will be better understood by examples and frequent use than by rules.

I. The simplest method and the most convenient one for those who have acquired some dexterity by frequent exercise consists in decomposing M or more generally a multiple of M into two parts, $kM = a + b$ (both parts can be positive or one positive and the other negative). The product of these two taken with the opposite sign will be a residue of M; for $-ab \equiv a^2 \equiv b^2$ (mod. M) and so $-abRM$. The numbers a, b should be taken in such a way that their product is divisible by a large square and their quotient is small or at least separable into factors which are not too large. This can be done without difficulty. It is especially to be recommended that a be a square or double a square or triple a square etc., which differs from M by a small number or at least by a number that can be resolved into suitable factors. Thus, e.g., $997331 = 999^2 - 2 \cdot 5 \cdot 67 = 994^2 + 5 \cdot 11 \cdot 13^2 = 2 \cdot 706^2 + 3 \cdot 17 \cdot 3^2 = 3 \cdot 575^2 + 11 \cdot 31 \cdot 4^2 = 3 \cdot 577^2 - 7 \cdot 13 \cdot 4^2 = 3 \cdot 578^2 - 7 \cdot 19 \cdot 37 = 11 \cdot 299^2 + 2 \cdot 3 \cdot 5 \cdot 29 \cdot 4^2 = 11 \cdot 301^2 + 5 \cdot 12^2$ etc. Thus we have the following residues: $2 \cdot 5 \cdot 67$, $-5 \cdot 11$, $-2 \cdot 3 \cdot 17$, $-3 \cdot 11 \cdot 31$, $3 \cdot 7 \cdot 13$, $3 \cdot 7 \cdot 19 \cdot 37$, $-2 \cdot 3 \cdot 5 \cdot 11 \cdot 29$. The last decomposition includes the residue $-5 \cdot 11$ which we already have. For the residues

$-3 \cdot 11 \cdot 31$, $-2 \cdot 3 \cdot 5 \cdot 11 \cdot 29$ we can substitute $3 \cdot 5 \cdot 31$, $2 \cdot 3 \cdot 29$ which result from their combination with $-5 \cdot 11$.

II. The second and third methods are derived from the fact that if two binary forms (A, B, C), (A', B', C') of the same determinant M, or $-M$ or more generally $\pm kM$ belong to the same genus, the numbers AA', AC', $A'C$ are residues of kM; this is not hard to see, since any characteristic number of one form, say m, is also a characteristic number of the other, and so mA, mC, mA', mC' are all residues of kM. If therefore (a, b, a') is a reduced form of the positive determinant M or the more general kM, and (a', b', a''), (a'', b'', a'''), etc. are forms in its period, they will be equivalent to it and certainly contained in the same genus. And the numbers aa', aa'', aa''', etc. will all be residues of M. We can compute a great number of forms in such a period with the help of the algorithm of article 187. Ordinarily the simplest residues will result from letting $a = 1$ and we can reject those that have factors that are too large. Here are the beginnings of the periods of the forms $(1, 998, -1327)$ and $(1, 1412, -918)$ whose determinants are 997331, 1994662:

(	1,	998,	−1327)	(	1,	1412,	−918)
(−1327,	329,		670)	(−918,	1342,		211)
(	670,	341,	−1315)	(	211,	1401,	−151)
(−1315,	974,		37)	(−151,	1317,		1723)
(	37,	987,	−626)	(	1723,	406,	−1062)
(−626,	891,		325)	(−1062,	656,		1473)
(	325,	734,	−1411)	(	1473,	817,	−901)
(−1411,	677,		382)	(−901,	985,		1137)
(	382,	851,	−715)			etc.	

Therefore the residues of the number 997331 are all the numbers $-1327, 670$, etc.; neglecting those that have factors too large, we have these: $2 \cdot 5 \cdot 67$, 37, 13, $-17 \cdot 83$, $-5 \cdot 11 \cdot 13$, $-2 \cdot 3 \cdot 17$, $-2 \cdot 59$, $-17 \cdot 53$; we have already found above the residue $2 \cdot 5 \cdot 67$ as well as $-5 \cdot 11$ which results from a combination of the third and the fifth.

III. Let C be any class different from the principal class of forms of a negative determinant $-M$ or more generally $-kM$ and let its period be $2C$, $3C$, etc. (art. 307). The classes $2C$, $4C$, etc. will belong to the principal genus; $3C$, $5C$, etc. to the same genus as C. If therefore (a, b, c) is the (simplest) form in C and (a', b', c') a form in some class of the period, say nC, either a' or aa' will be a residue of M according as n is even or odd (in the former case c' will also be a residue, in the latter case ac', ca', and cc'). The calculation of the period, i.e. of the simplest forms in its classes, is surprisingly easy when a is very small, especially when it $= 3$, which is always permissible when $kM \equiv 2$ (mod. 3). Here is the beginning of the period of the class which contains the form $(3, 1, 332444)$:

C	(3,	1,	332444)	$6C$	(729,	-209,	1428)
$2C$	(9,	-2,	110815)	$7C$	(476,	209,	2187)
$3C$	(27,	7,	36940)	$8C$	(1027,	342,	1085)
$4C$	(81,	34,	12327)	$9C$	(932,	-437,	1275)
$5C$	(243,	34,	4109)	$10C$	(425,	12,	2347)

After rejecting those that are not useful, we have the residues $3 \cdot 476$, 1027, 1085, 425 or (removing the quadratic factors) $3 \cdot 7 \cdot 17$, $13 \cdot 79$, $5 \cdot 7 \cdot 31$, 17. If we combine these judiciously with the eight residues found in II we get the twelve following, $-2 \cdot 3$, 13, $-2 \cdot 7$, 17, 37, -53, $-5 \cdot 11$, 79, -83, $-2 \cdot 59$, $-2 \cdot 5 \cdot 31$, $2 \cdot 5 \cdot 67$. The first six are the same as the ones we used in article 331. If we wish, we can add the residues 19 and -29, which we found in I; the others included there are dependent on the ones we have developed here.

▶ 333. The SECOND METHOD of resolving a given number M into factors depends on a consideration of the values of the expression $\sqrt{D}$ (mod. M) together with the following observations.

I. When M is a prime number or the power of an (odd) prime (which does not divide D), $-D$ will be a residue or a nonresidue of M according as M is contained in a form of the divisors or the nondivisors of $x^2 + D$. In the former case the expression $\sqrt{D}$ (mod. M) will have only two different divisors, which are opposite.

II. When M is composite, that is to say, it $= pp'p''$ etc. where the numbers p, p', p'', etc. are (different odd) primes (not dividing D) or powers of such numbers: $-D$ will be a residue of M only when it is a residue of each of the p, p', p'', etc., i.e. when all these numbers are contained in forms of the divisors of $x^2 + D$. Now if we designate the values of the expression $\sqrt{-D}$ relative to the moduli p, p', p'', etc. respectively by $\pm r$, $\pm r'$, $\pm r''$, etc., we will get all values of the same expression relative to the modulus M by deriving the numbers that are $\equiv r$ or $\equiv -r$ relative to p, those that are $\equiv r'$ or $\equiv -r'$ relative to p', etc. Their number will be $= 2^\mu$ where μ is the number of factors p, p', p'', etc. Now if these values are $R, -R, R', -R', R''$, etc. we see immediately that $R \equiv R$ relative to all the numbers p, p', p'', etc. but that $R \equiv -R$ relative to none of them. Thus M will be the greatest common divisor of the numbers M and $R - R$, and 1 is the greatest common divisor of M and $R + R$; but two values that are neither identical nor opposite, e.g. R and R', must be congruent relative to one or several of the numbers p, p', p'', etc. but not relative to all of them. Relative to the others we will have $R \equiv -R'$. Thus the product of the former will be the greatest common divisor of the numbers M and $R - R'$, and the product of the latter the greatest common divisor of M and $R + R'$. It follows from this that if we find all the greatest common divisors of M with the differences between the individual values of the expression $\sqrt{-D}$ (mod. M) and some given value, their complex will contain the numbers $1, p, p', p''$, etc. and all products of pairs and triples etc. of these numbers. *In this way, therefore, we can find the numbers p, p', p'', etc. from the values of that expression.*

Now since the method of article 327 reduces these values to the values of expressions of the form m/n (mod. M) with the denominator n relatively prime to M, it is not necessary for our present purposes to compute them. For the greatest common divisor of the number M and the difference between R and R', which correspond to m/n and m'/n', will obviously also be the greatest common divisor of the numbers M and $nn'(R - R')$, or of M and $mn' - m'n$, since the latter is congruent to $nn'(R - R')$ relative to the modulus M.

▶ 334. We can apply the preceding observations to our present problem in two ways; the first not only decides whether the given number M is prime or composite, but in the latter case also gives its factors; the second is superior in that it permits faster calculation, but unless it is repeated over and over again it does not produce the factors of composite numbers. It does however distinguish them from prime numbers.

I. Let us investigate the negative number $-D$ which is a quadratic residue of M. For this purpose the methods given in·I and II of article 332 can be used. In itself, the selection of the residue is arbitrary, nor is there any need here as in the preceding method that D be a small number. But the calculation will be shorter as the number of classes of binary forms contained in each properly primitive genus of the determinant $-D$ is smaller. Therefore it will be helpful to take residues that are contained among the 65 enumerated in article 303 if any such occur. Thus for $M = 997331$ the residue -102 will be the most suitable of all the negative residues given above. Now find all the different values of the expression $\sqrt{-D}$ (mod. M). If there are only two (opposite), M will certainly be a prime or the power of a prime. If there are many, say 2^μ, M will be composed of μ prime numbers or powers of primes. Their factors can be found by the method of the preceding article. Whether these factors are prime numbers or powers of primes can be determined directly, but the way in which the values of the expression $\sqrt{-D}$ are found will indicate all prime numbers which divide M. For if M is divisible by the square of a prime number π, the calculation will certainly produce one or more representations of the number $M = am^2 + 2bmn + cn^2$, in which the greatest common divisor of the numbers m, n is π (in such a way that $-D$ is also a residue of M/π^2). But when there is no representation for which m and n have a common divisor, this is certainly an indication that M is not divisible by a square, and so all the numbers p, p', p'', etc. are prime numbers.

Example. By the method given above we find that there are four values of the expression $\sqrt{-408}$ (mod. 997331) which coincide with the values of the expressions $\pm 1664/113$, $\pm 2824/3$; the greatest common divisors of 997331 with $3 \cdot 1664 - 113 \cdot 2824$ and $3 \cdot 1664 + 113 \cdot 2824$ or with 314120 and 324104 are 7853 and 127, so $997331 = 127 \cdot 7853$ as above.

II. Let us take a negative number $-D$ such that M is contained in a form of the divisors of $x^2 + D$; in itself it is arbitrary which number of this type is selected, but it is advantageous to have the number of classes in the genera of the determinant $-D$ as small as possible. There is no difficulty finding such a number; for among any number of numbers tried there are almost as many for which M is contained in the form of divisors as there are for which M is contained in the form of the nondivisors. It will therefore be proper to begin with the 65 numbers of article 303 (starting with

the largest) and if it happens that none of them is suitable (in
general this will happen only once in 16384 cases) we should pass
on to others in which only two classes are contained in each genus.
Then we should investigate the values of the expression $\sqrt{-D}$
(mod. M) and if we find any, the factors of M can be deduced
from it in the same manner as above; but if we find no values,
that is to say that $-D$ is a nonresidue of M, M will certainly be
neither a prime number nor a power of a prime number. If in this
case we want the factors themselves, we will have to repeat the
same operation, using other values for D, or try another method.

Thus, e.g., we find that 997331 is contained in a form of nondivisors of $x^2 + 1848$, $x^2 + 1365$, $x^2 + 1320$ but in a form of
divisors of $x^2 + 840$; for the values of the expression $\sqrt{-840}$
(mod. 997331) we get the expressions $\pm 1272/163$, $\pm 3288/125$
and from these we deduce the same factors as before. For more
examples consult article 328 where we first showed that 5428681
$= 307 \cdot 17683$; second that 4272943 is a prime number; third,
that 997331 is certainly composed of more than one prime
numbers.

The limits of the present work permit us to insert here only
the basic principles of each method of finding factors; we will
save for another occasion a more detailed discussion along with
auxiliary tables and other aids.

SECTION VII

EQUATIONS DEFINING SECTIONS OF A CIRCLE

▶ 335. Among the splendid developments contributed by modern mathematicians, the theory of circular functions without doubt holds a most important place. We shall have occasion in a variety of contexts to refer to this remarkable type of quantity, and there is no part of general mathematics that does not depend on it in some fashion. Since the most brilliant modern mathematicians by their industry and shrewdness have formulated for it an extensive discipline, we can hardly expect that any part of the theory, particularly the elements, can be significantly expanded. I will speak of the theory of trigonometric functions as related to arcs that are commensurable with the circumference, or of the theory of regular polygons. Only a small part of this theory has been developed so far, as the present section will make clear. The reader might be surprised to find a discussion of this subject in the present work which deals with a discipline apparently so unrelated; but the treatment itself will make abundantly clear that there is an intimate connection between this subject and higher Arithmetic.

The principles of the theory which we are going to explain actually extend much farther than we will indicate. For they can be applied not only to circular functions but just as well to other transcendental functions, e.g. to those which depend on the integral $\int [1/\sqrt{(1 - x^4)}]dx$ and also to various types of congruences. Since, however, we are preparing a substantial work on transcendental functions and since we will treat congruences at length as we continue our discussion of arithmetic, we have decided to consider only circular functions here. And although we could discuss them in all their generality, we reduce them to the simplest case in the following article, both for the sake of brevity and in order that the new principles of this theory may be more easily understood.

▶ 336. If we designate the circumference of the circle or four right angles by P, and if m, n are integers and n a product of relatively prime factors a, b, c, etc.: the angle $A = mP/n$ can be reduced by the methods of article 310 to the form $A = [(\alpha/a) + (\beta/b) + (\gamma/c) + \text{etc.}]P$, and the trigonometric functions corresponding to them can be deduced by known methods for the parts $\alpha P/a$, $\beta P/b$, etc. Therefore, since we can take a, b, c, etc. to be prime numbers or powers of prime numbers, it is sufficient to consider the division of the circle into parts whose number is a prime or the power of a prime, and we will immediately have a polygon of n sides from the polygons of a, b, c, etc. sides. However, we will restrict our discussion to the case where the circle is divided into an (odd) prime number of parts, especially for the following reason. It is clear that circular functions corresponding to the angle mP/p^2 are derived from functions belonging to mP/p by the solution of an equation of degree p. And from these by an equation of the same degree we can derive functions belonging to mP/p^3 etc. Therefore if we already have a polygon of p sides, to determine a polygon of p^λ sides we necessarily require the solution of $\lambda - 1$ equations of degree p. Even though the following theory could be extended to this case also, nevertheless we could not avoid so many equations of degree p because there is no way of reducing the number if p is prime. Thus, e.g., it will be shown below that a polygon of 17 sides can be constructed geometrically; but to get a polygon of 289 sides there is no way to avoid solving an equation of degree 17.

▶ 337. It is well known that the trigonometric functions of all the angles kP/n where k denotes in general all the numbers $0, 1, 2, \ldots$ $n - 1$, is expressed by the roots of equations of degree n. The *sine* involves the roots of equation (I):

$$x^n - \frac{1}{4}nx^{n-2} + \frac{1}{16}\frac{n(n-3)}{1 \cdot 2}x^{n-4} - \frac{1}{64}\frac{n(n-4)(n-5)}{1 \cdot 2 \cdot 3}x^{n-6}$$
$$+ \text{ etc. } \pm \frac{1}{2^{n-1}}nx = 0$$

the *cosine* involves the roots of equation (II):

$$x^n - \frac{1}{4}nx^{n-2} + \frac{1}{16}\frac{n(n-3)}{1 \cdot 2}x^{n-4} - \frac{1}{64}\frac{n(n-4)(n-5)}{1 \cdot 2 \cdot 3}x^{n-6}$$
$$+ \text{ etc. } \pm \frac{1}{2^{n-1}}nx - \frac{1}{2^{n-1}} = 0$$

and the *tangents* involve the roots of equation (III):

$$x^n - \frac{n(n-1)}{1 \cdot 2} x^{n-2} + \frac{n(n-1)(n-2)(n-3)}{1 \cdot 2 \cdot 3 \cdot 4} x^{n-4} - \text{etc.} \pm nx = 0$$

These equations are all true for any odd value of n, and equation II is true for even values also. If we set $n = 2m + 1$ they can be easily reduced to degree m; that is to say, for I and III by dividing on the left by x and substituting y for x^2. Equation II however includes the root $x = 1 (= \cos 0)$ and all the others are equal in pairs ($\cos P/n = \cos (n - 1)P/n$, $\cos 2P/n = \cos (n - 2)P/n$, etc.); thus the left side is divisible by $x - 1$ and the quotient will be a square. If we extract the square root, equation II is reduced to the following:

$$x^m + \frac{1}{2} x^{m-1} - \frac{1}{4}(m - 1)x^{m-2} - \frac{1}{8}(m - 2)x^{m-3}$$

$$+ \frac{1}{16} \frac{(m-2)(m-3)}{1 \cdot 2} x^{m-4} + \frac{1}{32} \frac{(m-3)(m-4)}{1 \cdot 2} x^{m-5} - \text{etc.} = 0$$

Its roots will be the cosines of the angles $P/n, 2P/n, 3P/n, \ldots mP/n$. There are no further reductions of these equations for the case where n is a prime number.

Nevertheless none of these equations is so tractable and so suitable for our purposes as $x^n - 1 = 0$. Its roots are intimately connected with the roots of the above. That is, if for brevity we write i for the imaginary quantity $\sqrt{-1}$, the roots of the equation $x^n - 1 = 0$ will be

$$\cos \frac{kP}{n} + i \sin \frac{kP}{n} = r$$

where for k we should take all the numbers $0, 1, 2, \ldots n - 1$. Therefore since $1/r = \cos kP/n - i \sin kP/n$ the roots of equation I will be $[r - (1/r)]/2i$ or $i(1 - r^2)/2r$; the roots of equation II, $[r + (1/r)]/2 = (1 + r^2)/2r$; finally the roots of equation III, $i(1 - r^2)/(1 + r^2)$. For this reason we build our investigation on a consideration of the equation $x^n - 1 = 0$, and presume that n is an odd prime number. In order not to interrupt the order of the investigation we will first consider the following lemma.

▶ 338. PROBLEM. *Given the equation*

$$(W)\ldots z^m + Az^{m-1} + \text{etc.} = 0$$

to find an equation (W') *whose roots are the λth power of the roots of equation* (W), *where λ is a given positive integral exponent.*

Solution. If we designate the roots of the equation W by a, b, c, etc., the roots of the equation W' will be $a^\lambda, b^\lambda, c^\lambda$, etc. By a well-known theorem of Newton, from the coefficients of equation W we can find the sum of any powers of the roots a, b, c, etc. Therefore, we want the sums

$$a^\lambda + b^\lambda + c^\lambda + \text{etc.,} \qquad a^{2\lambda} + b^{2\lambda} + c^{2\lambda} + \text{etc.,}$$

$$\text{up to } a^{m\lambda} + b^{m\lambda} + c^{m\lambda} + \text{etc.}$$

and by an inverse procedure according to the same theorem, the coefficients of the equation W' can be deduced. Q.E.F. At the same time it is clear that if all the coefficients of W are rational, all those in W' will also be rational. And by another method it can be proven that if all the former are integers, the latter will be integers also. We will not spend more time on this theorem here, since it is not necessary for our purpose.

▶ 339. The equation $x^n - 1 = 0$ (we will always presume that n is an odd prime number) has only one real root, $x = 1$; the remaining $n - 1$ roots which are given by the equation

$$x^{n-1} + x^{n-2} + \text{etc.} + x + 1 = 0$$

are all imaginary; we will denote their complex by Ω and the function

$$x^{n-1} + x^{n-2} + \text{etc.} + x + 1 = 0 \text{ by } X$$

If therefore r is any root in Ω, we will have $1 = r^n = r^{2n}$ etc. and in general $r^{en} = 1$ for any positive or negative integral value of e. Thus if λ, μ are integers which are congruent relative to n, we will have $r^\lambda = r^\mu$. But if λ, μ are noncongruent relative to n, then r^λ and r^μ will be unequal; for in this case we can find an integer v such that $(\lambda - \mu)v \equiv 1$ (mod. n) so $r^{(\lambda-\mu)v} = r$ and certainly $r^{\lambda-\mu}$ does not $= 1$. It is also clear that any power of r is also a root of the equation $x^n - 1 = 0$. Therefore, since the quantities $1(= r^0)$,

$r, r^2, \ldots r^{n-1}$ are all different, they will give us all the roots of the equation $x^n - 1 = 0$ and so the numbers $r, r^2, r^3, \ldots r^{n-1}$ will coincide with Ω. In general, therefore, Ω will coincide with $r^e, r^{2e}, r^{3e}, \ldots r^{(n-1)e}$ if e is any positive or negative integer not divisible by n. We have therefore

$$X = (x - r^e)(x - r^{2e})(x - \kappa^{3e}) \ldots (x - r^{(n-1)e})$$

and from this

$$r^e + r^{2e} + r^{3e} + \ldots + r^{(n-1)e} = -1$$

and

$$1 + r^e + r^{2e} + \ldots + r^{(n-1)e} = 0$$

If we have two roots such as r and $1/r$ $(=r^{n-1})$ or in general r^e and r^{-e}, we will call them *reciprocal* roots. Manifestly the product of two simple factors $x - r$ and $x - (1/r)$ is real and $= x^2 - 2x \cos \omega + 1$ so that the angle ω is equal either to the angle P/n or some multiple of it.

▶ 340. Since by designating one root of Ω by r we can express all roots of the equation $x^n - 1 = 0$ by powers of r, the product of several roots of this equation can be expressed by r^λ in such a way that λ is either 0 or positive and $< n$. Therefore if we let $\phi(t, u, v, \ldots)$ be a rational integral algebraic function of the unknowns t, u, v, etc. which is the sum of terms of the form $h t^\alpha u^\beta v^\gamma \ldots$: manifestly by substituting roots of the equation $x^n - 1 = 0$ for t, u, v, etc., say $t = a$, $u = b$, $v = c$, etc. then $\phi(a, b, c, \ldots)$ can be reduced to the form

$$A + A'r + A''r^2 + A'''r^3 + \ldots + A^v r^{n-1}$$

in such a way that the coefficients A, A', etc. (some of them can be missing and so $= 0$) are determined quantities. And all of these coefficients will be integers if all the coefficients in $\phi(t, u, v, \ldots)$, i.e. all the h, are integers. And if after this we substitute $a^2, b^2, c^2, \ldots$ for $t, u, v, \ldots$ respectively, each term $h t^\alpha u^\beta v^\gamma, \ldots$ which had been reduced to r^σ will now become $r^{2\sigma}$ and thus

$$\phi(a^2, b^2, c^2, \ldots) = A + A'r^2 + A''r^4 + A'''r^6 + \ldots + A^v r^{2n-2}$$

And in general for any integral value of λ,

$$\phi(a^\lambda, b^\lambda, c^\lambda, \ldots) = A + A'r^\lambda + A''r^{2\lambda} + \ldots + A^v r^{(n-1)\lambda}$$

This proposition is very important and is fundamental to the following discussion. It also follows from this that

$$\phi(1, 1, 1, \ldots) = \phi(a^n, b^n, c^n, \ldots) = A + A' + A'' + \ldots + A^\nu$$

and

$$\phi(a, b, c, \ldots) + \phi(a^2, b^2, c^2, \ldots) + \phi(a^3, b^3, c^3, \ldots) + \phi(a^n, b^n, c^n, \ldots)$$
$$= nA$$

This sum will be integral and divisible by n when all the coefficients in $\phi(t, u, v, \ldots)$ are integers.

▶ 341. THEOREM. *If the function X is divisible by the function of lower degree*

$$P = x^\lambda + Ax^{\lambda-1} + Bx^{\lambda-2} + \ldots + Kx + L$$

the coefficients $A, B, \ldots L$ cannot all be integers.

Demonstration. Let $X = PQ$ and $\mathfrak{P}$ the complex of the roots of the equation $P = 0$, $\mathfrak{Q}$ the complex of the roots of the equation $Q = 0$, so that Ω consists of $\mathfrak{P}$ and $\mathfrak{Q}$ taken together. Further let $\mathfrak{R}$ be the complex of reciprocal roots of $\mathfrak{P}$, $\mathfrak{S}$ the complex of reciprocal roots of $\mathfrak{Q}$ and let the roots which are contained in $\mathfrak{R}$ be roots of the equation $R = 0$ (this becomes $x^\lambda + (Kx^{\lambda-1}/L) + \text{etc.} + (Ax/L) + (1/L) = 0$ and let those that are contained in $\mathfrak{S}$ be roots of the equation $S = 0$. Manifestly if we take the roots $\mathfrak{R}$ and $\mathfrak{S}$ together we will get the complex Ω and $RS = X$. Now we must distinguish four cases.

I. When $\mathfrak{P}$ coincides with $\mathfrak{R}$ and consequently $P = R$. In this case obviously pairs of roots in $\mathfrak{P}$ will always be reciprocal and so P will be the product of $\lambda/2$ paired factors $x^2 - 2x\cos\omega + 1$; since such a factor $= (x - \cos\omega)^2 + \sin\omega^2$, it is clear that for any real value of x, P necessarily has a real positive value. Let the equations whose roots are the square, cubic, biquadratic, $\ldots n - 1$st powers of the roots in $\mathfrak{P}$ be respectively $P' = 0$, $P'' = 0$, $P''' = 0$, $\ldots P^\nu = 0$ and let the values of the functions $P, P', P'', \ldots P^\nu$ which are obtained by letting $x = 1$ be respectively $p, p', p'', \ldots p^\nu$. Then by what we have said before, p will be a positive quantity, and for a similar reason p', p'', etc. will also be positive. Since therefore p is the value of the function $(1 - t)(1 - u)(1 - v)$ etc. which is obtained by substituting for t, u, v, etc. the roots contained in $\mathfrak{P}$; p' the value of the same function obtained by substituting for t, u, v, etc. the squares of those roots etc.; and 0 its value when $t = 1$, $u = 1$, $v = 1$ etc.: the sum $p + p' + p'' \ldots p^v$ will be an integer divisible by n. Further the product $PP'P'' \ldots$ will be $= X^\lambda$ and so $pp'p'' \ldots = n^\lambda$.

Now if all the coefficients in P were rational, all of those in P', P'', etc. would also be rational by article 338. However by article 42 all these coefficients are necessarily integers. Thus p, p', p'', etc. would also be integers. And since their product is n^λ and their number is $n - 1 > \lambda$, some of them (at least $n - 1 - \lambda$) must $= 1$, and the others equal either to n or to a power of n. And if g of them $= 1$, the sum $p + p' +$ etc. will be $\equiv g$ (mod. n) and so certainly divisible by n. Thus our supposition is inconsistent.

II. When $\mathfrak{P}$ and $\mathfrak{R}$ do not coincide but contain some common roots, let $\mathfrak{T}$ be this complex and $T = 0$, the equation of which they are the roots. Then T will be the greatest common divisor of the functions P, R (as is clear from the theory of equations). However, pairs of roots in $\mathfrak{T}$ will be reciprocal and as we saw before not all the coefficients in T can be rational. This will certainly happen if all the members of P and thus also of R are rational, as one can see from the nature of the operation by which we find the greatest common divisor. Thus our supposition is absurd.

III. When $\mathfrak{Q}$ and $\mathfrak{S}$ either coincide or have common roots, we can show in exactly the same way that not all the coefficients in Q are rational; but they would be rational if all those in P were rational, so this is impossible.

IV. If $\mathfrak{P}$ has no root in common with $\mathfrak{R}$, and $\mathfrak{Q}$ none in common with $\mathfrak{S}$, all the roots $\mathfrak{P}$ would necessarily be found in $\mathfrak{S}$, and all the roots $\mathfrak{Q}$ in $\mathfrak{R}$. Therefore $P = S$ and $Q = R$, and so $X = PQ$ will be a product of P into R; i.e.

of $x^\lambda + Ax^{\lambda-1} \ldots + Kx + L$ into $x^\lambda + \dfrac{K}{L}x^{\lambda-1} \ldots + \dfrac{A}{L}x + \dfrac{1}{L}$

So letting $x = 1$, we have

$$nL = (1 + A \ldots + K + L)^2$$

Now if all coefficients in P were rational, and so by article 42 also integers, L, which must divide the last coefficient in X, i.e. unity, will necessarily $= \pm 1$ and so $\pm n$ would be a square. But since this is contrary to the hypothesis, the supposition is inconsistent.

By this theorem therefore it is clear that no matter how X is decomposed into factors, some of the coefficients at least will be irrational, and so cannot be determined except by an equation of a degree higher than unity.

▶ 342. It is not without some value to declare in a few words the purpose of the following discussions. We intend to resolve X *gradually* into more and more factors, and in such a way that their coefficients are determined by equations of as low an order as possible. In so doing we will finally come to simple factors or to the roots Ω. We will show that if the number $n - 1$ is resolved in any way into integral factors α, β, γ, etc. (we can assume each of them is prime), X can be resolved into α factors of $(n - 1)/\alpha$ dimensions with coefficients determined by an equation of degree α; each of these will be resolved into β others of $(n - 1)/\alpha\beta$ dimensions with the aid of an equation of degree β etc. Thus if we designate by v the number of factors α, β, γ, etc. the determination of the roots Ω is reduced to the solution of v equations of degree α, β, γ, etc. For example, for $n = 17$ where $n - 1 = 2 \cdot 2 \cdot 2 \cdot 2$, there will be four quadratic equations to solve; for $n = 73$ three quadratic and two cubic equations.

In what follows we will often have to consider powers of the root r whose exponents are again powers, and expressions of this sort are very hard to set up in type. Therefore for facility of expression we will use the following abbreviation. For r, r^1, r^3, etc. we will write [1], [2], [3], etc. and in general for r^λ where λ is any integer, we will write $[\lambda]$. Such expressions are not completely determined, but they will become so as soon as we take a specific root from Ω for r or [1]. In general $[\lambda], [\mu]$ will be equal or unequal according as λ, μ are congruent or noncongruent relative to the modulus n. Further $[0] = 1$; $[\lambda] \cdot [\mu] = [\lambda + \mu]$; $[\lambda]^v = [\lambda v]$; the sum $[0] + [\lambda] + [2\lambda] \ldots + [(n - 1)\lambda]$ is either 0 or n according as λ is not divisible or divisible by n.

▶ 343. If, for the modulus n, g is a number which we called a primitive root in Section III, the $n - 1$ numbers $1, g, g^2, \ldots g^{n-2}$ will be congruent to the numbers $1, 2, 3, \ldots n - 1$ relative to the modulus n. The order is not determined, but every number in one series will be congruent to one in the other. From this it follows immediately that the roots $[1], [g], [g^2], \ldots [g^{n-2}]$ coincide with Ω. By a similar argument the roots

$$[\lambda], [\lambda g], [\lambda g^2], \ldots [\lambda g^{n-2}]$$

will coincide with Ω when λ is any integer not divisible by n. Further since $g^{n-1} \equiv 1$ (mod. n) it is easy to see that the two roots $[\lambda g^\mu], [\lambda g^\nu]$ will be identical or different according as μ, ν are congruent or noncongruent relative to $n - 1$.

If therefore G is another primitive root, the roots $[1], [g] \ldots$ $[g^{n-2}]$ will also coincide with $[1], [G], \ldots [G^{n-2}]$ except for order. Further, if e is a divisor of $n - 1$, and we set $n - 1 = ef$, $g^e = h$, $G^e = H$, then the f numbers $1, h, h^2, \ldots h^{f-1}$ will be congruent to $1, H, H^2, \ldots H^{f-1}$ relative to n (without respect to order). For suppose that $G \equiv g^\omega$ (mod. n) and that μ is an arbitrary positive number $<f$ and that ν is the least residue of $\mu\omega$ (mod. f): then we will have $\nu e \equiv \mu\omega e$ (mod. $n - 1$) and so $g^{\nu e} \equiv g^{\mu\omega e} \equiv G^{\mu e}$ (mod. n) or $H^\mu \equiv h^\nu$; i.e. any number in the second series $1, H, H^2$, etc. will be congruent to a number in the series $1, h, h^2, \ldots$ and vice versa. Thus the f roots $[1], [h], [h^2], \ldots [h^{f-1}]$ will be identical with $[1], [H], [H^2], \ldots [H^{f-1}]$. In the same way it is easy to see that the more general series

$$[\lambda], [\lambda h], [\lambda h^2], \ldots [\lambda h^{f-1}] \quad \text{and} \quad [\lambda], [\lambda H], [\lambda H^2], \ldots [\lambda H^{f-1}]$$

coincide. We will designate the *aggregate* of f such roots, $[\lambda] + [\lambda h] + \text{etc.} + [\lambda h^{f-1}]$ by (f, λ). Since it is not changed by taking a different primitive root g, it must be considered as independent of g. And we will call the *complex* of the same roots the *period* (f, λ) and we disregard the order of the roots.[a] To indicate such a period it will be convenient to reduce each root to its simplest expression, that is, to substitute for the numbers $\lambda, \lambda h, \lambda h^2$, etc. their least residues relative to the modulus n. And if we wish we can order the terms of the period according to size.

For example, for $n = 19$, 2 is a primitive root and the period $(6, 1)$ consists of the roots $[1], [8], [64], [512], [4096], [32768]$ or

[a] In what follows we can also call the sum the numerical value of the period, or simply the period, when there is no fear of ambiguity.

[1], [7], [8], [11], [12], [18]. Similarly, the period (6, 2) consists of the roots [2], [3], [5], [14], [16], [17]. The period (6, 3) is identical with the preceding. The period (6, 4) contains the roots [4], [6], [9], [10], [13], [15].

▶ 344. We present immediately the following observations concerning periods of this type:

I. Since $\lambda h^f \equiv \lambda$, $\lambda h^{f+1} \equiv \lambda h$, etc. (mod. n), it is clear that (f, λ), $(f, \lambda h)$, $(f, \lambda h^2)$, etc. are composed of the same roots. In general therefore if we designate by $[\lambda']$ any root of (f, λ), this period will be completely identical with (f, λ'). If therefore two periods which have the same number of roots (we will call them *similar*) have one root in common, they will be identical. Therefore it cannot happen that two roots are contained together in a period and only one of them is found in another similar period. Further, if two roots $[\lambda]$; $[\lambda']$ belong to the same period of f terms, the value of the expression λ'/λ (mod. n) is congruent to some power of h; that is, we can presume that $\lambda' \equiv \lambda g^{ve}$ (mod. n).

II. If $f = n - 1$, $e = 1$ the period $(f, 1)$ will coincide with Ω. In the remaining cases Ω will be composed of the periods $(f, 1)$, (f, g), (f, g^2), ... (f, g^{e-1}). Therefore these periods will be completely different from one another and it is clear that any other similar period (f, λ) will coincide with one of these if $[\lambda]$ belongs to Ω; i.e. if λ is not divisible by n. The period $(f, 0)$ or (f, kn) is manifestly composed of f unities. It is also clear that if λ is any number nondivisible by n, the complex of e periods (f, λ), $(f, \lambda g)$, $(f, \lambda g^2)$ ... $(f, \lambda g^{e-1})$ will also coincide with Ω. Thus, e.g., for $n = 19$, $f = 6$, Ω will consist of the three periods (6, 1), (6, 2), (6, 4). Any other similar period, except (6, 0) can be reduced to one of these.

III. If $n - 1$ is the product of three positive numbers a, b, c, it is manifest that any period of bc terms is composed of b periods of c terms; for example (bc, λ) is composed of (c, λ), $(c, \lambda g^a)$, $(c, \lambda g^{2a})$ ... $(c, \lambda g^{ab-a})$. Thus these latter are said to be contained in the former. So for $n = 19$ the period (6, 1) consists of the three periods (2, 1), (2, 8), (2, 7). The first contains the roots r, r^{18}; the second r^8, r^{11}; the third r^7, r^{12}.

▶ 345. THEOREM. *Let (f, λ), (f, μ) be two similar periods which are identical or different, and let (f, λ) consist of the roots $[\lambda]$, $[\lambda']$, $[\lambda'']$,* *etc. Then the product of (f, λ) into (f, μ) will be the sum of f similar periods; that is to say*

$$= (f, \lambda + \mu) + (f, \lambda' + \mu) + (f, \lambda'' + \mu) + \text{etc.} = W$$

Demonstration. Let as above $n - 1 = ef$; g a primitive root for the modulus n and $h = g^e$. From what we have said above, the product we want will be

$$= [\mu] \cdot (f, \lambda) + [\mu h] \cdot (f, \lambda h) + [\mu h^2] \cdot (f, \lambda h^2) + \text{etc.}$$

and so

$$= [\lambda + \mu] \qquad + [\lambda h + \mu] \ldots \qquad + [\lambda h^{f-1} + \mu]$$

$$+ [\lambda h + \mu h] \qquad + [\lambda h^2 + \mu h] \ldots \qquad + [\lambda h^f + \mu h]$$

$$+ [\lambda h^2 + \mu h^2] + [\lambda h^3 + \mu h^2] \ldots + [\lambda h^{f+1} + \mu h^2] \text{ etc.}$$

This expression will contain all f^2 of the roots. And if we add the vertical columns together we will have

$$(f, \lambda + \mu) + (f, \lambda h + \mu) + \ldots + (f, \lambda h^{f-1} + \mu)$$

This expression coincides with W because by hypothesis the numbers $\lambda, \lambda', \lambda''$, etc. are congruent to $\lambda, \lambda h, \lambda h^2, \ldots \lambda h^{f-1}$ relative to the modulus n (we are not concerned with order here) and so also

$$\lambda + \mu, \quad \lambda' + \mu, \quad \lambda'' + \mu, \text{ etc.}$$

will be congruent to

$$\lambda + \mu, \quad \lambda h + \mu, \quad \lambda h^2 + \mu, \ldots \lambda h^{f-1} + \mu \qquad \text{Q.E.D.}$$

We add the following corollaries to this theorem:

I. If k is any integer, the product of $(f, k\lambda)$ into $(f, k\mu)$ will be

$$= (f, k(\lambda + \mu)) + (f, k(\lambda' + \mu)) + (f, k(\lambda'' + u)) + \text{etc.}$$

II. Since the single terms of W coincide either with the sum $(f, 0)$ which $= f$, or with one of the sums $(f, 1), (f, g), (f, g^2) \ldots$ (f, g^{e-1}), W can be reduced to the following form

$$W = af + b(f, 1) + b'(f, g) + b''(f, g^2) + \ldots + b^\varepsilon(f, g^{e-1})$$

where the coefficients a, b, b', etc. are positive integers (or some may even $= 0$). It is further clear that the product of $(f, k\lambda)$ into

$(f, k\mu)$ will then become

$$= af + b(f, k) + b'(f, kg) + \ldots + b^\varepsilon(f, kg^{e-1})$$

Thus, e.g., for $n = 19$ the product of the sum $(6, 1)$ into itself or the square of this sum will be $= (6, 2) + (6, 8) + (6, 9) + (6, 12) + (6, 13) + (6, 19) = 6 + 2(6, 1) + (6, 2) + 2(6, 4)$.

III. Since the product of the individual terms of W into a similar period (f, v) can be reduced to an analogous form, it is manifest that the product of three periods $(f, \lambda) \cdot (f, \mu) \cdot (f, v)$ can be represented by $cf + d(f, 1) \ldots + d^\varepsilon(f, g^{e-1})$ and the coefficients c, d, etc. will be integers and positive (or $= 0$) and for any integral value of k we have

$$(f, k\lambda) \cdot (f, k\mu) \cdot (f, kv) = cf + d(f, k) + d'(f, kg) + \text{etc.}$$

This theorem can be extended to the product of any number of similar periods, and it does not matter whether these periods are all different or partly or all identical.

IV. It follows from this that if in any rational integral algebraic function $F = \phi(t, u, v, \ldots)$ we substitute for the unknowns t, u, v, etc. respectively the similar periods $(f, \lambda), (f, \mu), (f, v)$, etc., its value will be reducible to the form

$$A + B(f, 1) + B'(f, g) + B''(f, g^2) \ldots + B^\varepsilon(f, g^{e-1})$$

and the coefficients A, B, B', etc. will all be integers if all the coefficients in F are integers. But if afterward we substitute $(f, k\lambda), (f, k\mu), (f, kv)$, etc. for t, u, v, etc. respectively, the value of F will be reduced to $A + B(f, k) + B'(f, kg) + \text{etc.}$

▶ 346. THEOREM. *If we suppose that λ is a number not divisible by n, and if for brevity we write p for (f, λ), any other similar period (f, μ) which has μ not divisible by n can be reduced to a form*

$$\alpha + \beta p + \gamma p^2 + \ldots + \theta p^{e-1}$$

where the coefficients α, β, etc. are determined rational quantities.

Demonstration. Let us designate by p', p'', p''', etc. the periods $(f, \lambda g), (f, \lambda g^2), (f, \lambda g^3)$, etc. on up to $(f, \lambda g^{e-1})$. Their number will be $e - 1$ and one of them will necessarily coincide with (f, μ). We immediately have the equation

$$0 = 1 + p + p' + p'' + p''' + \text{etc.} \dots \text{(I)}$$

Now if according to the rules of the preceding article we form the powers of p up to p^{e-1}, we will have $e - 2$ other equations

$$0 = p^2 + A + ap + a'p' + a''p'' + a'''p''' + \text{etc.} \dots \text{(II)}$$

$$0 = p^3 + B + bp + b'p' + b''p'' + b'''p''' + \text{etc.} \dots \text{(III)}$$

$$0 = p^4 + C + cp + c'p' + c''p'' + c'''p''' + \text{etc.} \dots \text{(IV) etc.}$$

All the coefficients A, a, a', etc.; B, b, b', etc.; etc. will be integers and as follows immediately from the preceding article, completely independent of λ; that is, the same equations will hold no matter what value we give to λ. This remark can also be extended to equation I as long as λ is not divisible by n. Let us suppose that $(f, \mu) = p'$; for it is easy to see that if (f, μ) coincides with any other period of the p'', p''', etc. the following line of argument can be used in a completely analogous way. Since the number of equations I, II, III, etc. is $e - 1$, the quantities p'', p''', etc. whose number is $= e - 2$ can be eliminated from them by known methods. The resulting equation (Z) will be free from them:

$$0 = \mathfrak{A} + \mathfrak{B}p + \mathfrak{C}p^2 + \text{etc.} + \mathfrak{M}p^{e-1} + \mathfrak{N}p'$$

This can be done in such a way that all the coefficients $\mathfrak{A}, \mathfrak{B}, \dots \mathfrak{N}$ are integers and certainly not all $= 0$. Now if we do not have $\mathfrak{N} = 0$, it follows that p' can be determined as the theorem demands. It remains therefore to show that we cannot have $\mathfrak{N} = 0$.

Suppose that $\mathfrak{N} = 0$. The equation Z becomes $\mathfrak{M}p^{e-1} + \text{etc.} + \mathfrak{B}p + \mathfrak{A} = 0$. Since this cannot be higher than degree $e - 1$, it cannot satisfy more than $e - 1$ different values of p. But since the equations from which Z is deduced are independent of λ, it follows that Z does not depend on λ and so it will hold, no matter what integer not divisible by n is taken for λ. Therefore this equation will be satisfied by any of the sums $(f, 1), (f, g), (f, g^2), \dots (f, g^{e-1})$, and it follows immediately that not all these sums can be unequal but at least two of them must be equal. Let one of these two equal sums contain the roots $[\zeta], [\zeta'], [\zeta'']$, etc. and the other the roots $[\eta]$, $[\eta'], [\eta'']$, etc. We will suppose (this is legitimate) that all the numbers ζ, ζ', ζ'', etc., η, η', η'', etc. are positive and $< n$. Manifestly all

will be different and none of them $=0$. We will designate by Y
the function

$$x^\zeta + x^{\zeta'} + x^{\zeta''} + \text{etc.} - x^\eta - x^{\eta'} - x^{\eta''} - \text{etc.}$$

Its highest term cannot exceed x^{n-1} and $Y = 0$ if we set $x = [1]$.
Thus Y will have a factor $x - [1]$ in *common* with the function
denoted by X in the preceding. It is easy to show that this would
be absurd. For if Y and X have a common factor, the greatest
common divisor of the functions X, Y (it cannot have dimension
$n - 1$ because Y is divisible by x) would have all of its coefficients
rational. This would follow from the nature of the operation involved in finding the greatest common divisor of two functions
whose coefficients are all rational. But in article 341 we showed
that X cannot have a factor with rational coefficients of dimension
less than $n - 1$. Therefore the supposition that $\mathfrak{N} = 0$ cannot be
consistent.

Example. For $n = 19$, $f = 6$ we have $p^2 = 6 + 2p + p' + 2p'$.
Since $0 = 1 + p + p' + p''$ we deduce that $p' = 4 - p^2$, $p'' = -5 - p + p^2$. Therefore

$$(6, 2) = 4 - (6, 1)^2, \qquad (6, 4) = -5 - (6, 1) + (6, 1)^2$$

$$(6, 4) = 4 - (6, 2)^2, \qquad (6, 1) = -5 - (6, 2) + (6, 2)^2$$

$$(6, 1) = 4 - (6, 4)^2, \qquad (6, 2) = -5 - (6, 4) + (6, 4)^2$$

▶347. THEOREM. *If* $F = \phi(t, u, v, \ldots)$ *is an invariable[b] rational
integral algebraic function in* f *unknowns* t, u, v, *etc. and if we substitute for these the* f *roots contained in the period* (f, λ), *by the
rules of article 340 the value of* F *is reduced to the form*

$$A + A'[1] + A''[2] + \text{etc.} = W$$

and the roots of this expression which belong to the same period of f
terms will have equal coefficients.

Demonstration. Let $[p]$, $[q]$ be two roots belonging to the same

---

[b] Invariable functions are those in which all the unknowns are contained in the same
way or, more clearly, functions which are not changed no matter how the unknowns are
permuted; such are, e.g., the sum of the unknowns, their product, the sum of the products
of pairs of them, etc.

period and suppose that p, q are positive and less than n. We must show that $[p]$ and $[q]$ have the same coefficient in W. Let $q \equiv pg^{ve}$ (mod. n); and let the roots contained in (f, λ) be $[\lambda], [\lambda'], [\lambda'']$, etc. where we suppose that $\lambda, \lambda', \lambda''$, etc. are positive and less than n; finally, let the least positive residues of the numbers $\lambda g^{ve}, \lambda' g^{ve}$, $\lambda'' g^{ve}$, etc. relative to the modulus n be μ, μ', μ'', etc. Manifestly they will be identical with the numbers $\lambda, \lambda', \lambda''$, etc., although the order may be transposed. From article 340 it is clear that

$$\phi([\lambda g^{ve}], [\lambda' g^{ve}], [\lambda'' g^{ve}], \ldots) = (I)$$

is reduced to

$$A + A'[g^{ve}] + A''[2g^{ve}] + \text{etc.} \qquad \text{or to}$$

$$A + A'[\theta] + A''[\theta'] + \text{etc.} = (W')$$

Here θ, θ', etc. are the least residues of the numbers $g^{ve}, 2g^{ve}$, etc. relative to the modulus n and so we see that $[q]$ has the same coefficient in (W') as $[p]$ has in (W). If we manipulate the expression (I) we will get the same thing we get from manipulating the expression $\phi([\mu], [\mu'], [\mu''], \text{etc.})$ because $\mu \equiv \lambda g^{ve}, \mu' \equiv \lambda' g^{ve}$, etc. (mod. n). Now this last expression produces the same result as $\phi([\lambda], [\lambda'], [\lambda''], \text{etc.})$ since the numbers μ, μ', μ'', etc. differ only in order from the numbers $\lambda, \lambda', \lambda''$, etc. and this does not matter in an invariable function. Thus W' is completely identical with W and so the root $[q]$ will have the same coefficient in W as $[p]$. Q.E.D.

We see therefore that W can be reduced to the form

$$A + a(f, 1) + a'(f, g) + a''(f, g^2) \ldots + a^{\varepsilon}(f, g^{e-1})$$

and the coefficients $A, a, \ldots a^{\varepsilon}$ will be determined quantities and integers if all the rational coefficients in F are integers. Thus, e.g., if $n = 19$, $f = 6$, $\lambda = 1$ and the function ϕ designates the sum of products of the unknowns taken two by two, its value is reduced to $3 + (6, 1) + (6, 4)$.

If after this the roots of another period $(f, k\lambda)$ are substituted for t, u, v, etc., the value of F will become

$$A + a(f, k) + a'(f, kg) + a''(f, kg^2) + \text{etc.}$$

▶ 348. In the equation

$$x^f - \alpha x^{f-1} + \beta x^{f-2} - \gamma x^{f-3} \ldots = 0$$

the coefficients α, β, γ, etc. are invariable functions of the roots; that is α is the sum of all of them, β is the sum of their products taken two at a time, γ the sum of their products taken three at a time, etc. Therefore in the equation whose roots are the ones contained in the period (f, λ), the first coefficient will $= (f, \lambda)$ and each of the others can be reduced to the form

$$A + a(f, 1) + a'(f, g) \ldots + a^\varepsilon(f, g^{e-1})$$

with all the A, a, a', etc. integers. It is further evident that the equation whose roots are the roots contained in any other period $(f, k\lambda)$ can be derived from the latter by substituting (f, k) for $(f, 1)$ in each of the coefficients, (f, kg) for (f, g), and in general (f, kp) for (f, p). In this way therefore we can assign e equations $z = 0$, $z' = 0$, $z'' = 0$, etc. whose roots will be the roots contained in $(f, 1), (f, g), (f, g^2)$, etc. as soon as we know the e sums $(f, 1)$, (f, g), (f, g^2), etc. or rather as soon as we find any *one* of them. This is true because, by article 346, all the rest can be deduced rationally from one of them. This done, the function X will be resolved into e factors of degree f, for manifestly the product of the functions z, z', z'', etc. will $= X$.

   Example. For $n = 19$ the sum of all the roots in the period $(6, 1)$ $= (6, 1) = \alpha$; the sum of their products taken two at a time $= 3 + (6, 1) + (6, 4) = \beta$; similarly, the sum of the products taken three at a time $= 2 + 2(6, 1) + (6, 2) = \gamma$; the sum of the products taken four at a time $= 3 + (6, 1) + (6, 4) = \delta$; the sum of products taken five at a time $= (6, 1) = \varepsilon$; the product of all of them $= 1$. Thus the equation

$$z = x^6 - \alpha x^5 + \beta x^4 - \gamma x^3 + \delta x^2 - \varepsilon x + 1 = 0$$

will contain all the roots included in $(6, 1)$. And if we substitute $(6, 2), (6, 4), (6, 1)$ for $(6, 1), (6, 2), (6, 4)$ respectively in the coefficients α, β, γ, etc. we will get the equation $z' = 0$ which will contain the roots of $(6, 2)$. And if the same permutation is applied again we will have the equation $z'' = 0$ containing the roots of $(6, 4)$, and the product $z z' z'' = X$.

▶ 349. It is often more convenient, especially when f is a large number, to deduce the coefficients β, γ, etc. from the sums of the powers of the roots by Newton's theorem. Thus the sum of the squares of roots contained in (f, λ) is $= (f, 2\lambda)$, the sum of the cubes is $=(f, 3\lambda)$, etc. If we write q, q', q'', etc. for (f, λ), $(f, 2\lambda)$, $(f, 3\lambda)$, etc. we will have

$$\alpha = q, \qquad 2\beta = \alpha q - q', \qquad 3\gamma = \beta q - \alpha q' + q'', \text{ etc.}$$

Here by article 345 the product of two periods is to be converted immediately into a sum of periods. Thus in our example, if we write p, p', p'' respectively for $(6, 1)$, $(6, 2)$, $(6, 4)$ we will have $q, q', q'', q''', q'''', q''''' $ respectively $= p, p', p', p'', p', p''$; thus

$$\alpha = p, \qquad 2\beta = p^2 - pp' = 6 + 2p + 2p''$$

$$3\gamma = (3 + p + p'')p - pp' + p' = 6 + 6p + 3p'$$

$$4\delta = (2 + 2p + p')p - (3 + p + p'')p' + pp' - p''$$

$$= 12 + 4p + 4p'', \text{ etc.}$$

However, it is sufficient to compute half the coefficients in this way, for it is not difficult to prove that the last are equal to the first in inverse order; that is, the last $=1$, the second last $=\alpha$, the third last $=\beta$, etc.; or, another way, the last can be derived from the first by substituting for $(f, 1)$, (f, g), etc. the periods $(f, -1)$, $(f, -g)$, etc. or $(f, n - 1)$, $(f, n - g)$, etc. The former case holds when f is even, the latter when f is odd. The last coefficient, however, will always $=1$. The basis for this is established by the theorem of article 79, but for the sake of brevity we will not dwell on the argument.

▶ 350. THEOREM. *Let* $n - 1$ *be the product of the three positive integers* α, β, γ *and let the period* $(\beta\gamma, \lambda)$ *which has* $\beta\gamma$ *terms be composed of* β *lesser periods of* γ *terms,* (γ, λ), (γ, λ'), (γ, λ''), *etc. Let us suppose further that in a function of* β *unknowns just as in article 347, that is in* $F = \phi(t, u, v, \ldots)$, *we substitute the sums* (γ, λ), (γ, λ'), (γ, λ''), *etc. for the unknowns* t, u, v, *etc. respectively and that according to the rules of the article 345.IV its value is reduced to*

$$A + a(\gamma, 1) + a'(\gamma, g) \ldots + a^\zeta(\gamma g^{\alpha\beta - \alpha}) \ldots + a^\theta(\gamma, g^{\alpha\beta - 1}) = W$$

then I say that if F *is an invariable function, the periods in* W *which are contained in the same period of the* $\beta\gamma$ *terms (i.e. in general the periods* (γ, g^μ) *and* $(\gamma, g^{\alpha\nu + \mu})$ *where* ν *is any integer) will have the same coefficients.*

Demonstration. Since the period $(\beta\gamma, \lambda g^\alpha)$ is identical with $(\beta\gamma, \lambda)$, the lesser periods $(\gamma, \lambda g^\alpha)$, $(\gamma, \lambda' g^\alpha)$, $(\gamma, \lambda'' g^\alpha)$, etc. which comprise the former, necessarily coincide with those that comprise the latter, although in a different order. And if we suppose that F will be transformed into W' by substituting the former quantities for t, u, v, etc., respectively, W' will coincide with W. But by article 347 we have

$$W' = A + a(\gamma, g^\alpha) + a'(\gamma, g^{\alpha+1})\ldots + a^\zeta(\gamma, g^{\alpha\beta})\ldots + a^\theta(\gamma, g^{\alpha\beta+\alpha-1})$$

$$= A + a(\gamma, g^\alpha) + a'(\gamma, g^{\alpha+1})\ldots + a^\zeta(\gamma, 1)\ldots + a^\theta(\gamma, g^{\alpha-1})$$

so this expression must coincide with W and the first, second, third, etc. coefficients in W (beginning with a) must coincide with the $(\alpha + 1)$st, the $(\alpha + 2)$nd, the $(\alpha + 3)$rd, etc. And we conclude in general that the coefficients of the periods (γ, g^μ), $(\gamma, g^{\alpha+\mu})$, $(\gamma, g^{2\alpha+\mu})$ $\ldots(\gamma, g^{\nu\alpha+\mu})$, which are the $\mu + 1$st, the $\alpha + \mu + 1$st, the $2\alpha + \mu + 1$st$\ldots\nu\alpha + \mu + 1$st, must coincide with one another.   Q.E.D.

This it is clear that W can be reduced to the form

$$A + a(\beta\gamma, 1) + a'(\beta\gamma, g)\ldots + a^\varepsilon(\beta\gamma, g^{\alpha-1})$$

with all the coefficients A, a, etc. integers when all the coefficients in F are integers. Suppose after this we substitute in F in place of the unknowns, β periods of γ terms which constitute another period of $\beta\gamma$ terms, for example those contained in $(\beta\gamma, \lambda k)$ which are $(\gamma, \lambda k)$, $(\gamma, \lambda' k)$, $(\gamma, \lambda'' k)$, etc. Then the resulting value will be $A + a(\beta\gamma, k) + a'(\beta\gamma, gk)\ldots + a^\varepsilon(\beta\gamma, g^{\alpha-1}k)$.

It is obvious that the theorem can also be extended to the case where $\alpha = 1$ or $\beta\gamma = n - 1$. In this case *all* the coefficients in W will be equal, and W will be reduced to the form $A + a(\beta\gamma, 1)$.

▶ 351. Now keeping the terminology of the preceding article, it is clear that the individual coefficients of the equation whose roots are the β sums (γ, λ), (γ, λ'), (γ, λ''), etc. can be reduced to a form like

$$A + a(\beta\gamma, 1) + a'(\beta\gamma, g)\ldots + a^\varepsilon(\beta\gamma, g^{\alpha-1})$$

and the numbers A, a, etc. will all be integers. And we can derive from this the equation whose roots are the β periods of γ terms contained in another period $(\beta\gamma, k\lambda)$ if in every coefficient we substitute $(\beta\gamma, k\mu)$ for every period $(\beta\gamma, \mu)$. If therefore $\alpha = 1$ all β

periods of γ terms will be determined by an equation of degree β, and each of the coefficients will be of the form $A + a(\beta\gamma, 1)$. As a result, *they will all be known quantities* because $(\beta\gamma, 1) = (n - 1, 1)$ $= -1$. If $\alpha > 1$, the coefficients of the equation whose roots are all the periods of γ terms contained in a given period of $\beta\gamma$ terms will be known quantities as long as all the numerical values of all α periods of $\beta\gamma$ terms are known. The calculation of the coefficients of these equations will often be much easier, especially when β is not very small, if first we calculate the sums of the powers of the roots and deduce from these the coefficients by the theorem of Newton, just as we did above in article 349.

Example 1. For $n = 19$ we want the equation whose roots are the sum $(6, 1), (6, 2), (6, 4)$. If we designate these roots by p, p', p'', etc. respectively and the equation we want by

$$x^3 - Ax^2 + Bx - C = 0$$

we get

$$A = p + p' + p'', \qquad B = pp' + pp'' + p'p'', \qquad C = pp'p''$$

and then

$$A = (18, 1) = -1$$

and

$$pp' = p + 2p' + 3p'', \quad pp'' = 2p + 3p' + p'', \quad p'p'' = 3p + p' + 2p''$$

so

$$B = 6(p + p' + p'') = 6(18, 1) = -6$$

and finally

$$C = (p + 2p' + 3p'')p'' = 3(6, 0) + 11(p + p' + p'') = 18 - 11 = 7$$

therefore the equation we want is

$$x^3 + x^2 - 6x - 7 = 0$$

Using the other method, we have

$$p + p' + p'' = -1$$
$$p^2 = 6 + 2p + p' + 2p'', \qquad p'p' = 6 + 2p' + p'' + 2p,$$
$$p''p'' = 6 + 2p'' + p + 2p'$$

therefore

$$p^2 + p'p' + p''p'' = 18 + 5(p + p' + p'') = 13$$

and similarly

$$p^3 + p'^3 + p''^3 = 36 + 34(p + p' + p'') = 2$$

Thus by Newton's theorem the same equation will be derived as before.

II. For $n = 19$ we want the equation whose roots are the sums $(2, 1), (2, 7), (2, 8)$. If we designate them by q, q', q'' we find

$$q + q' + q'' = (6, 1), \qquad qq' + qq'' + q'q'' = (6, 1) + (6, 4),$$

$$qq'q'' = 2 + (6, 2)$$

and so, keeping the same signs as in the preceding, the equation we want will be

$$x^3 - px^2 + (p + p'')x - 2 - p' = 0$$

The equation whose roots are the sums $(2, 2), (2, 3), (2, 5)$ contained in $(6, 2)$ can be deduced from the preceding by substituting p', p'', p for p, p', p'', respectively, and if we make the same substitution once again we will get the equation whose roots are the sums $(2, 4), (2, 6), (2, 9)$ contained in $(6, 4)$.

▶ 352. The preceding theorems contain along with their corollaries the basic principles of the whole theory, and the method of finding the values of the roots Ω can now be treated in a few words.

First we must take a number g which is a primitive root for the modulus n and find the least residues of the powers of g up to g^{n-2} relative to the modulus n. Resolve $n - 1$ into factors, and indeed into prime factors if we want to reduce the problem to equations of the lowest possible degree. Let these (the order is arbitrary) be $\alpha, \beta, \gamma, \ldots \zeta$ and set

$$\frac{n-1}{\alpha} = \beta\gamma \ldots \zeta = a, \qquad \frac{n-1}{\alpha\beta} = \gamma \ldots \zeta = b, \text{ etc.}$$

Distribute all the roots Ω into α periods of a terms, each of these again into β periods of b terms, and each of these again into γ periods, etc. According to the preceding article we want an equation (A) of degree α, whose roots are the α sums of a terms whose values can be determined by solving this equation.

But here a difficulty arises because it seems to be uncertain which sum should be made equal to which root of the equation (A); that is, which root should be denoted by $(a, 1)$, which by (a, g), etc. We can solve this difficulty in the following way. We can designate by $(a, 1)$ any root at all of the equation (A); for since any root of this equation is the sum of a roots of Ω, and it is completely arbitrary which root of Ω is denoted by [1], we will be free to assume that [1] expresses any of the roots which constitute a given root of equation (A) and that this root of equation (A) will be $(a, 1)$. The root [1] will not yet be completely determined, but, even if it remains entirely arbitrary or indefinite, we can adopt for [1] any of the roots that make up $(a, 1)$. As soon as $(a, 1)$ is determined, all the remaining sums of a terms can be deduced from it (art. 346). Thus it is clear that we need find only one root by this resolution. We can also use the following less direct method for the same purpose. Take for [1] a definite root; i.e. let $[1] = \cos kP/n + i \sin kP/n$ with the integer k taken arbitrarily but in such a way that it is not divisible by n. When this is done [2], [3], etc. will also determine definite roots, and the sums $(a, 1)$, (a, g), etc. will designate definite quantities. Now if these quantities are calculated from a table of sines with just enough precision so that one can decide which are the larger and which the smaller, there will be no doubt left as to how to distinguish the individual roots of the equation (A).

When in this way we have found all α sums of a terms, we will investigate by the methods of the preceding article the equation (B) of degree β, whose roots are the β sums of b terms contained in $(a, 1)$. The coefficients of this equation will all be known quantities. Therefore, since it is arbitrary which of the $a = \beta b$ roots contained in $(a, 1)$ is denoted by [1], any given root of equation (B) can be expressed by $(b, 1)$ because it is licit to suppose that one of the b roots of which it is composed is denoted by [1]. We will investigate therefore any one root of the equation (B) by its resolution. Let it $= (b, 1)$ and derive from it by article 346 all the remaining sums of b terms. In this way we have at the same time a method of confirming the calculation, since the sums of b terms which belong to the same periods of a terms is known. In some cases it is just as easy to form $\alpha - 1$ other equations of degree β, whose roots are respectively the individual β sums of b terms contained in the remaining periods of a terms (a, g), (a, g^2), etc. and to investigate *all* roots by the resolution of these equations and of the equation B. Then in the same way as above with the help of a table of sines we can decide which are the periods of b terms to which the individual roots found in this way are equal. But to help in this judgment various other devices can be used which cannot be fully explained here. One of them, however, for the case where $\beta = 2$ is especially useful and can be explained more briefly by illustration than by rule. We will use it in the following examples.

After we have found the values of all the $\alpha\beta$ sums of b terms in this way, we can use a similar method to determine by equations of degree γ all the $\alpha\beta\gamma$ sums of c terms. That is, we can *either* get *one* equation of degree γ whose roots, according to article 350, are the γ sums of c terms contained in $(b, 1)$, and by solving this find a root and let it $=(c, 1)$ and finally from this by the methods of article 346 deduce all the remaining sums; *or* in a similar way evolve the $\alpha\beta$ equations of degree γ whose roots are respectively the γ sums of c terms contained in the individual periods of b terms. We can solve all these equations for all their roots and determine the order of the roots with the help of the table of sines as we did above. However, for $\gamma = 2$ we can use the device we will demonstrate below.

If we continue in this way we will finally have all the $(n - 1)/\zeta$ sums of ζ terms; and if we find by the methods of article 348 the equation of degree ζ whose roots are the ζ roots of Ω contained in $(\zeta, 1)$, all its coefficients will be known quantities. And if we solve for any one of its roots, we can let it $=[1]$, and its powers will give us all the other roots Ω. If we prefer, we could solve for *all* the roots of this equation. Then by the solution of the other $[(n - 1)/\zeta] - 1$ equations of degree ζ, which contain respectively all the ζ roots in each of the remaining periods of ζ terms, we can find all the remaining roots Ω.

It is clear, however, that as soon as the first equation (A) is solved, or as soon as we have the values of all the α sums of a terms, we will also have the resolution of the function X into α factors of a dimensions, by article 348. Further, after solving equation (B) or after finding the values of all the $\alpha\beta$ sums of b terms, each of those factors will be resolved again into β factors, and so X will be resolved into $\alpha\beta$ factors of b dimensions etc.

▶ 353. *Example number one for* $n = 19$. Since here we have $n - 1 = 3 \cdot 3 \cdot 2$ finding the roots Ω is reduced to the solution of two cubic and one quadratic equation. This example is more easily understood because for the most part the necessary operations have already been discussed above. If we take the number 2 as the primitive root g, the least residues of its powers will produce the following (the exponents of the powers are written in the first line and the residues in the second):

0. 1. 2. 3. 4. 5. 6. 7. 8. 9. 10. 11. 12. 13. 14. 15. 16. 17

1. 2. 4. 8. 16. 13. 7. 14. 9. 18. 17. 15. 11. 3. 6. 12. 5. 10

From this, by articles 344, 345, we can easily find the following distribution of all the roots Ω into three periods of six terms and of each of these into three periods of two terms:

$$
\Omega = (18, 1)
\begin{cases}
(6, 1) \begin{cases} (2, 1) \; \ldots [1], [18] \\ (2, 8) \; \ldots [8], [11] \\ (2, 7) \; \ldots [7], [12] \end{cases} \\[2ex]
(6, 2) \begin{cases} (2, 2) \; \ldots [2], [17] \\ (2, 16) \ldots [3], [16] \\ (2, 14) \ldots [5], [14] \end{cases} \\[2ex]
(6, 4) \begin{cases} (2, 4) \; \ldots [4], [15] \\ (2, 13) \ldots [6], [13] \\ (2, 9) \; \ldots [9], [10] \end{cases}
\end{cases}
$$

The equation (A) whose roots are the sums $(6, 1), (6, 2), (6, 4)$ is found to be $x^3 + x^2 - 6x - 7 = 0$ and one of the roots is -1.2218761623. If we call this root $(6, 1)$ we have

$$(6, 2) = \quad 4 - (6, 1)^2 \qquad = 2.5070186441$$
$$(6, 4) = -5 - (6, 1) + (6, 1)^2 = -2.2851424818$$

Thus X is resolved into three factors of 6 dimensions, if these values are substituted in article 348.

The equation (B) whose roots are the sums $(2, 1), (2, 7), (2, 8)$ gives the equation

$$x^3 - (6, 1)x^2 + [(6, 1) + (6, 4)]x - 2 - (6, 2) = 0$$

or

$$x^3 + 1.2218761623x^2 - 3.5070186441x - 4.5070186441 = 0$$

One root is -1.3545631433 which we will call $(2, 1)$. By the method of article 346 we find the following equations (for brevity we will write q for $(2, 1)$).

$(2, 2) = q^2 - 2$, $(2, 3) = q^3 - 3q$, $(2, 4) = q^4 - 4q^2 + 2$,

$(2, 5) = q^5 - 5q^3 + 5q$, $(2, 6) = q^6 - 6q^4 + 9q^2 - 2$,

$$(2, 7) = q^7 - 7q^5 + 14q^3 - 7q$$

$$(2, 8) = q^8 - 8q^6 + 20q^4 - 16q^2 + 2$$

$$(2, 9) = q^9 - 9q^7 + 27q^5 - 30q^3 + 9q$$

In the present case these equations can be solved more easily as follows than by the methods of article 346. If we suppose that

$$[1] = \cos\frac{kP}{19} + i\sin\frac{kP}{19}$$

we have

$$[18] = \cos\frac{18kP}{19} + i\sin\frac{18kP}{19} = \cos\frac{kP}{19} - i\sin\frac{kP}{19}$$

and so

$$(2, 1) = 2\cos\frac{kP}{19}$$

and in general

$$[\lambda] = \cos\frac{\lambda kP}{19} + i\sin\frac{\lambda kP}{19} \quad \text{and} \quad \text{so}$$

$$(2, \lambda) = [\lambda] + [18\lambda] = [\lambda] + [-\lambda] = 2\cos\frac{\lambda kP}{19}$$

Therefore if $q/2 = \cos\omega$, we will have $(2, 2) = 2\cos 2\omega$, $(2, 3) = 2\cos 3\omega$ etc., and the same formulae as above will be derived from known equations for the cosines of multiple angles. Now from these formulae we derive the following numerical values:

$(2, 2) = -0.1651586909$	$(2, 6) = 0.4909709743$
$(2, 3) = 1.5782810188$	$(2, 7) = -1.7589475024$
$(2, 4) = -1.9727226068$	$(2, 8) = 1.8916344834$
$(2, 5) = 1.0938963162$	$(2, 9) = -0.8033908493$

The values of $(2, 7), (2, 8)$ can also be found from equation (B) of which they are the two remaining roots. And the doubt as to *which* of these roots is $(2, 7)$ and which $(2, 8)$ can be removed either by an approximate calculation according to the formulae given above or by means of sine tables. A cursory reference shows us that $(2, 1) = 2\cos\omega$ by letting $\omega = 7P/19$ and so we have

$$(2, 7) = 2\cos\tfrac{49}{19}P = 2\cos\tfrac{8}{19}P, \quad \text{and}$$
$$(2, 8) = 2\cos\tfrac{56}{19}P = 2\cos\tfrac{1}{19}P$$

Similarly we can find the sums $(2, 2), (2, 3), (2, 5)$ also by the equation

$$x^3 - (6, 2)x^2 + [(6, 1) + (6, 2)]x - 2 - (6, 4) = 0$$

whose roots they are, and the uncertainty as to which roots correspond to which sums can be removed in exactly the same way as before. Finally the sums $(2, 4), (2, 6), (2, 9)$ can be found by the equation

$$x^3 - (6, 4)x^2 + [(6, 2) + (6, 4)]x - 2 - (6, 1) = 0$$

[1] and [18] are the roots of the equation $x^2 - (2, 1)x + 1 = 0$. One of them

$$= \tfrac{1}{2}(2, 1) + i\sqrt{[1 - \tfrac{1}{4}(2, 1)^2]} = \tfrac{1}{2}(2, 1) + i\sqrt{[\tfrac{1}{2} - \tfrac{1}{4}(2, 2)]}$$

and the other

$$= \tfrac{1}{2}(2, 1) - i\sqrt{[\tfrac{1}{2} - \tfrac{1}{4}(2, 2)]}$$

and the numerical values will be

$$= -0.6772815716 \pm 0.7357239107\,i$$

The sixteen remaining roots can be found either from the powers of one or the other of these roots or by solving eight of the other similar equations. To decide which root has the positive sign for its imaginary part and which the negative we can use the method above or sine tables or the device that will be explained in the following example. In this way we will find the following values with the upper sign corresponding to the first root and the lower sign to the second root:

$$[1] \text{ and } [18] = -0.6772815716 \pm 0.7357239107\,i$$

$$[2] \text{ and } [17] = -0.0825793455 \mp 0.9965844930\,i$$

$$[3] \text{ and } [16] = 0.7891405094 \pm 0.6142127127\,i$$

$$[4] \text{ and } [15] = -0.9863613034 \pm 0.1645945903\,i$$

$$[5] \text{ and } [14] = 0.5469481581 \mp 0.8371664783\,i$$

$$[6] \text{ and } [13] = 0.2454854871 \pm 0.9694002659\,i$$

$$[7] \text{ and } [12] = -0.8794737512 \mp 0.4759473930\,i$$

$$[8] \text{ and } [11] = 0.9458172418 \mp 0.3246994692\,i$$

$$[9] \text{ and } [10] = -0.4016954247 \pm 0.9157733267\,i$$

▶ 354. *Second example for* $n = 17$. Here $n - 1 = 2 \cdot 2 \cdot 2 \cdot 2$ so the calculation will be reduced to four quadratic equations. For the primitive root we will take the number 3. The least residues of its powers relative to the modulus 17 are the following:

0. 1. 2. 3. 4. 5. 6. 7. 8. 9. 10. 11. 12. 13. 14. 15

1. 3. 9. 10. 13. 5. 15. 11. 16. 14. 8. 7. 4. 12. 2. 6

From this we derive the following distributions of the complex Ω into two periods of eight terms, four of four terms, eight of two terms:

$$\Omega = (16, 1) \begin{cases} (8, 1) \begin{cases} (4,\ 1) \begin{cases} (2,\ 1) \dots [1], [16] \\ (2, 13) \dots [4], [13] \end{cases} \\ (4,\ 9) \begin{cases} (2,\ 9) \dots [8], [\ 9] \\ (2, 15) \dots [2], [15] \end{cases} \end{cases} \\ (8, 3) \begin{cases} (4,\ 3) \begin{cases} (2,\ 3) \dots [3], [14] \\ (2,\ 5) \dots [5], [12] \end{cases} \\ (4, 10) \begin{cases} (2, 10) \dots [7], [10] \\ (2, 11) \dots [6], [11] \end{cases} \end{cases} \end{cases}$$

The equation (A) whose roots are the sums $(8, 1), (8, 3)$ is found by the rules of article 351 to be $x^2 + x - 4 = 0$. Its roots are

$$-(1/2) + (\sqrt{17}/2) = 1.5615528128$$

and

$$-(1/2) - (\sqrt{17}/2) = -2.5615528128$$

We will set the former $=(8, 1)$ so the latter necessarily $=(8, 3)$.

The equation (B) whose roots are the sums $(4, 1)$ and $(4, 9)$ is $x^2 - (8, 1)x - 1 = 0$. Its roots are

$$\tfrac{1}{2}(8, 1) \pm \tfrac{1}{2}\sqrt{[4 + (8, 1)^2]} = \tfrac{1}{2}(8, 1) \pm \tfrac{1}{2}\sqrt{[12 + 3(8, 1) + 4(8, 3)]}$$

We will set $(4, 1)$ equal to the quantity which has the positive radical sign and whose numerical value is 2.0494811777. Thus the quantity with the negative radical sign whose numerical value is -0.4879283649 will be expressed by $(4, 9)$. The remaining sums of four terms $(4, 3)$ and $(4, 10)$, can be calculated in two ways. *First*, by the method of article 346 which gives the following formulae when we abbreviate $(4, 1)$ by the letter p:

$$(4, \ 3) = -\tfrac{3}{2} + 3p - \tfrac{1}{2}p^3 \quad = \quad 0.3441507314$$
$$(4, 10) = \tfrac{3}{2} + 2p - p^2 - \tfrac{1}{2}p^3 = -2.9057035442$$

The same method gives the formula $(4, 9) = -1 - 6p + p^2 + p^3$ and from it we get the same value as above.

The *second* method allows us to determine the sums $(4, 3)$, $(4, 10)$ by solving the equation of which they are the roots. The equation is $x^2 - (8, 3)x - 1 = 0$. Its roots are

$$\tfrac{1}{2}(8, 3) \pm \tfrac{1}{2}\sqrt{[4 + (8, 3)^2]} \quad \text{or} \quad \tfrac{1}{2}(8, 3) + \tfrac{1}{2}\sqrt{[12 + 4(8, 1) + 3(8, 3)]}$$
$$\text{and} \quad \tfrac{1}{2}(8, 3) - \tfrac{1}{2}\sqrt{[12 + 4(8, 1) + 3(8, 3)]}$$

And we can remove the doubt as to *which* root should be expressed by $(4, 3)$ and which by $(4, 10)$ by the following device which we mentioned in article 352. Calculate the product of $(4, 1) - (4, 9)$ into $(4, 3) - (4, 10)$. It $= 2(8, 1) - 2(8, 3)$.[c] Manifestly the value of this expression is positive $= +2\sqrt{17}$ and, since the first factor of

[c] The real basis of this device is the fact that we can foresee in advance that the product does not contain sums of four terms but only sums of eight terms. The trained mathematician can easily grasp the reason for this. For the sake of brevity we shall omit it.

the product, $(4, 1) - (4, 9) = +\sqrt{12} + 3(8, 1) + 4(8, 3)$, is positive, the other factor, $(4, 3) - (4, 10)$, must also be positive. Therefore $(4, 3)$ is equal to the first root which has the positive sign in front of the radical, and $(4, 10)$ is equal to the second root. From these will result the same numerical values as above.

Having found all the sums of four terms, we proceed to the sums of two terms. Equation (C) whose roots are $(2, 1), (2, 13)$ and contained in $(4, 1)$ will be $x^2 - (4, 1)x + (4, 3) = 0$. Its roots are

$$\tfrac{1}{2}(4, 1) \pm \tfrac{1}{2}\sqrt{[-4(4, 3) + (4, 1)^2]}$$

or

$$\tfrac{1}{2}(4, 1) \pm \tfrac{1}{2}\sqrt{[4 + (4, 9) - 2(4, 3)]}$$

When we take the positive radical quantity whose value $= 1.8649444588$, we will set it $= (2, 1)$ and so $(2, 13)$ will be equal to the other whose value is $= 0.1845367189$. If the remaining sums of two terms are investigated by the method of article 346, we can use the same formulae for $(2, 2), (2, 3), (2, 4), (2, 5), (2, 6), (2, 7), (2, 8)$ as we did in the preceding example for similar quantities, that is to say, $(2, 2)$ [or $(2, 15)$] $= (2, 1)^2 - 2$ etc. But if it seems preferable to find them in pairs by solving a quadratic equation. For $(2, 9), (2, 15)$ we have the equation $x^2 - (4, 9)x + (4, 10) = 0$ whose roots are

$$\tfrac{1}{2}(4, 9) \pm \tfrac{1}{2}\sqrt{[4 + (4, 1) - 2(4, 10)]}$$

We can determine which sign to use in the same way as above. Calculating the product of $(2, 1) - (2, 13)$ into $(2, 9) - (2, 15)$ we get $-(4, 1) + (4, 9) - (4, 3) + (4, 10)$. Since this is negative and the factor $(2, 1) - (2, 13)$ is positive, $(2, 9) - (2, 15)$ must be negative and we should use the upper positive sign for $(2, 15)$ and the lower negative sign for $(2, 9)$. From this we find that $(2, 9) = -1.9659461994$, $(2, 15) = 1.4780178344$. Then, since in calculating the product of $(2, 1) - (2, 13)$ into $(2, 3) - (2, 5)$ we get the positive quantity $(4, 9) - (4, 10)$, the factor $(2, 3) - (2, 5)$ must be positive. And by a calculation like the one above we find

$$(2, 3) = \tfrac{1}{2}(4, 3) + \tfrac{1}{2}\sqrt{(4 + (4, 10) - 2(4, 9))} = \quad 0.8914767116$$
$$(2, 5) = \tfrac{1}{2}(4, 3) - \tfrac{1}{2}\sqrt{(4 + (4, 10) - 2(4, 9))} = -0.5473259801$$

Finally by completely analogous operations we have

$$(2, 10) = \tfrac{1}{2}(4, 10) - \tfrac{1}{2}\sqrt{(4 + (4, 3) - 2(4, 1))} = -1.7004342715$$

$$(2, 11) = \tfrac{1}{2}(4, 10) + \tfrac{1}{2}\sqrt{(4 + (4, 3) - 2(4, 1))} = -1.2052692728$$

It remains now to get down to the roots Ω themselves. Equation
(*D*) whose roots are [1] and [16] gives us $x^2 - (2, 1)x + 1 = 0$.
The roots of this are

$$\tfrac{1}{2}(2, 1) \pm \tfrac{1}{2}\sqrt{[(2, 1)^2 - 4]} \quad \text{or rather} \quad \tfrac{1}{2}(2, 1) \pm \tfrac{1}{2}i\sqrt{[4 - (2, 1)^2]}$$

or

$$\tfrac{1}{2}(2, 1) \pm \tfrac{1}{2}i\sqrt{[2 - (2, 15)]}$$

We will take the upper sign for [1], the lower for [16]. We can get
the fourteen remaining roots either from the powers of [1] or by
solving seven quadratic equations, each of which will give us two
roots, and the uncertainty about the signs of the radical quantities
can be removed by the same device we used above. Thus [4] and
[13] are the roots of the equation $x^2 - (2, 13)x + 1 = 0$ and so
equal to

$$\tfrac{1}{2}(2, 13) \pm \tfrac{1}{2}i\sqrt{[2 - (2, 9)]}$$

By calculating the product of [1] − [16] into [4] − [13] however
we get $(2, 5) - (2, 3)$, a real negative quantity. Therefore, since
[1] − [16] is $+i\sqrt{(2 - (2, 15))}$, i.e. the product of the imaginary *i*
into a real *positive* quantity, [4] − [13] must also be the product of
i into a real *positive* quantity because $i^2 = -1$. As a result we will
take the upper sign for [4] and the lower sign for [13]. Similarly
for the roots [8] and [9] we find

$$\tfrac{1}{2}(2, 9) \pm \tfrac{1}{2}i\sqrt{[2 - (2, 1)]}$$

so, since the product of [1] − [16] into [8] − [9] is $(2, 9) - (2, 10)$
and negative, we must take the upper sign for [8], the lower sign
for [9]. If we then compute the remaining roots we will obtain the
following numerical values, where the upper sign is to be taken
for the first root, the lower sign for the second:

$$[1], [16] \ldots \quad 0.9324722294 \pm 0.3612416662\, i$$
$$[2], [15] \ldots \quad 0.7390089172 \pm 0.6736956436\, i$$
$$[3], [14] \ldots \quad 0.4457383558 \pm 0.8951632914\, i$$

[4], [13]...   0.0922683595 ± 0.9957341763 i

[5], [12]... −0.2736629901 ± 0.9618256432 i

[6], [11]... −0.6026346364 ± 0.7980172273 i

[7], [10]... −0.8502171357 ± 0.5264321629 i

[8], [9]... −0.9829730997 ± 0.1837495178 i

What precedes can suffice for solving the equation $x^n - 1 = 0$ and so also for finding the trigonometric functions corresponding to the arcs that are commensurable with the circumference. But this subject is so important that we cannot conclude without indicating some of the observations that illustrate the argument, as well as examples that are related to it or depend on it. Among such we will especially select those that can be solved without a lot of apparatus that depends on other disciplines and we will consider them only as *examples* of this vast theory which must be considered in great detail later on.

▶ 355. Since n is always presumed to be odd, 2 will appear among the factors of $n - 1$, and the complex Ω will be composed of $(n - 1)/2$ periods of two terms. Such a period $(2, \lambda)$ will consist of the roots $[\lambda]$ and $[\lambda g^{(n-1)/2}]$ where as above g represents any primitive root for the modulus n. But $g^{(n-1)/2} \equiv -1$ (mod. n) and so $\lambda g^{(n-1)/2} \equiv -\lambda$ (see art. 62) and $[\lambda g^{(n-1)/2}] = [-\lambda]$. Therefore if we suppose that $[\lambda] = \cos kP/n + i \sin kP/n$, and $[-\lambda] = \cos kP/n - i \sin kP/n$, we will have the sum $(2, \lambda) = 2 \cos kP/n$. At this point we only draw the conclusion that the value of any sum of two terms is a real quantity. Since any period which has an even number of terms $= 2a$ and can be decomposed into a periods of two terms, it is clear in general that the value of any sum which has an even number of terms is always a real quantity. Therefore if in article 352 among the factors α, β, γ, etc. we save two until the end, all the operations will be done on real quantities until we come to a sum of two terms, and the imaginaries will be introduced when we pass from the sums to the roots themselves.

▶ 356. We should give special attention to the auxiliary equations by which we determine for any value of n the sum that forms the complex Ω. They are connected in a surprising way with the most recondite properties of the number n. Here we will restrict ourselves to the two following cases. *First*, the quadratic equation whose roots are sums of $(n - 1)/2$ terms, *second*, in the case where $n - 1$ has the factor 3, we will consider the cubic equation whose roots are sums of $(n - 1)/3$ terms.

If for brevity we write m for $(n - 1)/2$ and designate by g some primitive root for the modulus n, the complex Ω will consist of two periods $(m, 1)$ and (m, g). The former will contain the roots $[1], [g^2], [g^4], \ldots [g^{n-3}]$, the latter the roots $[g], [g^3], [g^5], \ldots [g^{n-2}]$. Let us suppose that the least positive residues of the numbers $g^2, g^4, \ldots g^{n-3}$ relative to the modulus n are, disregarding order, R, R', R'', etc. and that the residues of $g, g^3, g^5, \ldots g^{n-2}$ are N, N', N'', etc. Then the roots of which $(m, 1)$ consists, coincide with $[1], [R], [R'], [R'']$, etc. and the roots of the period (m, g) with $[N], [N'], [N'']$, etc. It is clear that all the numbers $1, R, R', R''$, etc. are *quadratic residues* of the number n. Since they are all different and less than n, and since their number is $= (n - 1)/2$ and so equal to the number of all positive residues of n that are less than n, these residues will coincide completely with those numbers. All the numbers N, N', N'', etc. are different from each other and from the numbers $1, R, R'$, etc. and together with these exhaust all the numbers $1, 2, 3, \ldots n - 1$. It follows that the numbers N, N', N'', etc. must coincide with all the positive *quadratic non-residues* of n that are less than n. Now if we suppose that the equation whose roots are the sums $(m, 1), (m, g)$ is

$$x^2 - Ax + B = 0$$

we have

$$A = (m, 1) + (m, g) = -1, \qquad B = (m, 1) \cdot (m, g)$$

The product of $(m, 1)$ into (m, g) by article 345

$$= (m, N + 1) + (m, N' + 1) + (m, N'' + 1) + \text{etc.} = W$$

and so will be reduced to a form $\alpha(m, 0) + \beta(m, 1) + \gamma(m, g)$. To determine the coefficients α, β, γ we observe *first* that $\alpha + \beta + \gamma = m$ (because the number of sums in $W = m$); *second*, that $\beta = \gamma$ (this follows from article 350 since the product $(m, 1) \cdot (m, g)$ is an invariable function of the sums $(m, 1), (m, g)$ of which the larger sum $(n - 1, 1)$ is composed); *third*, since all the numbers $N + 1, N' + 1, N'' + 1$, etc. are contained within the limits 2 and $n + 1$

exclusively, it is clear that *either* no sum in W can be reduced to
$(m, 0)$ and so $\alpha = 0$ when the number $n - 1$ does not occur among
the numbers N, N', N'', etc. *or* that one sum, say (m, n) can be
reduced to $(m, 0)$ and so $\alpha = 1$ when $n - 1$ does occur among the
numbers N, N', N'', etc. In the former case therefore we will have
$\alpha = 0$, $\beta = \gamma = m/2$, in the latter $\alpha = 1$, $\beta = \gamma = (m - 1)/2$. And
it follows that since the numbers β and γ must be integers, the
former case will hold, that is, $n - 1$ (or, what is the same thing,
-1) will not be found among the nonresidues of n when m is
even or n is of the form $4k + 1$. The latter case will hold, that is,
$n - 1$ or -1 will be a nonresidue of n whenever m is odd or n of
the form $4k + 3$.[d] Now since $(m, 0) = m$, $(m, 1) + (m, g) = -1$ the
product we seek will be $= m/2$ in the former case and $= (m + 1)/2$
in the latter. Thus the equation in the former case will be
$x^2 + x - (n - 1)/4 = 0$ with roots $-(1/2) \pm (\sqrt{n}/2)$, in the latter
$x^2 + x + [(n + 1)/4]$ with roots $-(1/2) \pm (i\sqrt{n}/2)$.

Let $\mathfrak{R}$ stand for all the positive quadratic residues of n that are
less than n and $\mathfrak{N}$ for all the nonresidues. Then no matter which
root of Ω is chosen for [1], the difference between the sums
$\Sigma[\mathfrak{R}]$ and $\Sigma[\mathfrak{N}]$ will be $= \pm\sqrt{n}$ for $n \equiv 1$ and $= \pm i\sqrt{n}$ for
$n \equiv 3$ (mod. 4). And it follows that if k is any integer not divisible
by n we will have

$$\sum \cos\frac{k\mathfrak{R}P}{n} - \sum \frac{k\mathfrak{N}P}{n} = \pm\sqrt{n}$$

and

$$\sum \sin\frac{k\mathfrak{R}P}{n} - \sum \sin\frac{k\mathfrak{N}P}{n} = 0$$

for $n \equiv 1$ (mod. 4). On the other hand for $n \equiv 3$ (mod. 4) the first
difference will $= 0$ and the second $= \pm\sqrt{n}$. These theorems are so
elegant that they deserve special note. We observe that the upper
signs always hold when for k we take unity or a quadratic residue
of n and the lower when k is a nonresidue. These theorems remain

[d] In this way we have given a new demonstration of the theorem which says that -1 is a
residue of all prime numbers of the form $4k + 1$ and a nonresidue of all those of the form
$4k + 3$. Above (art. 108, 109, 262) we proved it in several different ways. If it is preferable
to presuppose this theorem, there will be no need for the distinction between the two cases
because β and γ will already be integers.

the same or rather increase in elegance when they are extended to composite values of n. But these matters are on a higher level of investigation, and we will reserve their consideration for another occasion.

▶ 357. Let the equation of degree m whose roots are the m roots contained in the period $(m, 1)$ be the following:

$$x^m - ax^{m-1} + bx^{m-2} - \text{etc.} = 0$$

Here $z = 0$ and $a = (m, 1)$ and each of the remaining coefficients b etc. will be of the form $\mathfrak{A} + \mathfrak{B}(m, 1) + \mathfrak{C}(m, g)$ with $\mathfrak{A}, \mathfrak{B}, \mathfrak{C}$ integers (art. 348). If we denote by z' the function into which z is transformed, and if for $(m, 1)$ we everywhere substitute (m, g) and for (m, g) we substitute (m, g^2) or what is the same thing $(m, 1)$, then the roots of the equation $z' = 0$ will be the roots contained in (m, g) and the product

$$zz' = \frac{x^n - 1}{x - 1} = X$$

Therefore z can be reduced to a form $R + S(m, 1) + T(m, g)$ where R, S, T will be integral functions of x with all their coefficients integers. Having done this we will have

$$z' = R + S(m, g) + T(m, 1)$$

And if for brevity we write p and q for $(m, 1)$ and (m, g) respectively

$$2z = 2R + (S + T)(p + q) - (T - S)(p - q)$$
$$= 2R - S - T - (T - S)(p - q)$$

and similarly

$$2z' = 2R - S - T + (T - S)(p - q)$$

and if we set

$$2R - S - T = Y, \qquad T - S = Z$$

we will have $4X = Y^2 - (p - q)^2 Z^2$ and since $(p - q)^2 = \pm n$

$$4X = Y^2 \mp nZ^2$$

The upper sign will hold when n is of the form $4k + 1$, the lower when it is of the form $4k + 3$. This is the theorem we promised (art. 124) to prove. It is easy to see that the two terms of highest degree in the function Y will always be $2x^m + x^{m-1}$ and the highest in the function Z, x^{m-1}. The remaining coefficients, all of which will be integers, will vary according to the nature of the number n and cannot be given a general analytic formula.

Example. For $n = 17$, by the rules of article 348 the equation whose roots are the eight roots contained in $(8, 1)$ will be

$$x^8 - px^7 + (4 + p + 2q)x^6 - (4p + 3q)x^5 + (6 + 3p + 5q)x^4$$
$$-(4p + 3q)x^3 + (4 + p + 2q)x^2 - px + 1 = 0$$

therefore

$$R = x^8 + 4x^6 + 6x^4 + 4x^2 + 1$$
$$S = -x^7 + x^6 - 4x^5 + 3x^4 - 4x^3 + x^2 - x$$
$$T = 2x^6 - 3x^5 + 5x^4 - 3x^3 + 2x^2$$

and

$$Y = 2x^8 + x^7 + 5x^6 + 7x^5 + 4x^4 + 7x^3 + 5x^2 + x + 2$$
$$Z = x^7 + x^6 + x^5 + 2x^4 + x^3 + x^2 + x$$

Here are some other examples:

n	X	Z
3	$2x + 1$	1
5	$2x^2 + x + 2$	x
7	$2x^3 + x^2 - x - 2$	$x^2 + x$
11	$2x^5 + x^4 - 2x^3 + 2x^2 - x - 2$	$x^4 + x$
13	$2x^6 + x^5 + 4x^4 - x^3 + 4x^2$ $+ x + 2$	$x^5 + x^3 + x$
19	$2x^9 + x^8 - 4x^7 + 3x^6 + 5x^5$ $- 5x^4 - 3x^3 + 4x^2 - x - 2$	$x^8 - x^6 + x^5 + x^4 - x^3 + x$
23	$2x^{11} + x^{10} - 5x^9 - 8x^8$ $- 7x^7 - 4x^6 + 4x^5 + 7x^4$ $+ 8x^3 + 5x^2 - x - 2$	$x^{10} + x^9 - x^7 - 2x^6 - 2x^5$ $- x^4 + x^2 + x$

▶ 358. We proceed now to a consideration of the cubic equations which for the case where n is of the form $3k + 1$ determine the three sums of $(n - 1)/3$ terms which compose the complex Ω. Let g be any primitive root for the modulus n and $(n - 1)/3 = m$ which will be an even integer. Then the three sums that compose

Ω will be $(m, 1), (m, g), (m, g^2)$ for which we will write p, p', p'' respectively. It is clear that the first contains the roots $[1], [g^3]$, $[g^6], \ldots [g^{b-4}]$, the second the roots $[g], [g^4], \ldots [g^{n-3}]$, and the third the roots $[g^2], [g^5], \ldots [g^{n-2}]$. Let us suppose that the equation we want is

$$x^3 - Ax^2 + Bx - C = 0$$

We will have

$$A = p + p' + p'', \qquad B = pp' + p'p'' + pp'', \qquad C = pp'p''$$

and $A = -1$. Let the least positive residues of the numbers $g^3, g^6, \ldots g^{n-4}$ relative to the modulus n and disregarding order, be $\mathfrak{A}, \mathfrak{B}, \mathfrak{C}$, etc., and $\mathfrak{R}$ this complex with the number 1 added. Similarly let $\mathfrak{A}', \mathfrak{B}', \mathfrak{C}'$, etc. be the least residue of the numbers $g, g^4, g^7, \ldots g^{n-3}$ and $\mathfrak{R}'$ their complex; finally let $\mathfrak{A}'', \mathfrak{B}'', \mathfrak{C}''$, etc. be the least residues of $g^2, g^5, g^8, \ldots g^{n-2}$ and $\mathfrak{R}''$ their complex. Thus all the numbers in $\mathfrak{R}, \mathfrak{R}', \mathfrak{R}''$ will be different and will coincide with $1, 2, 3, \ldots n - 1$. First of all we must observe here that the number $n - 1$ must be in $\mathfrak{R}$, since it is easy to see that it is a residue of $g^{3m/2}$. It also follows from this that the two numbers $h, n - h$ will always be found in the *same* one of the three complexes $\mathfrak{R}, \mathfrak{R}', \mathfrak{R}''$, for if one of them is a residue of the power g^λ, the other will be a residue of the power $g^{\lambda + (3m/2)}$ or of $g^{\lambda - (3m/2)}$ if $\lambda > 3m/2$. We will denote by $(\mathfrak{RR})$ the numbers of numbers in the series $1, 2, 3, \ldots n - 1$ which belong to $\mathfrak{R}$ by themselves and when increased by unity; $(\mathfrak{RR}')$ will be the number of numbers in the same series, which are contained in $\mathfrak{R}$ themselves but are in $\mathfrak{R}'$ when increased by unity. It will be immediately obvious what is the meaning of the notation $(\mathfrak{RR}''), (\mathfrak{R}'\mathfrak{R}), (\mathfrak{R}'\mathfrak{R}'), (\mathfrak{R}'\mathfrak{R}''), (\mathfrak{R}''\mathfrak{R})$, $(\mathfrak{R}''\mathfrak{R}'), (\mathfrak{R}''\mathfrak{R}'')$. Having done this, I say *first* that $(\mathfrak{RR}') = (\mathfrak{R}'\mathfrak{R})$. For if we suppose that h, h', h'', etc. are all the numbers of the series $1, 2, 3, \ldots n - 1$ which are themselves in $\mathfrak{R}$ but with $h + 1, h' + 1, h'' + 1$, etc. in $\mathfrak{R}'$ and if we suppose that the number of these is $= \mathfrak{RR}'$, then it is clear that all the numbers $n - h - 1$, $n - h' - 1, n - h'' - 1$, etc. are contained in $\mathfrak{R}'$ and the next larger numbers $n - h, n - h'$, etc. in $\mathfrak{R}$; and since there are $(\mathfrak{R}'\mathfrak{R})$ such numbers in all, we certainly cannot have $(\mathfrak{R}'\mathfrak{R}) < (\mathfrak{RR}')$. We can then show that it is not possible to have $(\mathfrak{RR}') < (\mathfrak{R}'\mathfrak{R})$ so these numbers are necessarily equal. In exactly the same way we can show that $(\mathfrak{R}\mathfrak{R}'') = (\mathfrak{R}''\mathfrak{R})$, $(\mathfrak{R}'\mathfrak{R}'') = (\mathfrak{R}''\mathfrak{R}')$. *Second*, since any number in $\mathfrak{R}$ with the exception of the largest one $n - 1$ must be followed by the next larger one in $\mathfrak{R}$ or in $\mathfrak{R}'$ or in $\mathfrak{R}''$, the sum $(\mathfrak{R}\mathfrak{R}) + (\mathfrak{R}\mathfrak{R}') + (\mathfrak{R}\mathfrak{R}'')$ must be equal to the number of all numbers in $\mathfrak{R}$ diminished by unity, that is $= m - 1$. For a similar reason

$$(\mathfrak{R}'\mathfrak{R}) + (\mathfrak{R}'\mathfrak{R}') + (\mathfrak{R}'\mathfrak{R}'') = (\mathfrak{R}''\mathfrak{R}) + (\mathfrak{R}''\mathfrak{R}') + (\mathfrak{R}''\mathfrak{R}'') = m$$

With these preliminaries, by the rules of article 345 we evolve the product pp' into $(m, \mathfrak{A}' + 1) + (m, \mathfrak{B}' + 1) + (m, \mathfrak{C}' + 1) + $ etc. This expression is easily reduced to $(\mathfrak{R}'\mathfrak{R})p + (\mathfrak{R}'\mathfrak{R}')p' + (\mathfrak{R}'\mathfrak{R}'')p''$. By article 345.I we can get from this the product $p'p''$ by substituting for $(m, 1), (m, g), (m, g^2)$ respectively the quantities (m, g), $(m, g^2), (m, g^3)$, i.e. p', p'', p respectively for p, p', p''. Thus we have $p'p'' = (\mathfrak{R}'\mathfrak{R})p' + (\mathfrak{R}'\mathfrak{R}')p'' + (\mathfrak{R}'\mathfrak{R}'')p$. Similarly $p''p = (\mathfrak{R}'\mathfrak{R})p'' + (\mathfrak{R}'\mathfrak{R}')p + (\mathfrak{R}'\mathfrak{R}'')p'$. From this we get immediately

$$B = m(p + p' + p'') = -m$$

In a manner similar to that by which pp' was developed, we can also reduce pp'' to $(\mathfrak{R}''\mathfrak{R})p + (\mathfrak{R}''\mathfrak{R}')p' + (\mathfrak{R}''\mathfrak{R}'')p''$. And since this expression must be identical with the preceding, we will necessarily have $(\mathfrak{R}''\mathfrak{R}) = (\mathfrak{R}'\mathfrak{R}')$ and $(\mathfrak{R}''\mathfrak{R}'') = (\mathfrak{R}'\mathfrak{R})$. Now if we let

$$(\mathfrak{R}'\mathfrak{R}'') = (\mathfrak{R}''\mathfrak{R}') = a, \qquad (\mathfrak{R}''\mathfrak{R}'') = (\mathfrak{R}'\mathfrak{R}) = (\mathfrak{R}\mathfrak{R}') = b$$

$$(\mathfrak{R}'\mathfrak{R}') = (\mathfrak{R}''\mathfrak{R}) = (\mathfrak{R}\mathfrak{R}'') = c$$

we will have $m - 1 = (\mathfrak{R}\mathfrak{R}) + (\mathfrak{R}\mathfrak{R}') + (\mathfrak{R}\mathfrak{R}'') = (\mathfrak{R}\mathfrak{R}) + b + c$. And since $a + b + c = m, (\mathfrak{R}\mathfrak{R}) = a - 1$. Thus the nine unknown quantities are reduced to three a, b, c or rather, since $a + b + c = m$, to two. Finally it is clear that the square p^2 becomes $(m, 1 + 1) + (m, \mathfrak{A} + 1) + (m, \mathfrak{B} + 1) + (m, \mathfrak{C} + 1) + $ etc. Among the terms of this expression we have (m, n) which reduces to $(m, 0)$ or to m and the remaining terms reduce to $(\mathfrak{R}\mathfrak{R})p + (\mathfrak{R}\mathfrak{R}')p' + (\mathfrak{R}\mathfrak{R}'')p''$ so we have $p^2 = m + (a - 1)p + bp' + cp''$.

As a result of all this we have the following reductions:

$$
\begin{aligned}
p^2 &= m + (a - 1)p + bp' + cp'' \\
pp' &= bp + cp' + ap'' \\
pp'' &= cp + ap' + bp'' \\
p'p'' &= ap + bp' + cp''
\end{aligned}
$$

along with the conditional equation

$$a + b + c = m \tag{I}$$

and we know besides that these numbers are integers. As a result we have

$$
\begin{aligned}
C = p \cdot p'p'' &= ap^2 + bpp' + cpp'' \\
&= am + (a^2 + b^2 + c^2 - a)p + (ab + bc + ac)p' \\
&\quad + (ab + bc + ac)p''
\end{aligned}
$$

But since $pp'p''$ is an invariable function of p, p', p'', the coefficients by which it is multiplied in the preceding expression are necessarily equal (art. 350) and we have the new equation

$$a^2 + b^2 + c^2 - a = ab + bc + ac \tag{II}$$

and from this we get $C = am + (ab + bc + ac)(p + p' + p'')$ or (on account of (I) and the fact that $p + p' + p'' = -1$)

$$C = a^2 - bc \tag{III}$$

Now even though C depends on three unknowns and there are only two equations, nevertheless with the help of the condition that a, b, c be integers, they will suffice to completely determine C. To show this we express equation (II) as

$$
\begin{aligned}
12a + 12b + 12c + 4 &= 36a^2 + 36b^2 + 36c^2 - 36ab - 36ac \\
&\quad - 36bc - 24a + 12b + 12c + 4
\end{aligned}
$$

By (I), the left-hand side becomes $= 12m + 4 = 4n$. The right-hand side reduces to

$$(6a - 3b - 3c - 2)^2 + 27(b - c)^2$$

or if we write k for $2a - b - c$, to $(3k - 2)^2 + 27(b - c)^2$. Thus the number $4n$ (i.e. the quadruple of any prime of the form $3m + 1$) can be represented by the form $x^2 + 27y^2$. This can, of course, be deduced without any difficulty from the general theory of binary forms, but it is remarkable that such a decomposition is consistent with the values of a, b, c. Now the number $4n$ can always be decomposed in only one way into the product of a square and 27

times another square. We show this as follows.[e] If we suppose
that
$$4n = t^2 + 27u^2 = t't' + 27u'u'$$

we have *first*

$$(tt' - 27uu')^2 + 27(tu' + t'u)^2 = 16n^2$$

second

$$(tt' + 27uu')^2 + 27(tu' - t'u)^2 = 16n^2$$

third

$$(tu' + t'u)(tu' - t'u) = 4n(u'u' - u^2)$$

From the third equation it follows that n, since it is a prime
number, divides one of the numbers $tu' + t'u, tu' - t'u$. From the
first and the second, however, it is clear that each of these numbers
is less than n, so the one which n divides is necessarily $=0$. Therefore $u'u - u^2 = 0$ and $u'u' = u^2$ and $t't' = t^2$; i.e. the two decompositions are the same. Now suppose that the decomposition of $4n$
into a square and 27 times a square is known (this can be done
by the direct method of Section V or the indirect method of
art. 323, 324). We will then have $4n = M^2 + 27N^2$ and the
squares $(3k - 2)^2, (b - c)^2$ will be determined, and we will have
two equations in place of equation (II). But clearly not only the
square $(3k - 2)^2$ but its root $3k - 2$ will be determined. Because
it must either $= +M$ or $= -M$ the ambiguity is removed easily.
For since k must be an integer, we will have $3k - 2 = +M$ or
$= -M$ according as M is of the form $3z + 1$ or $3z + 2$.[f] Now
since $k = 2a - b - c = 3a - m$ we will have $a = (m + k)/3$,
$b + c = m - a = (2m - k)/3$ and so

$$C = a^2 - bc = a^2 - \tfrac{1}{4}(b + c)^2 + \tfrac{1}{4}(b - c)^2$$
$$= \tfrac{1}{9}(m + k)^2 - \tfrac{1}{36}(2m - k)^2 + \tfrac{1}{4}N^2 = \tfrac{1}{12}k^2 + \tfrac{1}{3}km + \tfrac{1}{4}N^2$$

and thus we have found all the coefficients of the equation.

---

[e] This proposition can be proved much more directly from the principles of Section V.

[f] Manifestly M cannot be of the form $3z$ because otherwise $4n$ would be divisible by 3.
With regard to the ambiguity as to whether $b - c$ must $=N$ or $= -N$ it is unnecessary to
consider the question here, and by the nature of the case it cannot be determined because
it depends on the selection of the primitive root g. For some primitive roots the difference
$b - c$ will be positive, for others negative.

Q.E.F. This formula will be much simpler if we substitute for N^2 its value from the equation $(3k - 2)^2 + 27N^2 = 4n = 12m + 4$. After calculation we get

$$C = \tfrac{1}{9}(m + k + 3km) = \tfrac{1}{9}(m + kn)$$

The same value can be reduced to $(3k - 2)N^2 + k^3 - 2k^2 + k - km + m$. And although this expression is less useful, it shows immediately that C must certainly be an integer, since it is even.

Example. For $n = 19$ we have $4n = 49 + 27$, so $3k - 2 = +7$, $k = 3$, $C = (6 + 57)/9 = 7$ and the equation we want is $x^3 + x^2 - 6x - 7 = 0$ as we saw above (art. 351). Similarly, for $n = 7, 13, 31, 37, 43, 61, 67$ the value of k is respectively $1, -1, 2, -3, -2, 1, -1$ and $C = 1, -1, 8, -11, -8, 9, -5$.

Although the problem we have solved in this article is rather intricate, we did not wish to omit it because of the elegance of the solution and because it gave occasion for using various devices that are fruitful also in other discussions.[g]

▶ 359. The preceding discussion had to do with the *discovery* of auxiliary equations. Now we will explain a very remarkable property concerning their *solution*. Everyone knows that the most eminent geometers have been ineffectual in the search for a general solution of equations higher than the fourth degree, or (to define the search more accurately) for the REDUCTION OF MIXED EQUATIONS TO PURE EQUATIONS. And there is little doubt that this problem does not so much defy modern methods of analysis as that it proposes the impossible (cf. what we said on this subject in *Demonstratio nova*, art. 9[1]). Nevertheless it is certain that there are innumerable mixed equations of every degree which admit a reduction to pure equations, and we trust that geometers will find it gratifying if we show that our equations are always of this kind. But because of the length of this discussion we will

---

[g] *Corollary.* Let ε be the root of the equation $x^3 - 1 = 0$ and we will have $(p + \varepsilon p' + \varepsilon^2 p'')^3 = n(M + N\sqrt{-27})/2$. Let $M/\sqrt{4n} = \cos \phi$, $N\sqrt{27}/\sqrt{4n} = \sin \phi$ and as a result

$$p = -\tfrac{1}{3} + \tfrac{2}{3}\cos \tfrac{1}{3}\phi\sqrt{n}; \qquad M \equiv +1 \pmod{3}; \qquad 1 \equiv M(1 \cdot 2 \cdot 3 \ldots m)^3 \pmod{n}$$

Setzt man $3x + 1 = y$, *so wird die Gleichung* $y^3 - 3ny - Mn = 0$ (If we let $3x + 1 = y$, we then have $y^3 - 3ny - Mn = 0$).

[1] This is Gauss' doctoral dissertation. Its full title is *Demonstratio nova theorematis Omnem Functionem Algebraicam Rationalem Integram unis variabilis in Factores Reales primi vel secundi gradus resolvi posse*, Helmstedt, 1799.

present here only the most important principles necessary to show the possibility of our claim and reserve for another time a more complete consideration worthy of this argument. We will first present some general observations about the roots of the equation $x^e - 1 = 0$ which also embrace the case where e is a composite number.

I. These roots are given (as is known from elementary text-books) by $\cos kP/e + i \sin kP/e$ where for k we take the e numbers $0, 1, 2, 3, \ldots e - 1$ or any others that are congruent to these relative to the modulus e. One root, for $k = 0$ or for any k divisible by e will $= 1$. For any other value of k there will be a root that is different from 1.

II. Since $(\cos kP/e + i \sin kP/e)^\lambda = \cos \lambda kP/e + i \sin \lambda kP/e$, it is clear that if R is such a root corresponding to a value of k which is relatively prime to e, then in the series R, R^2, R^3, etc. the eth term $= 1$, and all the antecedent values are different from 1. It follows immediately that all e of the quantities $1, R, R^2, R^3, \ldots R^{e-1}$ are unequal and, since they all satisfy the equation $x^e - 1 = 0$, they will give all the roots of this equation.

III. Under the same assumption the sum

$$1 + R^\lambda + R^{2\lambda} \ldots + R^{\lambda(e-1)} = 0$$

for any value of the integer λ not divisible by e. For it is $= (1 - R^{\lambda e})/(1 - R^\lambda)$ and the numerator of this fraction $= 0$, but the denominator is not $= 0$. When λ is divisible by e, the sum obviously $= e$.

▶ 360. Let n, as always, be a prime number, g a primitive root for the modulus n, and $n - 1$ the product of three positive integers α, β, γ. For brevity we will consider together the cases where α or $\gamma = 1$. When $\gamma = 1$, we can replace the sums $(\gamma, 1), (\gamma, g)$, etc. by the roots $[1], [g]$, etc. Suppose therefore that of all the α sums of $\beta\gamma$ terms $(\beta\gamma, 1), (\beta\gamma, g), (\beta\gamma, g^2), \ldots (\beta\gamma, g^{\alpha-1})$ are known and that we want to find the sums of γ terms. We have reduced the operation above to a mixed equation of degree β. Now we will show how to solve it by a pure equation of the same degree. For brevity for the sums

$$(\gamma, 1), (\gamma, g^\alpha), (\gamma, g^{2\alpha}), \ldots (\gamma, g^{\alpha\beta - \alpha})$$

which are contained in $(\beta\gamma, 1)$, we will write $a, b, c, \ldots m$ respectively. And for the sums

$$(\gamma, g), (\gamma, g^{\alpha+1}), \ldots (\gamma, g^{\alpha\beta-\alpha+1})$$

contained in $(\beta\gamma, g)$ we will write $a', b', \ldots m'$. And for

$$(\gamma, g^2), (\gamma, g^{\alpha+2}), \ldots (\gamma, g^{\alpha\beta-\alpha+2})$$

we will write $a'', b'', \ldots m''$, etc. until we come to those that are contained in $(\beta\gamma, g^{\alpha-1})$.

I. Let R be an undefined root of the equation $x^\beta - 1 = 0$ and let us suppose that the power of β degree of the function

$$t = a + Rb + R^2c + \ldots + R^{\beta-1}m$$

is, according to the rules of article 345,

$$N + Aa + Bb + Cc \ldots + Mm$$
$$+ A'a' + B'b' + C'c' \ldots + M'm'$$
$$+ A''a'' + B''b'' + C''c'' \ldots + M''m''$$
$$+ \text{etc.} \qquad\qquad = T$$

where all the coefficients N, A, B, A', etc. are rational integral functions of R. Let us also suppose that the β power of two other functions

$$u = R^\beta a + Rb + R^2c \ldots + R^{\beta-1}m$$
$$u' = b + Rc + R^2d \ldots + R^{\beta-2}m + R^{\beta-1}a$$

become respectively U and U'. It is easy to see from article 350 that since u' results from commuting the sums $a, b, c, \ldots m$ with $b, c, d, \ldots a$ that

$$U' = N + Ab + Bc + Cd \ldots + Ma$$
$$+ A'b' + B'c' + C'd' \ldots + M'a'$$
$$+ A''b'' + B''c'' + C''d'' \ldots + M''a''$$
$$+ \text{etc.}$$

It is also clear that, since $u = Ru'$, we will have $U = R^\beta U'$. And since $R^\beta = 1$ the corresponding coefficients in U and U' will be equal. Finally, since t and u differ only in so far as a is multiplied into t by unity and into u by R^β, all the corresponding coefficients (i.e. those that multiply the same sums) in T and U will be equal, and so also the corresponding coefficients in T and U'. Therefore $A = B = C$ etc. $= M$; $A' = B' = C'$ etc.; $A'' = B''$ $= C''$ etc.; etc. so T is reduced to a form like

$$N + A(\beta\gamma, 1) + A'(\beta\gamma, g) + A''(\beta\gamma, g^2) + \text{etc.}$$

where the individual coefficients N, A, A', etc. are of the form

$$pR^{\beta-1} + p'R^{\beta-2} + p''R^{\beta-3} + \text{etc.}$$

in such a way that p, p', p'', etc. are given integers.

II. If we take for R a determined root of the equation $x^\beta - 1 = 0$ (we suppose that we already have its solution) and in such a way that no power less than the β power is equal to unity, T will also be a determined quantity, and from it we can derive t by the pure equation $t^\beta - T = 0$. But since this equation has β roots which are $t, Rt, R^2t, \ldots R^{\beta-1}t$, there can be a doubt as to which root should be chosen. This is arbitrary, however, because we must remember that after all the sums of $\beta\gamma$ terms are determined, the root [1] is defined only in that one of the $\beta\gamma$ roots contained in $(\beta\gamma, 1)$ must be denoted by this symbol. So it is entirely arbitrary which of the β sums making up $(\beta\gamma, 1)$ is designated by a. And if after one of these sums is expressed by a we suppose that $t = \mathfrak{T}$, it is easy to see that the sum we now designate by b can be changed to a, and what was formerly $c, d, \ldots a, b$ now becomes $b, c, \ldots m, a$, and the value of t is now $= \mathfrak{T}/R = \mathfrak{T}R^{\beta-1}$. Similarly, if we now decide to let a equal the sum which in the beginning was c, the value of t becomes $\mathfrak{T}R^{\beta-2}$ and so on. Thus t can be considered equal to any of the quantities $\mathfrak{T}, \mathfrak{T}R^{\beta-1}, \mathfrak{T}R^{\beta-2}$, etc., i.e. to any root of the equation $x^\beta - T = 0$ according as we let one or another of the sums in $(\beta\gamma, 1)$ be expressed by $(\gamma, 1)$. Q.E.D.

III. After the quantity t has been determined in this way, we must investigate the $\beta - 1$ others which result from t by substituting for R successively $R^2, R^3, R^4, \ldots R^\beta$, that is, by finding

$$t' = a + R^2b + R^4c \ldots + R^{2\beta-2}m$$

$$t'' = a + R^3b + R^6c \ldots + R^{3\beta-3}m, \quad \text{etc.}$$

We have the last of these already because it manifestly $= a +$

$b + c \ldots + m = (\beta\gamma, 1)$. The others can be found in the following way. We use the methods of article 345 to find the product $t^{\beta-2}t'$ just as we found t^β in I. Then we use a method just like the preceding to show that from this we get a form

$$\mathfrak{R} + \mathfrak{A}(\beta\gamma, 1) + \mathfrak{A}'(\beta\gamma, g) + \mathfrak{A}''(\beta\gamma, g^2) \text{ etc.} = T'$$

Here $\mathfrak{R}$, $\mathfrak{A}$, $\mathfrak{A}'$, etc. are rational integral functions of R and so T' is a known quantity and $t' = T't^2/T$. In exactly the same way we can find T'' by calculating the product $t^{\beta-3}t''$. This expression will have a similar form and because its value is known we can derive the equation $t'' = T''t^3/T$. Then t''' can be found from the equation $t''' = T'''t^4/T$ where again T''' is a known quantity, etc.

This method would not be applicable if we had $t = 0$ for then $T = T' = T''$ etc. $= 0$. But it can be shown that this is impossible, although the demonstration is so long that we must omit it here. There are also some special artifices for converting the fractions T'/T, T''/T, etc. into rational *integral* functions of R and some shorter methods in the case where $\alpha = 1$ for finding the values of t', t'', etc. but we cannot consider them here.

IV. Finally, as soon as we have found t, t', t'', etc. by observation III of the preceding article we have immediately that $t + t' + t''$ + etc. $= \beta a$. This gives us the value of a and from this, by article 346, we can derive the values of all the remaining sums of γ terms. The values of b, c, d, etc. can also be found from the following equations, as a little investigation will show:

$$\beta b = R^{\beta-1}t + R^{\beta-2}t' + R^{\beta-3}t'' + \text{etc.}$$

$$\beta c = R^{2\beta-2}t + R^{2\beta-4}t' + R^{2\beta-6}t'' + \text{etc.}$$

$$\beta d = R^{3\beta-3}t + R^{3\beta-6}t' + R^{3\beta-9}t'' + \text{etc.,} \quad \text{etc.}$$

Among the great number of observations that we could make concerning the preceding discussion we will emphasize only one. With regard to the solution of the pure equation $x^\beta - T = 0$, it is clear that in many cases T has the imaginary value $P + iQ$ so the solution depends partly on the division of an angle (whose tangent $= Q/P$), partly on the division of a ratio [unity to $\sqrt{(P^2 + Q^2)}$] into β parts. And it is remarkable (we will not pursue this subject here) that the value of $\sqrt[\beta]{(P^2 + Q^2)}$ can always be expressed *rationally* by already known quantities. Thus, except for the

extraction of a square root, the *only* thing required for a solution
is the division of the angle, e.g. for $\beta = 3$ only the trisection of an
angle.

Finally, since nothing prevents us from setting $\alpha = 1, \gamma = 1$ and
so $\beta = n - 1$, it is evident that the solution of the equation
$x^n - 1 = 0$ can immediately be reduced to the solution of a pure
equation $x^{n-1} - T = 0$ of degree $n - 1$. Here T is determined by
the roots of the equation $x^{n-1} - 1 = 0$. As a result the division
of the whole circle into n parts requires, *first*, the division of the
whole circle into $n - 1$ parts; *second*, the division into $n - 1$ parts
of another arc which can be constructed as soon as the first
division is accomplished; *third*, the extraction of one square root,
and it can be shown that this is always $\sqrt{n}$.

▶ 361. It remains to examine more closely the connection
between the roots Ω and the trigonometric functions of the angles
$P/n, 2P/n, 3P/n \ldots (n - 1)P/n$. The method we used for finding
the roots of Ω (unless we consult sine tables, but this would be
less direct) leaves uncertain *which* roots correspond to the *individual* angles; i.e. which root $= \cos P/n + i \sin P/n$, which
$= \cos 2P/n + i \sin 2P/n$, etc. But this uncertainty can be easily
removed by reflecting that the cosines of the angles $P/n, 2P/n,$
$3P/n, \ldots (n - 1)P/2n$ are continually decreasing (provided we pay
attention to signs) and that the sines are positive. On the other
hand the angles $(n - 1)P/n, (n - 2)P/n, (n - 3)P/n, \ldots (n + 1)P/n$
have the same cosines as the above, but the sines are negative
although they have the same absolute value. Therefore of the
roots Ω the two that have the largest real parts (they are equal to
each other) correspond to the angles $P/n, (n - 1)P/n$. The former
has the imaginary quantity i positive, the latter negative. Of the
remaining $n - 3$ roots, those that have the largest real part
correspond to the angles $2P/n, (n - 2)P/n$, and so forth. As soon
as the root to which the angle P/n corresponds is known, those
that correspond to the remaining angles can be determined from
this one because, if we suppose that it $= [\lambda]$, the roots $[2\lambda], [3\lambda],$
$[4\lambda]$, etc. will correspond to the angles $2P/n, 3P/n, 4P/n$, etc. Thus
in the example in article 353 we see that the root corresponding
to the angle $p/19$ must be $[11]$, and $[8]$ to the angle $18P/19$.
Similarly the roots $[3], [16], [14], [5]$, etc. will correspond to the
angles $2P/19, 17P/19, 3P/19, 16P/19$, etc. In the example of article

354 the root [1] will correspond to the angle $P/17$, [2] to the angle $2P/17$, etc. In this way the cosines and sines of the angles P/n, $2P/n$, etc. will be completely determined.

▶ 362. With regard to the remaining trigonometric functions of these angles, there are those that can be derived from the corresponding sines and cosines by ordinary well-known methods. Thus secants and tangents can be found by dividing unity and the sine, respectively, by the cosine; cosecants, and cotangents by dividing unity and the cosine by the sine. But it will often be much more useful to obtain the same quantities with the help of the following formulae by addition alone and no divisions. Let ω be any one of the angles P/n, $2P/n$, ... $(n - 1)P/n$ and let $\cos\omega + i\sin\omega = R$ so that R will be one of the roots Ω, then

$$\cos\omega = \frac{1}{2}\left(R + \frac{1}{R}\right) = \frac{1 + R^2}{2R},$$

$$\sin\omega = \frac{1}{2i}\left(R - \frac{1}{R}\right) = \frac{i(1 - R^2)}{2R}$$

And from this

$$\sec\omega = \frac{2R}{1 + R^2} \qquad \tan\omega = \frac{i(1 - R^2)}{1 + R^2},$$

$$\text{cosec} = \frac{2Ri}{R^2 - 1}, \qquad \cotan\omega = \frac{i(R^2 + 1)}{R^2 - 1}$$

Now we will show how to transform the numerators of these four fractions so that they will be divisible by the denominators.

I. Since $R = R^{n+1} = R^{2n+1}$ we have $2R = R + R^{2n+1}$. This expression is divisible by $1 + R^2$ since n is an odd number. So we have

$$\sec\omega = R - R^3 + R^5 - R^7 \ldots + R^{2n-1}$$

and so (since $\sin\omega = -\sin(2n - 1)\omega$, $\sin 3\omega = -\sin(2n - 3)\omega$ etc. we have $\sin\omega - \sin 3\omega + \sin 5\omega \ldots + \sin(2n - 1)\omega = 0$)

$$\sec\omega = \cos\omega - \cos 3\omega + \cos 5\omega \ldots + \cos(2n - 1)\omega$$

or finally (since $\cos\omega = \cos(2n - 1)\omega$, $\cos 3\omega = \cos(2n - 3)\omega$, etc.)

$$= 2(\cos\omega - \cos 3\omega + \cos 5\omega \ldots \mp \cos(n - 2)\omega) \pm \cos n\omega$$

the upper or lower sign to be taken according as n is of the form $4k + 1$ or $4k + 3$. Obviously this formula can also be expressed as

$$\sec \omega = \pm [1 - 2\cos 2\omega + 2\cos 4\omega \ldots \pm 2\cos (n - 1)\omega]$$

II. Similarly by substituting $1 - R^{2n+2}$ for $1 - R^2$ we have

$$\tan \omega = i(1 - R^2 + R^4 - R^6 \ldots - R^{2n})$$

or (since $1 - R^{2n} = 0$, $R^2 - R^{2n-2} = 2i \sin 2\omega$, $R^4 - R^{2n-4} = 2i \sin 4\omega$, etc.)

$$\tan \omega = 2[\sin 2\omega - \sin 4\omega + \sin 6\omega \ldots \mp \sin (n - 1)\omega]$$

III. Since $1 + R^2 + R^4 \ldots + R^{2n-2} = 0$, we have

$$n = n - 1 - R^2 - R^4 \ldots R^{2n-2}$$

$$= (1 - 1) + (1 - R^2) + (1 - R^4) \ldots + (1 - R^{2n-2})$$

and each of its terms is divisible by $1 - R^2$. So

$$\frac{n}{1 - R^2} = 1 + (1 + R^2) + (1 + R^2 + R^4) \ldots + (1 + R^2 + R^4 \\ \ldots + R^{2n-4})$$

$$= (n - 1) + (n - 2)R^2 + (n - 3)R^4 \ldots + R^{2n-4}$$

Multiplying by 2, and subtracting the quantity

$$0 = (n - 1)(1 + R^2 + R^4 \ldots + R^{2n-2})$$

and again multiplying by R we have

$$\frac{2nR}{1 - R^2} = (n - 1)R + (n - 3)R^3 + (n - 5)R^5 \ldots - (n - 3)R^{2n-3} \\ - (n - 1)R^{2n-1}$$

and from this we get immediately

$$\operatorname{cosec} \omega = \frac{1}{n}[(n - 1)\sin \omega + (n - 3)\sin 3\omega \ldots \\ - (n - 1)\sin (2n - 1)\omega]$$

$$= \frac{2}{n}[(n - 1)\sin \omega + (n - 3)\sin 3\omega + \text{etc.} + 2\sin (n - 2)\omega]$$

This formula can also be expressed as

$$\operatorname{cosec} \omega = -\frac{2}{n}[2\sin 2\omega + 4\sin 4\omega + 6\sin 6\omega \ldots \\ + (n - 1)\sin (n - 1)\omega]$$

IV. If we multiply the value of $n/(1 - R^2)$ given above by $1 + R^2$ and subtract the quantity

$$0 = (n - 1)(1 + R^2 + R^4 \ldots + R^{2n-2})$$

we have

$$\frac{n(1 + R^2)}{1 - R^2} = (n - 2)R^2 + (n - 4)R^4 + (n - 6)R^6 \ldots$$
$$- (n - 2)R^{2n-2}$$

and from this it immediately follows that

$$\operatorname{cotan} \omega = \frac{1}{n}[(n - 2)\sin 2\omega + (n - 4)\sin 4\omega + (n - 6)\sin 6\omega \ldots$$
$$- (n - 2)\sin(n - 2)\omega]$$

$$= \frac{2}{n}[(n - 2)\sin 2\omega + (n - 4)\sin 4\omega \ldots + 3\sin(n - 3)\omega$$
$$+ \sin(n - 1)\omega]$$

and this formula can also be expressed as

$$\operatorname{cotan} \omega = -\frac{2}{n}[\sin \omega + 3\sin 3\omega \ldots + (n - 2)\sin(n - 2)\omega]$$

▶ 363. When $n - 1 = ef$, the function X can be resolved into e factors of degree f as soon as we know the values of all the e sums of f terms (art. 348). In the same way, if we suppose that $Z = 0$ is an equation of degree $n - 1$ whose roots are the sines or any other trigonometric functions of the angles $P/n, 2P/n \ldots (n - 1)P/n$, the function Z can be resolved into e factors of degree f in the following way.

Let Ω consist of the e periods of f terms, $(f, 1) = P, P', P''$, etc.; the period P of the roots $[1], [a], [b], [c]$, etc.; P' of the roots $[a'], [b'], [c']$, etc.; P'' of the roots $[a''], [b''], [c'']$, etc., etc. Let the angle ω correspond to the root $[1]$, and thus the angles $a\omega, b\omega$, etc. to the roots $[a], [b]$, etc., the angles $a'\omega, b'\omega$, etc. to the roots $[a'], [b']$, etc., the angles $a''\omega, b''\omega$, etc. to the roots $[a''], [b'']$, etc. It is easy to see that all these angles taken together coincide with respect to the trigonometric functions[h] with the angles P/n, $2P/n, 3P/n, \ldots (n - 1)P/n$. Now if we denote the function we are considering by the character ϕ prefixed to the angle, and if we

[h] Two angles coincide in this respect if their difference is equal to the circumference or to a multiple of it. We can say that they are *congruent relative to the circumference* if we want to use the term congruence in an extended sense.

let $= Y$ the product of the e factors

$$x - \phi\omega, \qquad x - \phi a\omega, \qquad x - \phi b\omega, \text{ etc.}$$

and the product of the factors $x - \phi a'\omega, x - \phi b'\omega$, etc. $= Y'$, the product of $x - \phi a''\omega, x - \phi b''\omega$, etc. $= Y''$ etc.: then necessarily the product $YY'Y''\ldots = Z$. It remains now to show that all the coefficients in the functions Y, Y', Y'', etc. can be reduced to the form

$$A + B(f, 1) + C(f, g) + D(f, g^2)\ldots + L(f, g^{e-1})$$

When we have done this, manifestly all of them will be known as soon as we know the values of all the sums of f terms. We show this in the following way.

Just as $\cos\omega = ([1]/2) + ([1]^{n-1}/2)$, $\sin\omega = -(i[1]/2) + (i[1]^{n-1}/2)$ so by the preceding article all the remaining trigonometric functions of the angle ω can be reduced to the form $\mathfrak{A} + \mathfrak{B}[1] + \mathfrak{C}[1]^2 + \mathfrak{D}[1]^3 +$ etc., and it is not difficult to see that the function of the angle $k\omega$ then becomes $= \mathfrak{A} + \mathfrak{B}[k] + \mathfrak{C}[k]^2 + \mathfrak{D}[k]^3 +$ etc. where k is any integer. Now since the individual coefficients in Y are invariable rational integral functions of $\phi\omega$, $\phi a\omega$, $\phi b\omega$, etc., if we substitute their values for these quantities, the individual coefficients will become invariable rational integral functions of $[1]$, $[a]$, $[b]$, etc. Therefore by article 347 they are reduced to the form $A + B(f, 1) + C(f, g) +$ etc. The coefficients in Y', Y'', etc. can also be reduced to similar forms.   Q.E.D.

▶ 364. We add a few observations concerning the problem of the preceding article.

I. The individual coefficients in Y' are functions of roots contained in the period P' [we can let it $=(f, a')$] just as the functions of the roots in P are the corresponding coefficients in Y. It is clear from article 347 therefore that we can derive Y' from Y by substituting everywhere in Y the quantities (f, a'), $(f, a'g)$, $(f, a'g^2)$, etc. for $(f, 1)$, (f, g), (f, g^2), etc. respectively. And Y'' can be derived from Y by substituting everywhere in Y (f, a''), $(f, a''g)$, $(f, a''g^2)$, etc. for $(f, 1)$, (f, g), (f, g^2), etc. respectively etc. Therefore as soon as we have the function Y, the remaining Y', Y'', etc. follow easily.

II. Let us suppose that

$$Y = x^f - \alpha x^{f-1} + \beta x^{f-2} - \text{etc.}$$

where the coefficients α, β, etc. are respectively the sum of the roots of the equation $Y = 0$, i.e. of the quantities $\phi\omega$, $\phi a\omega$, $\phi b\omega$, etc., the sum of their products taken two by two, etc. But often these coefficients will be found much more easily by a method similar to that of article 349, that is by calculating the sum of the roots $\phi\omega$, $\phi a\omega$, $\phi b\omega$, etc., the sum of their squares, cubes, etc. and deducing from this by Newton's theorem the coefficients we want. Whenever ϕ designates the tangent, secant, cotangent, or co-secant we have still other methods of abbreviating the process, but we cannot consider them here.

III. The case where f is an even number merits special considera-tion for then each of the periods P, P', P'', etc. will be composed of $f/2$ periods of two terms. Let P consist of the numbers $(2, 1)$, $(2, a), (2, b), (2, c)$, etc. The numbers $1, a, b, c$, etc. and $n - 1$, $n - a, n - b, n - c$, etc. taken together will coincide with the numbers $1, a, b, c$, etc. or at least (this comes to the same thing) will be congruent to them relative to the modulus n. But $\phi(n - 1)\omega = \pm\phi\omega$, $\phi(n - a)\omega = \pm\phi a\omega$ etc. the upper signs to be taken when ϕ designates the cosine or secant, the lower when ϕ designates the sine, tangent, cotangent, or cosecant. It follows from this that in the two previous cases the factors of which Y is composed will be equal two by two, and thus Y is a square and will $= y^2$ if we suppose that y is equal to the product of

$$x - \phi\omega, \qquad x - \phi a\omega, \qquad x - \phi b\omega, \text{ etc.}$$

In the same cases the remaining function Y', Y'', etc. will be squares, and if we suppose that P' is composed of $(2, a'), (2, b')$, $(2, c')$, etc.; P'' of $(2, a''), (2, b''), (2, c'')$, etc., etc., the product of $x - \phi a'\omega, x - \phi b'\omega, x - \phi c'\omega$, etc. $= y'$, the product of $x - \phi a''\omega$, $x - \phi b''\omega$, etc. $= y''$, etc., then $Y' = y'y'$, $Y'' = y''y''$, etc.; and the function Z will also be a square (cf. above, art. 337) and its root will be equal to the product of y, y', y'', etc. But clearly y', y'', etc. can be derived from y just as we said that Y, Y'' are derived from Y (cf. I). Further, the individual coefficients in y can also be reduced to the form

$$A + B(f, 1) + C(f, g) + \text{ etc.}$$

because the sums of the individual powers of the roots of the equation $y = 0$ are equal to half the powers of the equation $Y = 0$ and thus are reducible to such a form. In the four latter cases however Y will be the product of the factors

$$x^2 - (\phi\omega)^2, \qquad x^2 - (\phi a\omega)^2, \qquad x^2 - (\phi b\omega)^2, \text{ etc.}$$

and thus of the form

$$x^f - \lambda x^{f-2} + \mu x^{f-4} - \text{ etc.}$$

It is clear that the coefficients λ, μ, etc. can be deduced from the sums of squares, biquadrates, etc. of the roots, $\phi\omega$, $\phi a\omega$, $\phi b\omega$, etc. And the same thing is true for the functions Y', Y'', etc.

Example I. Let $n = 17$, $f = 8$ and let ϕ designate the cosine. Then we will have

$$Z = (x^8 + \tfrac{1}{2}x^7 - \tfrac{7}{4}x^6 - \tfrac{3}{4}x^5 + \tfrac{15}{16}x^4 + \tfrac{5}{16}x^3 - \tfrac{5}{32}x^2 - \tfrac{1}{32}x + \tfrac{1}{256})^2$$

and thus $\sqrt{Z}$ will be resolved into two factors y, y' of degree four. The period $P = (8, 1)$ consists of $(2, 1), (2, 9), (2, 13), (2, 15)$ so y will be a product of the factors

$$x - \phi\omega, \qquad x - \phi 9\omega, \qquad x - \phi 13\omega, \qquad x - \phi 15\omega$$

Substituting $([k]/2) + ([n - k]/2)$ for $\phi k\omega$ we find that

$$\phi\omega + \phi 9\omega + \phi 13\omega + \phi 15\omega = (8, 1)/2$$

$$(\phi\omega)^2 + (\phi 9\omega)^2 + (\phi 13\omega)^2 + (\phi 15\omega)^2 = 2 + [(8, 1)/4]$$

Thus the sum of the cubes is $= [3(8, 1)/8] + [(8, 3)/8]$ and the sum of the biquadrates is $= [1/2] + [5(8, 1)/16]$. So by Newton's theorem the coefficients in y will be

$$y = x^4 - \tfrac{1}{2}(8, 1)x^3 + \tfrac{1}{4}[(8, 1) + 2(8, 3)]x^2 - \tfrac{1}{8}[(8, 1) + 3(8, 3)]x$$
$$+ \tfrac{1}{16}[(8, 1) + (8, 3)]$$

and y' is derived from y by interchanging $(8, 1)$ and $(8, 3)$. Therefore if we substitute for $(8, 1), (8, 3)$ the values $-(1/2) + (\sqrt{17}/2)$, $-(1/2) - (\sqrt{17}/2)$ we get

$$y = x^4 + (\tfrac{1}{4} - \tfrac{1}{4}\sqrt{17})x^3 - (\tfrac{3}{8} + \tfrac{1}{8}\sqrt{17})x^2 + (\tfrac{1}{4} + \tfrac{1}{8}\sqrt{17})x - \tfrac{1}{16}$$

$$y' = x^4 + (\tfrac{1}{4} + \tfrac{1}{4}\sqrt{17})x^3 - (\tfrac{3}{8} - \tfrac{1}{8}\sqrt{17})x^2 + (\tfrac{1}{4} - \tfrac{1}{8}\sqrt{17})x - \tfrac{1}{16}$$

Similarly $\sqrt{Z}$ can be resolved into four factors of degree two. The first will be $(x - \phi\omega)(x - \phi 13\omega)$, the second $(x - \phi 9\omega)(x - \phi 15\omega)$, the third $(x - \phi 3\omega)(x - \phi 5\omega)$, the fourth $(x - \phi 10\omega)(x - \phi 11\omega)$, and all the coefficients in these factors can be expressed by the four sums $(4, 1)$, $(4, 9)$, $(4, 3)$, $(4, 10)$. Manifestly the product of the first factor into the second will be y, the product of the third into the fourth y'.

Example II. If with everything else the same, we suppose that ϕ stands for the sine so that

$$Z = x^{16} - \tfrac{17}{4}x^{14} + \tfrac{119}{16}x^{12} - \tfrac{221}{32}x^{10} + \tfrac{935}{256}x^8 - \tfrac{561}{512}x^6$$
$$+ \tfrac{357}{2048}x^4 - \tfrac{51}{4096}x^2 + \tfrac{17}{65536}$$

is to be resolved into two factors of degree 8 which we designate y, y', then y will be a product of four double factors

$$x^2 - (\phi\omega)^2, \qquad x^2 - (\phi 9\omega)^2, \qquad x^2 - (\phi 13\omega)^2, \qquad x^2 - (\phi 15\omega)^2$$

Now since

$$\phi k\omega = -\tfrac{1}{2}i[k] + \tfrac{1}{2}i[n - k]$$

we have

$$(\phi k\omega)^2 = -\tfrac{1}{4}[2k] + \tfrac{1}{2}[n] - \tfrac{1}{4}[2n - 2k] = \tfrac{1}{2} - \tfrac{1}{4}[2k] - \tfrac{1}{4}[2n - 2k]$$

Thus the sum of the squares of the roots $\phi\omega, \phi 9\omega, \phi 13\omega, \phi 15\omega$ will be $2 - ((8, 1)/4)$, the sum of their fourth powers $= (3/2) - (3(8, 1)/16)$, the sum of their sixth powers $= (5/4) - (9(8, 1)/64) - ((8, 3)/64)$, the sum of their eighth powers $(35/32) - (27(8, 1)/256) - ((8, 3)/32)$. As a result we have

$$y = x^8 - (2 - \tfrac{1}{4}(8, 1))x^6 + (\tfrac{3}{2} - \tfrac{5}{16}(8, 1) + \tfrac{1}{8}(8, 3))x^4$$
$$- (\tfrac{1}{2} - \tfrac{9}{64}(8, 1) + \tfrac{5}{64}(8, 3))x^2 + \tfrac{1}{16} - \tfrac{5}{256}(8, 1) + \tfrac{3}{256}(8, 3)$$

and y' is determined from y by interchanging $(8, 1)$, $(8, 3)$, so by substituting the values of these sums we get

$$y = x^8 - (\tfrac{17}{8} - \tfrac{1}{8}\sqrt{17})x^6 + (\tfrac{51}{32} - \tfrac{7}{32}\sqrt{17})x^4 - (\tfrac{17}{32} - \tfrac{7}{64}\sqrt{17})x^2$$
$$+ \tfrac{17}{256} - \tfrac{1}{64}\sqrt{17}$$

$$y' = x^8 - (\tfrac{17}{8} + \tfrac{1}{8}\sqrt{17})x^6 + (\tfrac{51}{32} + \tfrac{7}{32}\sqrt{17})x^4 - (\tfrac{17}{32} + \tfrac{7}{64}\sqrt{17})x^2$$
$$+ \tfrac{17}{256} + \tfrac{1}{64}\sqrt{17}$$

Thus Z can be resolved into four factors whose coefficients can be expressed as the sum of four terms. The product of two of them will be y, the product of the other two y'.

▶ 365. Thus by the preceding discussions we have reduced the division of the circle into n parts, if n is a prime number, to the solution of as many equations as there are factors in the number $n - 1$. The degree of the equations is determined by the size of the factors. Whenever therefore $n - 1$ is a power of the number 2, which happens when the value of n is 3, 5, 17, 257, 65537, etc. the sectioning of the circle is reduced to quadratic equations only, and the trigonometric functions of the angles P/n, $2P/n$, etc. can be expressed by square roots which are more or less complicated (according to the size of n). Thus in these cases the division of the circle into n parts or the inscription of a regular polygon of n sides can be accomplished by geometric constructions. Thus, e.g., for $n = 17$, by article 354, 361 we get the following expression for the cosine of the angle $P/17$:

$$-\tfrac{1}{16} + \tfrac{1}{16}\sqrt{17} + \tfrac{1}{16}\sqrt{[34 - 2\sqrt{17}]} + \tfrac{1}{8}\sqrt{[17 + 3\sqrt{17} - \sqrt{(34 - 2\sqrt{17}) - 2\sqrt{(34 + 2\sqrt{17})}]}}$$

The cosine of multiples of this angle will have a similar form, but the sine will have one more radical sign. It is certainly astonishing that although the geometric divisibility of the circle into three and five parts was already known in Euclid's time, nothing was added to this discovery for 2000 years. And all geometers had asserted that, except for those sections and the ones that derive directly from them (that is, division into 15, $3 \cdot 2^\mu$, $5 \cdot 2^\mu$, and 2^μ parts), there are no others that can be effected by geometric constructions. But it is easy to show that if the prime number $n = 2^m + 1$, the exponent m can have no other prime factors except 2, and so it is equal to 1 or 2 or a higher power of the number 2. For if m were divisible by any odd number ζ (greater than unity) so that $m = \zeta\eta$, then $2^m + 1$ would be divisible by $2^\eta + 1$ and so necessarily composite. All values of n, therefore, that can be reduced to quadratic equations, are contained in the form $2^{2^\nu} + 1$. Thus the five numbers 3, 5, 17, 257, 65537 result from letting $\nu = 0, 1, 2, 3, 4$ or $m = 1, 2, 4, 8, 16$. But the geometric division of the circle cannot be accomplished for *all* numbers contained in the formula but only for those that are prime. Fermat was misled by his induction and affirmed that all numbers contained in this form are necessarily prime, but the distinguished Euler first noticed that this rule is erroneous for $v = 5$ or $m = 32$, since the number $2^{32} + 1 = 4294967297$ involves the factor 641.

Whenever $n - 1$ implies prime factors other than 2, we are always led to equations of higher degree, namely, to one or more cubic equations when 3 appears once or several times among the prime factors of $n - 1$, to equations of the fifth degree when $n - 1$ is divisible by 5, etc. WE CAN SHOW WITH ALL RIGOR THAT THESE HIGHER-DEGREE EQUATIONS CANNOT BE AVOIDED IN ANY WAY NOR CAN THEY BE REDUCED TO LOWER-DEGREE EQUATIONS. The limits of the present work exclude this demonstration here, but we issue this warning lest anyone attempt to achieve geometric constructions for sections other than the ones suggested by our theory (e.g. sections into 7, 11, 13, 19, etc. parts) and so spend his time uselessly.

▶ 366. If a circle is to be cut into a^α parts where a is a prime number, manifestly this can be done geometrically when $a = 2$ but not for any other value of a if $\alpha > 1$, for then besides the equations required for the division into a parts, there will necessarily be $\alpha - 1$ others of degree a to be solved, and these cannot be avoided in any way or reduced. Therefore in general the degree of the necessary equations can be known from the prime factors of the number $(a - 1)a^{\alpha - 1}$ (including also the case where $\alpha = 1$).

Finally if the circle is to be cut into $N = a^\alpha b^\beta c^\gamma \ldots$ parts where a, b, c, etc. are unequal prime numbers, it suffices to effect divisions into $a^\alpha, b^\beta, c^\gamma$, etc. parts (art. 336). So in order to know the degree of the equations necessary for this purpose, we must consider the prime factors of the numbers

$$(a - 1)a^{\alpha - 1}, \qquad (b - 1)b^{\beta - 1}, \qquad (c - 1)c^{\gamma - 1}, \text{ etc.}$$

or, what comes to the same thing, the factors of their product. We remark that this product indicates the number of numbers relatively prime to N and less than it (art. 38). Geometrically therefore this division can be accomplished only when this number is a power of 2. But when the factors include primes other than 2, say p, p', etc. then equations of degree p, p', etc. cannot be avoided.

In general therefore in order to be able to divide the circle geo-metrically into N parts, N must be 2 or a higher power of 2, *or a* prime number of the form $2^m + 1$, *or* the product of several prime numbers of this form, *or* the product of one or several such primes into 2 or a higher power of 2. In brief, it is required that N imply no odd prime factor that is not of the form $2^m + 1$ nor any prime factor of the form $2^m + 1$ more than once. The following are the 38 values of N below 300:

2, 3, 4, 5, 6, 8, 10, 12, 15, 16, 17, 20, 24, 30, 32, 34, 40, 48, 51, 60, 64,

68, 80, 85, 96, 102, 120, 128, 136, 160, 170, 192, 204, 240, 255, 256,

257, 272.

ADDITIONAL NOTES

▶ *Art.* 28. The solution of the indeterminate equation $ax = by \pm 1$ was not first accomplished by the illustrious Euler (as stated in this section) but by a geometer of the seventeenth century, Bachet de Meziriac, the celebrated editor and commentator of Diophantus. It was the illustrious Lagrange who restored this honor to him (Appendix to Euler's *Algèbre*, p. 525,[1] where at the same time he indicates the nature of the method). Bachet published his discovery in the second edition of the book *Problèmes plaisants et délectables qui se font par les nombres* (1624). In the first edition (Lyon, 1612) which was the only one I saw, it was not included, although it was mentioned.

▶ *Art.* 151, 296, 297. The illustrious Legendre presented his demonstration again in his excellent work, *Essai d'une théorie des nombres* (p. 214 ff.[2]) but in such a way as to change nothing essential. So this method is still subject to all the objections contained in article 297. It is true that the theorem (on which one supposition is based) which states that any arithmetic progression $l, l + k, l + 2k$, etc. contains prime numbers if k and l do not have a common divisor, is given more fully here (p. 12 ff.[3]), but it does not yet seem to satisfy geometric rigor. But even if this theorem were fully demonstrated, the second supposition remains (that there are prime numbers of the form $4n + 3$ for which a given positive prime number of the form $4n + 1$ is a quadratic nonresidue) and I do not know whether this can be proven *rigorously* unless the fundamental theorem is *presumed*. But it must be remarked that Legendre did not tacitly assume this last supposition, nor did he ignore it (p. 221).

▶ *Art.* 288–293. The same argument presented here as a special application of the theory of ternary forms, and which seems to be so absolute with respect to rigor and generality that nothing more could be desired,

---

[1] Cf. p. 11, note.

[2] The reference is to *Seconde Partie: Propriétés Générales des Nombres*. The particular article is No. 8, "De la manière de déterminer x pour que $x^2 + a$ soit divisible par un nombre composé quelconque N."

[3] "Introduction. XXI—Contenant des Notions Générales sur les Nombres."

461

is treated much more fully by the illustrious Legendre in the third part of his work (pp. 321–400[a]). He uses principles and methods quite different from ours, but in this way he encounters many difficulties that hinder him from strengthening these remarkable theorems with a rigorous demonstration. He indicates these difficulties candidly but, unless I am mistaken, these can be more easily dealt with than the presupposition of the theorem that is cited in the note at the foot of p. 371 (the theorem begins, "In any arithmetic progression etc.").

▶ *Art.* 306.VIII. In the third chiliad of negative determinants there are 37 which are irregular; 18 of them have 2 as index of irregularity, the other 19 the index 3.

▶ *Art.* 306.X. We have lately been successful in solving fully the question proposed here. We will very soon publish this discussion in our continuation of this work. It illustrates brilliantly many parts of higher Arithmetic and Analysis. The same solution proves that the coefficient m in article 304 $= \gamma\pi = 2.3458847616$ where γ is the same quantity as in article 302 and π is the length of half a circle of radius 1.

---

[a] The reader hardly needs to be warned that our ternary forms are not to be confused with what Legendre calls *forme trinaire d'un nombre*. By this expression he means the decomposition of a number into three squares.

TABLE 1 (art. 58, 91)

		2 . 3 . 5 . 7 . 11	13 . 17 . 19 . 23 . 29	31 . 37 . 41 . 43 . 47	53 . 59 . 61 . 67 . 71	73 . 79 . 83 . 89
3	2	I				
5	2	1 . 3				
7	3	2 . 1 . 5				
9	2	1 . *. 5 . 4				
11	2	1 . 8 . 4 . 7				
13	6	5 . 8 . 9 . 7 . 11				
16	5	* . 3 . 1 . 2 . 1	3			
17	10	10 . 11 . 7 . 9 . 13	12			
19	10	17 . 5 . 2 . 12 . 6	13 . 8			
23	10	8 . 20 . 15 . 21 . 3	12 . 17 . 5			
25	2	1 . 7 . *. 5 . 16	19 . 13 . 18 . 11			
27	2	1 . *. 5 . 16 . 13	8 . 15 . 12 . 11			
29	10	11 . 27 . 18 . 20 . 23	2 . 7 . 15 . 24			
31	17	12 . 13 . 20 . 4 . 29	23 . 1 . 22 . 21 . 27			
32	5	* . 3 . 1 . 2 . 5	7 . 4 . 7 . 6 . 3	0		
37	5	11 . 34 . 1 . 28 . 6	13 . 5 . 25 . 21 . 15	27		
41	6	26 . 15 . 22 . 39 . 3	31 . 33 . 9 . 36 . 7	28 . 32		
43	28	39 . 17 . 5 . 7 . 6	40 . 16 . 29 . 20 . 25	32 . 35 . 18		
47	10	30 . 18 . 17 . 38 . 27	3 . 42 . 29 . 39 . 43	5 . 24 . 25 . 37		
49	10	2 . 13 . 41 . *. 16	9 . 31 . 35 . 32 . 24	7 . 38 . 27 . 36 . 23		
53	26	25 . 9 . 31 . 38 . 46	28 . 42 . 41 . 39 . 6	45 . 22 . 33 . 30 . 8		
59	10	25 . 32 . 34 . 44 . 45	23 . 14 . 22 . 27 . 4	7 . 41 . 2 . 13 . 53	28	
61	10	47 . 42 . 14 . 23 . 45	20 . 49 . 22 . 39 . 25	13 . 33 . 18 . 41 . 40	51 . 17	
64	5	* . 3 . 1 . 10 . 5	15 . 12 . 7 . 14 . 11	8 . 9 . 14 . 13 . 12	5 . 1 . 3	
67	12	29 . 9 . 39 . 7 . 61	23 . 8 . 26 . 20 . 22	43 . 44 . 19 . 63 . 64	3 . 54 . 5	
71	62	58 . 18 . 14 . 33 . 43	27 . 7 . 38 . 5 . 4	13 . 30 . 55 . 44 . 17	59 . 29 . 37 . 11	
73	5	8 . 6 . 1 . 33 . 55	59 . 21 . 62 . 46 . 35	11 . 64 . 4 . 51 . 31	53 . 5 . 58 . 50 . 44	
79	29	50 . 71 . 34 . 19 . 70	74 . 9 . 10 . 52 . 1	76 . 23 . 21 . 47 . 55	7 . 17 . 75 . 54 . 33	4
81	11	25 . *. 35 . 22 . 1	38 . 15 . 12 . 5 . 7	14 . 24 . 29 . 10 . 13	45 . 53 . 4 . 20 . 33	48 . 52
83	50	3 . 52 . 81 . 24 . 72	67 . 4 . 59 . 16 . 36	32 . 60 . 38 . 49 . 69	13 . 20 . 34 . 53 . 17	43 . 47
89	30	72 . 87 . 18 . 7 . 4	65 . 82 . 53 . 31 . 29	57 . 77 . 67 . 59 . 34	10 . 45 . 19 . 32 . 26	68 . 46 . 27
97	10	86 . 2 . 11 . 53 . 82	83 . 19 . 27 . 79 . 47	26 . 41 . 71 . 44 . 60	14 . 65 . 32 . 51 . 25	20 . 42 . 91 . 18

TABLE 2 (art. 99)

	−1	+2	+3	+5	+7	+11	+13	+17	+19	+23	+29	+31	+37	+41	+43	+47	+53	+59	+61	+67	+71	+73	+79	+83	+89	+97
3			—		—				—			—	—		—				—			—	—			—
5	—						—			—	—			—	—			—			—			—		—
7		—				—				—	—			—		—				—		—	—			
11	—			—							—			—			—				—			—		—
13	—		—						—	—					—					—			—			—
17	—				—		—	—						—						—	—			—		
19			—	—	—		—	—						—					—			—	—			
23	—		—		—		—						—							—			—		—	
29	—		—	—		—			—	—								—				—				
31		—	—		—		—			—							—			—				—		—
37	—		—	—	—								—				—					—	—			
41	—		—		—					—					—		—			—			—			
43	—		—		—	—			—			—				—				—		—	—			
47	—		—		—		—					—		—				—					—		—	—
53	—					—			—	—		—			—					—				—		
59		—	—	—			—			—					—					—			—		—	
61	—		—	—				—		—							—			—			—		—	
67	—						—	—						—			—			—		—	—		—	
71		—	—	—				—		—					—		—				—	—	—		—	
73	—			—				—							—					—	—		—		—	
79		—	—		—	—				—	—									—		—	—			—
83		—		—		—		—			—		—		—						—	—		—		—
89	—	—	—	—	—	—							—	—					—	—	—	—		—		—
97	—	—	—			—									—	—	—		—	—		—	—		—	—

TABLE 3 (art. 316)

3	(0)..3; (1)..6
7	(0)..142857
9	(0)..1; (1)..2; (2)..4; (3)..8; (4)..7; (5)..5
11	(0)..09; (1)..18; (2)..36; (3)..72; (4)..45
13	(0)..076923; (1)..461538
17	(0)..0588235294 117647
19	(0)..0526315789 47368421
23	(0)..0434782608 6956521739 13
27	(0)..037; (1)..074; (2)..148; (3)..296; (4)..592; (5)..185
29	(0)..0344827586 2068965517 24137931
31	(0)..0322580645 16129; (1)..5483870967 74193
37	(0)..027; (1)..135; (2)..675; (3)..378; (4)..891; (5)..459
	(6)..297; (7)..486; (8)..432; (9)..162; (10)..810; (11)..054
41	(0)..02439; (1)..14634; (2)..87804; (3)..26829; (4)..60975; (5)..65853; (6)..95121; (7)..70731
43	(0)..0232558139 5348837209 3; (1)..6511627906 9767441860 4
47	(0)..0212765957 4468085106 3829787234 0425531914 893617
49	(0)..0204081632 6530612244 8979591836 7346938775 51
53	(0)..0188679245 283; (1)..4905660377 358; (2)..7547169811 320; (3)..6226415094 339
59	(0)..0169491525 4237288135 5932203389 8305084745 7627118644 06779661
61	(0)..0163934426 2295081967 2131147540 9836065573 7704918032 7868852459
67	(0)..0149253731 3432835820 8955223880 597; (1)..1791044776 1194029850 7462686567 164
71	(0)..0140845070 4225352112 6760563380 28169; (1)..8732394366 1971830985 9154929577 46478
73	(0)..01369863; (1)..06849315; (2)..34246575; (3)..71232876; (4)..56164383
	(5)..80821917; (6)..04109589; (7)..20547945; (8)..02739726
79	(0)..0126582278 481; (1)..3670886075 949; (2)..6455696202 531
	(3)..7215189873 417; (4)..9240506329 113; (5)..7974683544 303
81	(0)..012345679; (1)..135802469; (2)..493827160; (3)..432098765; (4)..753086419; (5)..283950617
83	(0)..0120481927 7108433734 939759036I 4457831325 3
	(1)..6024096385 5421686746 9879518072 2891566265 0
89	(0)..0112359550 5617977528 0898876404 4943820224 7191
	(1)..3370786516 8539325842 6966292134 8314606741 5730
97	(0)..0103092783 5051546391 7525773195 8762886597 9381443298 9690721649 4845360824 7422680412
	3711340206 185567

GAUSS' HANDWRITTEN NOTES

[Art. 40. The lines, "If there is a third number $C, \ldots$ by letting $k\alpha = a$, $k\beta = b$, $\gamma = c$, $\lambda' = \mu$"]. If there is a third number C, let λ' be the greatest common divisor of the numbers λ, C and it will also be the greatest common divisor of the numbers A, B, C.[*] Determine the numbers k, γ in such a way that $k\lambda + \gamma C = \lambda'$. Then $k\alpha A + k\beta B + \gamma C = \lambda'$. If there is a fourth number D, let λ'' be the greatest common divisor of λ', D (it is easy to see that it will also be the greatest common divisor of A, B, C, D), and let $k'\lambda' + \delta D = \lambda''$. Then we have

$$kk'\alpha A + kk'\beta B + k'\gamma C + \delta D = \lambda''.$$

[Art. 114.] A more elegant demonstration goes as follows:

$$(a^{3n} - a^n)^2 = 2 + (a^{4n} + 1)(a^{2n} - 2)$$
$$(a^{3n} + a^n)^2 = -2 + (a^{4n} + 1)(a^{2n} + 2)$$

and so $\sqrt{2} \equiv \pm(a^{3n} - a^n)$ and $\sqrt{-2} \equiv \pm(a^{3n} + a^n)$ (mod. $8n + 1$).

[Art. 256. In part VI were indicated 16 positive determinants of form $8n + 5$ for which the number of properly primitive classes is three times greater than the number of improperly primitive classes, namely $37, \ldots, 573$. To these are added] $677, 701, 709, 757, 781, 813, 829, 877, 885, 901, 909, 925, 933, 973, 997$, giving 31 out of 125.

[Art. 301. The last sentence: "In this way the sum of the number of genera for the determinants -1 to -100 is found to

---

* Obviously λ' divides all the numbers A, B, C. If it were not the *greatest* common divisor, the greatest would be larger than λ'. Now since this greatest divisor divides A, B, C it also divides $k\alpha A + k\beta B + \gamma C$, that is λ' itself. Thus a larger number divides a smaller one. Q.E.A. This result can be even more easily established from art. 18.

467

be = 234.4, whereas it is actually 233."] From -1 to -3000, the table gives 11166, the formula 11167.9.

[Art. 336.] If all numbers of the $2^{2^m} + 1$ were prime numbers, then a sufficiently accurate approximation for the number of numbers in question (N) which are smaller than the given number M would be $\frac{1}{2}(\log M/\log 2)^2$.

[Art. 42. The theorem concerning the divisors of an algebraically whole rational function with integral coefficients] 1797, July 22.

[Art. 130. The words, "Now that we have rigorously ... some prime less than itself."] We discovered this proof April 8, 1796.

[Art. 131.] We discovered the fundamental theorem by induction in March 1795. We found our first proof, the one contained in this section, April 1796.

[Art. 133. The words, "We begin now a more general investigation. Consider any two odd, signed numbers P and Q which are relatively prime,"] April 29, 1796.

[Art. 145. The words, "Furthermore the theorems ... new method of demonstrating them,"] Feb. 4, 1797.

[Sect. V. The heading, "Forms and Indeterminate Equations of the Second Degree,"] From here on, June 22, 1796.

[Art. 234. The words, "we will go on to another very important subject, the *composition* of forms,"] These investigations were begun in the autumn of 1798.

[Art. 262. The words. "From this principle ... pertaining to the residues -1, $+2$, -2,"] The principles of this method were first discovered July 27, 1796, but it was refined and really reduced to its present form in the year 1800.

[Art. 266. The words, "But there are very many beautiful ... brief digressions into this theory,"] February 14, 1799.

[Art. 272. The words, "First we will show ... for any given determinant,"] February 13, 1800.

[Art. 287. III. The words, "In a similar way ... Unless we are very much mistaken these theorems etc.,"] First proven in the month of April 1798.

[Art. 302. The words, "The average number of classes however increases very regularly,"] A first intimation of this in the beginning of 1799.

[Art. 306. X. The words, "We observe finally ... and the determinant to which it belongs,"] Everything we desired turned out so well as to leave nothing more to be desired. November 30–December 3, 1800.

[Art. 365.] March 30, 1796, we discovered that a circle is geometrically divisible into 17 parts.

LIST OF SPECIAL SYMBOLS

b/a (mod. m)	13
ϕA	20
ψd	33
$\sqrt[n]{A}$ (mod. p)	39
aRb	88
aNb	88
$[x, y]$	95
(a, b, c)	108
$(m; k)$	198
Rp	223
Np	223
1, 4	223
1 *and* 7, 8	223
(g, h)	227
$\sqrt{(ax^2 + 2bxy + cy^2)}$ (mod. m)	227
$\sqrt{(a, b, c)}$ or $\sqrt{F}$ (mod. m)	227
$\sqrt{M(a, b, c)}$ or $\sqrt{MF}$ (mod. m)	227
$M \cdot (a, b, c)$ or MF	227
$\begin{pmatrix} a, a', a'' \\ b, b', b'' \end{pmatrix}$	293
$\begin{pmatrix} \alpha & \beta & \gamma \\ \alpha' & \beta' & \gamma' \\ \alpha'' & \beta'' & \gamma'' \end{pmatrix}$	294
$[\lambda]$	414
(f, λ)	415

DIRECTORY OF TERMS

Modulus, 1
Dimension of a Prime Factor, 7
Fermat's Theorem, 32
Equivalent Indices Relative to the Modulus
$p - 1$, 37
Primitive Roots, 37
Wilson's Theorem, 50
The Fundamental Theorem on Quadratic
Residues, 88
Congruence: 1, 7, 9; algebraic, 9; degree of
algebraic, 9; root of, 9; solution of, 9;
solvable and unsolvable, 9; transcenden-
tal, 9; base of, 37; nonpure congruence,
106; method of exclusion for solving
$x^2 - A$ (mod. m), 383; an excluding num-
ber for solving $x^2 - A$ (mod. m), 384
Noncongruence, 1, 7
Residues, 1; least and absolutely least, 2;
least negative and least positive, 2; quadratic, 64; associated, 72; biquadratic, 76
Nonresidues, 1; quadratic nonresidues, 64
Forms of the Second Degree, 108
Forms of the Second, Third, Fourth Degree,
etc., 292
Binary, Ternary, Quaternary Forms, etc.,
292
Binary Forms of the Second Degree: forms
of divisor of $x^2 - A$, 101; forms of nondivisors of $x^2 - A$, 101; determinant of a
form, 109; representation of, 109; one
form contained in another, 111; implied
in another, 111; transformed into another,
111; properly and improperly contained
in another, 112; similar, dissimilar, proper, and improper transformations of one
form into another, 112; equivalent forms,
112; properly and improperly equivalent
forms, 113;
neighboring forms, 115; neighbor by

the first and by the last part, 115; opposite forms, 115; reduced forms, 135, 152,
191; reduced form with a negative determinant, 135; reduced form with a nonquadratic positive determinant, 152;
period of a form, 156; associated periods,
158; simplest in a class of forms, 213;
positive and negative forms, 216; derived,
217; primitive, 217; form derived from
a primitive, 217; properly and improperly
primitive forms, 217; form derived from a
properly and improperly primitive form,
218;
principal form, 225; a form as a quadratic residue of a number, 227; composite
form, 233; transformable, 233; taking a
form directly, and inversely, 236; resolving a form into others, 256; degree of a
form, 292; representation of binary forms
by ternary forms, 311; improper representation of binary forms by ternary
forms, 322
Classes of Forms of Second Degree, 213;
classes with given determinants, 139, 195;
representing form, 213; simplest form,
213; opposite classes, 215; opposite to
themselves, 216; ambiguous class of
forms, 216; negative, 216; neighboring,
216; positive, 216; primitive, 217; class
derived from a primitive class, 217; properly and improperly primitive class, 218;
class derived from a properly and improperly primitive class, 218; complete character of a class of forms, 224; principal
class, 225; class composed of two given
classes, 264; composition of classes, 264;
duplication and triplication of a class, 266
Genus (Genera) of Forms of the Second
Degree, 220, 224; genera of properly

471

primitive classes, 224; principal genus, 225; composition of genera of forms, 263; genus composed of two or more genera, 263; number of classes in a genus, 268

Orders of Classes of Forms, 218; number of classes in different orders, 253; composition of orders of forms, 260; simplest form in an order, 266; number of ambiguous classes in an order, 277

Ternary Forms of the Second Degree, 292; adjoint of a ternary form, 293; contained in another, 294; determinant of, 294; implying another, 294; substitution for the transformation of one ternary form into another, 294; adjoint of a substitution, 296; transposition of a substitution, 296; equivalent ternary forms, 296; classes of ternary forms, 297; definite and indefinite ternary forms, 298; positive and negative ternary forms, 298; representation of a number by a ternary form, 311; adjoint of a representation of a ternary form, 312; proper and improper representation by a ternary form, 313; adjoint of the representation by a ternary form, 313; adjoint of the representation of a binary form by a ternary form, 313; improper representation of a binary form by a ternary form, 322

Determinant of a Form of the Second Degree, 109; regular and irregular determinants, 368; exponent of irregularity, 369; double or multiple indices of a determinant, 370

Numbers and Representation of Numbers by a Form: a number belonging to an exponent, 34; index of a number in a congruence, 37; primitive roots of a number in a congruence, 37; a primitive root as base of a congruence, 37; associate numbers relative to a modulus, 51; forms of divisors and of nondivisors of $x^2 - A$, 100; representation of a number by a form, 109; representation of a number by a form belonging to a value of the expression $\sqrt{(b^2 - ac)}$, 110; representation belonging to different, to opposite, and to the same numbers, 111; a form as a quadratic residue of a number, 227; representation of numbers by ternary forms, 311; proper and improper representation of a number by a form, 313

Value of the Expression $\sqrt{(b^2 - ac)}$ (mod. m), 110

A Representation Belonging to the Value of the Expression $\sqrt{(b^2 - ac)}$, 110

Value of the Expression

$\sqrt{(ax^2 - 2bxy - cy^2)}$ (mod. m), 227

Value of the Expression $\sqrt{(a, b, c)}$ OR $\sqrt{F}$ (mod. m), 227

Value of the Expression $\sqrt{M(a, b, c)}$ OR $\sqrt{MF}$ (mod. m), 227; equivalent and different values of the expression $\sqrt{M(a, b, c)}$ (mod. m), 229

THE YALE PAPERBOUNDS

Y–1   LIBERAL EDUCATION AND THE DEMOCRATIC IDEAL by A. Whitney Griswold
Y–2   A TOUCH OF THE POET by Eugene O'Neill
Y–3   THE FOLKLORE OF CAPITALISM by Thurman Arnold
Y–4   THE LOWER DEPTHS AND OTHER PLAYS by Maxim Gorky
Y–5   THE HEAVENLY CITY OF THE EIGHTEENTH-CENTURY PHILOSOPHERS by Carl Becker
Y–6   LORCA by Roy Campbell
Y–7   THE AMERICAN MIND by Henry Steele Commager
Y–8   GOD AND PHILOSOPHY by Etienne Gilson
Y–9   SARTRE by Iris Murdoch
Y–10 AN INTRODUCTION TO THE PHILOSOPHY OF LAW by Roscoe Pound
Y–11 THE COURAGE TO BE by Paul Tillich
Y–12 PSYCHOANALYSIS AND RELIGION by Erich Fromm
Y–13 BONE THOUGHTS by George Starbuck
Y–14 PSYCHOLOGY AND RELIGION by C. G. Jung
Y–15 EDUCATION AT THE CROSSROADS by Jacques Maritain
Y–16 LEGENDS OF HAWAII by Padraic Colum
Y–17 AN INTRODUCTION TO LINGUISTIC SCIENCE by E. H. Sturtevant
Y–18 A COMMON FAITH by John Dewey
Y–19 ETHICS AND LANGUAGE by Charles L. Stevenson
Y–20 BECOMING by Gordon W. Allport
Y–21 THE NATURE OF THE JUDICIAL PROCESS by Benjamin N. Cardozo
Y–22 PASSIVE RESISTANCE IN SOUTH AFRICA by Leo Kuper
Y–23 THE MEANING OF EVOLUTION by George Gaylord Simpson
Y–24 PINCKNEY'S TREATY by Samuel Flagg Bemis
Y–25 TRAGIC THEMES IN WESTERN LITERATURE edited by Cleanth Brooks
Y–26 THREE STUDIES IN MODERN FRENCH LITERATURE by J. M. Cocking, Enid Starkie, and Martin Jarrett-Kerr
Y–27 WAY TO WISDOM by Karl Jaspers
Y–28 DAILY LIFE IN ANCIENT ROME by Jérôme Carcopino
Y–29 THE CHRISTIAN IDEA OF EDUCATION edited by Edmund Fuller
Y–30 FRIAR FELIX AT LARGE by H. F. M. Prescott
Y–31 THE COURT AND THE CASTLE by Rebecca West
Y–32 SCIENCE AND COMMON SENSE by James B. Conant
Y–33 THE MYTH OF THE STATE by Ernst Cassirer
Y–34 FRUSTRATION AND AGGRESSION by John Dollard et al.
Y–35 THE INTEGRATIVE ACTION OF THE NERVOUS SYSTEM by Sir Charles Sherrington
Y–38 POEMS by Alan Dugan
Y–39 GOLD AND THE DOLLAR CRISIS by Robert Triffin
Y–40 THE STRATEGY OF ECONOMIC DEVELOPMENT by Albert O. Hirschman
Y–41 THE LONELY CROWD by David Riesman
Y–42 LIFE OF THE PAST by George Gaylord Simpson
Y–43 A HISTORY OF RUSSIA by George Vernadsky

Y-44 THE COLONIAL BACKGROUND OF THE AMERICAN REVOLUTION by *Charles M. Andrews*

Y-45 THE FAMILY OF GOD by *W. Lloyd Warner*

Y-46 THE MAKING OF THE MIDDLE AGES by *R. W. Southern*

Y-47 THE DYNAMICS OF CULTURE CHANGE by *Bronislaw Malinowski*

Y-48 ELEMENTARY PARTICLES by *Enrico Fermi*

Y-49 SWEDEN: THE MIDDLE WAY by *Marquis W. Childs*

Y-50 JONATHAN DICKINSON'S JOURNAL edited by *Evangeline Walker Andrews and Charles McLean Andrews*

Y-51 MODERN FRENCH THEATRE by *Jacques Guicharnaud*

Y-52 AN ESSAY ON MAN by *Ernst Cassirer*

Y-53 THE FRAMING OF THE CONSTITUTION OF THE UNITED STATES by *Max Farrand*

Y-54 JOURNEY TO AMERICA by *Alexis de Tocqueville*

Y-55 THE HIGHER LEARNING IN AMERICA by *Robert M. Hutchins*

Y-56 THE VISION OF TRAGEDY by *Richard B. Sewall*

Y-57 MY EYES HAVE A COLD NOSE by *Hector Chevigny*

Y-58 CHILD TRAINING AND PERSONALITY by *John W. M. Whiting and Irvin L. Child*

Y-59 RECEPTORS AND SENSORY PERCEPTION by *Ragnar Granit*

Y-61 LONG DAY'S JOURNEY INTO NIGHT by *Eugene O'Neill*

Y-62 JAY'S TREATY by *Samuel Flagg Bemis*

Y-63 SHAKESPEARE: A BIOGRAPHICAL HANDBOOK by *Gerald Eades Bentley*

Y-64 THE POETRY OF MEDITATION by *Louis L. Martz*

Y-65 SOCIAL LEARNING AND IMITATION by *Neal E. Miller and John Dollard*

Y-66 LINCOLN AND HIS PARTY IN THE SECESSION CRISIS by *David M. Potter*

Y-67 SCIENCE SINCE BABYLON by *Derek J. de Solla Price*

Y-68 PLANNING FOR FREEDOM by *Eugene V. Rostow*

Y-69 BUREAUCRACY by *Ludwig von Mises*

Y-70 JOSIAH WILLARD GIBBS by *Lynde Phelps Wheeler*

Y-71 HOW TO BE FIT by *Robert Kiphuth*

Y-72 YANKEE CITY by *W. Lloyd Warner*

Y-73 WHO GOVERNS? by *Robert A. Dahl*

Y-74 THE SOVEREIGN PREROGATIVE by *Eugene V. Rostow*

Y-75 THE PSYCHOLOGY OF C. G. JUNG by *Jolande Jacobi*

Y-76 COMMUNICATION AND PERSUASION by *Carl I. Hovland, Irving L. Janis, and Harold H. Kelley*

Y-77 IDEOLOGICAL DIFFERENCES AND WORLD ORDER edited by *F. S. C. Northrop*

Y-78 THE ECONOMICS OF LABOR by *E. H. Phelps Brown*

Y-79 FOREIGN TRADE AND THE NATIONAL ECONOMY by *Charles P. Kindleberger*

Y-80 VOLPONE edited by *Alvin B. Kernan*

Y-81 TWO EARLY TUDOR LIVES edited by *Richard S. Sylvester and Davis P. Harding*

Y-82 DIMENSIONAL ANALYSIS by *P. W. Bridgman*

Y-83 ORIENTAL DESPOTISM by *Karl A. Wittfogel*

Y-84 THE COMPUTER AND THE BRAIN by *John von Neumann*
Y-85 MANHATTAN PASTURES by *Sandra Hochman*
Y-86 CONCEPTS OF CRITICISM by *René Wellek*
Y-87 THE HIDDEN GOD by *Cleanth Brooks*
Y-88 THE GROWTH OF THE LAW by *Benjamin N. Cardozo*
Y-89 THE DEVELOPMENT OF CONSTITUTIONAL GUARANTEES OF LIBERTY by *Roscoe Pound*
Y-90 POWER AND SOCIETY by *Harold D. Lasswell and Abraham Kaplan*
Y-91 JOYCE AND AQUINAS by *William T. Noon, S.J.*
Y-92 HENRY ADAMS: SCIENTIFIC HISTORIAN by *William Jordy*
Y-93 THE PROSE STYLE OF SAMUEL JOHNSON by *William K. Wimsatt, Jr.*
Y-94 BEYOND THE WELFARE STATE by *Gunnar Myrdal*
Y-95 THE POEMS OF EDWARD TAYLOR edited by *Donald E. Stanford*
Y-96 ORTEGA Y GASSET by *José Ferrater Mora*
Y-97 NAPOLEON: FOR AND AGAINST by *Pieter Geyl*
Y-98 THE MEANING OF GOD IN HUMAN EXPERIENCE by *William Ernest Hocking*
Y-99 THE VICTORIAN FRAME OF MIND by *Walter E. Houghton*
Y-100 POLITICS, PERSONALITY, AND NATION BUILDING by *Lucian W. Pye*
Y-101 MORE STATELY MANSIONS by *Eugene O'Neill*
Y-102 MODERN DEMOCRACY by *Carl L. Becker*
Y-103 THE AMERICAN FEDERAL EXECUTIVE by *W. Lloyd Warner, Paul P. Van Riper, Norman H. Martin, Orvis F. Collins*
Y-104 POEMS 2 by *Alan Dugan*
Y-105 OPEN VISTAS by *Henry Margenau*
Y-106 BARTHOLOMEW FAIR edited by *Eugene M. Waith*
Y-107 LECTURES ON MODERN IDEALISM by *Josiah Royce*
Y-108 SHAKESPEARE'S STAGE by *A. M. Nagler*
Y-109 THE DIVINE RELATIVITY by *Charles Hartshorne*
Y-110 BEHAVIOR THEORY AND CONDITIONING by *Kenneth W. Spence*
Y-111 THE BREAKING OF THE DAY by *Peter Davison*
Y-112 THE GROWTH OF SCIENTIFIC IDEAS by *W. P. D. Wightman*
Y-113 THE PASTORAL ART OF ROBERT FROST by *John F. Lynen*
Y-114 STATE AND LAW: SOVIET AND YUGOSLAV THEORY by *Ivo Lapenna*
Y-115 FUNDAMENTAL LEGAL CONCEPTIONS by *Wesley Newcomb Hohfeld*
Y-116 MANKIND EVOLVING by *Theodosius Dobzhansky*
Y-117 THE AUTOBIOGRAPHY OF BENJAMIN FRANKLIN edited by *Leonard W. Labaree et al.*
Y-118 THE FAR EASTERN POLICY OF THE UNITED STATES by *A. Whitney Griswold*
Y-119 UTOPIA edited by *Edward Surtz, S.J.*
Y-120 THE IMAGINATION OF JEAN GENET by *Joseph H. McMahon*
Y-121 SCIENCE AND CRITICISM by *Herbert J. Muller*
Y-122 LYRIC POETRY OF THE ITALIAN RENAISSANCE *L. R. Lind, collector*
Y-123 TRANSFORMING TRADITIONAL AGRICULTURE by *Theodore W. Schultz*
Y-124 FACTS AND VALUES by *Charles L. Stevenson*
Y-125 THE AGE OF JOHNSON edited by *Frederick W. Hilles*
Y-126 THE LIBERAL TEMPER IN GREEK POLITICS by *Eric A. Havelock*
Y-127 PROTEIN METABOLISM IN THE PLANT by *Albert Charles Chibnall*

Y-128 RUSSIAN FOREIGN POLICY *edited by Ivo J. Lederer*

Y-129 SCHOOLS AND SCHOLARSHIP *edited by Edmund Fuller*

Y-130 THE COLONIAL PERIOD OF AMERICAN HISTORY, VOLUME 1 *by Charles M. Andrews*

Y-131 THE COLONIAL PERIOD OF AMERICAN HISTORY, VOLUME 2 *by Charles M. Andrews*

Y-132 THE COLONIAL PERIOD OF AMERICAN HISTORY, VOLUME 3 *by Charles M. Andrews*

Y-133 THE COLONIAL PERIOD OF AMERICAN HISTORY, VOLUME 4 *by Charles M. Andrews*

Y-134 DIRECT USE OF THE SUN'S ENERGY *by Farrington Daniels*

Y-135 THE ECONOMICS OF SOVIET PLANNING *by Abram Bergson*

Y-136 CENTRAL PLANNING *by Jan Tinbergen*

Y-137 INNOCENCE AND EXPERIENCE *by E. D. Hirsch, Jr.*

Y-138 THE PROBLEM OF GOD *by John Courtney Murray, S.J.*

Y-139 THE IMPERIAL INTELLECT *by A. Dwight Culler*

Y-140 THE SHAPE OF TIME *by George Kubler*

Y-141 DREAM BARKER *by Jean Valentine*

Y-142 AN AFRICAN BOURGEOISIE *by Leo Kuper*

Y-143 SOUTHERN NEGROES, 1861–1865 *by Bell Irvin Wiley*

Y-144 MILLHANDS AND PREACHERS *by Liston Pope*

Y-145 FACES IN THE CROWD *by David Riesman*

Y-146 THE PHILOSOPHY OF SYMBOLIC FORMS, VOLUME 1: LANGUAGE *by Ernst Cassirer*

Y-147 THE PHILOSOPHY OF SYMBOLIC FORMS, VOLUME 2: MYTHICAL THOUGHT *by Ernst Cassirer*

Y-148 THE PHILOSOPHY OF SYMBOLIC FORMS, VOLUME 3: THE PHENOMENOLOGY OF KNOWLEDGE *by Ernst Cassirer*

Y-149 THE NEGRO NOVEL IN AMERICA *by Robert A. Bone*

Y-150 SOCIAL JUDGMENT *by Muzafer Sherif and Carl I. Hovland*

Y-151 COMMUNITY POWER AND POLITICAL THEORY *by Nelson W. Polsby*

Y-152 THE MORALITY OF LAW *by Lon L. Fuller*

Printed in the United States
5540